Introduction to Stellar Winds

This long-awaited graduate textbook, written by two pioneers of the field, is the first to provide a comprehensive introduction to the observations, theories and consequences of stellar winds. The rates of mass loss and the wind velocities are explained from basic physical principles. This textbook also includes chapters clearly explaining the formation and evolution of interstellar bubbles, and the effects of mass loss on the evolution of high- and low-mass stars. Each topic is introduced in a simple way to explain the basic processes and then developed to provide a solid foundation for understanding current research.

This authoritative textbook is designed for advanced undergraduate and graduate students and researchers seeking an understanding of stellar winds and, more generally, supersonic flows from astrophysical objects. It is based on courses taught in Europe and the US over the past 20 years and includes 70 problems for coursework or self-study.

Henny J. G. L. M. Lamers is Professor of Astrophysics and Space Research at the University of Utrecht.

Joseph P. Cassinelli is Professor of Astronomy at the University of Wisconsin, Madison.

Introduction to Stellar Winds

HENNY J. G. L. M. LAMERS

and

JOSEPH P. CASSINELLI

PUBLISHED BY THE PRESS SYNDICATE OF THE UNIVERSITY OF CAMBRIDGE
The Pitt Building, Trumpington Street, Cambridge, United Kingdom

CAMBRIDGE UNIVERSITY PRESS
The Edinburgh Building, Cambridge CB2 2RU, UK www.cup.cam.ac.uk
40 West 20th Street, New York, NY 10011-4211, USA www.cup.org
10 Stamford Road, Oakleigh, Melbourne 3166, Australia
Ruiz de Alarcón 13, 28014 Madrid, Spain

First published 1999

Printed in the United Kingdom at the University Press, Cambridge

Typeface Monotype Times 10/13pt *System* LaTeX [EPC]

A catalogue record of this book is available from the British Library

Library of Congress Cataloguing in Publication data
Lamers, Henny J. G. L. M., 1941–
Introduction to stellar winds / Henny J. G. L. M. Lamers and Joseph P. Cassinelli.
p. cm.
Includes bibliographical references and index.
ISBN 0 521 59398 0 (hardback). – ISBN 0 521 59565 7 (pbk.)
1. Stellar winds. 2. Mass loss (Astrophysics)
I. Lamers, Henny J. G. L. M. 1941– Cassinelli, Joseph P. 1940– .
II. Title.
QB461.L34 1999
523.8′5–dc21 98-15182 CIP

ISBN 0 521 59398 0 hardback
ISBN 0 521 59565 7 paperback

To the love in our lives:

to Trudi and Mary

Contents

Preface

During the last thirty years astronomers have discovered that nearly all stars are losing mass in the form of stellar winds through a major fraction of their lives. This mass loss affects their evolution from their origin to their death. It also leads to spectacular interactions between the supersonic stellar winds and the interstellar medium in the form of planetary nebulae and ring nebulae and in the form of interstellar bubbles and superbubbles. The return of matter from stars into the interstellar medium and the formation of bubbles and superbubbles changes the chemical composition of the galaxies and affects their kinematical properties.

Literature in this field has grown tremendously over the past three decades. On the one hand this is due to the advance of spectroscopic observations over the full range of the spectrum and to the enormous improvements in image resolution from ground based telescopes and the Hubble Space Telescope which results in spectacular images of the nebulae formed by stellar winds. On the other hand it is the result of many theoretical studies to explain the basic mechanisms for stellar winds and the interactions with their surroundings. Many reviews have been published that give an overview of specific aspects of stellar winds or mass loss from stars.

Our colleagues and students made us aware that there is a need for a book that not only gives an overview of the studies of winds and their effect on stellar evolution and on the galaxies, but that also discusses the basic principles of stellar winds and mass loss: the many types of observations and the methods to derive the properties of the winds; the mechanisms that have been proposed to explain the winds; the effects of the winds on their surroundings and the changes in stellar evolution due to mass loss. As a result of these requests, we decided to join our efforts in writing such a book. We started in 1989 when

Lamers occupied the Brittingham Chair of the University of Wisconsin for a period of seven months. Since then, we have worked together on the book almost every summer in either Utrecht or Madison or Baltimore.

The purpose of this book is to provide a basic but thorough description of the observations of stellar winds, the theories of stellar winds and the effects of winds and mass loss on the interstellar environment and on the stellar evolution. In each of the theoretical chapters we first discuss in very simple terms the basic physical mechanisms, and then we continue to derive the theory in physical and mathematical terms. We end most chapters by describing the status of the research at the present time. We have aimed at a level of the book that can be understood by students in astrophysics or physics at the undergraduate level.

The book can be divided into four parts:

- In part one we describe the historical development of the studies of stellar winds (Chapter 1). Few astronomers are aware of the important role that has been played by the very early observations of novae in the gradual realization that stars can lose mass. Nor is it generally known that a firm basis for interpreting observational properties of winds was developed already in the 1930s.
- Part two describes the observations of stellar winds in different wavelength regions and by different types of spectroscopic or photometric methods (Chapter 2). For each type of observation we give the basic diagnostic method how the mass loss rate from a star and the velocity of its wind can be derived. At the end of this chapter we give an overview of the mass loss rates and the wind velocities of the different types of hot and cool stars.
- Part three, the larger part of the book, discusses the different stellar wind theories. We begin with two simple chapters (3 and 4) that describe the basic underlying physics of all stellar wind models in which the gas is driven from a low, highly subsonic velocity at the photosphere, to a supersonic velocity at large distance. Much of this was developed by Eugene Parker and colleagues around 1960 to explain the observations of the solar wind. We then describe the six mechanisms that have been proposed for explaining the stellar winds of different types of stars (Chapters 5 through 10). The last chapter of this part (Chapter 11) describes the two theories that have been proposed for the formation of outflowing disks from hot stars. We were both involved in the development of these theories. Our goal in each chapter is to explain the distinguishing features

of each wind driving mechanism, and then to develop and use the fundamental equations to derive formulae for the mass loss rates and wind velocities as a function of the stellar properties. This enables us to predict and understand the types of stars or the evolution phase of the stars for which each mechanism can drive the stellar wind. By comparing the predicted properties of the various mass loss mechanisms, we can point out the strength or the shortcomings of each mass loss mechanism in a way that has not been done before.

– Part four discusses the effects of stellar winds on the interstellar medium (Chapter 12) and on the stellar evolution (Chapter 13). From an observational perspective the most dramatic consequences of winds are produced by the interaction of winds with the surrounding interstellar medium. There shocks lead to the production of planetary nebulae and to expanding and extended wind bubbles which are observable in galaxies. The dominant effect of the winds on the underlying star is the stripping of the outermost layers. In the most extreme cases this can lead to a loss of more than half of the star's original mass, and thereby expose and eject enriched nuclear processed material that was in the deep interior of the star.

The book can be used as a textbook for teaching stellar winds or mass loss. To as great an extent as possible, each chapter stands alone and can be taught or studied separately. The exceptions are Chapters 3 and 4 where the basic fundamentals of all stellar wind theories are explained. At the urging of the publisher, we have also included a set of problems, which we hope will be useful guides for understanding the chapters.

We have both been involved in stellar wind studies since the 1960s. Cassinelli started his research under the supervision of Jack Brandt in the development of models of the solar wind. Lamers was involved in the first wind studies based on ultraviolet spectra obtained with the S59 and the *Copernicus* satellite under the guidance of Kees de Jager. Our subsequent work with Anne Underhill, Don Morton, Jack Rogerson, Karl Heinz Böhm, John Castor and David Hummer provided us with the fundamentals of radiation transfer that are needed to understand the observations of winds from the most luminous stars. Together with many colleagues and students we have been active in this field ever since: studying observations of winds in X-rays, the ultraviolet, the optical, the infrared and at radio wavelengths. We have both been involved in theoretical explanations of stellar winds and of outflowing disks and we have been teaching extensively about stellar winds and stellar evolution.

In reading the completed book we can trace many of the ideas back to collaborations with former students including Lucky Achmad, Eric Bakker, Alex de Koter, Huib Henrichs, Lex Kaper, Jean Paul Koninx, Norman Trams, Jeroen van Gent, Robert Voors and Rens Waters in the Netherlands, and Jon Bjorkman, David Cohen, Glen Cooper, Bernhard Haisch, Lee Hartmann, Rico Ignace, Gordon Olson, Clint Poe, Daniel VandenBerk, Wayne Waldron and Mark Wolfire in Wisconsin. We also developed a better understanding of the subject by working closely with postdoctoral associates including David Abbott, You Hua Chu, David Friend, Joe Mac Farlane, Kenny Wood and Thierry Lanz.

Significant improvements in the book were made because of comments by Lucky Achmad, Philippe Eenens, Thomas Gaeng, Rico Ignace, Garik Israelian, John Mathis, Pat Morris, Robin Shelton, Jeroen van Gent, Kim Venn, Jorick Vink, Robert Voors, Bart Wakker and Wayne Waldron on drafts of the chapters. We thank Wendy Ashman for developing several schematic drawings for the book. We are especially indebted to Alex de Koter for critical comments on most of the book, to Wayne Waldron for his detailed checks and comments on the magnetically driven wind chapters, to Derck Massa for his persistent proddings that got this project underway, and to John Mathis for providing us with data on the grain properties in dust driven winds.

We are grateful to Sharon Pittman, Marion Wijburg, and secretaries at JILA for typing of several of the chapters, to Ingrid Kallick and Daniel VandeBerk for designing several figures and to Lucky Achmad, Alex de Koter, Sake Hogeveen, Stephan Jansen and Ed van der Zalm for efficient help with the computing systems.

We thank our wives Trudi Lamers and Mary Cassinelli for their love and encouragement throughout the many years that it took to bring this book to completion. Perhaps the best aspect of this endeavor was that it led the four of us to take many enjoyable bicycle rides through the beautiful forests and fields in the Netherlands and on roads through rolling hills of Wisconsin.

Henny Lamers and Joe Cassinelli
August 1997

1 Historical overview

Stellar winds are the continuous outflow of material from stars. The ejection of material plays a major role in the life cycle of stars. In the case of massive stars, the winds remove more than half of the star's original mass before the star explodes as a supernova. In this book we will explore the many mechanisms that can lead a star to eject matter in the form of a steady stellar wind. We will also discuss the interaction of winds with the interstellar medium of our galaxy, and the effects of mass loss on the evolution of a star. We start by giving in this chapter a brief overview of the historical development of the subject, especially focusing on the early observations and theoretical advances that led us to our current level of understanding.

1.1 Historical introduction

1.1.1 The early developments

The names 'solar wind' and 'stellar winds' were both coined by Eugene Parker (1958, 1960). However, the origins of the basic ideas regarding mass loss from stars arose long before that.

The earliest phase in the development of the subject concerns the realization that a few stars are like 'novae', in having spectra with very broad emission lines. Novae are sudden outbursts of light from certain types of stars, and the outbursts are also associated with the high speed ejection of material. Tycho Brahe's observation of a 'new star' or nova in 1572 marks the birth of stellar astronomy as a study of objects that are not perfect celestial objects, but rather ones that can change in interesting ways. We now realize that Tycho's new star was really a supernova, an extremely rare event. This outburst was visible even at noon in the daytime sky, and it remained visible as a night time object for another 16 months. The next such astronomical event occured 28 years later, when Blaeu (1600) discovered another 'new star' that later got the name P Cygni. We now know that this also was not what we would now call a nova, but rather a star with a persistent, strong stellar wind. P Cygni is now classified as a luminous blue variable star. After P Cygni flared up to third magnitude in 1600, it showed large brightness variations in the next sixty years, but after that it has been rather constant in brightness. Optical prism observations in the 19th century showed that other objects have prominent line profiles like those seen in the spectrum of P Cygni. These 'P Cygni' lines have an emission component that is shifted longward of line center, along with an absorption component that is shifted to the shortward side of line center. Such profiles have played an important role in our discovery of wind properties. Wolf and Rayet (1867) discovered a class of stars that have spectral lines like those seen in novae and in P Cygni, but the Wolf-Rayet stars do not give rise to outbursts, nor do they fade from view with time.

By the end of the 19th century spectroscopic and photographic techniques had improved to the extent that the profiles in the spectra of novae could be recorded and the shells could be directly photographed. Nova Aurigae 1891 was the first really well studied 'classical nova'. Campbell (1892) found that the lines in its spectrum had widths, $\Delta\lambda$, that are proportional to the central wavelength of the line. This result confirmed that the Doppler effect (for which $\Delta\lambda = \lambda v/c$) was responsible for the broadening. However, Doppler broadening could arise either from expansion or from some sort of 'turbulent' motion, so outflows were not yet proven. Photographs of novae later showed the presence of shells centered on the site of the outbursts. The diameters of the shells increased with time; so obviously they were expanding away from the central star. The photographs of novae shells confirmed the link between the P Cygni shaped line profiles and the outward expansion of gaseous envelopes.

The strong similarity between the P Cygni profiles of novae with those of Wolf-Rayet stars led Beals (1929) to propose that a 'continuous outflow' occurs from Wolf-Rayet stars. Soon thereafter, Chandrasekhar (1934b) developed a solid footing for interpreting P Cygni profiles as arising in expanding atmospheres. Kosirev (1934) used Chandrasekhar's continuum and line profile diagnostics, along with observational data to derive the mass loss rates and terminal velocity of the flow from a Wolf-Rayet star. Kosirev estimated that the star had a mass loss rate of 10^{-5} solar masses per year, and that the maximal expansion velocity was about 1000 kilometers per second. Both of these estimates agree within a factor three of the current values for this star! So by the mid-1930s, astronomers had achieved a fairly sophisticated understanding of the nature of rapidly expanding atmospheres of some hot stars.

In the case of cool stars, it was more difficult to find evidence for mass loss by expanding atmospheres. Adams and MacCormack (1935) found that strong resonance lines in very bright red supergiants have shortward shifted absorption features, but the implied expansion speed was small ($\approx$ 5 km s^{-1}), and well below the velocity of escape from the stars. Analyzing similar data, Spitzer (1939) concluded in favor of a 'fountain model' for the atmospheres of red supergiants. He suggested that radiation drives matter upward, but at some height the ionization stage changes and the radiative acceleration drops so that the matter falls back onto the star. Hence, in Spitzer's view, the observed line displacements did not indicate a loss of mass by the star. The crucial evidence for actual mass loss from red giant stars was developed by Deutsch (1956) from observations of the binary system α Her. The system consists of an M5 giant and a G5 dwarf. Deutsch noticed that the spectrum of the G5 star contained a set of shifted lines that are also present in the expansion of the M star envelope. He concluded that both stars are enveloped in gas that has been ejected by the M star, and he found that the G star serves only as a 'very convenient probe for the detection of the remoter parts of the circumstellar envelope', see Figure (1.1). Although the inferred expansion speeds were below the escape speed in the atmosphere of the M giant star, by the time the expansion had reached the radial distance of the G star, the flow speed was sufficient that the matter would escape from the system. Deutsch estimated that the M star is losing mass to the interstellar medium at a rate of about 10^{-7} $M_{\odot}$ yr^{-1}.

In addition to spectroscopic evidence, it was recognized from stellar interior considerations that mass loss must be occuring from stars. The mass loss was known to be necessary in order that stars with masses originally larger than the Chandrasekhar limit of 1.4 $M_{\odot}$, could evolve to the white dwarf phase (Wilson, 1960b). By the time the solar wind was observationally confirmed by the *Mariner 2* interplanetary probe, (Neugebauer and Snyder, 1962), mass loss from several classes of stars were known and Deutsch (1968) stated that mass loss was 'ubiquitous across the HR diagram'.

1.1.2 The forces driving mass loss

Several mechanisms for driving mass from a star have been proposed. For most of these we can trace their origins to solar studies.

Parker's original theory for the solar wind, in 1958, concerned *flows driven by gas pressure gradients.* However, observations from probes in interplanetary space showed evidence for a strong time dependence of solar wind properties. To explain these, *the Alfvén wave driven wind theory* was developed by Parker (1965), Alazraki and Couturier (1971) and Belcher (1971). To understand the effects of the co-rotating magnetic field on the wind, Weber and Davis (1967) developed *the magnetic rotator theory.* Their model showed that angular momentum can be transferred from a star by a wind. It also led to an explanation of the slow rotation rate of the sun; a long standing problem in astronomy.

Even the origins of *radiation driven wind theory* can be traced to solar studies. The first idea that atoms could be pushed out by the absorption of radiation came in a paper by Saha (1919), not long after the quantum theory was developed. Radiation was then realized to propagate as photons, each having a momentum $h\nu/c$. The absorption of the momentum in the radiation field could 'levitate' or drive matter upward. Milne (1924) and Johnson (1925) developed the basic equations for acceleration of matter by line radiation, and they suggested that radiation pressure could extend and heat the outer solar atmosphere.

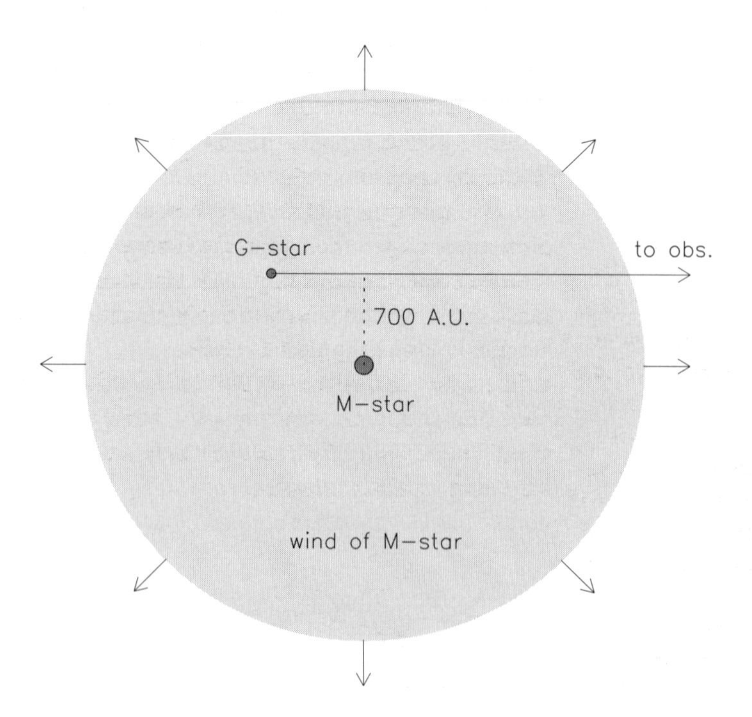

Figure 1.1 Deutsch's model of the α Herculis system. The G-type visual companion lies within the expanding envelope of the M5II star. The G star plays no appreciable role in the mass loss, but absorption lines in its spectrum provide a probe of the wind for the observer located off to the right. (After Deutsch, 1960)

The modern version of radiation driven wind theory was developed by Lucy and Solomon (1970), and Castor, Abbott and Klein (1975). The motivation for these radiation driven wind models was not solar observations, but rather the rocket observations of the UV spectra of a few O and B supergiants by Morton and his collaborators during the late 1960s. Figure (1.2) shows the spectrum of the hot supergiant ζ Ori which has several P Cygni lines, and wind-broadened lines, that indicated to Morton (1967) that the star has a mass loss rate of a few times 10^{-6} $M_{\odot}\,\mathrm{yr}^{-1}$ and a velocity of about 2500 $\mathrm{km\,s}^{-1}$.

The field of *stellar winds* became very lively after observations could be made in essentially all of the spectral regions that are not available from the ground. Rockets, balloons, and satellites launched since the mid 1960s have shown evidence for stellar winds at wavelength bands including X-rays, extreme-ultraviolet, ultraviolet, optical, infrared, millimeter and the radio region. The observations show evidence for phenomena ranging from masering in OH/IR stars, to X-ray and synchrotron emission in luminous hot stars.

1.2 Chronology of the development of the subject of stellar winds

In Appendix A we give a chronological list of observational discoveries and theoretical developments in the field of stellar winds. This list is far from complete, but it allows us to trace the development of specific topics discussed in this book.

We can identify five phases that have led to our current level of understanding of winds.

(I) The first phase (1572-1910) is the realization that *nova-like phenomena* indicate that stars eject mass, and that this ejection can be detected spectroscopically. This phase runs from the detection of the nova-like objects B Cas and P Cygni around 1600 to the photographic studies of the early 20th century.

(II) The second phase (1913-1947) involves the development of *diagnostics of mass loss* leading to a fairly good picture of the outflow from hot stars and a basic understanding of the physical conditions in the outer solar atmosphere.

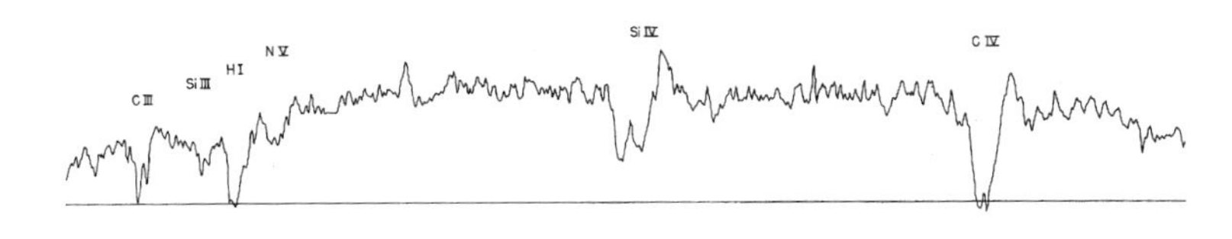

Figure 1.2 The spectrum of ζ Orionis O9.5Ia, obtained in one of the first rocket observations of stars by Morton (1967). The resonance lines of Si IV, at 1393.8Å, and C IV, at 1549.5Å, have P Cygni profiles. (From Morton, 1969)

(III) The third phase (1947-1967) contains the proof for *mass loss from cool giants* and the development of the subject of the *solar wind.*
(IV) The fourth phase (1967-1982) traces the rapid development of the subject of *stellar winds* made possible by observations with space instruments.
(V) The final phase (1980-present) is the current era which is concerned with *time dependent phenomena.* Effects such as rotation and pulsation can lead to non-spherical and variable winds. Winds from luminous stars or collections of stars in other galaxies are also current topics of interest.

1.3 Conclusions

It took many decades for astronomers to come to the realization that most classes of stars undergo some form of continuous mass loss. However, once observations could be made from space, the subject of *stellar winds*, as separate from the solar wind, developed rapidly. Among the space observatories that have been useful for studying winds are the *Copernicus* satellite, the *International Ultraviolet Explorer*, (*IUE*), the *Einstein* X-ray observatory, the *Infrared Astronomy Satellite* (*IRAS*), *Extreme Ultraviolet Explorer*, (*EUVE*), *Hubble Space Telescope*, and the *Infrared Space Observatory* (*ISO*).

Currently, global properties of stellar winds and mass loss are reasonably well understood for most types of stars. However, there are still many aspects of stellar winds that are uncertain. We do not really understand the variability of stellar winds, nor the processes that lead to variability. Magnetic fields can play an important role in affecting the mass loss and velocity structure of winds, but we know very little about strengths and geometrical distribution of the fields in stars. Pulsation plays an important role in initiating outflow from cool giant stars, but our understanding of the process is quite basic. The rotation of a star produces interesting, but poorly understood effects.

Almost all the knowledge that we have about stellar winds comes from the study of stars in our galaxy. Until recently there were only a few observations of stellar winds in other galaxies, simply because it is difficult to obtain the spectral and spatial resolution required for detailed studies. Ultraviolet spectra obtained with the *Hubble Space Telescope* of stars in the LMC, SMC and M31 show that the mass loss rates depend on metallicity, so that stars in lower metallicity galaxies have smaller mass loss rates than galactic stars of the same luminosity, temperature and mass. The evolution of stars depends crucially on their mass loss during various phases in their lives. Changes in the mass loss rate by even a factor two can drastically modify the evolution of a star. Therefore the evolution of massive stars in galaxies of low metallicity may be very different from that in our galaxy.

1.4 Suggested reading

Clerke, Agnes M. 1903, *Problems in Astrophysics* (London: Adams & Black).
(This book presents a fascinating review of the state of the then new field of 'astrophysics' at the end of the 19th century.)

de Jager, C. 1980, *The Brightest Stars* (Dordrecht: Reidel)
(A good overview of the classes of stars that have the largest mass loss rates. Such stars are located across the upper luminosity portion of the Hertzsprung Russell diagram.)

Goldberg, L. 1986, 'Mass Loss' in *The M-Type Stars* NASA SP-492, eds. H. R. Johnson & F. R. Querci

Parker, E. N. 1965, 'Dynamical Theory of the Solar Wind', *Space Science Reviews* **4**, 666.
(A review about early solar wind studies)

2 Observations of stellar winds

Stars emit not only radiation but also particles. The emission of particles is called *the stellar wind.*

The two most important parameters regarding a stellar wind that can be derived from the observations are the *mass loss rate* $\dot{M}$, which is the amount of mass lost by the star per unit time,† and the *terminal velocity* v_∞, which is the velocity of the stellar wind at a large distance from the star. By convention, the mass loss rate $\dot{M}$ is always positive and it is expressed in units of solar masses per year, with $1\ M_\odot\ \mathrm{yr}^{-1} = 6.303 \times 10^{25}\ \mathrm{g\ s}^{-1}$. A star with $\dot{M} = 10^{-6}\ M_\odot\ \mathrm{yr}^{-1}$, which is not an unusual value, loses an amount of mass equal to the total mass of the earth in three years. The terminal velocity v_∞ of a stellar wind ranges typically from about $10\ \mathrm{km\,s}^{-1}$ for a cool supergiant star to $3000\ \mathrm{km\,s}^{-1}$ for a luminous hot star.‡

The values of $\dot{M}$ and v_∞ are important because

(1) $\dot{M}$ describes how much material is lost by the star per unit of time. This is important for the evolution of the stars, because stars with high mass loss rates will evolve differently from those with low mass loss rates.

(2) Different stellar wind theories predict different mass loss rates and different terminal velocities for a star. So by comparing the observed values with the predictions we can learn which mechanism is responsible for the mass loss from a star.

(3) The gas that escapes from the star into space carries kinetic energy into the interstellar medium. The amount of kinetic energy that a stellar wind deposits into the interstellar medium per unit of

† The symbol $\dot{M}$, called 'M-dot', refers to the conventional use of the superscript 'dot' to indicate a time-derivative.

‡ The 'terminal' velocity should not be confused with the 'thermal' velocity of the gas in the wind, which is typically of the order of $10\ \mathrm{km\,s}^{-1}$.

time is $0.5\dot{M}v_\infty{}^2$. In order to study the effect of a stellar wind on its surrounding interstellar material, the values of $\dot{M}$ and v_∞ throughout the evolution of the star have to be known.

For a star with a stationary spherically symmetric wind, the mass loss rate is related to the density and the velocity at any point in the wind via the *equation of mass continuity*

$$\dot{M} = 4\pi r^2 \rho(r) v(r) \tag{2.1}$$

where r is the distance from the center of the star, ρ is the density and v is the velocity. This equation states that no material is destroyed nor created in the stellar wind, so the same amount of gas flows per second through a sphere at any radial distance r from the center of the star.

The gas that escapes from the outer layers of the stars is accelerated outwards from a small radial velocity ($v \leq 1$ km s^{-1}) at the photosphere of the star, to some high velocity at large distance from the star. At very large distance r, measured from the center of the star, the velocity asymptotically approaches the terminal velocity, $v_\infty = v(r \to \infty)$. The distribution of the velocity of the wind with radial distance from the star is called the *velocity law*, $v(r)$. Observations and models of stellar winds indicate that the velocity law can often be approximated by a *β-law*, which varies as

$$v(r) \simeq v_0 + (v_\infty - v_0)\left(1 - \frac{R_*}{r}\right)^\beta \tag{2.2}$$

This law describes a general increase of v with distance from v_0 at the photosphere ($r = R_*$) to v_∞ at large distance, with $v_0 \ll v_\infty$. The parameter β describes how 'steep' the velocity law is. For instance, hot stars have winds with a velocity law of $\beta \simeq 0.8$. These winds experience a fast acceleration and reach 80 percent of their terminal velocity already at $r = 4.1\ R_*$, i.e. 3.1 R_* above the stellar surface. Winds of cool stars accelerate more slowly, corresponding to larger values of β. Figure (2.1) shows the β-law for several values of β.

The velocity law can also be approximated by an alternative form of the β-law

$$v(r) \simeq v_\infty \left(1 - \frac{r_0}{r}\right)^\beta \tag{2.3}$$

with

$$r_0 = R_* \left\{1 - \left(\frac{v_0}{v_\infty}\right)^{1/\beta}\right\} \tag{2.4}$$

This form of the β-law has the advantage that it is easier to handle when it appears in integral equations, for instance in the expressions for the column density or the emission measure that we will use later. It is slightly different from the previous expression at small distances from the photosphere.

In this chapter we describe the several types of observations of stellar winds. For each type of observation we also give a short description of the analysis of the observations. We first discuss the different processes of line formation that can occur in stellar winds (§ 2.1) We discuss the observations of P Cygni profiles, atomic emission lines, the radio and infrared radiation from winds, molecular emission lines and the infrared radiation from dusty winds in §§ 2.2 through 2.6. In §§ 2.7 and 2.8 we summarize the mass loss rates and the terminal velocities of the winds of hot and cool stars that have been derived from the observations.

2.1 The formation of spectral lines in stellar winds

The spectral lines from stellar winds can easily be distinguished from the photospheric lines because of their large width or wavelength shift due to the outflowing motion of the gas in the wind. Wind lines can appear in absorption, in emission or as a combination of the two: a P Cygni profile. This depends on the efficiency of the creation or destruction of line photons in the stellar wind. We can distinguish

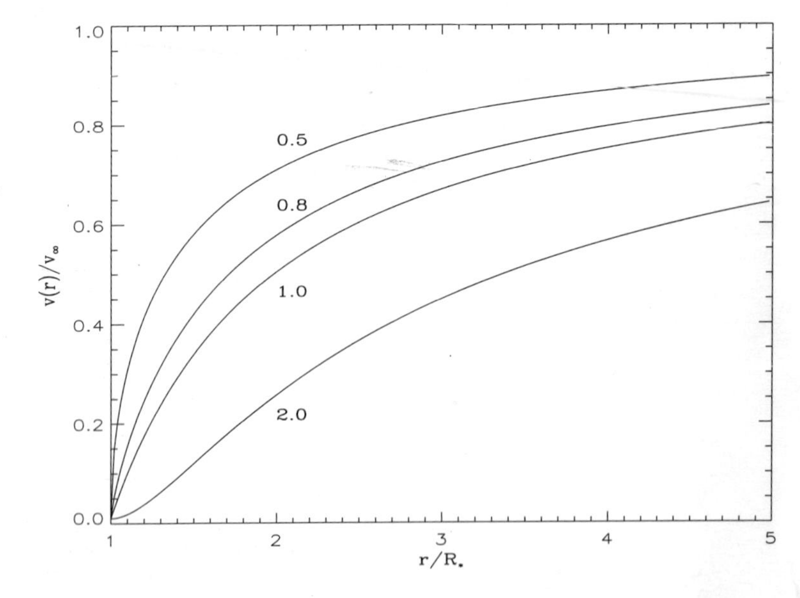

Figure 2.1 The β-type velocity laws for stellar winds for several values of β, for an assumed initial velocity of $w_0 = v_{sound}/v_\infty = 0.01$. The law for $\beta = 0.8$ agrees with the observations and the theory for the winds of O stars.

five processes for line formation in winds. Four of them are shown in Fig. (2.2).

2.1.1 Line scattering

If a photon emitted by the photosphere is absorbed by an atom in an atomic transition, it gives rise to photo-excitation of an electron of the atom. After a very short time the photon is re-emitted again by electron de-excitation to the same original level. In this case the emitted photon has almost the same frequency as the absorbed photon, apart from the Doppler shift due to the thermal motions of the atom. This process is called 'line scattering' because it seems that the photon was only scattered in another direction. It is important for line transitions of low atomic levels because the life time for spontaneous de-excitation of low excited levels is small (typically 10^{-10} to 10^{-9} seconds). If the line transition is from the ground state of the atom, the line is called a 'resonance line' and the scattering is called *'resonance scattering'*. Most of the observed P Cygni profiles are formed by resonance scattering.

2.1.2 Line emission by recombination

If an ion in a stellar wind collides with an electron it can recombine. The most likely recombination is directly to the ground state of the ion. However the ion can also recombine to an excited state. The resulting excited ion may then cascade downwards in the energy level diagram by photo-deexcitation. Each de-excitation results in the emission of a line photon, so this process creates line photons in a stellar wind. Lines from specific electron transitions, that have a high probability of being fed by recombination with subsequent photo-deexcitation,

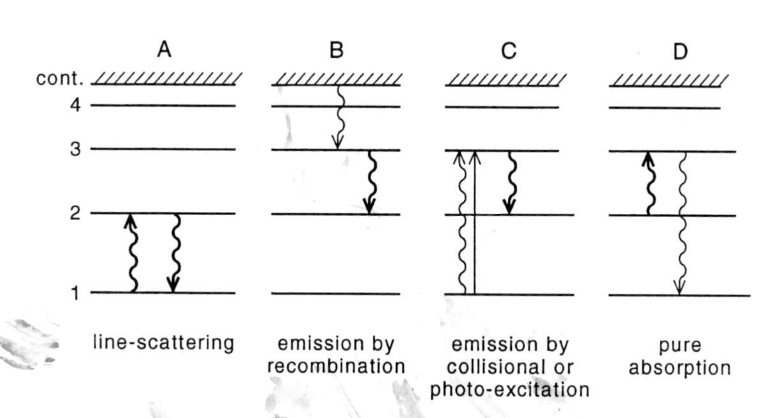

Figure 2.2 Different processes for line formation in stellar winds.

may thus appear in emission. This process is responsible for the Hα emission and the infrared emission lines in the winds of hot stars.

2.1.3 Line emission from collisional- or photo-excitation

Excitation of an atom by collisions from the ground state to an excited state can result in subsequent photo-deexcitation to a lower level. This process creates line-photons by converting kinetic energy into photon energy. This process is most efficient in hot plasmas where collisions are energetic and frequent. It is responsible for the formation of emission lines from hot chromospheres and coronae.

If the excitation is not by collisions but by photo-excitation from the ground state to a higher excited state, subsequent photo-deexcitation to another excited state will also create line photons. This process can give rise to emission lines from stellar winds. However, it is usually not important because photo-excitation from the ground state results mainly in resonance scattering.

2.1.4 Pure absorption

Photo-excitation of an excited atom into a higher excited state can be followed by spontaneous de-excitation to another (lower) level. This results in the destruction of the photons of the first transition and the creation of photons of the second transition. In practice, this process is not important for stellar winds because the vast majority of atoms in a stellar wind are in their ground state.

2.1.5 Masering by stimulated emission

If a photon traveling through a stellar wind hits an excited atom or molecule that can emit exactly the same photon by photo-deexcitation, the process of stimulated emission becomes important. The atom then de-excites to the lower level by emitting a photon with exactly the same frequency and *in the same direction* as the original photon. Instead of one, there will now be two similar photons travelling in the same direction. Repetition of this process can result in the creation of large numbers of line photons travelling in the same direction. This is called 'masering'. In order for masering to be efficient, the conditions in the wind have to be just right: a high fraction of the atoms or molecules should be in the excited upper level of the transition and there should be no velocity gradient in the direction of the travelling photons. (If there is a velocity gradient, the Doppler shift will inhibit the process). This process is responsible for the very strong and usually very narrow maser emission lines of abundant molecules in the winds of cool stars.

2.2 P Cygni profiles

The most sensitive indicators of mass loss from hot stars are the spectral lines due to atomic transitions from the ground state, called *resonance lines*, of abundant ions. Well-known examples are the resonance lines of C IV, N V and Si IV on the ultraviolet (UV) spectra of stars of types O to early-B, the resonance lines of C II in the UV spectra of stars of types late-B to A and the resonance lines of Mg II in the UV spectra of stars of types late-B to M. The large abundance of these ions combined with the large oscillator strengths of their atomic resonance transitions implies that they can produce an observable absorption line that shows the Doppler shift due to the outflow, even if the mass loss rate is small.

If the column density of the absorbing ions in the wind between the observer and the stellar photosphere is small, between 10^{13} and 10^{14} ions cm^{-2}, the resonance lines will produce a weak but observable absorption component. This absorption component is Doppler shifted to a shorter wavelength because it is formed in a region that is moving outward from the star and toward the observer. If the column density of the absorbing ions is high, say larger than about 10^{15} ions cm^{-2}, the lines will have so-called 'P Cygni profiles' consisting of a violet-shifted absorption component and a red-shifted† emission component. This is shown in Fig. (2.3).

The most important lines that show P Cygni profiles are listed in Table (2.1). Atlases of observed P Cygni profiles have been published by Snow and Morton (1976) and by Snow *et al.* (1994). The first mass loss determination based on a detailed analysis of P Cygni profiles was made by Lamers and Morton (1976).

Most of the observed P Cygni profiles of the UV resonance lines in the spectra of hot supergiants are due to scattering of photons rather than to absorption. These lines are from an electronic transition

† The terms 'violet-shifted' and 'red-shifted' will be used here to indicate Doppler shifts to shorter and longer wavelengths than the rest wavelength of the spectral line. We use these terms independent of the location of the line in the spectrum. So a 'violet-shift' refers to a shift to shorter wavelength even if the line is in the ultraviolet wavelength range.

between the ground state and the first excited state. After absorption with photo-excitation, the atom has a high probability of returning to the ground state via photo-deexcitation in the same transition. So the photon is not lost, but only scattered to a different direction.

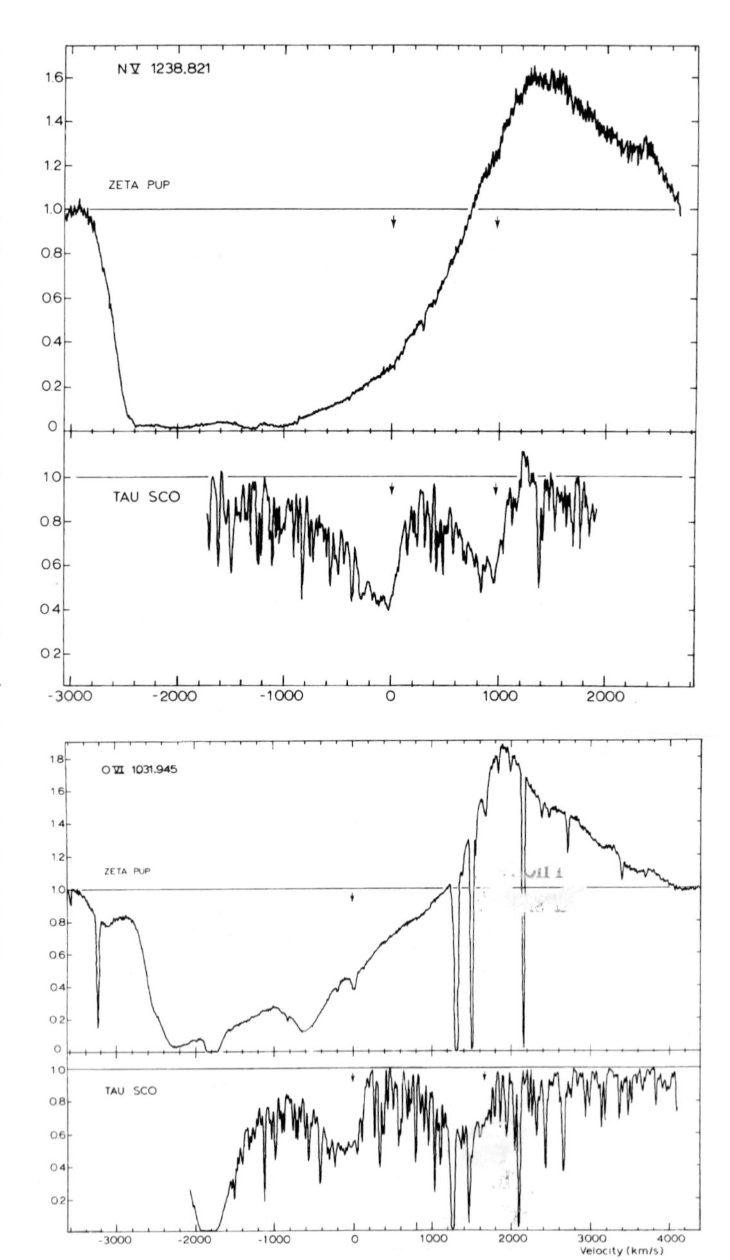

Figure 2.3 The P Cygni profiles of the N V doublet (upper) and the O VI doublet (lower) in the UV spectrum of ζ Pup (O4If) and τ Sco (B0V). The rest wavelengths are indicated by small arrows. The doublet lines blend into one strong P Cygni profile in the spectrum of ζ Pup, and they are seen separately in the spectrum of τ Sco. The profiles of τ Sco extend to $v \simeq -1500$ km s^{-1}, and those of ζ Pup to -2700 km s^{-1}. The spectrum of τ Sco also shows many narrow photospheric absorption lines. (From Lamers and Morton, 1976 and Lamers and Rogerson, 1978)

Table 2.1 *Some important ultraviolet lines that show the effects of mass loss*

Ion	Abundance	Ion.pot (eV)	λ (Å)	Exc.pot (eV)	g_ℓ	f	Notes
C II	3.7×10^{-4}	11.26	1334.532	0.00	2	0.128	
			1335.708	0.01	4	0.319	3 lines
C III	3.7×10^{-4}	24.38	1175.67	6.50	9	0.257	6 lines
C IV	3.7×10^{-4}	47.89	1548.195	0.00	2	0.191	
			1550.770	0.00	2	0.095	
N IV	1.1×10^{-4}	47.45	1718.551	16.20	3	0.179	
N V	1.1×10^{-4}	77.47	1238.821	0.00	2	0.157	
			1242.804	0.00	2	0.078	
O VI	6.8×10^{-4}	113.90	1031.928	0.00	2	0.130	
			1037.619	0.00	2	0.065	
Mg II	3.5×10^{-5}	7.65	2795.528	0.00	2	0.612	
			2802.705	0.00	2	0.305	
Si II	3.5×10^{-5}	8.15	1526.707	0.00	2	0.230	
			1533.431	0.00	2	0.229	
Si III	3.5×10^{-5}	16.35	1206.500	0.00	1	1.669	
Si IV	3.5×10^{-5}	33.49	1393.755	0.00	2	0.514	
			1402.770	0.00	2	0.255	
Fe II	2.5×10^{-5}	7.87	2585.876	0.00	10	0.065	
			2598.370	0.05	8	0.099	
			2599.396	0.00	10	0.224	

column 2: the abundance of the element relative to H by number.
column 3: the ionization potential of the next lower ion.
columns 5 and 6: the excitation potential and the statistical weight of the lower level of the transition.
column 7: f is the oscillator strength of the transition.

2.2.1 A qualitative explanation of P Cygni profiles

The formation of P Cygni profiles can be understood qualitatively by a simple model of a spherically symmetric outflowing wind in which the velocity increases outward, such as shown in Fig. (2.4). An outside observer can recognize four regions which contribute to the formation of a spectral line:
(a) the star (S) which emits a continuum, possibly with a photospheric absorption component at the rest wavelength λ_0 of the line.
(b) the 'tube' (F) in front of the stellar disk. The gas in F is moving to the observer with velocities between $v \simeq 0$ and v_∞.
(c) The 'tube' (O) that is occulted by the star. The gas in O is moving away from the observer, but the radiation from this region does not reach the observer.

(d) The region to the sides of the star that would be observed as a halo (H) around the star if the wind could be resolved spatially by the observer. The gas in region H has both positive and negative velocity components along the line of sight to the observer.

Figure (2.4) also shows the contribution of the different regions of a wind to the flux that gives rise to a P Cygni profile. The star S emits a continuum with a photospheric absorption line. The region F in front of the star scatters radiation from the star, that would have reached the observer if there were no wind, out of the line of sight to the

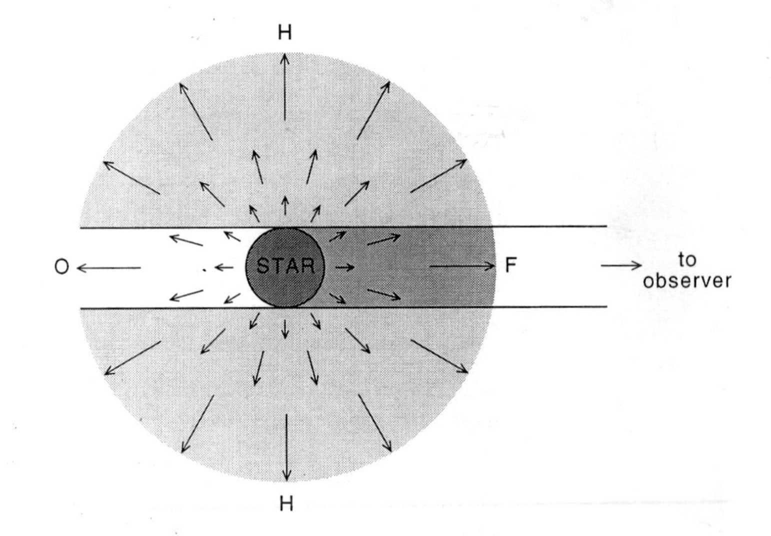

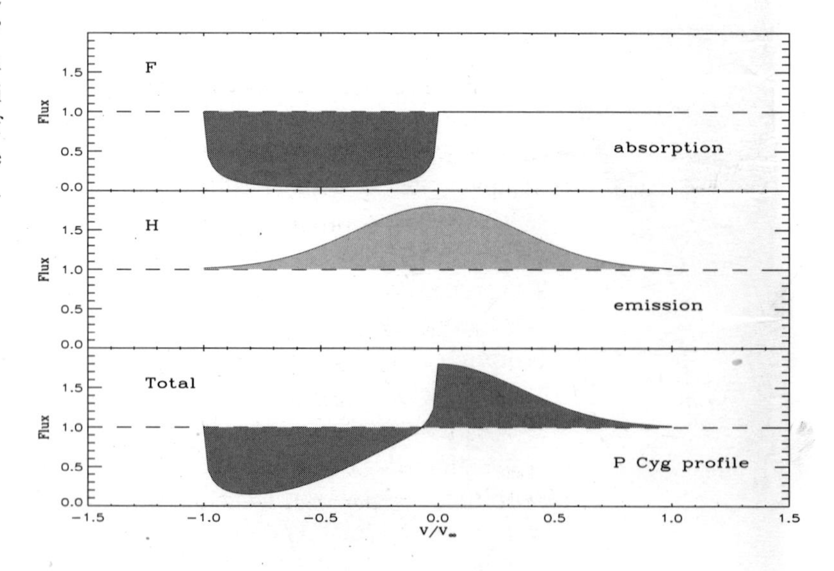

Figure 2.4 Top: the geometry of a stellar wind with outward increasing velocity. The observer can distinguish four regions: the star (S), the 'tube' in front of the star (F), the occulted 'tube' (O) behind the star, the 'halo' (H) surrounding the stellar disk. Bottom: the contributions from the star (the continuum) with the absorption from F and the emission from H. The P Cygni profile is the sum of these three contributions.

observer. This removal of photons produces a violet shifted absorption component with a Doppler shift between $-v_\infty$ and 0. This absorption does not reach zero flux because there is also some scattering into the line of sight in this region. The halo H scatters radiation from the photosphere in all directions. Some of this radiation is sent in the direction of the observer. This produces an emission component with a Doppler shift between $-v_\infty$ and v_∞, but with the largest contribution at Doppler shift 0. The net result, which is the simple addition of the contributions from the region in front of the star (S+F) and the halo, is a P Cygni profile.

2.2.2 A quantitative explanation of P Cygni profiles

An alternative way to understand the shape of the P Cygni profiles, that gives more quantitative insight, is shown in Figs. (2.5) and (2.6). Isotropic scattering of stellar photons in a geometrically and optically thin shell around the star at distance between r and $r + \Delta r$ with expansion velocity between v and $v + \Delta v$ produces a line profile that consists of a narrow violet-shifted absorption component and a *flat* emission component. The absorption component extends from Doppler velocity $-v$ to $-v(1 - \cos\Theta_*)$ (if the intrinsic broadening by thermal motions is neglected), where $\sin\Theta_* = R_*/r$. The emission component extends from $-v(1-\cos\Theta_*)$ to $+v(1-\cos\Theta_*)$ (see Sobolev, 1960). The strength of the absorption depends on the column density of the number of absorbing ions in the shell. The amount of emission is equal to the amount of the absorption minus the emission between Doppler velocities $+v$ and $+v(1 - \cos\Theta_*)$ that is occulted by the star. The upper part of Fig. (2.5) shows the geometry of the shell and the lower part shows the resulting lineprofile. A P Cygni profile formed by scattering in a wind with increasing velocity can be considered as the sum of the contributions of many shells of different velocities. This is shown in Fig. (2.6). The upper part of the figure shows the contributions of the shells: each shell producing a narrow violet shifted absorption and a wide and flat emission. The sum of the emission and the absorption at each Doppler velocity gives the resulting P Cygni profile.

With this explanation it is immediately clear how the P Cygni profile would react to changes in the density or the ionization in the wind. We give a few examples:

(a) The presence of a shell with an extra high density in the wind at velocity v_S will give rise to an extra absoption dip at Doppler velocity $-v_S$ and a smeared out emission component between about $-v_S$ and $+v_S$.

(b) If the ion that produces the line does not occur within the inner part of the wind where $v(r) < v_{\rm in}$, the shells with a low velocity will not contribute. This gives a profile with a violet shifted absorption between $-v_\infty$ and $-v_{\rm in}$, and an emission between $-v_{\rm in}$ and $+v_\infty$, that is flat between $-v_{\rm in}$ and $+v_{\rm in}$.
(c) If the ion that produces the line does not occur at large distance in the wind where $v(r) > v_{\rm out}$, the shells with higher velocity will not contribute. This gives a narrow P Cygni profile that does not extend to the terminal velocity.
(d) If the observed ion only occurs in the wind close to the star, the

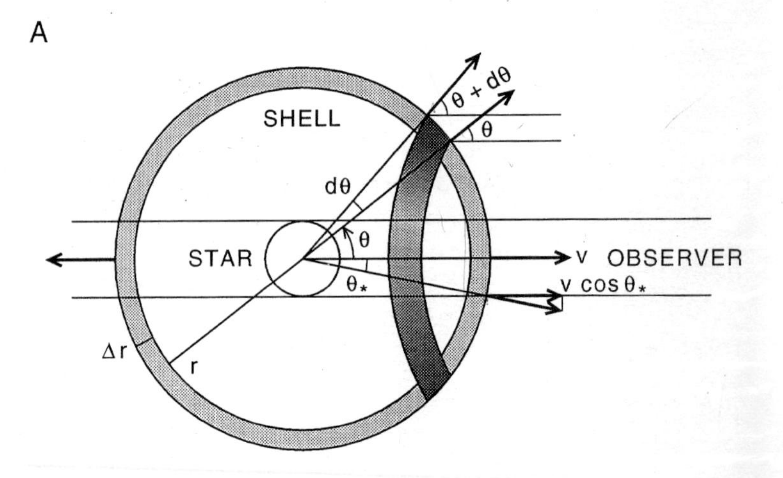

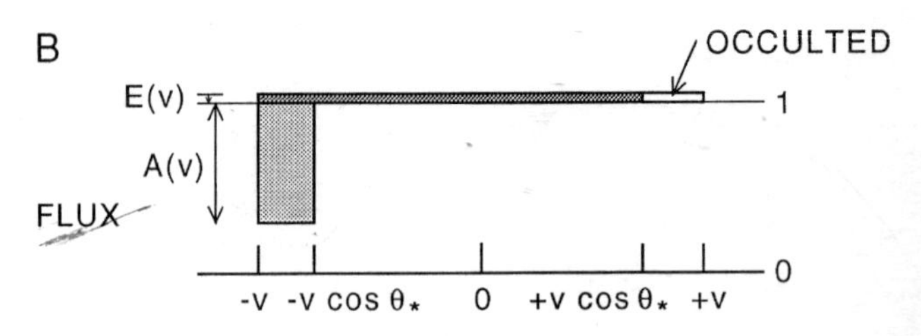

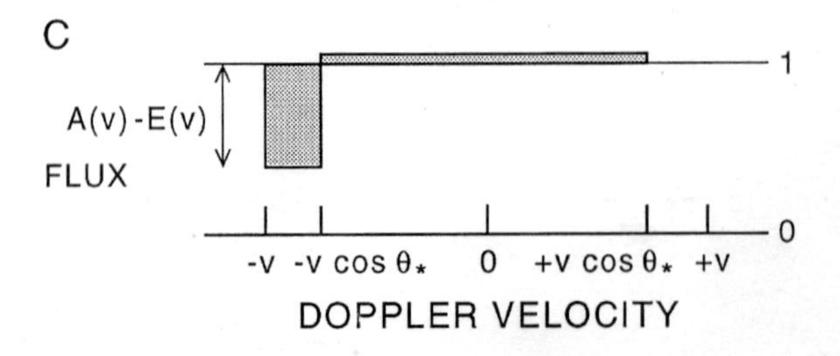

Figure 2.5 The profile of a thin scattering shell. A: the geometry of the shell. B: the resulting absorption, $A(v)$, and the flat emission $E(v)$. Part of this emission is occulted by the star. C: the resulting profile is the sum of the absorption and the emisson.

emission will be considerably smaller than the absorption because the occultation is significant.

Note that the cross-over from absorption to emission does not necessarily indicate the velocity of the star, but depends on the wavelength distribution of the absorption and the emission.

The ratio between the strength of the emission and absorption component depends on the size of the wind region where the scattering occurs relative to the size of the star. If the star emits a continuum

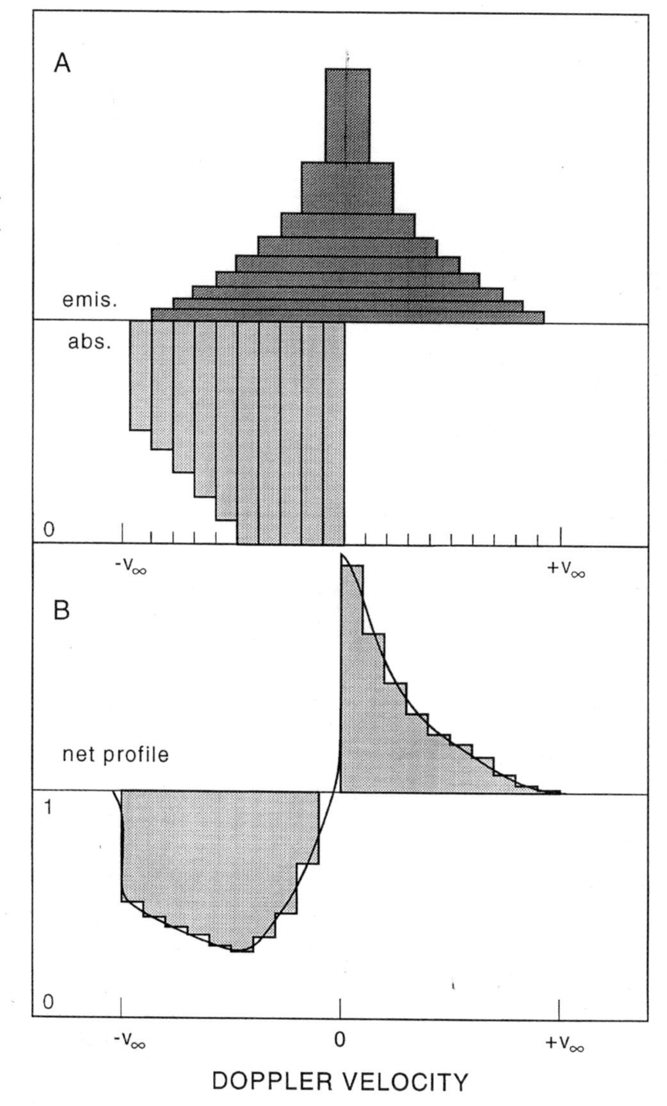

Figure 2.6 The profile of a scattering line in a wind with outward increasing velocity. Upper figure: the separate contributions of shells consist of narrow absorptions and flat wide emissions. Lower figure: the resulting P Cygni profile is the sum of all the contributions.

without an absorption component, the difference between the strength of the emission and the absorption component is only due to the photons that are lost by back-scattering into the star. If the star is small compared to the scattering region, little radiation is 'lost' and the emission will be about equal to the absorption. If the star is large compared to the size of the region in the wind where line scattering occurs, a considerable fraction of the photons are lost by back-scattering into the star. In that case the emission will be smaller than the absorption. In the limiting case of a thin scattering layer just above the photosphere half of the photons are scattered back into the star. If the photosphere emits only a continuum the emission component would thus be half as strong as the absorption component. The presence of a photospheric absorption component complicates this relation between the net absorption and the size of the scattering region and increases the strength of the absorption compared to that of the emission.

2.2.3 Mass loss studies of P Cygni profiles

The P Cygni profiles of UV lines in the spectra of early type stars have been studied quantitatively by various authors to derive information on the velocity law of the wind and the mass loss rates, e.g. Groenewegen *et al.* (1989), Howarth and Prinja (1989), Prinja *et al.* (1990), and Lamers *et al.* (1995).

For strongly saturated lines the profiles are mostly sensitive to the velocity law. So the velocity law and v_∞ can be derived most accurately from the profiles of saturated lines which have a steep violet absorption edge. The violet absorption edge of the saturated profiles reach the continuum at a Doppler velocity of $v_{\rm edge} \simeq -(v_\infty + 2v_{\rm turb})$, where $v_{\rm turb}$ is the turbulent velocity in the stellar wind at a distance of $r \gtrsim 10R_*$, where v_∞ is reached. The black (zero intensity) absorption trough of strongly saturated lines extend to a Doppler velocity of about $-v_\infty$ (see Fig. 2.3).

The mass loss rate can be derived from unsaturated P Cygni profiles. To this purpose the observed profiles are compared with predicted profiles for different radial distributions $n_i(r)$ of the observed ions in the wind. When the predicted and observed profiles match one another, the distribution of $n_i(r)$ is known. The ion density $n_i(r)$ can be converted into a density distribution $\rho(r)$ if the abundance and the ionization fraction of the observed ion is known in the wind.

$$n_i(r) = \frac{n_i(r)}{n_E(r)} \cdot \frac{n_E(r)}{n_H(r)} \cdot \frac{n_H(r)}{\rho(r)} \cdot \rho(r) = q_i(r) A_E \frac{n_H}{\rho} \frac{\dot{M}}{4\pi r^2 v(r)} \tag{2.5}$$

where $A_E = n_E/n_H$ is the abundance of the element with respect to H and $q_i = n_i/n_E$ is the fraction of ions in the right stage of ionization and excitation to produce the line. The ratio n_H/ρ depends on the composition of the wind and is $(1.36m_H)^{-1} = 4.43 \times 10^{23}$ atoms g^{-1} for solar composition. We have used the mass continuity equation (2.1) to express ρ in terms of $\dot{M}$. With $v(r)$ and $\rho(r)$ known, the mass loss rate can be derived. Thus we see that both the mass loss rates and terminal velocities of stellar winds can be derived from P Cygni profiles.

The terminal velocities and the velocity law can very well be derived from the P Cygni profiles (Prinja *et al.*, 1990; Lamers *et al.*, 1995). The determinations of mass loss rates from P Cygni profiles is much more complicated because it depends crucially on the adopted ionization fractions. These fractions are not well known for hot stars because of the sensitivity to non-LTE effects, the presence of shocks and the superionization (Groenewegen and Lamers, 1991; Pauldrach *et al.*, 1994; Lamers *et al.*, 1998).

2.3 Emission lines from winds

Stars with high mass loss rates, on the order of 10^{-6} $M_\odot$ yr^{-1} or higher, may show emission lines in their spectra. The best known example is the optical Hα line ($\lambda = 6562$ Å) in the spectra of O stars and B-type supergiants. The study of stellar winds by means of this line has the obvious advantage that the line can be observed from the ground (in contrast with the UV P Cygni profiles which require observations from satellites). Other examples of emission lines are the Paschen and Brackett lines of H in the near infrared and the He II lines at $\lambda = 1640$ and 4686 Å in the spectra of hot stars. The visual spectra of Wolf-Rayet stars (WR stars) are dominated by emission lines formed in their high density winds.

The strength and the profiles of the emission lines provide information about the mass loss rate and the velocity law of stellar winds. Most of the observed atomic emission lines are formed by recombination and thus have an emissivity proportional to ρ^2. This implies that the emission originates mainly from regions of high density, i.e. the lower layers of the wind close to the star, where most of the wind acceleration occurs.

Examples of emission lines are shown in Fig. (2.7). The emission lines are approximately symmetric around their rest wavelength with a typical FWHM of a few hundred km s^{-1}. This is smaller than the terminal velocity of the winds, which is typically about 2000 km s^{-1} for an O or an early-type B star, because the emission lines are formed in the lower layers of the winds where the density is high. The emission lines of WR stars have a FWHM comparable to the terminal velocity because these lines are formed over a very extended region of the wind due to the much higher density throughout the winds of these stars.

The general theory for the formation of emission lines in stellar winds was described by Sobolev (1960). The first reliable determinations of mass loss rates from the Hα line were made by Klein and Castor (1978) and drastically improved by Puls *et al.* (1995). The emission lines from the winds of Wolf-Rayet stars have been studied extensively, see e.g. Hamann *et al.* (1995) and references therein.

2.3.1 The determination of mass loss rates from optically thin emission lines

The large velocity gradients in stellar winds make it easy for line photons to escape. This is because in a spherical wind with an increasing outward velocity all the points are moving away from one another. This is shown in Fig. (2.8). As soon as a line photon, created by recombination, has traveled a distance $l > 2v_{\rm th}/(dv/dl)$, where $v_{\rm th}$ is the thermal velocity and (dv/dl) is the velocity gradient of the wind along the path of the photon, it is Doppler shifted with respect to the surrounding gas by more than $2v_{\rm th}$ and cannot be absorbed any more in the same line transition. So winds with $v_\infty \gg v_{\rm th}$ are optically thin for line radiation. This greatly simplifies the theory and the calculations of emission line profiles. The optically thin assumption is approximately valid for the Hα line and for the infrared recombination lines from the winds of early type stars.

If the wind is completely optically thin for a line transition, the mass loss rate can be derived directly from the luminosity L_l of the

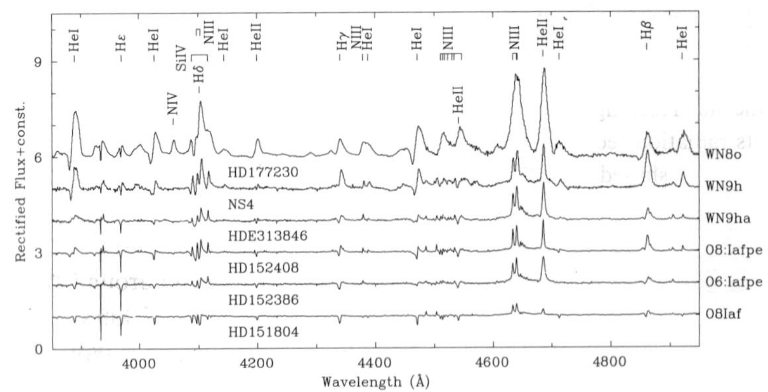

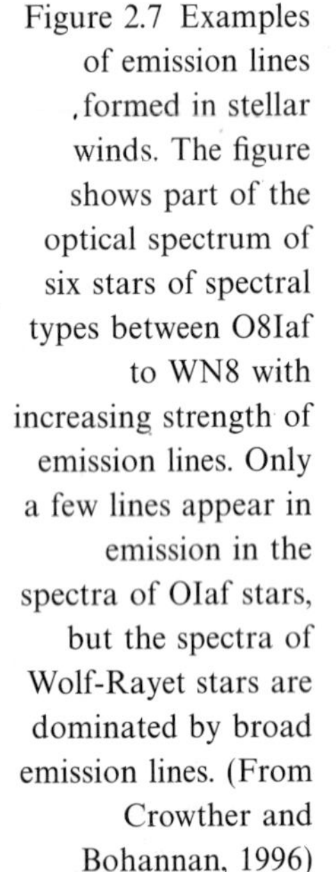
Figure 2.7 Examples of emission lines formed in stellar winds. The figure shows part of the optical spectrum of six stars of spectral types between O8Iaf to WN8 with increasing strength of emission lines. Only a few lines appear in emission in the spectra of OIaf stars, but the spectra of Wolf-Rayet stars are dominated by broad emission lines. (From Crowther and Bohannan, 1996)

line, i.e. the energy emitted per second by the star in the line. This quantity can be derived from the spectrum if the distance to the star is known.

Let $j_l(r)$ be the *line emissivity* of the wind, which is the amount of energy generated by the emission of line photons, in erg cm^{-3} s^{-1} at distance r from the star. The emissivity of H and He lines has been tabulated by Osterbrock (1989). The emissivity of the Hα line in an ionized wind of $5000 < T < 20\,000$ K is

$$j_l \simeq 3.56 \times 10^{-25}\, n_e\, n_p\, (T/10^4)^{-0.96} \tag{2.6}$$

erg s^{-1} cm^{-3}, where n_e and n_p are the electron and proton densities. (Case B: Osterbrock, 1989). The total line luminosity is simply the volume integral over the wind of the emissivity, apart from a correction for the fraction of the emitted photons that are intercepted by the star. So the line luminosity is

$$L_l = \int_{r_{\rm min}}^{\infty} 4\pi r^2 j_l(r)\{1 - W(r)\}dr \tag{2.7}$$

The integration starts at a distance $r_{\rm min}$, which is the distance where the optical depth of the wind reaches a value of about 1 for continuum

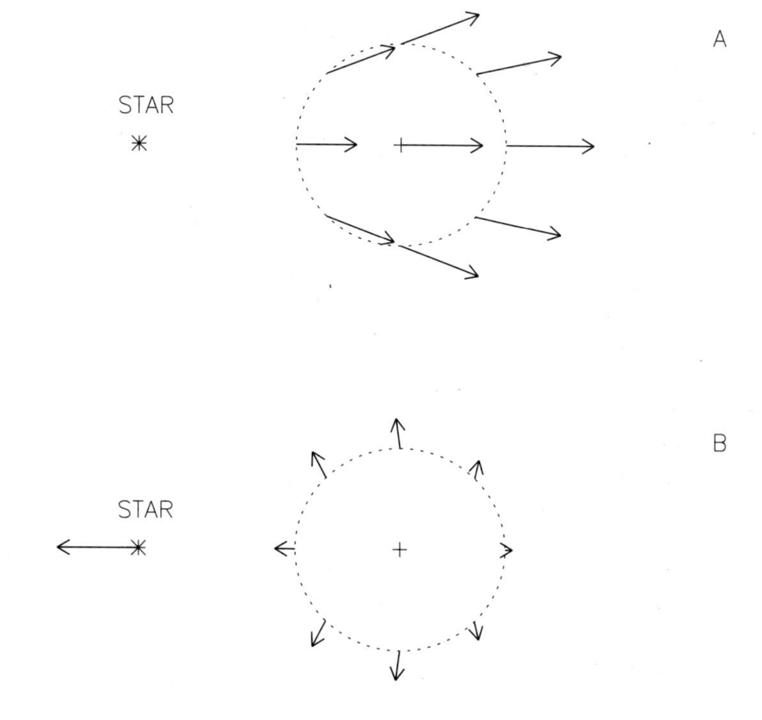

Figure 2.8 A: the velocity vectors in an expanding wind in and around a point, indicated by +. The star is indicated by an asterisk. B: the velocity vectors in the comoving frame relative to point +. An observer in + sees its surrounding and the star receding and its radiation red shifted.

radiation at the wavelength of the line. This is because line photons cannot escape if they are emitted in layers that are optically thick for continuum radiation at their wavelength. If the wind is optically thin for continuum radiation, one can take the radius of the sonic point $r_s \simeq R_*$ as the lower limit, because the velocity gradient of the wind is large at $r > r_s$ and small at $r < r_s$. So photons cannot easily escape from layers below the sonic point.

The factor $W(r)$ describes the probability that a photon, emitted at r in a random direction, is intercepted by the star, which has a radius $r_{\rm min}$ at the wavelength of the line. The factor $1 - W(r)$ is the probability that a photon emitted at r escapes. $W(r)$ is the fraction of the solid angle that is covered by the star, so

$$W(r) = \frac{1}{2}\left\{1 - \sqrt{1 - (r_{\rm min}/r)^2}\right\} \tag{2.8}$$

This factor is called the *geometrical dilution factor.* Photons emitted just above $r_{\rm min}$ have a fifty percent chance of disappearing into the optically thick layers below, so $W(r \rightarrow r_{\rm min}) = 0.5$. Photons emitted at a large distance have a much smaller probability of being intercepted by the star, because the stellar disk covers only a small fraction of the total solid angle of 4π steradians. It is easy to show that for a large distance $W(r) \simeq (r/2r_{\rm min})^2$ if $r \gg r_{\rm min}$.

The emissivity of recombination lines is proportional to the density squared, because it involves collisions between ions and electrons, and it is a function, $f_{\rm rec}(T)$, of temperature. Therefore one can write

$$j_l(r) = \rho^2(r) f_{\rm rec}(T) \tag{2.9}$$

The density in the wind can be expressed in terms of the mass loss rate and the velocity law via the mass continuity equation $\rho(r) = \dot{M}/4\pi r^2 v(r)$, Eq. (2.1). It is convenient to define the dimensionless distance parameter $x \equiv r/R_*$ and velocity $w \equiv v/v_\infty$, because the normalized velocity law $w(x)$ is about the same for most early type stars, despite large differences in R_* and v_∞. Substitution of Eqs. (2.9) and (2.8) into Eq. (2.7) gives a simple expression for the line luminosity

$$L_l = \frac{1}{8\pi R_*}\left(\frac{\dot{M}}{v_\infty}\right)^2 \int_{x_{\rm min}}^{\infty} \frac{f_{\rm rec}(T)}{wx^2}\left\{1 - \sqrt{1 - (x_{\rm min}/x)^2}\right\} dx \tag{2.10}$$

If the velocity law and the temperature structure of the wind are known the integral can be calculated and the line luminosity can be expressed in terms of the mass loss rate.

Equation (2.10) shows that the line luminosity of optically thin lines is proportional to $(\dot{M}/v_\infty)^2$ or $\dot{M} \sim v_\infty \sqrt{L_l}$. The dependence of $\dot{M}$ on $\sqrt{L_l}$ makes the line luminosity an accurate indicator of the mass loss rate, provided that the terminal velocity is known from the UV P Cygni profiles.

The emission of Hα has been used to derive mass loss rates from O stars by Klein and Castor (1978), Leitherer (1988), Scuderi *et al.* (1992), Lamers and Leitherer (1993) and Puls *et al.* (1995).

2.4 The infrared and radio excess from stellar winds

Stars with an ionized stellar wind emit an excess of continuum emission at long wavelengths, i.e. from the infrared (IR) to the radio region. The *excess flux* is measured relative to the flux expected from the photosphere of the star if it did not have a wind. The excess flux is due to free-free emission (Bremsstrahlung) from the wind. The free-free emission depends on the density and temperature structure of the winds. For a given temperature structure, the density structure can be derived from the energy distribution of the excess flux. If the velocity law is known, e.g. from the UV lines, the mass loss rate can be determined from the density and the velocity. (The radio or infrared excess itself does not give information about the velocity of the wind because it is continuum radiation so the Doppler effect cannot be used.)

The *radio excess* or *mm-excess* has been measured for only a few dozen early type stars, mainly by Abbott and colleagues (see e.g. Bieging *et al.*, 1989), by Leitherer *et al.* (1995, 1997) and Scuderi *et al.*, (1997). The number of stars is small because the free-free emission from stellar winds is very weak and only on the order of a few milli-Janskys (1 Jy $= 10^{-23}$ erg cm^{-2}s^{-1} Hz^{-1}) at a wavelength of 6 cm for a star at a distance of a kiloparsec with a mass loss rate of about 10^{-6} $M_\odot$ yr^{-1}.

The *infrared excess* has been measured for many more stars, mainly from the ground by Barlow and Cohen (1977), Abbott *et al.* (1984) and by the IRAS satellite (see e.g. Waters *et al.*, 1987) The IR free-free emission from stellar winds was first measured and explained in terms of ionized winds by Barlow and Cohen (1977). For normal O and B stars the IR-excess is generally small and on the order of only a few tenths of a magnitude at 5 to 10 μm. The accuracy of the measurement of the excess thus depends critically on the accuracy of the photospheric flux in the IR. The IR-excess is formed in the lower layers of the wind, typically below $r < 1.5R_*$ where the acceleration of the wind occurs. The determination of mass loss rates from the IR excess thus requires an accurate knowledge of the velocity law.

Figure (2.9) shows the energy distribution of the star P Cygni, a luminous blue variable of spectral type B1 Ia^+, from the visual to radio wavelength as an example of a typical energy distribution with infrared and radio excess. Notice that the observed flux decreases to longer wavelength but that the excess increases relative to the expected photospheric flux, which is the Rayleigh-Jeans tail of a Planck function: $F_\nu \sim \lambda^{-2}$. The flux in the radio region is expected to be a power law of slope $F_\nu \sim \nu^{0.60} \sim \lambda^{-0.60}$ or $F_\lambda \sim \lambda^{-1.4}$ (see below). This is a typical characteristic of free-free emission.

2.4.1 A simple qualitative explanation of the free-free emission

For a simple explanation of the infrared and radio excess due to free-free emission we adopt the zero-order solution of the radiative

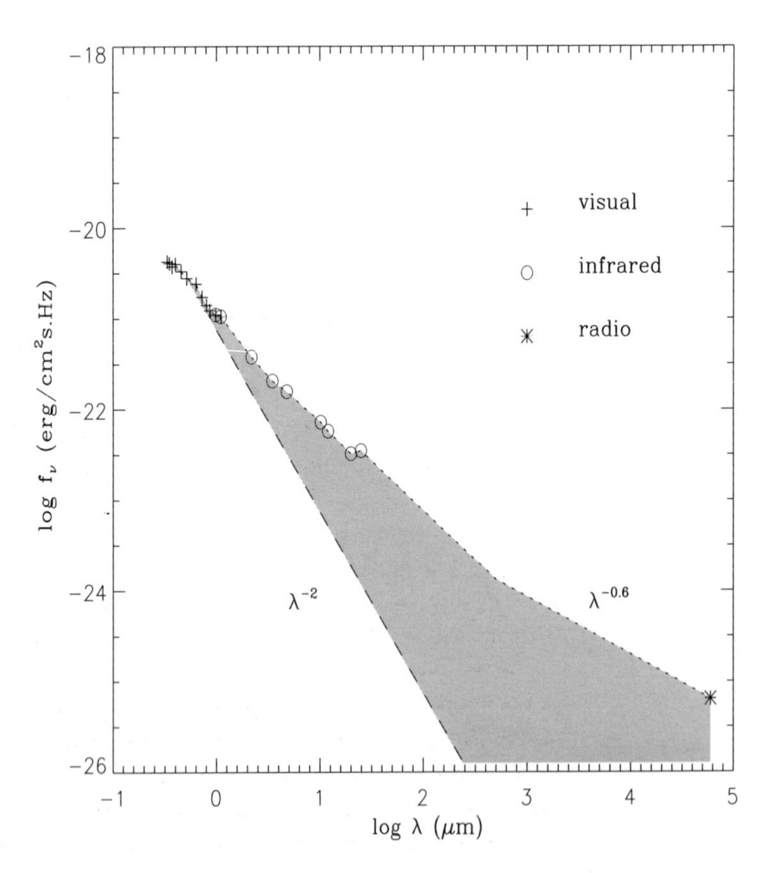

Figure 2.9 The energy distribution of the luminous blue variable star P Cygni. The dashed line is the expected spectrum of a hydrostatic photosphere without a wind, i.e. approximately a blackbody extrapolation of the visual flux. The grey area is the excess. (Data from Lamers *et al.*, 1983 and Waters and Wesselius, 1986)

transfer in a stellar atmosphere, i.e. the Eddington-Barbier relation

$$L_\nu \simeq 4\pi r^2(\tau_\nu = \tau_{\rm eff})\, \pi\, B_\nu(T(\tau_\nu = \tau_{\rm eff})) \tag{2.11}$$

where L_ν is the monochromatic luminosity in erg s^{-1}Hz^{-1}. This heuristic approach was first used by Cassinelli and Hartmann (1977). The effective optical depth, $\tau_{\rm eff}$, is 2/3 in a plane-parallel atmosphere and about 1/3 in an extended stellar wind. The free-free opacity increases to longer wavelengths as $\kappa_\nu \simeq \nu^{-2}$ (see below) and so the optical depth, $\tau_\nu = \int \kappa_\nu \rho dr$, along a line of sight into the wind also increases to longer wavelength as ν^{-2} or λ^2. This implies that the radius where $\tau = \tau_{\rm eff}$ in the wind increases with wavelength. So if we could observe a star with a stellar wind from a close distance, we would notice that the radius of the star increases with wavelength: an IR image would show a larger star than a visual image, and a radio image would show an even larger radius of about 10^2 R_*. We cannot observe the stars from a close distance, except the sun, so we cannot observe the increase in radius with wavelength. However we can observe the increase in monochromatic luminosity of the star with increasing wavelength that results from the increase in radius.

Compare the luminosity L_ν from a star with a wind, Eq. (2.11), with that expected from the star without a stellar wind

$$L_\nu^* \simeq 4\pi\, R_*^2\, \pi B_\nu(T_{\rm eff}) \tag{2.12}$$

where L_ν^* is the luminosity from the photosphere, and T_{eff} is the effective temperature of the star. The ratio between the flux from a star with a wind and without a wind is

$$\frac{L_\nu}{L_\nu^*} \simeq \frac{r^2(\tau_\nu = \tau_{\rm eff})}{R_*^2} \cdot \frac{B_{\nu\,\rm wind}(T(\tau_\nu = \tau_{\rm eff}))}{B_\nu(T_{\rm eff})} \tag{2.13}$$

This equation shows that a star with an ionized wind will have an *excess* radiation because $r(\tau_\nu = \tau_{\rm eff}) > R_*$, if $T_{\rm wind} \simeq T_{\rm eff}$. (The Planck function at long wavelength is not very sensitive to temperature and scales as $B_\nu \sim T$.) The excess will increase to longer wavelength because $r(\tau_\nu \simeq \tau_{\rm eff})$ moves outward with decreasing ν or increasing λ since $\kappa_\nu \sim \nu^{-2}$. This is shown schematically in Fig. (2.10).

From Eq. (2.13) we see that in principle the wind can also result in a *deficiency* of radiation if the wind is cold with $T_w \ll T_{\rm eff}$ so that the decrease of B_ν is larger than the increase of $r^2(\tau_\nu \simeq \tau_{\rm eff})$. The observations show that most early type stars with a wind have an IR or radio *excess*, so obviously the wind temperature is not much lower than the photospheric temperature. However there are some exceptions for supergiants with very massive winds.

2.4.2 The radio free-free emission

The free-free opacity or the free-free absorption coefficient per unit mass, $\kappa_\nu^{\rm ff}$ in cm^2 g^{-1}, of an ionized gas at long wavelength, where $h\nu < kT$, is

$$\kappa_\nu^{\rm ff} = 1.78 \times 10^{-2} Z^2 \, \nu^{-2} \, g_\nu \, T^{-3/2} \, \frac{n_i n_e}{\rho} \tag{2.14}$$

where Z^2 is the square of the rms charge of the atoms ($Z = 1$ for singly ionized gas), n_e and n_i are the electron and ion densities in cm^{-3}, ρ is the density in g cm^{-3} and g_ν is the gaunt factor which is approximately

$$g_\nu \simeq 10.6 + 1.90 \log T - 1.26 \log \nu Z \tag{2.15}$$

for radio waves (Allen, 1973, p. 102). In the range of $10\,000 < T < 50\,000$ and $0.1 < \lambda < 30$ cm the gaunt factor can be approximated by a power law

$$g_\nu \simeq 1.37 \; T^{0.135} \; \lambda^{0.084} \tag{2.16}$$

if λ is in cm.

For an ionized wind n_e and n_i are both proportional to the density, so we can write $n_e = \gamma_e n_i$, where γ_e is the number of free electrons per ion, and $n_i = \rho(\mu_i m_H)^{-1}$, where μ_i is the mean atomic mass of the ions

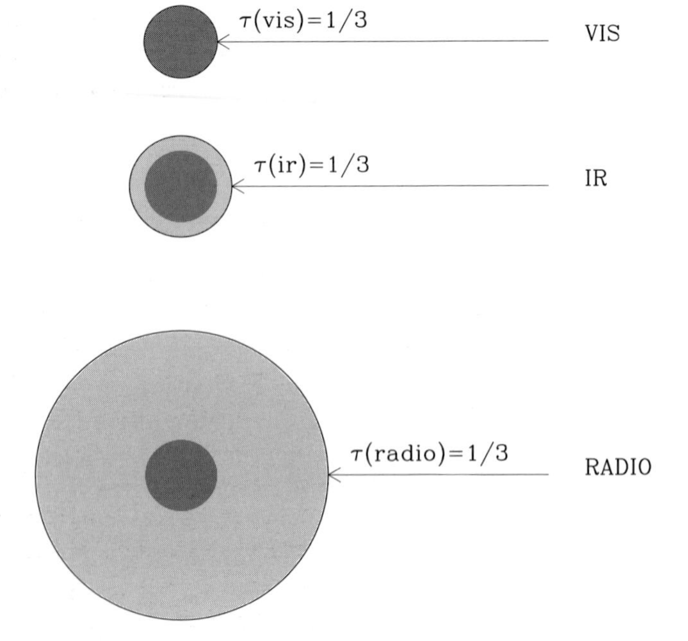

Figure 2.10 A schematic representation of the infrared and radio excess due to free-free emission from a stellar wind. The effective radius of the star, where $\tau \simeq 1/3$, increases to longer wavelengths.

in units of m_H. The density can be expressed in terms of the mass loss rate and the velocity law, $\rho = \dot{M}/4\pi r^2 v$. So the absorption coefficient per cm^3, $k_\nu = \kappa_\nu \rho$ in cm^{-1}, is

$$k_\nu \sim \left(\frac{Z^2\gamma_e}{\mu_i^2}\right) \nu^{-2}\, g_\nu\, T^{-3/2}\, R_*^{-4} \left(\frac{\dot{M}}{v_\infty}\right)^2 \frac{1}{x^4 w^2} \tag{2.17}$$

where the normalized distance and velocity are $x = r/R_*$ and $w = v(r)/v_\infty$, as before.

The free-free emission from a stellar wind in the radio region is formed at a large distance from the star where the wind has reached its terminal velocity. Therefore we can estimate the radio emission from an ionized wind by adopting a constant velocity, $v = v_\infty$ or $w = 1$, and a constant temperature, T_w.

The radial optical depth for free-free absorption of an isothermal stellar wind with constant velocity is

$$\tau_\nu = \int_r^\infty k_\nu\, dr \sim \frac{Z^2\gamma_e}{\mu_i^2} \nu^{-2}\, g_\nu\, T_w^{-3/2}\, R_*^{-3} \left(\frac{\dot{M}}{v_\infty}\right)^2 x^{-3} \tag{2.18}$$

This equation shows that the optical depth $\tau_\nu = \tau_{\rm eff} = 1/3$ is reached at a distance x where

$$x(\tau_\nu = \tau_{\rm eff}) \sim \left(\frac{Z^2\gamma_e}{\mu_i^2}\right)^{1/3} \nu^{-2/3}\, g_\nu^{1/3}\, T_w^{-1/2}\, R_*^{-1} \left(\frac{\dot{M}}{v_\infty}\right)^{2/3} \tag{2.19}$$

The Planck function in the radio region can be written in its Rayleigh-Jeans approximation $B_\nu \sim T\,\nu^2$ and so the luminosity of the wind at radio wavelengths is in the zero-order approximation (Eq. 2.11)

$$\begin{aligned} L_\nu &\simeq 4\pi R_*^2\, x^2(\tau_\nu \simeq \tau_{\rm eff})\, \pi B_\nu(\tau_\nu \simeq \tau_{\rm eff}) \\ &\sim \left(\frac{Z^2\gamma_e}{\mu_i^2}\right)^{2/3} \left(\frac{\dot{M}}{v_\infty}\right)^{4/3} \nu^{2/3} g_\nu^{2/3} \end{aligned} \tag{2.20}$$

Notice that the temperature of the wind and the radius of the star do not enter in this expression. If the weak dependence of the gaunt factor ν (Eq. 2.16) is taken into account, we find that the predicted energy distibution is a power law $L_\nu \sim \nu^{0.61}$ which is characteristic for the radio free-free emission from an isothermal wind with a constant velocity.

If the constants are taken into account and the slight dependence of the gaunt factor on frequency and temperature, the relation between mass loss and monochromatic radio flux becomes

$$\frac{\dot{M}}{v_\infty} = 0.095\, (f_\nu d^2)^{3/4} \left(\frac{\mu^2}{Z^2\gamma_e}\right)^{1/2} (\nu g_\nu)^{-1/2} \tag{2.21}$$

(Wright and Barlow, 1975), where f_ν is the measured radio flux in Janskys, d is the distance of the star in kpc, $\dot{M}$ is in $M_\odot\,\mathrm{yr}^{-1}$, v_∞ is in km s^{-1} and ν is in Hz. The mass loss rate is almost independent of the temperature of the wind, apart from the very weak temperature dependence of the gaunt factor. This is an advantage because the thermal structure of stellar winds is not well known.

Figure (2.11) shows the characteristic energy distribution of a star with free-free emission from an ionized wind. The stellar flux is assumed to be a Planck function. Up to a few microns the infrared energy distribution is the Rayleigh-Jeans tail of the photospheric flux. At longer wavelengths the free-free emission produces an excess flux. At $\lambda > 100\ \mu$m the flux varies as $\lambda^{-0.6}$ because the free-free emission originates from the constant velocity region of the wind. The shape of the energy distribution between about 3 and 100 μm depends on the temperature and density structure in the layers where the velocity increases outwards. Lamers and Waters (1984) have published energy distributions for a grid of stellar wind models.

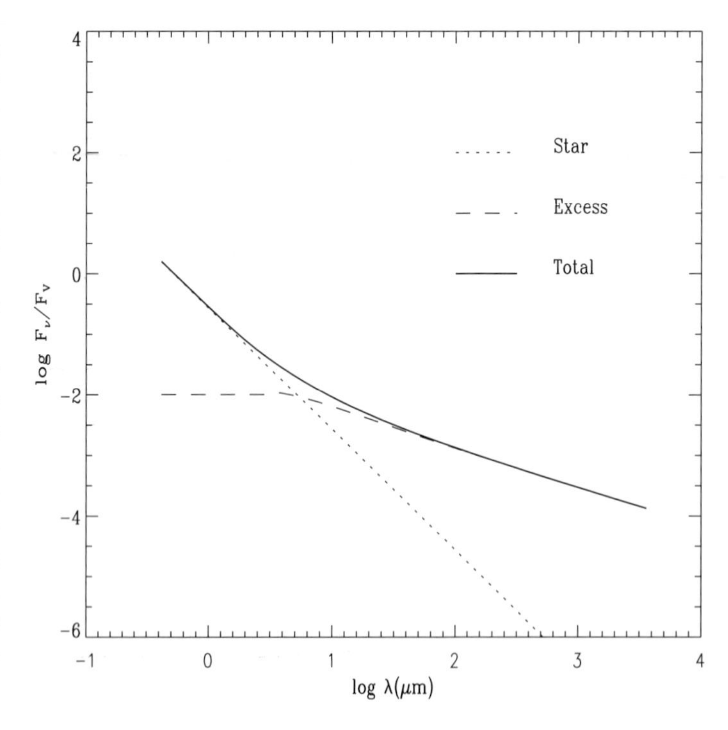

Figure 2.11 The characteristic energy distribution of a star of $R_* = 10\ R_\odot$, $T_{\mathrm{eff}} = 37500$ K, with free-free emission from a wind of $\dot{M} = 1 \times 10^{-5}$ $M_\odot.\mathrm{yr}^{-1}$, $T_w = 30000$ K, v_∞=2000 km s^{-1} and $\beta = 1.0$. The dotted line is the photospheric spectrum, the dashed line is the free-free emission and the thick line is the resulting spectrum. (Data from Lamers and Waters, 1984)

2.5 Molecular emission lines from the winds of cool stars

Winds from cool stars emit molecular emission lines. The main reason for the emission is the large geometrical extent of the winds of cool stars, compared to the size of the star itself. For instance, the CO line at 2.6 mm emitted by the wind of a red supergiant is formed over a region with a radius on the order of 10^4 stellar radii. The huge size of the line emitting region is the result of the high mass loss rate, typically of the order of 10^{-6} to 10^{-5} $M_{\odot}\,\mathrm{yr}^{-1}$, and the low outflow velocity of about 10 $\mathrm{km\,s}^{-1}$. These combine to give a relatively high density in the wind.

The profiles of molecular emission lines are either flat-topped (if the lines are optically thin) or parabolic (if the lines are optically thick). However, in some cases the lines are double peaked with strong spikes at the edges of the profiles. Such lines are formed by masering. Figure (2.12) shows some observed profiles.

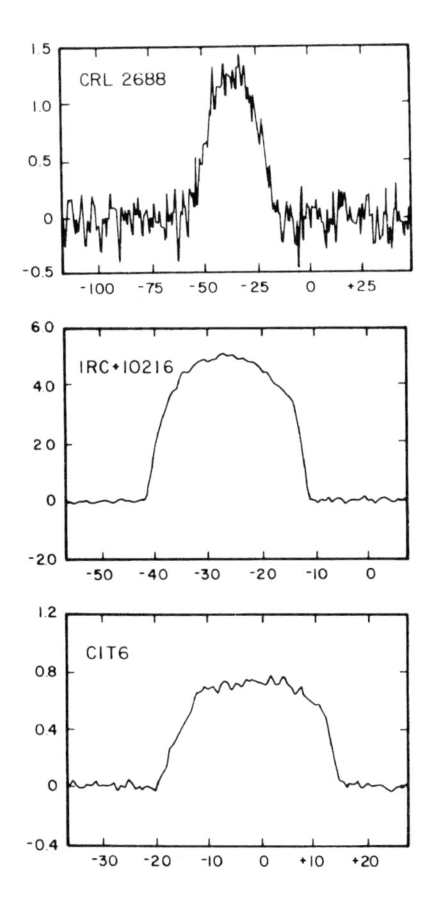

Figure 2.12 Observed CO $J = 1 \rightarrow 0$ profiles of three cool stars. CRL 2688 has a parabolic profile (optically thick), IRC+10216 has an intermediate profile (partially optically thick), and CIT6 has a flat-topped profile (optically thin). (From Knapp and Morris, 1985)

Table 2.2 *Some important molecular lines that show the effects of mass loss*

Molecule	frequ. (GHz)	λ (cm)	Transition	Pumping line (μm)
CO	230.7	0.130	(v,J)=(0,2) → (0,1)	4.6
	115.4	0.260	(v,J)=(0,1) → (0,0)	4.6
SiO	86.85	0.35	(v,J)=(0,2) → (0,1)	8
	43.42	0.69	(v,J)=(0,1) → (0,0)	8
OH	1.612	18.61	(J,F)=(3/2,1) → (3/2,2)	119,79,53,35
	1.665	18.02	(J,F)=(3/2,1) → (3/2,1)	119,79,53,35
	1.667	18.00	(J,F)=(3/2,2) → (3/2,2)	119,79,53,35

The observed molecular emission lines typically show lines with full widths of the order of 20 to 50 km s^{-1}, which indicate that the outflow velocities are only about 10 to 25 km s^{-1}. This is smaller than the escape velocity at the stellar surface, which is typically 50 to 100 km s^{-1} for cool supergiants. However, the molecular lines are formed at large distances from the star where the local escape velocity is much smaller than near the stellar surface. So, although the observed outflow velocities are smaller than $v_{\rm esc}(R_*)$, they still indicate mass loss. Table (2.2) lists some important molecular emission lines used for mass loss studies.

The most important molecular lines for the studies of mass loss from cool stars are the CO lines at 1.3 and 2.6 mm. These lines are most suitable for determining the mass loss rates of cool stars because they can be observed in winds of both O-rich and C-rich stars. In C-rich stars, where the ratio of C/O by number is larger than 1, all the O is in the form of CO. In O-rich stars, where O is more abundant than C, all the C is in the form of CO. Since CO is a very stable molecule, it can already form in the photospheres of very cool stars and persist throughout the wind up to very large distances, so the density of CO molecules in the winds of cool stars is strictly proportional to the density. Other molecules, such as OH, SiO, SO and SO_2, are very sensitive to the details of the chemistry of the wind and are less suitable for determining mass loss rates than the CO lines. For these reasons we concentrate this discussion on the CO lines.

2.5.1 The excitation of CO molecules in the winds of cool stars

The two most important excitation processes for CO molecules in the winds of luminous cool stars are the excitation of the rotational levels by collisions with H_2 molecules and the photo-excitation of the vibrational levels by IR-photons. This is called '*infrared pumping*'. Cool stars often have dust in their winds which emits infrared radiation (see § 2.6). The process of pumping is shown in Fig. (2.13).

A CO molecule in the lowest vibrational quantum state ($v = 0$) can be excited to a higher vibrational state by the absorption of an infrared photon, e.g. to $v = 1$. This transition occurs in the infrared at 4.6 μm for CO. The original quantum number has changed by $\Delta J = \pm 1$. After the absorption the molecule immediately re-emits an IR photon and de-excites back to the vibrational ground state $v = 0$. During this de-excitation J changes again by $\Delta J = \pm 1$, so that in the end the quantum number has changed by $J_{\rm end} - J_{\rm start} = +2$, 0 or -2. There is a higher probabilty that $J_{\rm end} - J_{\rm start} \geq 0$ because of the quantum-mechanical branching ratios for spontaneous de-excitation. So the excitation and de-excitation by absorption and emission of infrared photons will produce an average shift upward on the rotational ladder. The CO molecules can also be pumped via $v = 0 \rightarrow 2$ levels by IR photons at 2.3 μm.

Collisions can redistribute the populations over the rotational ladder and bring the distribution into a Boltzmann equilibrium. If the collisions are more important than the pumping by infrared photons, the distribution of the molecules over the different J levels will be approximately in LTE (local thermodynamic equilibrium) with respect to the temperature in the wind. If the pumping by the infrared pho-

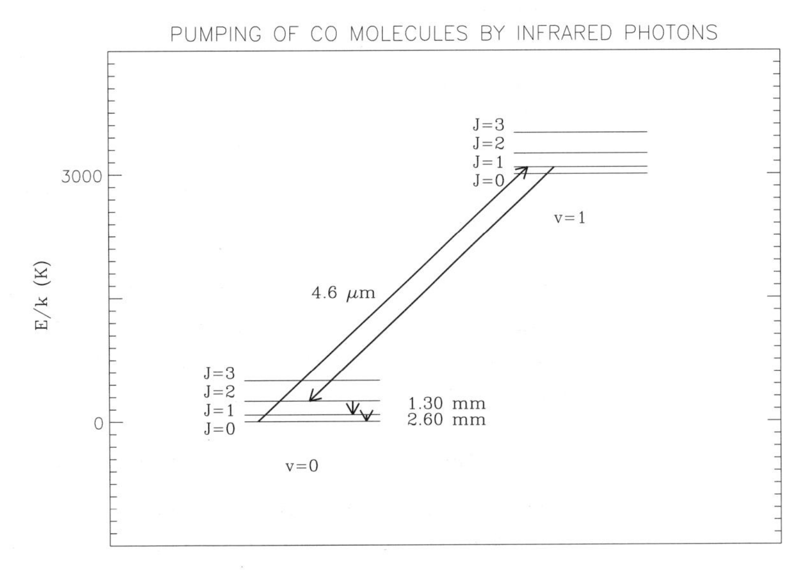

Figure 2.13 The population of the rotational levels of CO by the absorption and emission of IR photons emitted by dust. The figure only shows two vibrational levels (v=0 and 1) and the lowest rotational levels (J=0 to 3). The transitions obey the selection rule of $\Delta J = \pm 1$. The de-excitations in the rotational transitions of $v = 0$ produce emission lines at mm-wavelengths. (The energy differences between rotational levels are exaggerated.)

tons dominate, the distribution will deviate from LTE. If the pumping results in a population inversion, such that there are more molecules in a higher rotational state than in a lower state, masering will occur (see below).

2.5.2 Mass loss determinations from optically thin CO emission lines

The CO emission lines are formed over a very extended region of the winds of cool stars, so we can assume a constant outflow velocity in the line forming region. We describe the relation between the strength of the optically thin lines and the mass loss rate.

Consider a molecular emission line formed in a constant velocity wind by a transition between two rotational levels, $J_u \rightarrow J_l$, where u and l stand for upper and lower respectively. The number density of the molecules in the lower level of the transition, n_l, is proportional to the density in the wind

$$n_l(r) \sim \left(\frac{\dot{M}}{v}\right) r^{-2} \tag{2.22}$$

So $n_l \sim r^{-2}$ in a wind with constant velocity. The number density of CO molecules in the upper level, $n_u(r)$, relative to that of the lower level, can be expressed in terms of the *excitation temperature*, $T_{\rm exc}$, which is defined by the Boltzmann formula

$$\frac{n_u(r)}{n_l(r)} = \frac{g_u}{g_l} e^{-\chi/kT_{\rm exc}(r)} \tag{2.23}$$

where g_u and g_l are the statistical weights of the two rotational levels.

If pumping by IR photons is not important then $T_{\rm exc}$ is equal to the temperature of the gas in the wind, typically between 10^2 and 10^1 K. In case of efficient infrared pumping $T_{\rm exc}$ can go up to higher values of typically 10^2 K. In case of a population inversion, when $n_u/g_u > n_l/g_l$, $T_{\rm exc}$ becomes negative.

The emissivity j_l of line photons, in erg cm^{-3}s^{-1}, is equal to the number density of molecules in the upper state times the probability A_{ul} for spontaneous de-excitation per second, times the energy $h\nu$ of the emitted photons. So

$$\begin{aligned} j_l(r) &= n_u(r)\, A_{ul}\, h\nu \\ &\simeq \left(\frac{n_{\rm CO}}{\rho}\right) \left(\frac{\dot{M}}{4\pi r^2 v}\right) A_{ul}\, h\nu\, \frac{g_u}{g_l}\, e^{-\chi/kT_{\rm exc}(r)} \end{aligned} \tag{2.24}$$

If the line is optically thin and all emitted photons can escape, the luminosity of the line is (see Eq. 2.7)

$$L_l = \int_{R_*}^{r_{\rm max}} 4\pi r^2 j_l(r) dr$$
$$\simeq \left(\frac{n_{\rm CO}}{\rho}\right) \left(\frac{\dot{M}}{v}\right) A_{ul}\, h\nu\, \frac{g_u}{g_l}\, e^{-\chi/kT_{\rm exc}}\, r_{\rm max} \qquad (2.25)$$

if $T_{\rm exc}$ is constant and $r_{\rm max} \gg R_*$. Since the molecular lines are formed over a very extended wind, we can ignore the correction factor $1-W(r)$ for photons that are intercepted by the star (Eq. 2.7).

Notice that the integral diverges if the upper limit is set to $r = \infty$. This is a consequence of the fact that the emissivity is simply proportional to the density in the wind, and hence its volume integral is proportional to the total mass of the wind. Adopting an infinite radius is equivalent to assuming that the wind has been blowing for an infinite time, so the total mass of the wind would be infinite. In reality the mass in the wind from a cool star is limited by the fact that the high density winds of cool stars are only blowing during a relatively short time, typically less than 10^5 years. So the upper limit to the size of the wind should be set by the time that the wind is blowing at the present mass loss rate: $r_{\rm max} = v \times t$. This time estimate comes from stellar evolution calculations.

Equation (2.25) is only valid if the emission lines are optically thin. Fortunately this can be checked very simply on the basis of the shape of the line profile. It can be shown that the profiles of lines with an emissivity $j_l \sim \rho$, formed in winds with a constant velocity, are flat-topped if the lines are optically thin, and parabolic if they are optically thick (Sobolev, 1960). The flat topped profiles extend from Doppler velocity $-v$ to $+v$. So the velocity of the wind can be easily derived from the width of flat-topped molecular emission lines.

The proportionality constant between $\dot{M}$ and L_l in Eq. (2.25) can be calculated in a simple way if the level populations are in LTE and the temperature structure of the wind is known because $T_{\rm exc}(r) = T_w(r)$. However, if IR pumping is important, the calculation of the excitation temperature involves the solution of the radiative transfer equation for the IR line at 4.6 μm that is responsible for the pumping, and the statistical equilibrium equations for the calculation of the populations of the different rotational levels. Such calculations have been carried out by e.g. Morris (1980) for the CO lines, who also calculated the line profiles for a grid of stellar wind models. The CO $J = 1 \rightarrow 0$ line is optically thin for low mass loss rates of $\dot{M} \lesssim 10^{-6}\ M_\odot\,{\rm yr}^{-1}$. For higher mass loss rates the winds are no longer optically thin so Eq. (2.25) cannot be used. In that case the mass loss rate is derived from the fitting of the observed line profile to the predicted ones for different mass loss rates.

Figure (2.14) shows theoretical profiles of the CO $J = 1 \rightarrow 0$ line formed in a wind of constant velocity and different mass loss rates. The monochromatic fluxes are expressed in terms of the antenne temperature T_B in K.† For high mass loss rates, $\dot{M} \gtrsim 2 \times 10^{-5}$ $M_\odot\,\mathrm{yr}^{-1}$, the winds are optically thick for the line so they have a parabolic shape. For lower mass loss rates of $\dot{M} \simeq 2 \times 10^{-6}$ to 10^{-5} $M_\odot\,\mathrm{yr}^{-1}$ the winds are partially optically thick so the profiles are flat-topped parabolae. For low mass loss rates of $\dot{M} \lesssim 10^{-6}\ M_\odot\,\mathrm{yr}^{-1}$ the profiles become 'hollow' due to masering. We note that the profiles depend on the beam width of the telescope because the line emitting region of the winds is so large that the telescope only receives radiation from part of it.

Knapp and Morris (1985) derived an expression for the mass loss rate for optically thick CO $J = 1 \rightarrow 0$ lines

$$\dot{M} = 5 \times 10^{-16}\ T_B v^2 D^2 f_{CO}^{-0.85} \tag{2.26}$$

where T_B is the antenne temperature of the center of the line, v is in $\mathrm{km\,s^{-1}}$, D is the distance of the star in pc, and f_{CO} is the number abundance of CO relative to H_2, with $f_{CO} \simeq 2 \times 10^{-4}$. This relation is valid for $\dot{M} f_{CO}/v \gtrsim 1.3 \times 10^{-10}$. It has been used extensively to derive mass loss rates from AGB stars.

2.5.3 The profiles of maser lines

The absorption coefficient for a line transition is proportional to the number density of the atoms or molecules in the lower level, l, times a correction factor $\eta = \left[1 - n_u g_l / n_l g_u\right]$ for stimulated emission. If the pumping by IR photons results in a *population inversion*, then $\eta < 0$ and so the absorption coefficient becomes negative. The intensity of line radiation in a wind with a population inversion can be very high if the conditions are right. This can easily be seen from the equation for the intensity of a ray with initial intensity I_0 passing through a medium of optical depth τ : $I \simeq I_0 e^{-\tau}$. A negative value of τ results in an *increase* of the intensity rather than a decrease as in the normal situation. This is *masering*.

Let us consider masering on a molecular scale. In a normal situation, a photon can be absorbed in a line transition. This results in an

† The antenne temperature is the brightness temperature of an extended source (larger than the beam width of the telescope) that emits a monochromatic flux $B_\nu(T_B)$ per unit area per steradian, that would give the same flux as the observed source.

excitation of the molecule from the lower level, l, to the upper level u. There is also a probability that a similar photon meets an excited molecule in level u. In that case the passage of the photon triggers a de-excitation of the molecule from level u to l. This results in the stimulated emission of a photon with *the same energy and direction* as the one passing by. So now there are two photons in the same direction. In a normal situation there is a higher probability that photons are absorbed than created by stimulated emission. In case

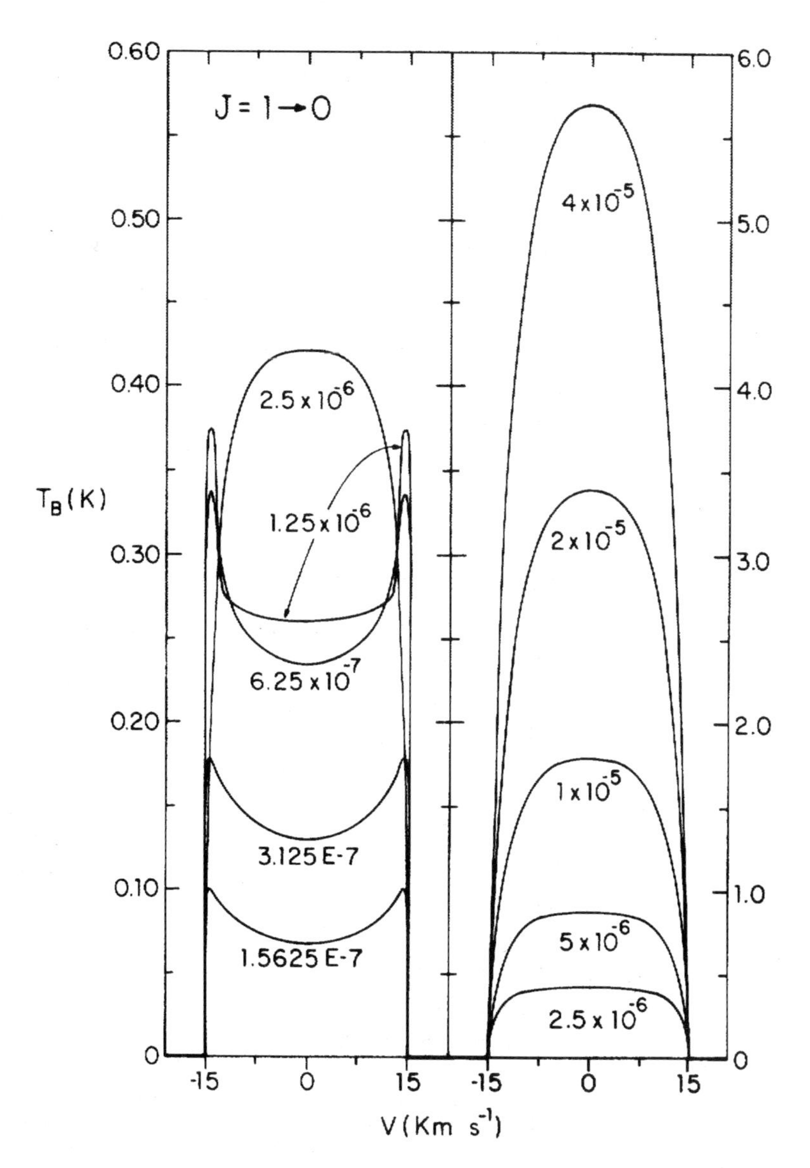

Figure 2.14 Predicted profiles of the CO $J = 1 \rightarrow 0$ line at 2.60 mm for winds around a star of $T_{\mathrm{eff}} = 2000$ K, $R_* = 700\ R_\odot$ and a wind velocity of 15 km s^{-1} and different mass loss rates. The mass loss rates are indicated. The profiles vary from parabolic shapes for optically thick winds to hollow shapes for low density winds with masering. (From Morris, 1980)

of a population inversion there is a higher probability for stimulated emission than for absorption. So the passing of line photons through a gas with population inversions results in an *increase* in intensity rather than in a decrease.

The masering of molecular lines is limited by two effects. If the density of the wind is so high that the rate of collisions with H_2 molecules dominates the other excitation and de-excitation rates, the level population thermalizes, i.e. it is forced to LTE values. This is called *quenching* of the maser. (This explains why maser profiles in Fig. 2.14 only occur for low mass loss rates.) If the rate of creation of photons by stimulated emission per unit volume, approaches the rate at which the population inversion is formed per unit volume, the masering reaches a maximum because the molecules cannot de-excite faster than they are excited. This is called *saturation* of the maser.

The profiles of maser lines can be calculated if the populations, $n_l(r)$ and $n_u(r)$, of the upper and lower levels of the line are known. In that case the (negative) optical depth can be calculated and the amplification of the intensity by a factor $e^{|\tau_\nu|}$ can be found for any frequency and any line of sight through the wind.

The maximum masering occurs along a line of sight that has the highest number of molecules *moving with the same velocity component*

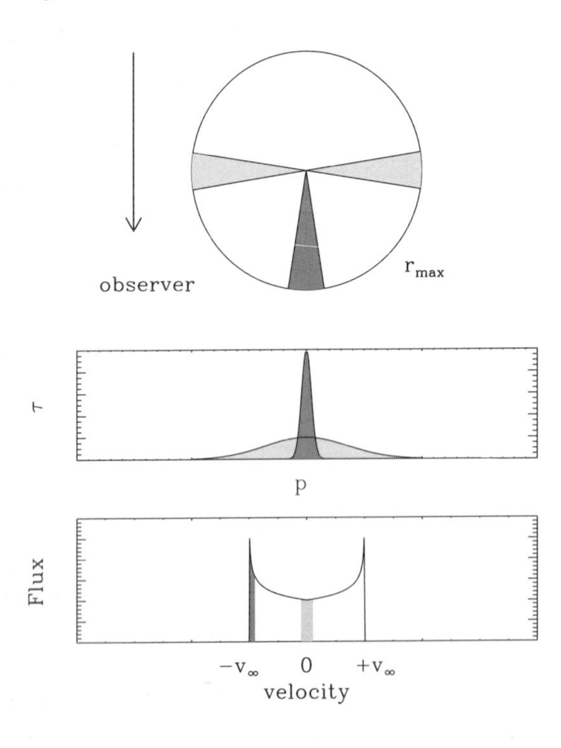

Figure 2.15 The formation of double peaked maser lines. Upper: the geometry with the regions where $-v < v_z < -v + \Delta$ (dark grey) and where $-\Delta < v_z < +\Delta$ (light grey). Middle: the optical depth $\tau(v_z = -v)$ (dark grey) and $\tau(v_z = 0)$ (light grey) as a function of the line of sight p. Lower: the profile of a maser with the flux from the grey regions indicated. The large value of $\tau(v_z = -v)$ gives a high flux at the edge of the profile.

to the observer. If there is a velocity gradient in the direction to the observer, masering can only occur over the limited distance, because the Doppler shift moves the photons out of the frequency reach of the absorption profile. This is equivalent to the statement that masering is most effective when the *monochromatic* optical depth is largest because the intensity varies as $I \sim e^{|\tau_\nu|}$. For a wind with a constant outflow velocity, v, the smallest velocity gradient along a line of sight occurs for lines of sight to the star or just passing the stellar disk. The gas in front of the star moves with a constant Doppler velocity $-v$ to the observer, and the gas almost behind the star moves with a constant Doppler velocity $+v$. So the monochromatic optical depth is largest at Doppler velocity $+v$ and $-v$. Therefore the intensity of molecular maser lines formed in a wind of constant outflow velocity is highest at the edges. This gives maser lines their characteristic double peaked profiles. See Fig. (2.15).

The terminal velocities of the winds of cool stars can easily be measured from the velocity separation, $2v$, of the peaks of the maser lines or from the width of the saturated parabolic lines or the unsaturated flat-topped lines.

2.6 The infrared and millimeter radiation from dust

Cool stars with high mass loss rates emit an excess of radiation at long wavelengths due to the radiation from the dust in their winds. This radiation is observed at infrared and mm wavelengths. The energy distribution of radiation by dust is very distinct from that of free-free radiation discussed before (§ 2.4) and shown in Fig. (2.9). Free-free emission produces an energy distribution which decreases to longer wavelengths. Dust emission produces an energy distribution with a characteristic bump in the IR, resembling a Planck function or a combination of Planck functions with $T \simeq 10^2$ to 10^3 K. The wavelength of the bump is related to the mean temperature of the dust.

If the column density of dust in the wind is small, the energy distribution consists of a stellar component of $T_* \simeq 2000$ to 3000 K in the visual and in the near IR and a dust component at longer wavelength. However, if the column density is large, the dust extinction can block the stellar radiation so that the energy distribution shows only the dust component at long wavelengths. This is the case for the 'OH/IR stars', so named because they show OH masering and large IR excesses.

Figure (2.16) shows the observed energy distributions of six cool stars with dust emission, from Bedijn (1987). The energy distribution of the first four stars, o Cet, R Cas, TX Cam and IK Tau, shows the photospheric spectrum at short wavelength of $1 \lesssim \lambda \lesssim 5$ μm, with a contribution by dust at longer wavelengths. The dust produces emission bands at 9.7 and 20 μm (see below). The last two stars, IRC+10011 and OH39.7+1.5, have such a high mass loss rate, that the dusty wind is optically thick for the stellar radiation at $\lambda \lesssim 10$ μm. The photospheric radiation is suppressed and the energy distibution peaks at longer wavelengths.

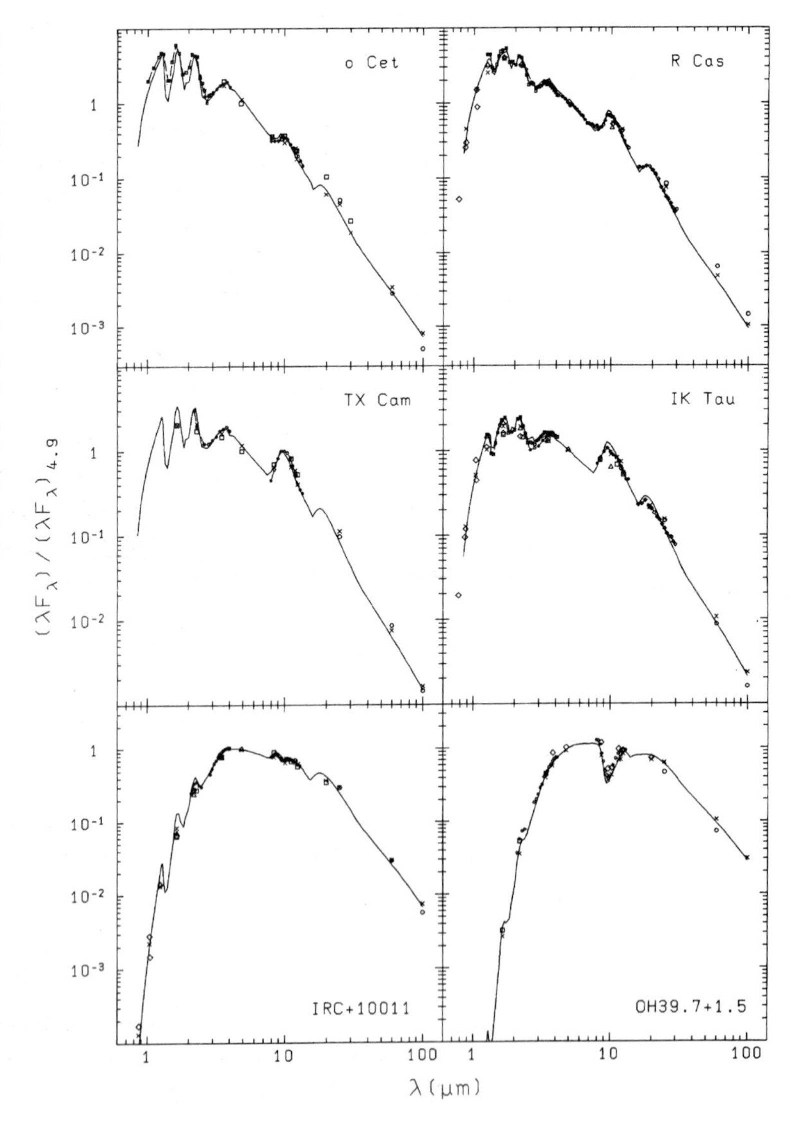

Figure 2.16 The observed energy distributions, in terms of λF_λ of six stars with dusty winds. The first four stars show the photospheric energy distribution plus a contribution by the wind. The lower two stars show only the emission by the dust. The features at 9.7 and 20 μm are due to silicates. (From Bedijn, 1987)

Apart from the broad component due to thermal dust radiation, the observations also show the presence of narrower dust features in absorption or in emission. The most prominent of these are the features at 3.1 μm due to ice, at 9.7 and 20 μm due to amorphous silicate in the winds of O-rich stars, at 3.3 μm due to polycyclic aromatic hydrocarbons (PAH) and at 11.15 μm due to SiC in the winds of C-rich stars (Roche, 1989).

The first studies of dust emission from winds of cool stars were made by Woolf and Ney (1969), Gehrz (1972) and Gehrz and Hackwell (1976). The *IRAS* and *ISO*-satellites have extended the number of observations very drastically and allowed both detailed and statistical studies of mass loss from cool stars (e.g. Jura, 1987; van der Veen, 1989; Habing, 1996).

In this section we describe the energy distibutions due to dust in a simple way and we discuss the methods used to derive the mass loss rates from the dust emission. Since dust radiates continuum emission and no narrow lines, its radiation cannot be used to derive the outflow velocity. Therefore the information about the velocity has to be derived independently from the width or profiles of molecular lines.

2.6.1 The temperature distribution of dust

The energy distribution from a dusty wind depends on the distribution of the dust through the wind, its temperature and its emissivity. The temperature of a dust grain is determined by the radiative equilibrium between the rate of radiative heating due to the absorption of stellar photons and the radiative cooling due to thermal emission, as described in detail § 7.4. The frequency dependence of the dust absorption coefficient, in the wavelength range where the star emits most of its radiation, can often be approximated by a power law of frequency $\kappa_\nu \sim \nu^p$. The parameter p depends on the nature of the dust, but it is usually $p \simeq 1$.

The temperature of the dust grains is independent of the grain size but it depends on the radiation from the star and the distance from the center of the star. In § 7.4 we will show that the dust temperature varies as

$$T_d(r) \;=\; T_* \; W(r)^{1/(4+p)} \;\simeq\; T_* \left(\frac{2r}{R_*}\right)^{-2/(4+p)} \tag{2.27}$$

(see Eq. 7.36) where T_* is the effective temperature of the star and $W(r)$ is the geometrical dilution factor (Eq. 2.8). This shows that T_d varies approximately as $r^{-2/5}$ if $p \simeq 1$.

Grains can only form at a distance where the equilibrium temperature is lower than the condensation temperature T_c of the grains, which is typically about 1500 K. This means that the distance where condensation can occur, r_c, can be derived from Eq. (2.27) by setting $T_d(r_c)$ equal to T_c. So the condensation distance r_c is given by

$$W(r_c) = \left(\frac{T_c}{T_*}\right)^{4+p} \quad \text{or} \quad r_c \simeq \frac{R_*}{2}\left(\frac{T_*}{T_c}\right)^{(4+p)/2} \tag{2.28}$$

The last approximation is valid if $T_c < 0.6\ T_*$. The condensation radii are typically $r_c \simeq 1.2$ to 4 R_* for cool stars.

2.6.2 The energy distribution of optically thin dusty winds

If there is a continuous outflow of gas from the star that condenses into dust at the condensation radius, then the inner radius of the dust will be $r_{\rm min} = r_c$. This is the case for red giants and AGB stars which have strong winds that are still ongoing. If, on the other hand, the star is observed some time after the termination of the wind phase, then the inner radius of the dust has moved out to $r_{\rm min} = v_d \times t_{\rm end}$, where v_d is the outflow velocity of the dust and $t_{\rm end}$ is the time since the end of the wind phase. This is the case for post-AGB stars, for which the high mass loss phase has stopped at the end of the AGB phase. It also applies to novae after their outbursts. The outer radius of the dust shell is at $r_{\rm max} = v_d \times t_{\rm begin}$, where $t_{\rm begin}$ is the time since the beginning of the wind phase.

Once the dust is formed at the condensation distance, it is not destroyed in the wind and keeps about the same distribution of sizes. This means that the dust density decreases outwards as r^{-2} because the winds of cool stars quickly reach their terminal velocity of about 10 to 25 km s^{-1}.

For dust of a single particle size of radius a, the mass loss rate in the form of dust can be expressed in terms of the continuity equation (2.1)

$$\dot{M}_d = \left(\frac{4\pi}{3} a^3 \bar{\rho}\ n_d(r)\right) 4\pi r^2\ v_d \tag{2.29}$$

where $\bar{\rho}$ is the mass density in the dust grains, which is 2.26 and 3.30 g cm^{-3} for silicates and graphite respectively (Draine and Lee, 1984). The first right hand term is the mass of a grain times the number density of the grains, so it is the density of the dust in g cm^{-3}. So the number density of the grains can be written as

$$n_d(r) = \frac{3}{16\pi^2} \frac{1}{< a^3 > \bar{\rho}} \frac{\dot{M}_d}{v_d} \frac{1}{r^2} \tag{2.30}$$

where $< a^3 >$ is the mean value of a^3 in the case of a distribution in size of the grains.

The emission of a single dust grain is the product of its surface times the efficiency factor Q_ν^A for emission times the Planck function at the dust temperature

$$\epsilon_d(\nu) = 4\pi a^2 Q_\nu^A B_\nu(T_d) \tag{2.31}$$

The efficiency factors for different types of grains are shown in § 7.3.

With the temperature structure, the dust density and the emissivity known, the IR radiation from an optically thin dust shell can be calculated easily. The luminosity of the dust shell is

$$\begin{aligned} L_\nu &= \int_{r_{\min}}^{r_{\max}} 4\pi r^2 n_d(r) < a^2 > Q_\nu^A B_\nu(T_d)\{1 - W(r)\}\, dr \\ &= \frac{3}{4\pi} \frac{< a^2 >}{< a^3 >} \frac{1}{\bar{\rho}} \frac{\dot{M}_d}{v_d} Q_\nu^A \int_{r_{\min}}^{r_{\max}} B_\nu(T_d)\{1 - W(r)\}\, dr \end{aligned} \tag{2.32}$$

where $< a^2 >$ is the mean value of a^2 in the case that the dust particles have a distribution in size. The factor $\{1 - W(r)\} \simeq 1$ corrects for the fraction of the photons that are intercepted by the star (Eq. 2.8).

We see that the luminosity of the dust radiation is proportional to $(\dot{M}_d/v_d)$ and to the distance integral of the Planck function. The shape of the energy distribution is given by the product of Q_ν^A and the integral over the Planck function. It can be calculated exactly with the temperature distribution of Eq. (2.27). Figure (2.17) shows theoretical energy distributions for an absorption coefficient $\kappa_\nu \sim \nu^p$ with $p = 1$ and with $T_* = 2500$ K, and for various values of $r_{\min}$.

Notice that the energy distribution consists of four parts.

(i): At the shortest wavelength the dust does not contribute, so the the energy distribution is approximately that of the star. In reality, the short wavelength flux of a star surrounded by a dusty wind is always *smaller* than the stellar flux because it is the absorbed short wavelength flux that heats the gas and is re-emitted in the infrared. The total flux of a star plus wind cannot be larger than that of the star itself.

(ii): At wavelengths of a few μm only the hottest grains contribute. This means that the peak in the dust energy distibution indicates the temperature of the hottest dust, $\lambda_{\max}(\mu\text{m}) \simeq 5100/T_{\max}$. If $r_{\min} = T_c \simeq 1000$ K the peak occurs at a wavelength of a few microns, but if the inner radius of the dust shell has moved out to larger distances the peak will move to a longer wavelength.

(iii): At wavelengths beyond the peak, the energy distribution is approximately a power law with a slope that depends on the value of p.

(iv): At very long wavelength $\lambda(\mu m) \gg 5100/T_d(r_{max})$, (not shown in Fig. 2.17) the radiation by the dust decreases to zero.

So, from the shape of the energy distribution from an optically thin dusty wind one can derive the minimum and maximum radius of the dust shell, and the spectral index p of the absorption coefficient. If the size of the dust grains and their absorption properties are known, the mass loss rate can be derived from the absolute value of the flux at any frequency. The models for optically thin dusty winds are valid for cool stars with mass loss rates less than about $5 \times 10^{-7}\ M_\odot\,\mathrm{yr}^{-1}$.

2.6.3 The energy distribution of optically thick dusty winds

The energy distributions of optically thick dusty winds have been calculated by many groups, e.g. Rowan Robinson *et al.* (1986), Bedijn (1987), and Ivezic and Elitzur (1995). See van der Veen and Olofsson (1990) for a review. The predictions by Bedijn (1987) are shown in Fig. (2.18).

The mass loss rate that can be derived from the IR energy distributions of dusty winds refers only to the mass lost in the form of dust. This is usually only a small fraction of the mass lost in the form of gas. To find the total mass loss rate one has to multiply the value of $\dot{M}_d$ with the gas/dust ratio μ of dusty winds. This ratio has been derived for a number of cool stars by comparing the mass loss rate, derived from molecular lines (gas mass loss), to the dust mass loss rate. One

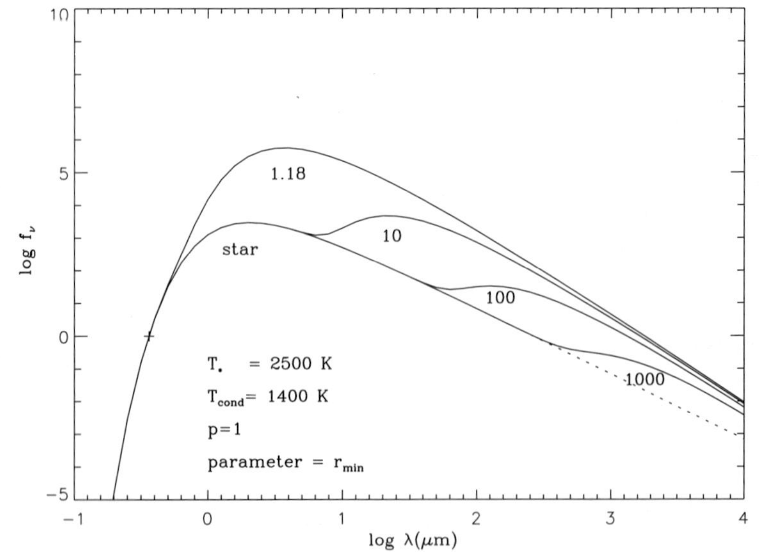

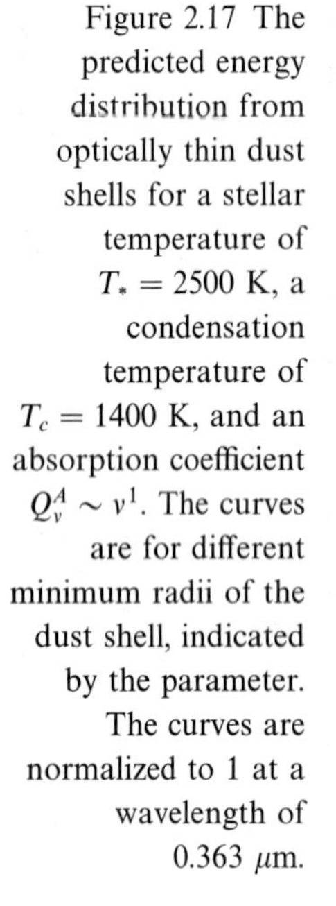
Figure 2.17 The predicted energy distribution from optically thin dust shells for a stellar temperature of $T_* = 2500$ K, a condensation temperature of $T_c = 1400$ K, and an absorption coefficient $Q_\nu^A \sim \nu^1$. The curves are for different minimum radii of the dust shell, indicated by the parameter. The curves are normalized to 1 at a wavelength of 0.363 μm.

generally assumes a gas/dust ratio of about 200 (e.g. Whitelock *et al.*, 1994). So the total mass loss rate is typically 200 times the dust mass loss rate.

Jura (1987) has derived a formula for calculating the mass loss rate from the flux at 60 μm as measured by the *IRAS* satellite

$$\dot{M} = 7.7\ 10^{-10}\ \mu\ v_{15}\ D^2\ L_4^{-0.5}\ F_{60}\ \sqrt{\lambda_{10}}\ \left(\frac{150}{\kappa_{60}}\right) \qquad M_\odot\ \mathrm{yr}^{-1} \quad (2.33)$$

where μ is the gas to dust ratio (usually assumed to be about 200), v_{15} is the wind velocity in units of 15 km s^{-1}, D is the distance of the star in kpc, L_4 is the luminosity of the star in $10^4\ L_\odot$, F_{60} is the colour-corrected *IRAS* flux at 60 μm in Jy, λ_{10} is the mean wavelength at which the circumstellar envelope emits in units of 10 μm and κ_{60} is the dust opacity at 60 μm in cm^2 per gram.

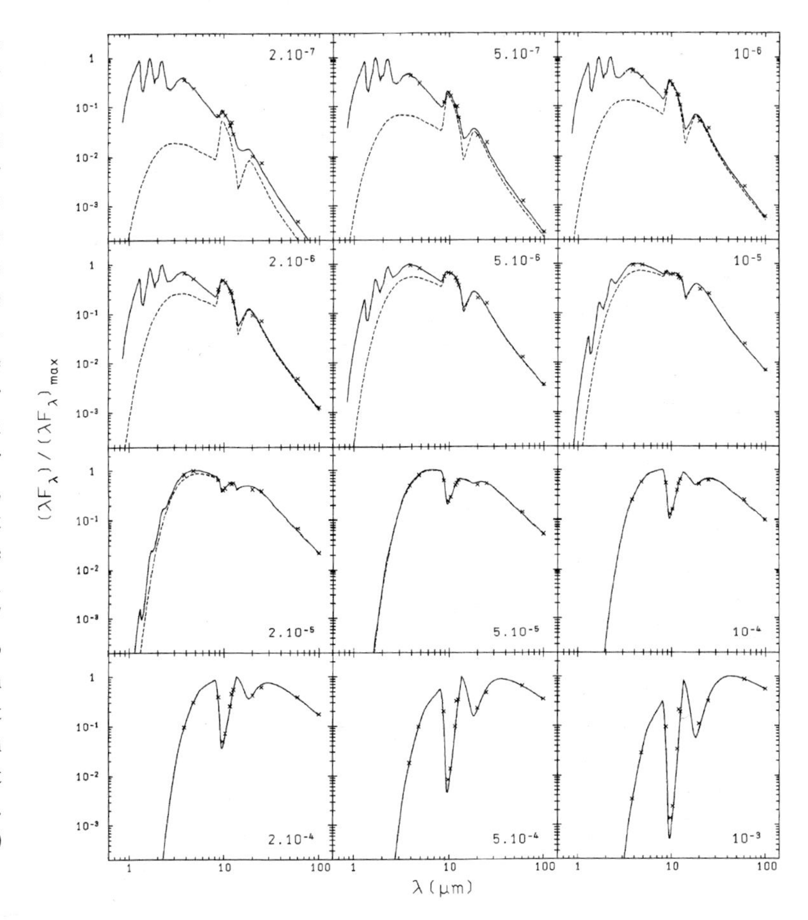

Figure 2.18 The predicted energy distributions, in terms of λF_λ of stars with dusty winds, for different mass loss rates indicated by the parameter. The dashed line is the dust shell spectrum. The full line is the spectrum of the star ($T_{\mathrm{eff}} = 3000$ K) with the dust shell. For $\dot{M} \gtrsim 2\ 10^{-5}\ M_\odot\ \mathrm{yr}^{-1}$, the star does not contribute to the spectrum any more. The features at 9.7 and 20 μm are due to silicates. They are in emission at $\dot{M} < 10^{-5}$ and in absorption at $\dot{M} > 10^{-5}\ M_\odot\ \mathrm{yr}^{-1}$. (From Bedijn, 1987)

2.7 Mass loss rates and terminal velocities of winds from hot stars

We briefly review the knowledge about mass loss rates and terminal velocities of stellar winds.

2.7.1 O and B stars

The most comprehensive studies of the mass loss rates from O and B supergiants are by Howarth and Prinja (1989), based on UV lines, Lamers and Leitherer (1993) based on compilations of radio data and Hα, and Puls *et al.* (1995), based on Hα. The results for the Galactic O and B type stars are shown in Fig. (2.19a) and those for the O stars in the Magellanic Clouds are shown in Fig. (2.19b).

The radiation driven wind theory predicts (see § 8.7) that the mass loss rate depends on the stellar parameters as

$$\dot{M} \sim L_*^{1/\alpha} M_{\rm eff}^{\frac{\alpha-1}{\alpha}} \tag{2.34}$$

and

$$v_\infty \sim v_{\rm esc} \sim \left(\frac{M_{\rm eff}}{R_*}\right)^{0.5} \tag{2.35}$$

where α is a force multiplier parameter, which is about 0.6 for OB stars (see § 8.6). The effective mass is the mass corrected for the radiative force due to electron scattering, $M_{\rm eff} = M_*(1 - \Gamma_e)$ with

$$\Gamma_e = \frac{\sigma_e L_*}{4\pi c G M_*}. \tag{2.36}$$

The electron scattering coefficient per unit mass σ_e depends on the chemical composition of the wind and the degree of ionization (see Eq. 8.93). Its value is about $\sigma_e \simeq 0.30$ cm^2 g^{-1} for the winds of hot stars. The mass of the star is the most uncertain parameter in Eqs. (2.34) and (2.35). Therefore it is better to use the *modified wind momentum*

$$\dot{M}\, v_\infty\, R_*^{0.5} \sim L_*^{1/\alpha} M_{\rm eff}^{\frac{\alpha-1}{\alpha}+\frac{1}{2}} \tag{2.37}$$

in the comparison between observed and predicted mass loss rates, because it is proportional to $M_{\rm eff}^{-0.17}$ if $\alpha = 0.60$ and $M_{\rm eff}^{-0.04}$ if $\alpha = 0.65$. This means that $\dot{M} v_\infty R_*^{0.5}$ is almost independent of the effective mass of the star (e.g. Kudritzki *et al.*, 1995). Figures (2.19a) and (2.19b) show this product as a function of L_*. The mean relation for the galactic stars in Fig. (2.19a) is

$$\log(\dot{M}\, v_\infty\, R_*^{0.5}) \;=\; -1.37 \;+\; 2.07 \log(L_*/10^6) \tag{2.38}$$

if $\dot{M}$ is in $M_\odot\,{\rm yr}^{-1}$, v_∞ is in km s^{-1}, R_* is in $R_\odot$ and L_* is in $L_\odot$.

The mass loss rates of O stars in the Large and Small Magellanic Clouds are compared to those of the Galactic stars in Fig. (2.19b). We can expect that the mass loss rates of the LMC and SMC stars are smaller, because the stars in these galaxies have a lower metallicity of about 0.5 and 0.1 times the solar value respectively. The radiation driven wind theory predicts that $\dot{M} v_\infty R_*^{0.5}$ scales with $Z^{1.0}$ (see § 8.6). Figure (2.19b) shows that the LMC and SMC stars are indeed generally below the mean relation of the galactic stars, but that the difference not only depends on metallicity but also on luminosity. Lamers and Cassinelli (1995) have derived equations that descibe the observed relations.

The terminal velocities of winds of O and B supergiants depend on the effective escape velocity at the stellar surface and on the effective temperature of the star. The effective escape velocity is the Newtonian escape velocity minus a correction for the radiation pressure due to electron scattering,

$$v_{\rm esc} = \sqrt{2(1 - \Gamma_e) G M_* / R_*} \tag{2.39}$$

The theory of radiation driven winds predict that the ratio $v_{\rm esc}/v_\infty$ depends on $T_{\rm eff}$. The result of a survey by Lamers *et al.* (1995) is shown in Fig. (2.20). The ratio $v_{\rm esc}/v_\infty$ is about 2.6 for stars with $T_{\rm eff} > 21\ 000$ K, drops steeply to 1.3 near $T_{\rm eff} = 21\ 000$ K, and jumps to a value of 0.7 for stars with $T_{\rm eff} < 10\ 000$ K. The jump near 21 000 K is due to the fact that the winds of the hotter stars are mainly driven by high ionization lines of C, N, O etc. in the Lyman continuum, whereas the winds of the stars with $10\ 000 < T_{\rm eff} < 20\ 000$ K are mainly driven by a large number of metal lines in the Lyman and the Balmer continuum (Pauldrach and Puls, 1990; Lamers and Pauldrach, 1991). The discontinuities near 21 000 and 10 000 K are called *bi-stability jumps*, because a star near this boundary could jump from one type of wind to another.

2.7.2 Wolf-Rayet stars

Wolf-Rayet stars are massive stars in a late evolution stage, when mass loss has stripped most of their H-rich envelope. Their luminosity is typically between 3×10^4 and 10^6 $L_\odot$, and their effective temperature is between about 30 000 and 100 000 K. (The effective temperature of WR stars is not well defined, because the continuum is formed in the wind and hence the radius depends on the wavelength). There are two types of WR stars: the N-rich WN stars and the C-rich WC stars (Conti, 1988; Lamers *et al.*, 1991). The spectrum of WR stars is characterized by emission lines formed in the wind.

The mass loss rates from WR stars have been derived from radio fluxes by Abbott *et al.* (1986) and Leitherer *et al.* (1995) and from the emission lines by many groups (e.g. Hillier, 1984; Schmutz *et al.*, 1989; Hamann *et al.*, 1995; Crowther *et al.*, 1995). Hamann *et al.* (1995)

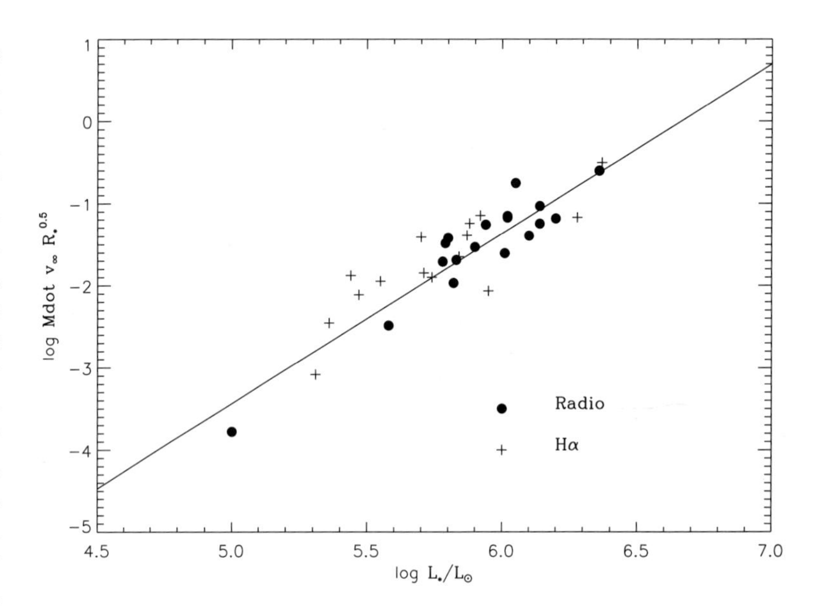

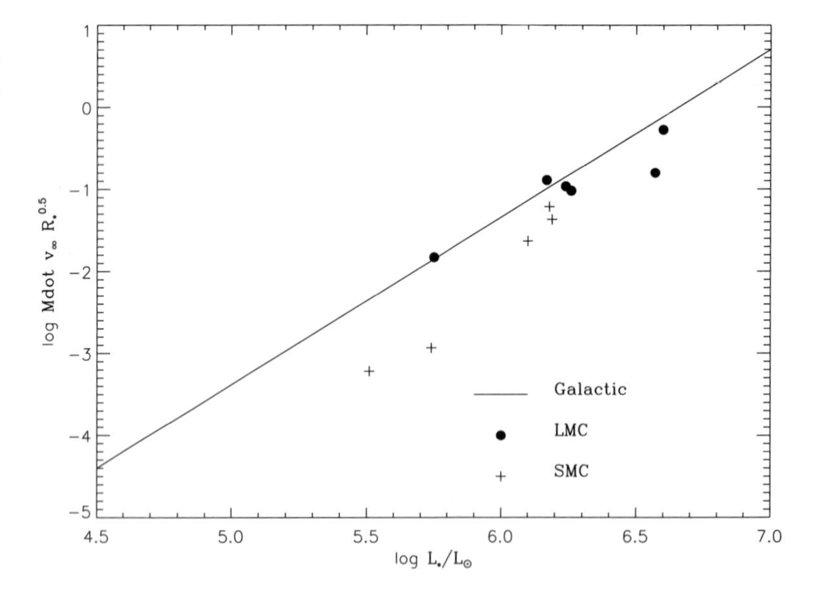

Figure 2.19 The modified wind momentum $\dot{M} v_\infty R_*^{0.5}$ versus L_* for O and B stars in the Galaxy (upper panel, a) and in the LMC and SMC (lower panel, b). The mean relation for the Galactic stars is given by Eq. (2.38). The wind momentum is lower for the LMC and SMC stars. The mass loss rates are derived from Hα by Puls *et al.* (1995), and from the radio flux by Abbott *et al.* (1981), Lamers and Leitherer (1993) and Leitherer *et al.* (1995).

has derived the mean stellar parameters, mass loss rates and terminal velocities of the N-rich WN stars. The wind momentum $\dot{M}v_\infty$ is plotted versus the photon momentum L_*/c in Fig. (2.21a). The figure shows that the wind momentum of WN stars is between 2 and 100 times the photon momentum. In Chapter 8 we will show that the radiation driven wind theory predicts a maximum wind momentum of $\dot{M}v_\infty = L_*/c$ in the limit that *all* photons from the star are scattered once in the wind. This is called the *single scattering limit* (see § 8.2 and Eq. 8.19). The WN stars obviously surpass this limit by a large factor. The mass loss rates of WC stars are similar to those of the WN stars and the wind momentum also surpasses the photon momentum by large factors (Hamann, 1995).

Figure (2.21b) shows the terminal velocity of the winds of WN stars versus the effective escape velocity, based on the parameters M_*, R_*, and L_* derived by Hamann *et al.* (1995). An electron scattering coefficient of $\sigma_e = 0.25$ cm^2 g^{-1} was adopted for the calculation of the effective escape velocity. The late-WN stars have smaller escape velocities than the early-WN stars because their stellar radius is larger. The terminal velocities are between 700 and 2500 km s^{-1}and the ratio $v_\infty/v_{\rm esc}$ is between 1.5 and 4 with no clear relation to spectral type.

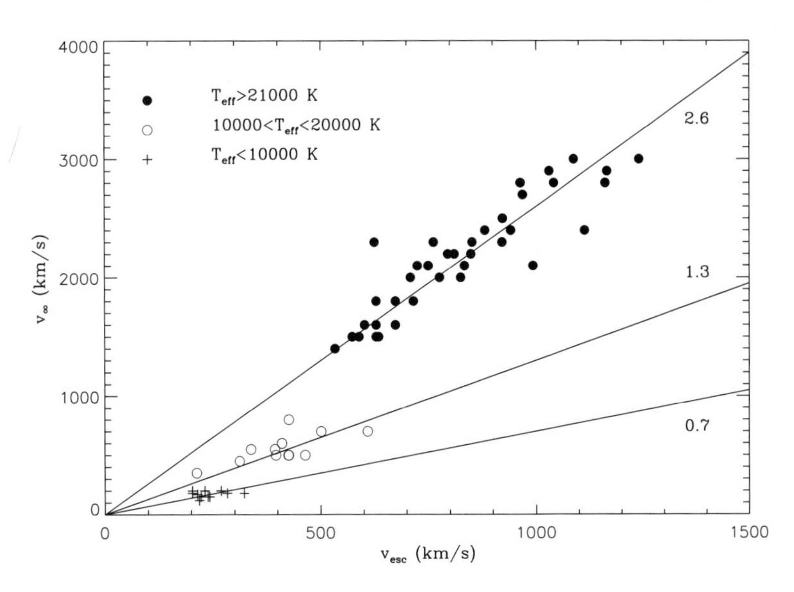

Figure 2.20 The terminal velocities of the winds, v_∞, versus the effective escape velocities $v_{\rm esc}$ for O, B and A stars. Symbols denote different temperature intervals: $T_{\rm eff} > 21000$ K (dots); $10000 < T_{\rm eff} < 20000$ K (circles); $5000 < T_{\rm eff} < 10000$ K (+). The uncertainty in v_∞ is typically 10 to 20 %. Notice the three linear relations. (Data from Lamers *et al.*, 1995)

2.7.3 Central stars of planetary nebulae

The ultraviolet spectra of central stars of planetary nebulae (CSPN) show P Cygni profiles of the resonance lines of C IV, N V and Si IV and sometimes also of lines from excited levels of N IV, O IV and O V. The effective temperatures of these stars is high, between about 30 000 K and 120 000 K and their luminosities are in the range of

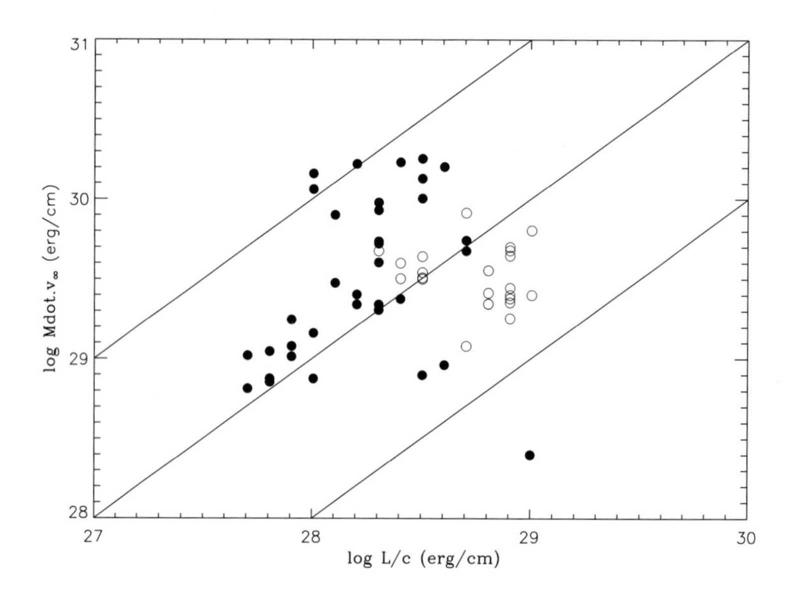

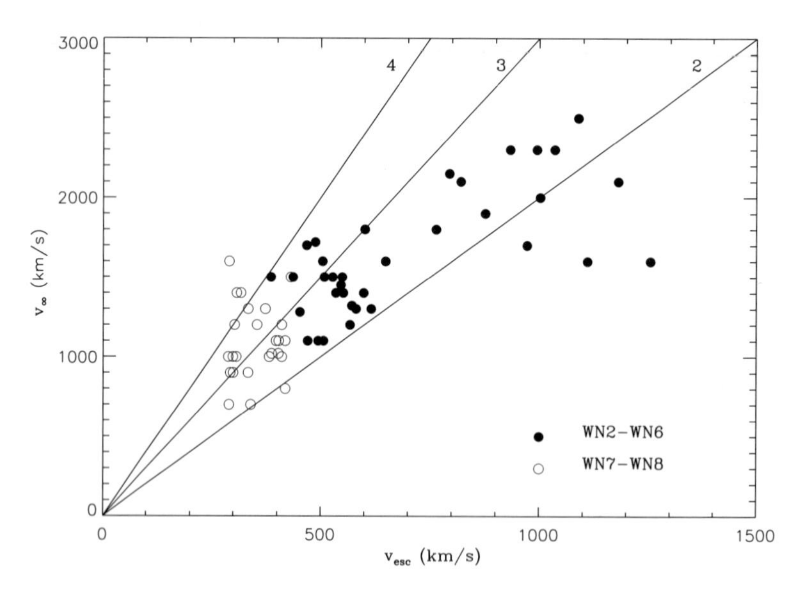

Figure 2.21 The wind momentum of WN stars as a function of the photon momentum. The wind momentum $\dot{M}v_\infty$ is typically 2 to 100 times larger than L_*/c (upper panel, a). The ratio between the terminal velocity and the escape velocity of the WN stars is between 1.5 and 4 (lower panel, b). (Based on data from Hamann *et al.*, 1995)

$3.5 < \log L_*/L_\odot < 4.3$. This implies radii of about 0.3 to 3 $R_\odot$. The masses are small and of the order of 0.5 to 0.6 $M_\odot$ because they are the progenitors of white dwarfs. The combination of low mass and small radii and high luminosity implies that the effective escape velocity at the stellar surface is between 200 and 800 km s^{-1}.

Mass loss rates for CSPN have been derived from the P Cygni profiles of the UV lines and from the visual emission lines by various groups; see Mendez *et al.* (1992), Perinotto (1993), Kudritzki *et al.* (1997) and references therein. The mass loss rates are small and of the order of 10^{-9} to 10^{-7} $M_\odot$ yr^{-1}. Yet, these small mass loss rates produce strong P Cygni profiles because the radii of the stars are small. The optical depth of the P Cygni profiles scales roughly with the column density in the wind and inversely with the terminal velocity, so $\tau \sim \dot{M}/(R_* v_\infty{}^2)$ (§ 8.4). A typical CSPN with $\dot{M} = 10^{-8}$ $M_\odot$ yr^{-1}, $v_\infty = 3000$ km s^{-1} and $R_* = R_\odot$ will have about the same optical depth of the P Cygni profiles as a typical O supergiant of $R_* = 50R_\odot$ with $v_\infty = 2000$ km s^{-1} and $\dot{M} = 2 \times 10^{-5}$ $M_\odot$ yr^{-1}.

Figure (2.22) shows the modified wind momentum (see Eq. 2.37) of the CSPN plotted versus the stellar luminosity. The modified wind momentum of the CSPN suggests a separation into two groups. If the CSPN are compared with the normal OB stars, with mass loss rates derived from the radio data, it turns out that the modified wind momentum of the lower group falls on the extrapolation of the relation

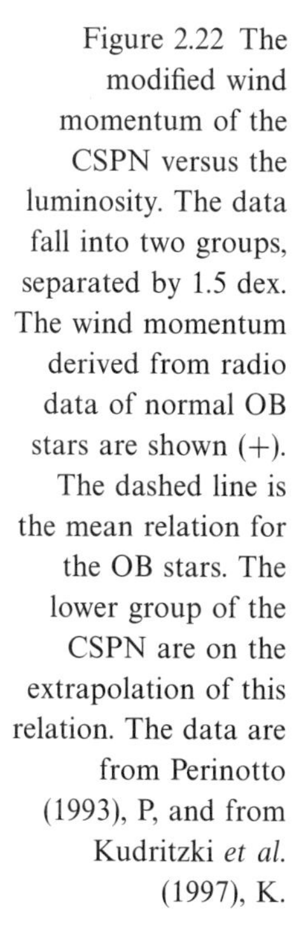

Figure 2.22 The modified wind momentum of the CSPN versus the luminosity. The data fall into two groups, separated by 1.5 dex. The wind momentum derived from radio data of normal OB stars are shown (+). The dashed line is the mean relation for the OB stars. The lower group of the CSPN are on the extrapolation of this relation. The data are from Perinotto (1993), P, and from Kudritzki *et al.* (1997), K.

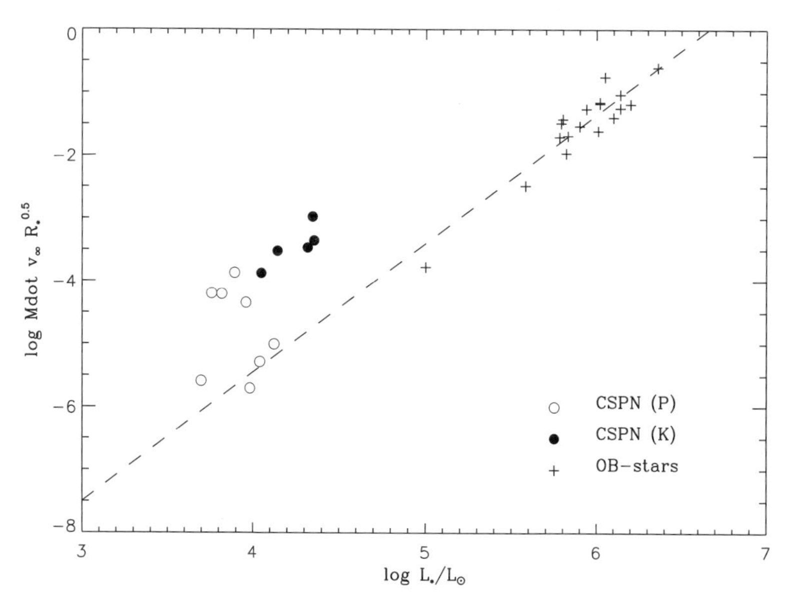

for the OB stars, Eq. (2.38) and Fig. (2.19a), but that the modified wind momentum of the other group is about 1.5 dex higher.

The P Cygni profiles of the UV lines show that the terminal velocities of their stellar winds are between 500 and 4000 km s^{-1}, which is about 3 to 5 times the escape velocity. This suggests that the winds of CSPN are driven by radiation pressure in spectral lines, similar to the winds of OB stars.

Figure (2.23) shows the terminal velocities of stellar winds of 8 CSPN versus the photospheric escape velocity. The data are from the reviews by Perinotto (1993) and by Kudritzki *et al.* (1997). We assumed a mass of 0.6 $M_{\odot}$ and an electron scattering coefficient of $\sigma_e = 0.29$ cm^2 g^{-1} (corresponding to a fully ionized wind with a He/H ratio of 0.15) for the calculation of the effective escape velocity. The mean relation is

$$v_{\infty} \simeq 4.4 \times v_e \tag{2.40}$$

This is higher than the mean ratio of 2.6 for the normal O stars (see Fig. 2.20).

2.8 Mass loss rates and terminal velocities of winds from cool stars

2.8.1 Red giants and supergiants

The mass loss rates from red giants and supergiants have been derived from Hα, from shifted chromospheric lines, from UV lines of red

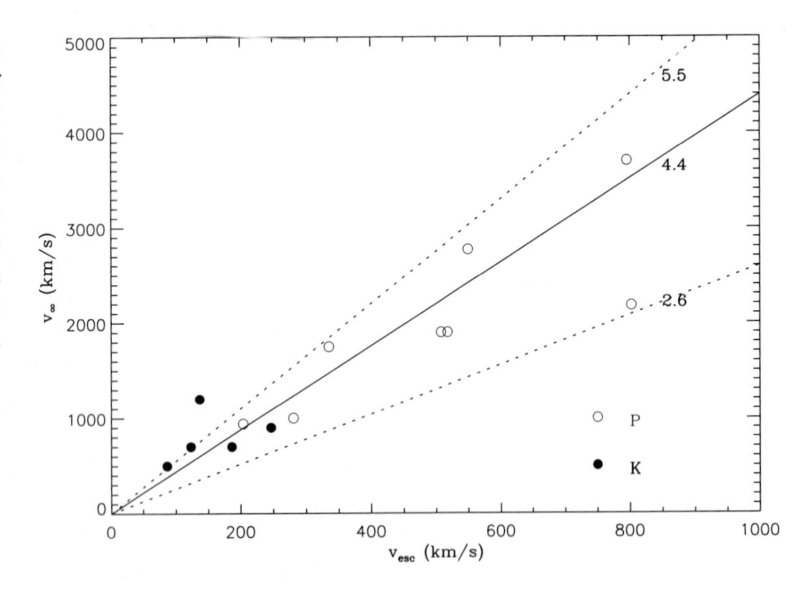

Figure 2.23 The terminal velocity of the winds of CSPN versus the effective escape velocity at the stellar surface. The mean relation is $v_{\infty} \simeq 4.4\ v_e$. The data are from Perinotto (1993), P, and from Kudritzki *et al.* (1997), K.

Table 2.3 *Mass loss determinations of red supergiants in binary systems*

Star	Type	M_*	R_*	$\log L_*$	$\dot{M}$	v_∞	v_e	v_∞/v_e
22 Vul	G3 II-Ib	4.3	40	2.99	6×10^{-9}	160	202	0.78
31 Cyg	K4 Ib	6.2	202	3.91	4×10^{-8}	80	108	0.74
ζ Aur	K4 Ib	8.3	140	3.41	6×10^{-9}	40	150	0.27
32 Cyg	K5 Iab	8.0	188	3.82	3×10^{-8}	60	127	0.47
δ Sge	M2 II	8.0	140	3.43	2×10^{-8}	28	147	0.19
α Sco	M1.5 Iab-Ib	18.0	575	4.68	1×10^{-6}	17	109	0.16

M_*, R_* and L_* are in solar units; $\dot{M}$ is in $M_\odot\,\mathrm{yr}^{-1}$; v_∞ is in $\mathrm{km\,s}^{-1}$.

supergiants with a hot companion. The data are reviewed by Dupree (1986) and Dupree and Reimers (1987).

Half a dozen red supergiants of types G3 to M2 with luminosity class between II and Iab have reliable mass loss rates, masses, radii and luminosities. These are the red supergiants with early type companion stars. In this case the mass of the red supergiant can be derived from the orbital motion of the two stars, and the mass loss rate from the red supergiant can be derived accurately by observing the lines from the wind of the cool star in absorption against the hot star. This method has been applied by Reimers and colleagues to six systems (see Dupree and Reimers, 1987). The data for these stars are listed in Table (2.3). The mass loss rates are between 10^{-9} and 10^{-6} $M_\odot\,\mathrm{yr}^{-1}$. From these data Reimers (1975) derived a relation between the mass loss rate and the product $L_* R_*/M_*$ (see Fig. 2.24).

This 'Reimers relation' can be written as

$$\dot{M} = 4 \times 10^{-13} \eta_R \frac{(L_*/L_\odot)(R_*/R_\odot)}{M_*/M_\odot} \tag{2.41}$$

in $M_\odot\,\mathrm{yr}^{-1}$, where $1/3 < \eta_R < 3$ is an empirical correction factor for different types of stars. So $\eta_R = 1$ for the red supergiants. This relation has been widely used in the literature even for other cool stars such as AGB stars, for which it was not derived. For comparison: the mass loss rate of the sun is 2×10^{-14} $M_\odot\,\mathrm{yr}^{-1}$. Since M_*/R_* is proportional to v_{esc}^2, the Reimers relation implies that the potential energy of the wind of red supergiants, $0.5\dot{M}v_{\mathrm{esc}}^2$, is proportional to the luminosity.

The terminal velocity of the winds of the red supergiants listed in Table (2.3) are between 17 and 160 $\mathrm{km\,s}^{-1}$ and the ratio $v_\infty/v_{\mathrm{esc}}$ varies between 0.16 and 0.78. Both quantities decrease with increasing luminosity.

The mass loss rates of red giants of types K and M and luminosity class II and III are plotted versus spectral type in Fig. (2.25). $\dot{M}$ of class II stars increases from about 3 10^{-10} at K0II to 10^{-7} $M_{\odot}\,\mathrm{yr}^{-1}$ at M5II. The mass loss rate of class III stars also increase to later types, but there is a large scatter. Part of this scatter is due to the fact that the stellar parameters of the individual red giants are poorly known. The figure suggests that the mass loss rate of class III stars does not only depend on spectral type. The mass loss rates predicted by the 'Reimers relation' with $\eta_R = 1$ are indicated by dashed lines. The adopted stellar parameters are from Lang (1992) with $M_* = 3M_{\odot}$ for class II stars and $M_* = 1.2M_{\odot}$ for class III stars. These predicted relations show approximately the same general trends as the observed ones, but there is an offset of about 1 to 0.5 dex for class II stars.

The terminal velocities of the winds of K and M giants scatter between about 10 and 200 km s^{-1} with no obvious correlation to the stellar parameters (Dupree, 1986).

2.8.2 Asymptotic giant branch stars, OH/IR stars and Miras

Asymptotic giant granch (AGB) stars are the late evolution stages of low and intermediate mass stars when the star has a degenerate core, a He-burning and a H-burning shell, and a very extended convective H-rich envelope. A typical AGB star has a radius of about 3×10^{13} cm $= 430\ R_{\odot}\ \simeq 2\ AU$, a luminosity of $6.10^3\ L_{\odot}$, $T_{\mathrm{eff}} \simeq 2500$ K, and a mass between 0.6 and 6 $M_{\odot}$. The small mass and the large radius result in a very low surface gravity of about 10^{-1} cm s^{-2}. The small

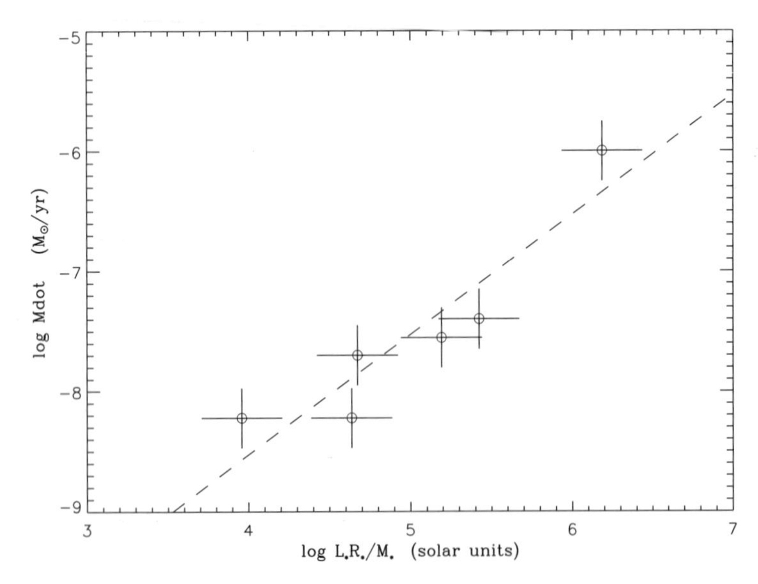

Figure 2.24 The empirical 'Reimers relation' between the mass loss rates of red supergiants and the ratio L_*R_*/M_*. The mean relation is log $\dot{M} = -12.40 + \log(L_*R_*/M_*)$. (Data from Dupree and Reimers, 1987)

escape velocity of about 40 km s^{-1} makes it easy for the outer layers to escape, if there is a driving mechanism. There are two mechanisms that can drive the winds of AGB stars: pulsation and radiation pressure on dust. The observed high mass loss rates of the AGB stars are probably due to a combination of both mechanisms: pulsation in an atmosphere of low gravity results in a large scale density height which facilitates the formation of dust. The radiation pressure on the dust and the interactions of the dust grains with the gas then results in a high mass loss rate. (This mechanism is discussed in detail in Chapter 7.)

The pulsation of the AGB stars can be observed in the form of photometric variations. Stars that show variations with periods longer than about 100 days and amplitudes of $\Delta V \gtrsim 2.5^m$ are called Miras. If the mass loss rate is so high that the dusty wind obscures the star and the stellar radiation is reprocessed by the dust, the star will become an infrared source. The high mass loss then also gives rise to strong OH maser lines. The star is then an OH/IR star. The properties of AGB stars, their winds and circumstellar envelopes have been reviewed extensively by Habing (1996).

The mass loss rates from OH/IR stars and from AGB stars have been derived from molecular lines and from the dust emission. This results in two different mass loss rates: one for the the dust and one for the gas. Studies of stars for which both $\dot{M}_{\text{gas}}$ and $\dot{M}_{\text{dust}}$ can be

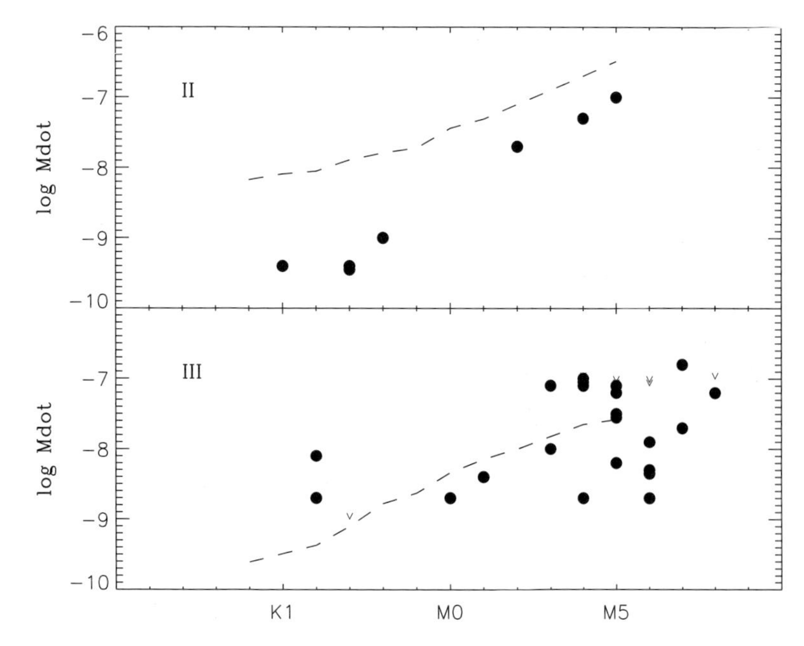

Figure 2.25 The mass loss rates of red giants of class II (upper) and III (lower) are plotted versus spectral type, from K1 to K5 and from M0 to M8. The arrows indicate upper limits. The dashed lines show the predicted Reimers relation. (Based on data from Dupree, 1986)

derived independently have shown that the gas/dust ratio in mass is typically about 200 (e.g. Whitelock *et al.*, 1994), so most of the wind escapes in the form of gas. Using this 'canonical' gas/dust ratio, the mass loss rate of many more stars has been derived from the dust radiation only.

Mass loss rates of AGB stars have been derived from the IR energy distributions, i.e. from the dust radiation by many groups (see the reviews by van der Veen and Olofsson, 1990 and by Habing, 1996). It turns out that the $K - [12]$ magnitude scales very nicely with the mass loss rate, where [12] is the magnitude of the star measured with the 12 μm filter of the *IRAS* satellite (Whitelock *et al.*, 1994).

The mass loss rates of the pulsating variables with a mass of $M_* \simeq 2.5\ M_\odot$ are correlated with the pulsational period P. Figure (2.26) shows the relation between $\dot{M}$ and P for Galactic Mira stars of spectral types M and S and for Galactic pulsating OH/IR stars. The mass loss rates are derived from the CO emission lines and from the 60 μm flux. There is a strong correlation between P and $\log \dot{M}$ for stars with periods less than about 600 days. For longer periods the mass loss seems to saturate to some constant value of about 10^{-4} $M_\odot\,\mathrm{yr}^{-1}$.

The mean empirical relation, derived by Vassiliades and Wood (1993) for periods less than 600 days is

$$\log \dot{M} \;=\; -11.4 \;+\; 0.0123 \times P(\mathrm{days}) \tag{2.42}$$

in $M_\odot\,\mathrm{yr}^{-1}$. This shows that the mass loss rate of AGB stars increases exponentially with P, until it reaches the very high value of 10^{-4} $M_\odot\,\mathrm{yr}^{-1}$ for the OH/IR stars. This very high mass loss rate is referred to as the 'superwind phase'. Eq. (2.42) is for pulsating stars with $M_* \simeq 2.5\ M_\odot$.

Vassiliades and Wood (1993) have argued that the relation should be different for AGB stars more massive than 2.5 $M_\odot$. They propose the following relation

$$\log \dot{M} = -11.4 + 0.0123 \times \left\{ P(\mathrm{days}) - 100 \left(\frac{M_*}{M_\odot} - 2.5 \right) \right\} \tag{2.43}$$

for $M_* > 2.5\ M_\odot$. With these two equations, (2.42) and (2.43), the mass loss rates of low and intermediate mass stars can be calculated during their AGB phase. The period P is assumed to be the fundamental period, which depends on the stellar parameters as

$$\log P(\mathrm{days}) = -2.07 + 1.94 \times \log \left(\frac{R_*}{R_\odot} \right) - 0.9 \times \log \left(\frac{M_*}{M_\odot} \right) \tag{2.44}$$

(Wood, 1990). Alternative expressions for the mass loss rates of AGB stars have been given by Blöcker (1995).

The terminal velocity of the AGB winds are small and of the order of about 5 to 25 km s^{-1} (e.g. Loup *et al.*, 1993).

2.9 Conclusions

In this chapter we have discussed the various ways to detect and measure mass loss from stars. These methods cover the full spectral energy range from the UV to the radio region. The methods that use spectral lines, e.g. P Cygni profiles from UV resonance lines and optical emission lines for hot stars and molecular lines for cool stars, have the advantage that they provide information about both the velocity (from Doppler shifts or line profiles) and the density or column density (from the line strengths or the line profiles). The methods that use continuum radiation to study mass loss, e.g. the IR or radio free-free emission for hot stars and the dust for cool stars, provide information about the amount of gas or dust in the star, but not on the velocity. In order to derive mass loss rates from continuum radiation one needs independent velocity information from line profiles.

The methods that use line profiles are generally sensitive to the assumptions about the thermal structure of the wind, because the line absorption or line emission is usually strongly dependent on the ionization and excitation conditions. The methods that use continuum radiation are generally less sensitive to the details of the wind structure.

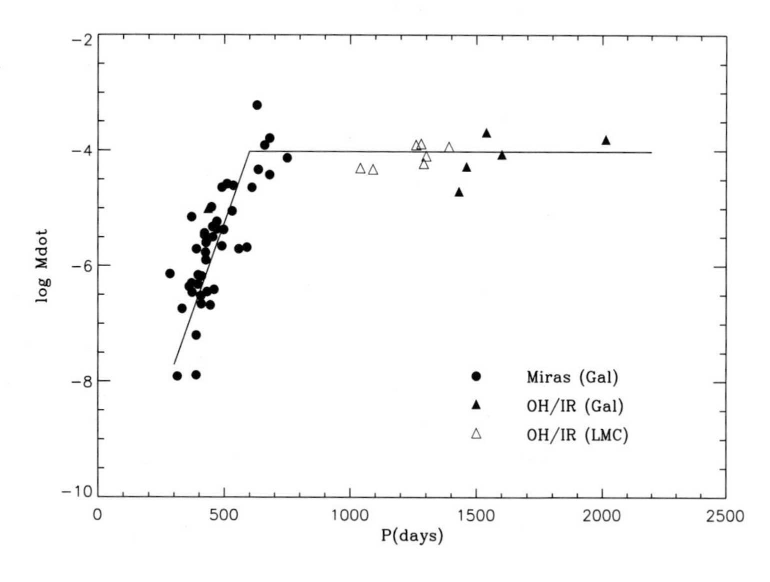

Figure 2.26 The mass loss rates of Mira variables and OH/IR stars as a function of their pulsational period. Dots refer to Miras of types M, S and C. Triangles refer to OH/IR stars in the Galaxy (filled) and the LMC (open). The dashed line shows the mean relation for $P < 600$ days. (Data from Vassiliades and Wood, 1993)

Therefore the best way to determine mass loss rates from stars is to use both kinds of observations. For the hot stars the most reliable mass loss rates are derived from radio continuum measurements combined with velocity information from the P Cygni profiles in the UV. For the cool stars the most accurate mass loss rates are derived from the continuum dust emission in the infrared combined with the velocity information from the CO lines in the radio region.

The mass loss rates from hot stars in various phases of evolution, e.g. the OB stars and the central stars of planetary nebulae show a rather tight correlation with the stellar luminosity. The terminal velocities of the winds of hot stars are typically a few times the effective escape velocity. Both arguments strongly suggest that the dominant mass loss mechanism for hot stars is radiation pressure in numerous spectral lines. This mechanism works via the transfer of *momentum* from the radiation to the gas. The high momentum of the winds of Wolf-Rayet stars cannot (yet?) be explained by the radiation driven wind models.

The mass loss rates from cool stars show a much less clear picture. There is as yet no clear relation between the mass loss rate of solar type stars or red giants and the stellar parameters. For the AGB stars there is a clear correlation between the mass loss rate and the pulsation period. In this case the mass loss rate is probably driven by a combination of two effects: pulsations result in extended envelopes with large density scaleheights and the subsequent dust formation results in a very large radiation pressure on the dust grains which share their momentum with the gas by dust-gas collisions.

2.10 Suggested reading

Cassinelli, J.P. & Lamers, H.J.G.L.M. 1987 'Winds from hot young stars', in *Exploring the universe in the ultraviolet with the IUE satellite*, eds. Y. Kondo *et al.* (Dordrecht: Reidel), p. 139
(A review of mass loss from hot stars)

Dupree, A.K. & Reimers, D. 1987 'Mass loss from cool stars' in *Exploring the universe in the ultraviolet with the IUE satellite*, eds. Y. Kondo *et al.* (Dordrecht: Reidel), p. 321
(A review of mass loss from cool stars)

Habing, H.J. 1996 'Circumstellar envelopes of asymptotic giant branch stars', *A & A Reviews* **7**, 97
(A review of mass loss and evolution of AGB stars)

Hamann, W.R. 1995 'Model atmospheres and spectral analysis of Wolf-Rayet stars', in *Wolf-Rayet stars: binaries, colliding winds, evolution*, eds K.A. van der Hucht and P.M. Williams (Dordrecht: Kluwer) p. 105
(A review of mass loss from WR stars)

Kudritzki, R.P., Mendez, R.H., Puls, J. & McCarthy, J.K. 1997 'Mass loss from central stars of planetary nebulae', in *Planetary Nebulae, IAU Symp. 180*, eds. H.J. Habing and H.J.G.L.M. Lamers (Dordrecht: Kluwer), p. 64
(A review of mass loss from CSPN stars)

Snow, T.P., Lamers, H.J.G.L.M., Lindholm, D.M. & Odell, A.P. 1994 'An atlas of ultraviolet P Cygni profiles', *Ap. J. Sup.* **95**, 163

Sobolev, V.V. 1960 *Moving envelopes of stars* (Cambridge MA: Harvard Univ. Press)
(The basic theory of line formation in stellar winds)

3 Basic concepts: isothermal winds

The purpose of this chapter is to describe and explain some of the fundamental properties of the stellar wind models. This is done by deriving the equations for idealized simple winds. For these simple models the equation of motion can be solved easily so that the velocity and density structures of the wind are known. The solutions show how the velocities and densities depend on the forces in the wind. They also show that the mass loss rate of a stationary wind model is uniquely determined by the solution of the equations, i.e., given the lower boundary conditions in the wind and the forces and energy gains and losses, a physically realistic solution exists for only one specific value of the mass loss rate. The simple solutions discussed in this chapter show how this so-called *critical solution* depends on the forces and the energy of the wind. Although only simplified models are considered in this chapter, the conclusions are qualitatively valid for the more complicated and detailed models which will be described in later chapters.

Section 3.1 describes the simplest possible model of an isothermal wind in which gas pressure provides the outward force. In §§ 3.2, 3.3 and 3.4 the effects of additional forces in isothermal wind models are considered; first as simple analytic expressions, such as a force which varies as r^{-2}, or as $v\ dv/dr$, and later in more general terms. Section 3.5 describes the analogy between stellar winds and rocket nozzles. Section 3.6 gives the conclusions.

3.1 Isothermal winds with gas pressure only

In this section we describe the structure of an isothermal stellar wind in which the gas is subject to only two forces: the inward directed gravity and the outward directed gradient of the gas pressure. We will show that the momentum equation has many solutions, depending on the boundary conditions but only one of them, the *critical solution* starts subsonic at the lower boundary of the wind and reaches supersonic velocities at large distance. The mass loss rate is fixed by the critical solution.

3.1.1 The momentum equation and the critical point

For a time-independent stellar wind with a constant mass loss rate, the amount of gas passing through any sphere of radius r is constant. This is expressed in the *equation of mass conservation*

$$\dot{M} = 4\pi r^2 \rho(r) v(r) = 4\pi F_m \tag{3.1}$$

where $F_{\rm m}$ is the mass flux from the star per steradian.

The motion of the gas in a stellar wind is described by Newton's law, $f = m \times a$ or $f = \rho \; dv/dt$ if f is the force per unit volume and f/ρ is the force per unit mass. The velocity and location of a unit of mass that is accelerated in the wind depends on distance r and time t, so the velocity gradient in Newton's law is

$$\frac{dv(r,t)}{dt} = \frac{\partial v(r,t)}{\partial t} + \frac{\partial v(r,t)}{\partial r}\frac{dr(t)}{dt} = v(r)\frac{dv}{dr} \tag{3.2}$$

In a stationary, i.e. time-independent, wind the velocity at a given distance does not change with time so $\delta v(r,t)/\delta t = 0$. If there are no other forces acting on the wind than the gas pressure and gravity the equation of motion is

$$v\frac{dv}{dr} + \frac{1}{\rho}\frac{dp}{dr} + \frac{GM_*}{r^2} = 0 \tag{3.3}$$

This equation describes the motion or the momentum of the gas in a stationary stellar wind. It is usually called the *momentum equation.* The first term of the momentum equation is the acceleration, which is produced by the pressure gradient (second term) and the gravity (third term). The energy equation is simply $T(r) = T =$ constant since the thermal structure is assumed to be somehow maintained at a constant temperature. This is a basic assumption of this chapter. Non-isothermal winds are discussed in the next chapter. If the flow behaves like a perfect gas then

$$p = \mathscr{R}\rho T/\mu \tag{3.4}$$

where $\mathscr{R}$ is the gas constant and μ is the mean atomic weight of the particles expressed in units of $m_{\rm H}$. The mean atomic weight μ is assumed to be constant, with a value of $\mu = 0.602$ for solar composition material. The force due to the pressure gradient can be written as

$$\frac{1}{\rho}\frac{dp}{dr} = \mathscr{R}\mu\frac{dT}{dr} + \frac{\mathscr{R}T}{\mu\rho}\frac{d\rho}{dr} = \left(\frac{\mathscr{R}T}{\mu}\right)\frac{1}{\rho}\frac{d\rho}{dr} \tag{3.5}$$

for an isothermal wind.

The density gradient can be expressed in a velocity gradient by (3.1)

$$\frac{1}{\rho}\frac{d\rho}{dr} = -\frac{1}{v}\frac{dv}{dr} - \frac{2}{r} \tag{3.6}$$

Substituting (3.6) and (3.5) into (3.3) yields

$$v\frac{dv}{dr} + \frac{\mathscr{R}T}{\mu}\left\{-\frac{1}{v}\frac{dv}{dr} - \frac{2}{r}\right\} + \frac{GM_*}{r^2} = 0 \tag{3.7}$$

or

$$\frac{1}{v}\frac{dv}{dr} = \left\{\frac{2a^2}{r} - \frac{GM_*}{r^2}\right\} / \left\{v^2 - a^2\right\} \tag{3.8}$$

where

$$a = (\mathscr{R}T/\mu)^{1/2} \tag{3.9}$$

is the isothermal speed of sound, which is constant in an isothermal wind. The lower boundary condition of Eq. (3.8) is the bottom of the isothermal region, located at r_0 where $v(r_0) = v_0$. In general r_0 is about the photospheric radius or a bit larger if the star is to be surrounded by an isothermal corona.

The momentum equation (3.8) has a singularity at the point where $v(r) = a$. We will show below that this singularity is extremely important, because it implies that the mass loss rate is fixed. Let us first consider some of the immediate consequences of Eq. (3.8) for the structure of the wind.

The numerator contains an r^{-1} and an r^{-2} term. This numerator goes to zero at a distance

$$r = r_c \equiv GM_*/2a^2 \tag{3.10}$$

This is the *critical distance*, or the distance of the *critical point*. Such a distance exists in the isothermal wind at $r_c > r_0$ if

$$GM_*/2a^2 > r_0 \quad \text{or} \quad GM_*/2r_0 \equiv \frac{v_{\text{esc}}^2(r_0)}{4} > a^2. \tag{3.11}$$

If the critical point were located at $r_c < r_0$, it would not be in the isothermal region, but in the underlying photosphere or in the transition region. In that case the assumption of an isothermal wind made in this chapter would not be valid. The velocity gradient at the critical distance will be zero, because the numerator equals zero, unless

$v(r_c) = a$. Similarly, the velocity gradient at the distance where $v = a$ will be $\pm\infty$, because the denominator $= 0$, unless $r = r_c$ when $v = a$. So the only solution which can have a positive velocity gradient at all distances is the one that goes through the critical point. This is the *critical solution* for which

$$v(r_c) = a \qquad \text{at} \qquad r_c = \frac{GM_*}{2a^2} \tag{3.12}$$

So we find that at the critical point

$$v(r_c) = a = \frac{v_{\rm esc}(r_c)}{2} \tag{3.13}$$

where $v_{\rm esc} = \sqrt{2GM_*/r_c}$ is the escape velocity at the critical point. The point in the wind where $v(r) = a$ is called the *sonic point*. The point where $v(r) = v_{\rm esc}(r)$ is called the *escape point*. The point where the numerator of the momentum equation goes to zero is referred to as the *Parker point* after Eugene Parker, who described and solved the equations for the solar wind in 1958. In an isothermal wind the critical point coincides with the sonic point, but this is not necessarily true for other wind models that we will consider. The critical solution is transonic, because it starts subsonic at small distances and reaches a supersonic velocity at large distances.

The topology of the solutions of Eq. (3.8) is shown in Fig. (3.1) for various initial velocities $v(r_0)$. In the region of $r < r_c$, where the numerator of Eq. (3.8) is negative, the velocity gradient is positive if $v(r) < a$ and negative if $v(r) > a$. Similarly, at $r > r_c$, where the numerator of Eq. (3.8) is positive, the velocity gradient will be negative if $v(r) < a$ and positive if $v(r) > a$. Curve 1 is the critical solution which starts subsonic, goes through the critical point and reaches supersonic values farther outwards. Curve 2 also passes through the critical point, but it starts supersonically. Curve 3 is the solution in the case that the lower boundary velocity v_0 is too small to reach the critical point, so it remains subsonic with a maximum at r_c. Curve 4 is the solution for a supersonic starting value of v_0. Curves 5 and 6 are mathematical solutions which do not have a physical meaning because they are multi-valued. They show that $dv/dr \to \infty$ at $v = a$ in agreement with Eq. (3.8), unless the numerator is zero.

The slope of the critical solution at r_c can be found by applying de l'Hopital's rule (see Appendix 3) to Eq. (3.8), with $f(r)$ and $g(r)$ as the numerator and the denominator of Eq. (3.8). De l'Hopital's rule results in

$$\frac{1}{v}\left(\frac{dv}{dr}\right)_{r_c} = \left\{-\frac{2a^2}{r_c^2} + \frac{2GM_*}{r_c^3}\right\}\left\{\frac{2v\,dv}{dr}\right\}_{r_c}^{-1} = \frac{a^2}{r_c^2}\left(\frac{v\,dv}{dr}\right)_{r_c}^{-1} \tag{3.14}$$

This gives

$$\left(\frac{dv}{dr}\right)_{r_c} = \frac{\pm 2a^3}{GM_*} \qquad (3.15)$$

The positive or negative sign is a result of the fact that de l'Hopital's rule results in an expression for $(dv/dr)^2$. A singularity which has two equal positive and negative gradients at the critical point is called an X-type singularity because of its topology (see Fig. 3.1).

The preceeding discussion has shown that there is only one solution which starts subsonic and ends supersonic. This critical solution occurs for only one particular value of the velocity at the lower boundary: v_0(crit). This implies that an isothermal envelope with given density ρ_0 at its bottom can only produce a transonic wind if

$$\dot{M} = 4\pi r_0^2 \rho_0 v_0(\text{crit}) \qquad (3.16)$$

This is a very important result which shows that an isothermal wind with a given lower boundary (ρ_0, T_0 and gravity) can reach supersonic velocities for only one specific value of the mass loss rate! At first glance this may seem surprising, because one might have expected intuitively that an isothermal wind can sustain a range of mass loss rates, each with its corresponding velocity law provided by the forces from the gas pressure and the gravity. This, however is not the case, because the wind should not only satisfy the conservation

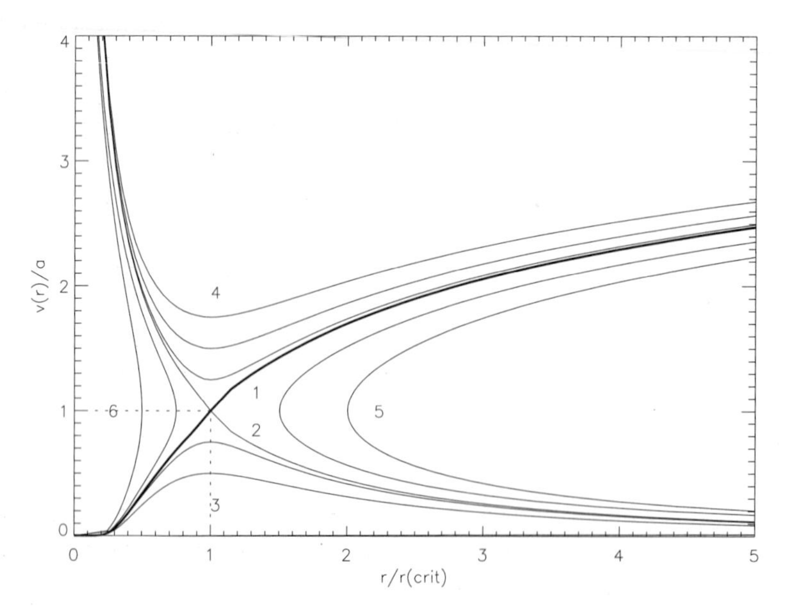

Figure 3.1 Solutions of the momentum equation of an isothermal stellar wind with gas pressure and gravity in terms of v/a versus r/r_c. For this particular case $r_c = 5\ R_*$. The different curves are described in the text. Curve 1 (thick) is the transonic solution with increasing velocity through the critical point r_c where $v = a$.

of momentum and mass but also the conservation of energy. We will show this below.

3.1.2 The energy of the wind

In the case of an isothermal wind, discussed here, the conservation of energy is hidden in the condition $T(r) = T$. Consider the total energy of the outflowing matter per unit mass

$$e(r) = \frac{v^2}{2} - \frac{GM_*}{r} + \frac{5}{2}\frac{\mathscr{R}T}{\mu} \tag{3.17}$$

in which the first two terms are the kinetic and potential energy and the last term is the *enthalpy* (see Appendix 2), which is the sum of the internal energy, $3\mathscr{R}T/2\mu$, and the potential for work by adiabatic expansion $\mathscr{R}T/\mu$. At the critical point where $v^2/2 = a^2/2$ and $GM_*/r_c = 2a^2$ the sum of the kinetic, potential and internal energy is exactly zero. We will see later that this condition is specific to the isothermal case and does not apply in general at the critical point. The energy equation (3.17) can be written as

$$e(r) = e(r_0) + \frac{(v^2 - v_0^2)}{2} + \frac{GM_*}{r_0}\left(1 - \frac{r_0}{r}\right) \tag{3.18}$$

This shows that the energy of the isothermal wind is not constant, but that an energy equal to the sum of the two last terms has to be added throughout the wind to keep it isothermal. This makes the isothermal approximation seem artificial and unrealistic. However, it turns out that winds in which the heating and cooling are mainly due to radiative processes may be close to isothermal.

At the bottom of the isothermal wind, where $v^2 < a^2 < v_{esc}^2$, the energy is negative and of the order of the potential energy

$$e(r_0) \simeq -\frac{GM_*}{r_0} \tag{3.19}$$

if $a^2 \ll v_{esc}^2$. At large distances where $v^2 > a^2$ and $v^2 > GM_*/r$ the energy is positive and of the order of the kinetic energy

$$e(r \to \infty) \simeq \frac{v^2}{2} \tag{3.20}$$

if $v \gg a$. The difference between $e(r_0)$ and $e(\infty)$ has been added to the wind to overcome the gravitational well, provide it with kinetic energy and keep it isothermal.

3.1.3 The velocity and density distribution

To show the effect that the atmospheric expansion has on the density distribution of the wind, it is interesting to compare its density structure with that of a hydrostatic envelope of the same gravity and temperature. To this purpose we first derive the velocity law of the wind. The momentum equation (3.8) has an analytic solution which can easily be found by expressing the first two terms of Eq. (3.7) as $d(v^2/2)/dr$ and $-a^2 d(\ln v)/dr$. The solution is

$$\frac{v^2}{2} - a^2 \ln(v) = 2a^2 \ln(r) + \frac{GM_*}{r} + \text{constant} \tag{3.21}$$

The constant is fixed by the condition $v(r_c) = a$ at the critical point. This yields the solution

$$v \exp\left(-\frac{v^2}{2a^2}\right) = a\left(\frac{r_c}{r}\right)^2 \exp\left\{-\frac{2r_c}{r} + \frac{3}{2}\right\} \tag{3.22}$$

with $r_c = GM_*/2a^2$. Equation (3.22) is derived by dividing Eq. (3.21) by a^2 and taking the exponents on both sides.

The initial velocity at the lower boundary of the isothermal region can be derived by applying Eq. (3.22) at r_0. At the bottom of a gravitationally bound subsonic wind with $v_0 \ll a < v_{\rm esc}$ one finds

$$\begin{aligned} v_0 &\simeq a\left(\frac{r_c}{r_0}\right)^2 \exp\left\{-\frac{2r_c}{r_0} + \frac{3}{2}\right\} \\ &= a\left(\frac{v_{\rm esc}(r_0)}{2a}\right)^2 \exp\left\{-\frac{v_{\rm esc}^2(r_0)}{2a^2} + \frac{3}{2}\right\} \end{aligned} \tag{3.23}$$

This is an interesting result because it shows that an isothermal wind that is driven by gas pressure can only become supersonic if the initial velocity has the very specific value of Eq. (3.23). This velocity, together with the density ρ_0 at the lower boundary r_0 of the isothermal region, sets the mass loss rate of the star. We will derive the value of $\dot{M}$ below.

The velocity throughout the wind can also be expressed as

$$\frac{v}{v_0} \exp\left(-\frac{v^2}{2a^2}\right) = \left(\frac{r_0}{r}\right)^2 \exp\left\{\frac{GM_*}{a^2}\left(\frac{1}{r_0} - \frac{1}{r}\right)\right\} \tag{3.24}$$

At large distances where $r \gg r_0$, the velocity law approaches

$$v(r \to \infty) \simeq 2a\sqrt{\ln(r/r_0)}, \tag{3.25}$$

which increases infinitely. This is a consequence of the assumption that the wind is isothermal up to very large distances. It requires the continuous addition of energy and the resulting gas pressure then accelerates the wind indefinitely. Clearly this is an unrealistic situation.

In reality the winds are approximately isothermal only up to a certain distance and the velocity does not increase significantly beyond that distance.

The density structure is given by the mass continuity equation (3.1) which yields

$$\frac{\rho}{\rho_0} \exp\left\{+\frac{1}{2}\left(\frac{v_0 \rho_0 r_0^2}{a \rho r^2}\right)^2\right\} = \exp\left\{-\frac{GM_*}{a^2}\left(\frac{1}{r_0} - \frac{1}{r}\right)\right\} \quad (3.26)$$

This equation can be solved numerically to give $\rho(r)/\rho_0$. The result is shown in Fig. (3.2) for a wind model with a temperature such that $a^2 = 0.05 GM_*/r_0$, which implies a critical point at $r_c = 10 r_0$. Let us now compare this with the density distribution of a hydrostatic atmosphere. In a static atmosphere the density is given by the hydrostatic equation

$$\frac{1}{\rho}\frac{dp}{dr} + \frac{GM_*}{r^2} = 0 \quad (3.27)$$

which transforms, with Eq. (3.5), into

$$\frac{r^2}{\rho}\frac{d\rho}{dr} = -\frac{GM_*}{a^2} \quad (3.28)$$

if the atmosphere is isothermal. The solution is

$$\frac{\rho(r)}{\rho_0} = \exp\left\{-\frac{GM_*}{a^2}\left(\frac{1}{r_0} - \frac{1}{r}\right)\right\} = \exp\left\{-\frac{(r - r_0)}{\mathscr{H}_0}\frac{r_0}{r}\right\} \quad (3.29)$$

where

$$\mathscr{H}_0 = \mathscr{R}T/\mu g_0 \quad \text{with} \quad g_0 = GM_*/r_0^2 \quad (3.30)$$

is the density scale height at the bottom of the isothermal region.

The density structure (3.29) of the hydrostatic region is very similar to that of an isothermal wind. In fact, Eq. (3.26) is equal to Eq. (3.29) at the bottom of a wind if v_0 is highly subsonic. The density structure of an isothermal wind is compared with the hydrostatic case in Fig. (3.2). This figure shows that the densities are very similar, except close to the critical point where the wind density drops below that of a hydrostatic model. The difference at the critical or sonic point is exactly a factor exp (−0.5) because $v_0 \rho_0 r_0^2 = a \rho(r_c) r_c^2$.

The close agreement between the hydrostatic and wind density structure in the subsonic region where $v \leq 0.5a$ is due to the fact that the term $v dv/dr$ in the momentum equation (3.3) is much smaller than the pressure gradient term. In other words: *the structure of the subsonic region is mainly determined by the hydrostatic density structure and not by the velocity law!*

3.1.4 The mass loss rate

The good agreement between the wind density and the hydrostatic density in the subsonic region of an isothermal wind allows an easy estimate of the mass loss rates of isothermal winds driven by gas pressure. Suppose a star with radius R_* and mass M_* has a corona of temperature T_c which has a base density ρ_0. Assuming that the coronal temperature is about constant, the density in the subsonic region varies about exponentially with the scale height (3.29) out to r_c where $v_{\rm esc} = 2a$. The mass loss rate can then be estimated from the distance, density and velocity at the critical point.

$$\dot{M} = 4\pi\rho_c\, a\, r_c^2 \approx 4\pi\rho_0\, a\, r_0^2 \left(\frac{r_c}{r_0}\right)^2 \exp\left\{-\frac{r_c - r_0}{\mathcal{H}_0} \cdot \frac{r_0}{r_c}\right\} \tag{3.31}$$

The mass loss rate for an exactly isothermal wind, given by Eq. (3.1) at the critical point with the density given by Eq. (3.28) and the correction factor exp (-0.5) is

$$\begin{aligned}\dot{M} &= 4\pi\rho_0\, a\, r_0^2 \left\{\frac{v_{\rm esc}(r_0)}{2a}\right\}^2 \exp\left\{-\frac{(r_c - r_0)}{\mathcal{H}_0}\frac{r_0}{r_c} - \frac{1}{2}\right\} \\ &= 4\pi\rho_0\, a\, r_0^2 \left\{\frac{v_{\rm esc}(r_0)}{2a}\right\}^2 \exp\left\{-\frac{v_{\rm esc}^2(r_0)}{2a^2} + \frac{3}{2}\right\}\end{aligned} \tag{3.32}$$

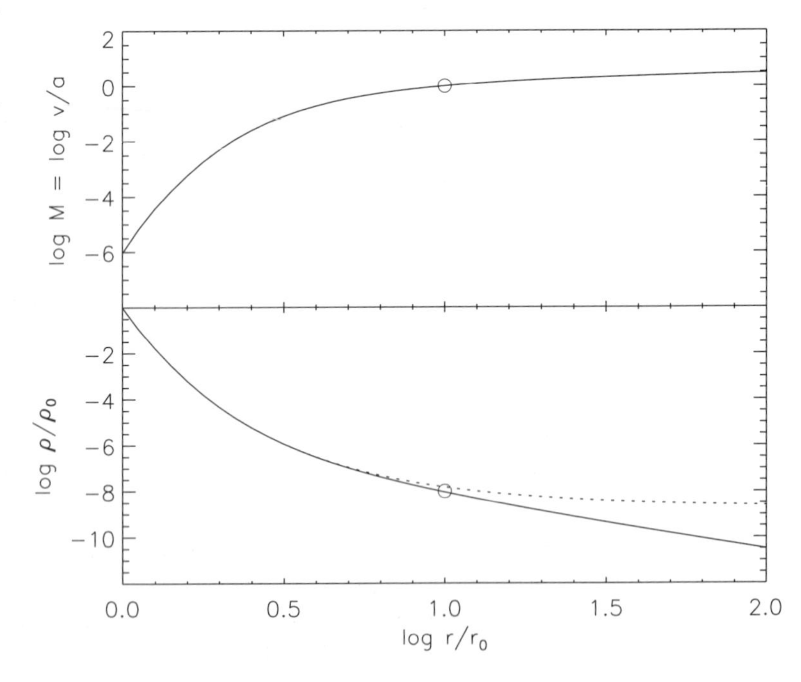

Figure 3.2 The velocity, in terms of Mach number ($M = v/a$), and the density, in terms of the value at the lower boundary r_0, as a function of r/r_0. The location of the critical point is indicated by a circle. The dashed line shows the density distribution of a hydrostatic atmosphere with the same temperature. The two density distributions are very similar in the subsonic part of the flow.

Table 3.1 *Characteristics of isothermal winds with a density at the lower boundary of* $\rho_0 = 10^{-14}$ g/cm^3

M_* ($M_\odot$)	R_* ($R_\odot$)	v_{esc} (km/s)	T (K)	a (km/s)	$\mathscr{H}_0$ (R_*)	r_c (R_*)	$\frac{r_c - R_*}{\mathscr{H}_0}$	$\dot{M}$ ($M_\odot$/yr)
1	1	617.5	1.10^5	37.2	$7.3\ 10^{-3}$	68.7	$9.3\ 10^3$	$1.2\ 10^{-68}$
			3.10^5	64.5	$2.2\ 10^{-2}$	22.9	$1.0\ 10^3$	$1.5\ 10^{-28}$
			1.10^6	117.7	$7.3\ 10^{-2}$	6.9	$8.0\ 10^1$	$1.6\ 10^{-14}$
			3.10^6	203.9	$2.2\ 10^{-1}$	2.3	5.9	$8.2\ 10^{-11}$
			5.10^6	263.2	$3.6\ 10^{-1}$	1.4	1.1	$4.0\ 10^{-10}$
1	100	61.7	3.10^3	6.4	$2.2\ 10^{-2}$	22.9	$1.0\ 10^3$	$1.5\ 10^{-25}$
			1.10^4	11.8	$7.3\ 10^{-2}$	6.9	$8.1\ 10^1$	$1.6\ 10^{-11}$
			3.10^4	20.4	$2.2\ 10^{-1}$	2.3	5.9	$8.2\ 10^{-8}$
			5.10^4	26.3	$3.6\ 10^{-1}$	1.4	1.1	$4.0\ 10^{-7}$
10	10	617.5	1.10^5	37.2	$7.3\ 10^{-3}$	68.7	$9.3\ 10^3$	$1.2\ 10^{-66}$
			3.10^5	64.5	$2.2\ 10^{-2}$	22.9	$1.0\ 10^3$	$1.5\ 10^{-26}$
			1.10^6	117.7	$7.3\ 10^{-2}$	6.9	$8.0\ 10^1$	$1.6\ 10^{-12}$
			3.10^6	203.9	$2.2\ 10^{-1}$	2.3	5.9	$8.2\ 10^{-9}$
			5.10^6	263.2	$3.6\ 10^{-1}$	1.4	1.1	$4.0\ 10^{-8}$
10	1000	61.7	3.10^3	6.4	$2.2\ 10^{-2}$	22.9	$1.0\ 10^3$	$1.5\ 10^{-23}$
			1.10^4	11.8	$7.3\ 10^{-2}$	6.9	$8.1\ 10^1$	$1.6\ 10^{-9}$
			3.10^4	20.4	$2.2\ 10^{-1}$	2.3	5.9	$8.2\ 10^{-6}$
			5.10^4	26.3	$3.6\ 10^{-1}$	1.4	1.1	$4.0\ 10^{-5}$

1 $M_\odot$/yr = 6.303 10^{25} g/s, $\mu = 0.60$

Estimates of the mass loss for a few characteristic stars are given in Table (3.1). Notice the extreme sensitivity of the mass loss rate to the ratio v_{esc}/a, because the density at the critical point decreases exponentially as $v_{esc}(r_0)^2/2a^2$ or as $r_0/\mathscr{H}_0$. The predicted mass loss rate of a solar type star with a corona of 1×10^6 K and $\rho_0 = 10^{-14}$ g cm^{-3} is 1.6×10^{-14} $M_\odot$ yr^{-1}. This is in reasonable agreement with the observed rate of 2×10^{-14} $M_\odot$yr^{-1}.

3.1.5 Conclusions

This section has shown that the equation describing the velocity of isothermal winds has many solutions depending on the initial conditions at the bottom of the wind. There is only one critical solution for which the velocity increases from subsonic at the bottom to supersonic far out. This velocity passes through the critical point where $v(r_c) = a = v_{esc}(r_c)/2$, and implies one particular value of the initial velocity v_0 at the lower boundary r_0 of the isothermal region. If the density at r_0 is fixed, the mass loss rate is fixed by $\dot{M} = 4\pi\rho_0 v_0 r_0^2$. The total energy increases from negative at the base of the wind to positive in the supersonic region, so the flow requires the input of energy into the wind. This energy input is needed to keep the flow isothermal and it is this energy that is tranferred into kinetic energy of the wind via the gas pressure. In the subsonic region of the wind the density structure is very similar to that of a hydrostatic atmosphere, with a difference of only a factor exp (− 0.5) at the critical point.

It is important to realize that the location of the critical point and the value of the mass loss rate depend only on the conditions between the bottom of the isothermal region and the critical point. They do not depend on the conditions beyond the critical point but only require that the velocity remains larger than a at $r > r_c$.

3.2 Isothermal winds with an $f \sim r^{-2}$ force

The previous section has shown that the velocity structure of an isothermal wind without additional forces will only become supersonic for one particular mass loss rate because the solution of the momentum equation has to pass through a critical point. One might wonder about the effect of additional forces on the wind, such as, e.g., radiation pressure or wave pressure. Will the critical point remain at the same location? Will an extra outward force in the wind increase the velocity or increase the mass loss or both? How does the effect depend on the location in the wind where the force is applied? In particular, does the application of a force in the subsonic region have the same effect as a force applied in the supersonic region? These questions will be discussed in this section.

The wind is again assumed to be isothermal, similar to the previous section. This simplifies the problems because the energy equation is reduced to $T(r) = T$ and it allows the isolation of the effects of the forces on the winds by comparison with the isothermal model with only gas pressure discussed in the previous section.

This section deals with simple forces of the form $f \sim r^{-2}$. Such a force can be produced by radiation pressure due to optically thin lines or continua such as electron scattering or dust scattering. This is because the radiative flux F varies as r^{-2} and thus the radiative acceleration is $g_{\rm rad} = \kappa_F F(r)/c = \kappa_F F(R_*)(r/R_*)^{-2}/c \sim r^{-2}$ where κ_F is the flux-mean opacity. The last equality is only valid if κ_F is independent of distance.

The presence of a positive r^{-2} force which is smaller than the acceleration of gravity and which acts throughout the whole wind will obviously have the same effect as a reduction of the gravity or the mass of the star by a constant factor. In that case the mass loss rate and the velocity can simply be solutions of Eqs. (3.22) and (3.32). However, if the force operates only in the lower part of the wind or only in the upper part, the solutions will be different.

The assumption of a r^{-2} force which acts only at restricted distances may seem very artificial. However, there are winds which resemble such simple models. For instance the dust which forms at a certain distance from red supergiants or Miras produces a sudden switch-on of a radiative force which varies as r^{-2} in dust-driven wind models. What is more important to our discussion is that these simplified models illustrate an important property of winds: the different reaction to forces applied in the subcritical and the supercritical regions.

The momentum equation of an isothermal wind with an additional positive force $f = Ar^{-2}$ is

$$v\frac{dv}{dr} = -\frac{1}{\rho}\frac{dp}{dr} - \frac{GM_*}{r^2} + \frac{A}{r^2} \tag{3.33}$$

In this equation A is a positive constant in the region where the force operates and zero outside that region. Expressing the pressure gradient in terms of the velocity gradient by means of the mass continuity equation (3.1), with (3.5) and (3.6), results in the momentum equation

$$\frac{1}{v}\frac{dv}{dr} = \left\{\frac{2a^2}{r} - \frac{GM_*}{r^2} + \frac{A}{r^2}\right\} / \left\{v^2 - a^2\right\} \tag{3.34}$$

The equation has a similar structure as in the case of an isothermal wind with gas pressure only, Eq. (3.8). In particular it has the property of a critical equation for which the velocity gradient dv/dr can only be positive throughout the wind if the numerator and the denominator both reach zero at the same point. This implies that there is one critical solution for one particular value of the initial velocity v_0 and one mass loss rate if the lower boundary conditions of the isothermal region (ρ_0, T, r_0) are known. This critical solution depends on the value of A and on the region where $A \neq 0$.

3.2.1 An $f \sim r^{-2}$ force throughout the wind

Suppose that $A \neq 0$ throughout the whole wind. This represents the case for an ionized wind with radiation pressure due to electron scattering. Then the momentum equation (3.34) reduces to that of a wind without an extra force, Eq. (3.8), if the stellar mass is replaced by an effective mass defined as

$$M_{\text{eff}} = M_* - \frac{A}{G} = M_*(1-\Gamma) \tag{3.35}$$

with

$$\Gamma = A/GM_* \tag{3.36}$$

If $\Gamma < 1$, which means that $M_{\text{eff}} > 0$, the conditions of the critical point are the same as in Eq. (3.12). In this case, the velocity at the critical point, $v(r_c) = a$, remains the same as before, but the critical point is now closer to the star by a factor $(1-\Gamma)$. This can result in a considerable increase in the mass loss rate (Eq. 3.32) for two reasons: the density scale height $\mathscr{H}_0$ in the subcritical region increases (because g_0 decreases), so the outward density decrease in the subcritical region is slower (Eq. 3.29), and secondly the critical point is closer to r_0. Both effects result in a considerable increase of ρ_c/ρ_0 compared to the case of gas pressure only, and in an increase of $\dot{M}$.

Now consider what happens if the force increases in strength. When Γ increases from zero to positive values, the critical point moves closer in since

$$r_c(\Gamma) = \frac{GM_*(1-\Gamma)}{2a^2} \tag{3.37}$$

until Γ reaches a maximum value such that r_c reaches the bottom of the isothermal region at r_0. This occurs at

$$\Gamma_{\text{max}} = 1 - \frac{2a^2 r_0}{GM_*} = 1 - \left\{\frac{2a}{v_{\text{esc}}(r_0)}\right\}^2 \tag{3.38}$$

with $\Gamma_{\text{max}} \simeq 1$ if $2a \ll v_{\text{esc}}\ (r_0)$. If $\Gamma = \Gamma_{\text{max}}$ the numerator of the momentum equation (3.34) is zero at r_0. If $\Gamma > \Gamma_{\text{max}}$ and the velocity at r_0 is subsonic, $v_0 < a$, then the velocity gradient will be negative for $r > r_0$ and the velocity will *decrease* outward, rather than increase. This is due to the fact that the velocity gradient in the subsonic region decreases when the force increases.

It may seem surprising that a large outward directed force in the subsonic region results in a *deceleration* ($dv/dr < 0$) rather than an *acceleration* of the subsonic flow. However, this can be understood by remembering the conclusion in the previous section that the density structure of the subsonic region is mainly determined by the *hydrostatic equilibrium* condition and not by the velocity law. So the application of a force in the subsonic region which counteracts gravity increases the density scale height, and produces a slower outward decrease of ρ. Because the density is related to the velocity by the mass continuity equation, the application of a force translates into a smaller velocity gradient. If the net force in the subsonic region were directed outwards the density would have to increase outwards if the atmosphere was to be hydrostatic. This increase of density corresponds to a decrease of the velocity.

Figure (3.3) shows the effect of increasing $\Gamma = A/GM_*$ in the range of $0 < \Gamma < \Gamma_{max}$ on the density and velocity structure. The upper figure shows the velocity in terms of the Mach number $M = v/a$ and the lower figure shows the resulting density structure normalized to ρ_0. The temperature of the wind is such that $a^2 = 0.05GM_*/r_0$. The curves are for different values of Γ. The mass loss rate for a fixed value of ρ_0 scales as $\dot{M} \sim v_0$. Notice the very rapid increase of $M(x = 1) = v_0/a$ and of the resulting mass loss rate by orders of magnitude if Γ increases from 0 to 0.8. The figure shows that an increase of the outward force results in a *smaller acceleration* but in a *higher velocity* of the wind at any distance. This is because the initial velocity v_0 has to increase with increasing Γ in order to obtain a transonic solution through the critical point.

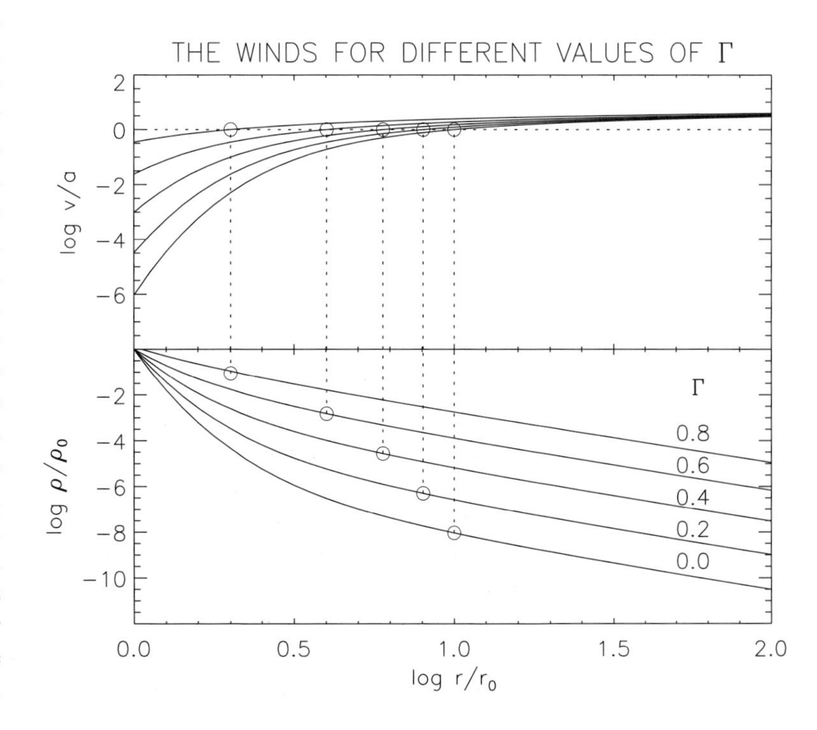

Figure 3.3 The velocity structure, expressed in the Mach number $M = v/a$, and the density structure as a function of distance r/r_0, for an isothermal wind with an extra outward force of the type $f = \Gamma(GM_*/r^2)$. The location of the critical point is indicated by a circle at each of the curves. Notice the very rapid increase of $v(r_0)$ and of the resulting mass loss rate, $\dot{M} \sim v_0$, by orders of magnitude when Γ increases.

3.2.2 An $f \sim r^{-2}$ force from a certain distance onwards

Now consider the effect of a r^{-2} force which acts beyond a certain distance r_d, so $\Gamma = 0$ for $r < r_d$ and $0 < \Gamma < 1$ for $r > r_d$. It resembles the winds of cool stars for which radiation pressure on dust becomes important at the distance where the dust is formed. It will be shown that the effect of such a force on the velocity and the mass loss rate depends very critically on the location of r_d. For simplicity we assume that whenever Γ is positive it has the same constant value of $\Gamma = 0.5$.

The different cases are shown in Fig. (3.4). The temperature of the wind model is such that $\mathscr{R}T/\mu = a^2 = (GM_*/r_0)/2\sqrt{(10)}$. So the critical point is at $\log r_c/R_* = 0.5$ in the absence of the extra force. As the location of the critical point depends on Γ, we will indicate it as $r_c(\Gamma)$. The dependence of $\dot{M}$ in the five cases can be judged from the velocity laws by realizing that the velocity at the critical point is $v_c = a$ in all five cases and that $\dot{M}$ is proportional to $v(r_0)$ because $\rho(r_0)$ is a fixed boundary condition. The top figure shows the location of $r_c(0)$ in case of $\Gamma = 0$ (Eq. 3.12). The bottom figure shows the location of $r_c(\Gamma)$ in case of $\Gamma = 0.5$ throughout the wind (Eq. 3.37). These are the two extreme cases.

If $r_d > r_c(0)$ neither the structure of the subsonic region nor the location of the critical point is affected by Γ, so the mass loss rate will be the same as in the case of $\Gamma = 0$. This shows a very important characteristic of stellar winds: *an outward force applied to the wind above the critical point does not affect the mass loss rate.* We will show

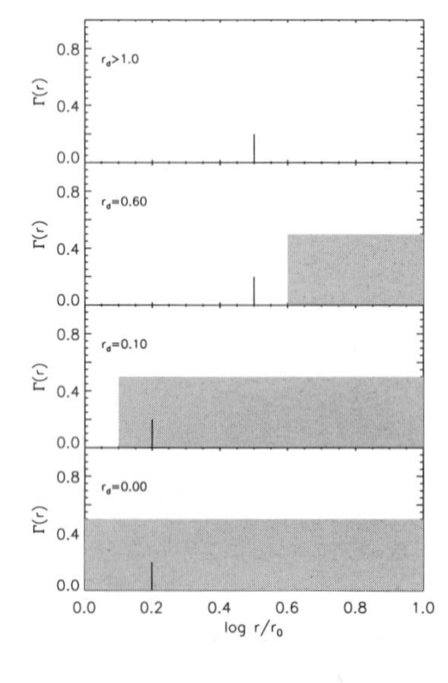

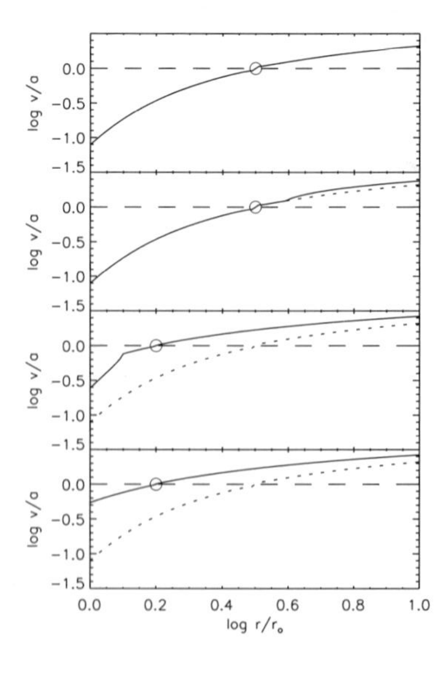

Figure 3.4 The effect of an outward force $f(r) = \Gamma(r)GM_*/r^2$ on the velocity. The left hand side shows the various distributions of $\Gamma(r)$ and the right hand side shows the resulting wind velocity. The wind velocity for $\Gamma(r) = 0$ is shown by a dotted line. The location of the critical point is indicated by a circle (right) or by a tickmark (left). Notice the changes in the location of the critical point and in the mass loss rate $\dot{M} \sim v(r_0)$ if $\Gamma(r) > 0$ in the subsonic region.

later in Chapter 4 that the same is true for the addition of energy above the critical point. The velocity in the supersonic regions at $r > r_d$ will be larger however than in the case of $\Gamma = 0$, because the numerator of Eq. (3.34) is larger when $\Gamma > 0$ and the denominator of Eq. (3.34) is positive. So increasing Γ above the critical point will result in a steeper velocity law and larger velocities at $r > r_d$. Recall that in the supersonic region the value of Γ may exceed 1.

If $r_d < r_c(\Gamma)$ the critical point occurs at $r_c(\Gamma)$. This is due to the fact that the location of r_c is given by the *local* condition that the numerator of Eq. (3.34) is zero. The mass loss depends on the value of Γ through the whole subcritical region $r_0 < r < r_c(\Gamma)$ because the velocity law in this region is affected by $\Gamma(r)$ according to Eq. (3.34). A positive value of $\Gamma > 0$ implies a smaller velocity gradient and a smaller density gradient at $r_d < r < r_c(0)$ than in the case of $\Gamma = 0$. This smaller density gradient results in a higher value of ρ at $r_c(\Gamma)$ and thus a higher mass loss rate.

If the value of Γ becomes positive in the region between $r_c(\Gamma)$ and $r_c(0)$ then the location of the critical point depends sensitively on the shape of the $\Gamma(r)$ function. If Γ were a mathematical stepfunction the numerator of Eq. (3.34) would not have a zero-point in that case. This can easily be verified by applying the critical point condition (3.37). If $\Gamma = 0$ then $r_c = r_c(0)$, but $\Gamma \neq 0$ at $r_c(0)$. Similarly if $\Gamma \neq 0$ then $r_c = r_c(\Gamma)$, but $\Gamma = 0$ at $r_c(\Gamma)$. In this case the numerator reaches zero only when Γ is not a step function, but a slower function of r in the range of $r_c(\Gamma) < r < r_c(0)$. The critical point of Eq. (3.35) is then given by the solution of the condition

$$\frac{2a^2}{r_c} - \frac{GM_*}{r_c^2}\{1 - \Gamma(r_c)\} = 0 \tag{3.39}$$

for

$$\frac{r_c}{1 - \Gamma(r_c)} = \frac{GM_*}{2a^2} \tag{3.40}$$

The critical point will thus be located in the region where $\Gamma(r)$ increases. The mass loss will be higher than in the case of $\Gamma = 0$ because the velocity of $v = a$ is reached closer to the star than $r_c(0)$ and thus at higher density than in the case of $\Gamma = 0$.

3.2.3 The mass loss rate for constant Γ

The velocity and density structure in the subcritical part of the wind are only affected by Γ if $\Gamma > 0$ in the subcritical region. In this case the mass loss rate is also affected by Γ.

If a force with $0 < \Gamma < \Gamma_{\text{max}}$ is applied in the subsonic part of the wind the velocity gradient will be smaller but the velocities will be larger than without such a force, and the critical point moves closer to the base of the isothermal region. This is because the velocity reaches the sound speed at the critical point and since the velocity gradient in the subsonic region decreases with increasing Γ, the velocity in the subsonic region must increase. This results in a drastic increase of $\dot{M}$ because the density scale height in the subsonic region becomes larger and the critical point moves closer in. An estimate of the increase of $\dot{M}$ can be obtained from Eq. (3.32). If the r^{-2} force operates throughout the subsonic region then $v_{\text{esc}}^2 \to v_{\text{esc}}^2(1-\Gamma)$, $r_c \to r_c(1-\Gamma)$ and $\mathscr{H} \to \mathscr{H}/(1-\Gamma)$ and so the mass loss rate increases by a factor

$$\frac{\dot{M}}{\dot{M}(\Gamma=0)} = (1-\Gamma)\cdot\exp\left\{\frac{\Gamma\, r_0}{\mathscr{H}(\Gamma=0)}\right\} \tag{3.41}$$

The maximum increase, which corresponds to Γ_{max}, is

$$\frac{\dot{M}(\Gamma_{\text{max}})}{\dot{M}(\Gamma=0)} = \frac{2\mathscr{H}(\Gamma=0)}{r_0}\exp\left\{\frac{r_0}{\mathscr{H}(\Gamma=0)} - \frac{3}{2}\right\} \tag{3.42}$$

in which case the critical point coincides with the bottom of the isothermal region, and $\dot{M}(\Gamma_{\text{max}}) = 4\pi r_0^2 a \rho_0$. Some estimates of $\dot{M}(\Gamma)$ for different stars are given in Table (3.2). Notice the strong dependence of $\dot{M}$ on Γ.

3.2.4 Conclusions

The examples shown here have demonstrated some important consequences of an outward r^{-2} force in the wind.

1. The value of Γ must be less than Γ_{max} (Eq. 3.38) in the subsonic part of the flow, because the structure of this region is mainly determined by hydrostatic equilibrium and an isothermal region cannot be hydrostatic if $\Gamma > \Gamma_{\text{max}}$. So the outward force in the subsonic region should not exceed gravity. In the supersonic region the force may exceed gravity, $\Gamma > 1$, which will result in a steeper velocity law.

2. If a force with $0 < \Gamma < \Gamma_{\text{max}}$ is applied in the subcritical part of the wind the velocity gradient will be smaller but the velocities will be higher than in the absence of the force. The critical point moves closer to the star and the mass loss rate will increase.

3. If a force with $\Gamma > 0$ is only applied in the supercritical region it will not affect the mass loss rate, but it will result in higher supersonic velocities.

4. If an r^{-2} force increases outwards such that $\Gamma(r)$ increases from 0

Table 3.2 *Estimates of mass loss rates for isothermal winds with radiation pressure in the continuum*

M_* ($M_\odot$)	R_* ($R_\odot$)	T (K)	Γ	$\mathscr{H}_0$ (R_*)	r_c (R_*)	$\frac{r_c - R_*}{\mathscr{H}_0}$	$\dot{M}$ ($M_\odot/yr$)
1	1	3.10^5	0	$2.2\ 10^{-2}$	22.9	$1.0\ 10^3$	$1.5\ 10^{-28}$
			0.5	$4.4\ 10^{-2}$	11.5	$2.4\ 10^2$	$5.6\ 10^{-19}$
			0.9	$2.2\ 10^{-1}$	2.3	5.9	$8.8\ 10^{-12}$
1	1	3.10^6	0	$2.2\ 10^{-1}$	22.9	5.9	$8.2\ 10^{-28}$
			0.5	$4.4\ 10^{-1}$	1.15	$3.4\ 10^{-1}$	$4.0\ 10^{-10}$
			0.9	2.2	<1		
1	100	3.10^3	0	$2.2\ 10^{-2}$	22.9	$1.0\ 10^3$	$1.5\ 10^{-25}$
			0.5	$4.4\ 10^{-2}$	11.5	$2.4\ 10^2$	$5.6\ 10^{-16}$
			0.9	$2.2\ 10^{-1}$	2.3	5.9	$8.8\ 10^{-9}$
1	100	3.10^4	0	$2.2\ 10^{-1}$	22.9	5.9	$8.2\ 10^{-8}$
			0.5	$4.4\ 10^{-1}$	1.15	$3.4\ 10^{-1}$	$4.0\ 10^{-7}$
			0.9	2.2	<1		
10	10	3.10^5	0	$2.2\ 10^{-2}$	22.9	$1.0\ 10^3$	$1.5\ 10^{-26}$
			0.5	$4.4\ 10^{-2}$	11.5	$2.4\ 10^2$	$5.6\ 10^{-17}$
			0.9	$2.2\ 10^{-1}$	2.3	5.9	$8.8\ 10^{-10}$
10	10	3.10^6	0	$2.2\ 10^{-1}$	2.3	5.9	$8.2\ 10^{-9}$
			0.5	$4.4\ 10^{-1}$	1.15	$3.4\ 10^{-1}$	$4.0\ 10^{-8}$
			0.9	2.2	<1		
10	1000	3.10^3	0	$2.2\ 10^{-2}$	22.9	$1.0\ 10^3$	$1.5\ 10^{-23}$
			0.5	$4.4\ 10^{-2}$	11.5	$2.4\ 10^2$	$5.6\ 10^{-14}$
			0.9	$2.2\ 10^{-1}$	2.3	5.9	$8.8\ 10^{-7}$
10	1000	3.10^4	0	$2.2\ 10^{-1}$	2.3	5.9	$8.2\ 10^{-6}$
			0.5	$4.4\ 10^{-1}$	1.15	$3.4\ 10^{-1}$	$4.0\ 10^{-5}$
			0.9	2.2	<1		

to Γ in the region where $r_c(\Gamma) < r_d < r(0)$, the critical point will occur in the region where the force increases as given by Eq. (3.39).

This last situation resembles the mass loss from dust driven winds. The resulting mass loss rate can be estimated by assuming a nearly hydrostatic density distribution, Eq. (3.29), between r_0 and the distance r_d where the dust is formed.

$$\dot{M} \simeq 4\pi r_d^2\ a\ \rho_0\ \exp\left\{-\frac{(r_d - r_0)}{\mathscr{H}(r_0)} \cdot \frac{r_0}{r_d}\right\} \tag{3.43}$$

For a typical red supergiant with $r_0 \simeq R_* \simeq 10^3 R_\odot$, $T_{\rm wind} \simeq T_{\rm eff} \simeq 3.10^3$ K, $M_* \simeq 20 M_\odot$ and $\rho_0 \simeq \rho(\tau \simeq 1) \simeq 10^{-10}$ g cm^{-3}, the dust forms at $T \sim 10^3$ K which is at $r_d \simeq 4R_*$. The atmospheric scale height is $\mathscr{H}_0 = 7.10^{-3}\ R_* = 5.10^{11}$ cm. This results in an estimate of $\dot{M} \sim 10^{-50} M_\odot/\mathrm{yr}$. This very small value is due to the fact that the critical point r_d is located about 6.10^2 scale heights above the photosphere where the density is very low. If the scale height were increased by about a factor 8 from 5.10^{11} cm to 4.10^{12} cm, the mass loss rate would increase to 10^{-5} $M_\odot/\mathrm{yr}$, i.e., by 45 orders of magnitude! This shows the sensitivity of the mass loss rate to the scale height of the subsonic region. We will discuss mass loss from dust driven winds in detail in Chapter 7.

3.3 Isothermal winds with an $f \sim v(dv/dr)$ force

As a second example of a force which can be expressed in a simple analytical form we consider the radiation pressure due to optically thick spectral lines in the wind. In the limit of large velocity gradients in the wind the intrinsic width of the absorption profile can be neglected compared to the Doppler shifts due to the flow velocities. This leads to the so-called Sobolev approximation, which we will discuss in more detail in Chapter 8. The amount of radiative momentum absorbed per cm^3 and per second in the wind due to one optically thick line of negligible intrinsic width is $F_\nu(r) \cdot \Delta\nu/c$, where $F_\nu(r) \simeq F_\nu(R_*) \cdot (R_*/r)^2$ is the monochromatic flux at distance r and $\Delta\nu$ is the Doppler shift due to the velocity gradient over a distance of 1 cm, so $\Delta\nu = (\nu_0/c)(dv/dr)\Delta r$ with $\Delta r = 1$ cm. The force is proportional to the momentum of the absorbed radiation per unit volume, so $f_{\mathrm{rad}} \sim r^{-2}(dv/dr)$ and per unit mass it is $f_{\mathrm{rad}} \sim \rho^{-1} r^{-2}(dv/dr) \sim v(dv/dr)$. So the radiation pressure due to an optically thick line gives a force which is proportional to $v(dv/dr)$.

3.3.1 The momentum equation and the critical point

The momentum equation of an isothermal wind with such a force is

$$v\frac{dv}{dr} = -\frac{1}{\rho}\frac{dp}{dr} - \frac{GM_*}{r^2} + B \cdot v\left(\frac{dv}{dr}\right) \tag{3.44}$$

where B is a positive constant which depends on the number of optically thick lines and on the flux at their frequencies. In this simple case we assume B to be constant. The pressure gradient can be expressed in a velocity gradient by means of the mass continuity equation as in Eq. (3.6), which gives

$$\frac{1}{v}\frac{dv}{dr} = \left\{\frac{2a^2}{r} - \frac{GM_*}{r^2}\right\} / \left\{v^2(1-B) - a^2\right\} \tag{3.45}$$

The radiative acceleration, i.e. the last term in Eq. (3.44), produces an extra factor $1 - B$ in the denominator of the momentum equation compared to the isothermal wind without an additional force, Eq. (3.8). It is easy to see that Eq. (3.45) is exactly similar to the momentum equation (3.8) if we transform $v(r)$ into $v'(r) = v(r)(1 - B)^{-1/2}$ for $B < 1$. This implies that the properties at the critical point derived from Eq. (3.8) can be applied directly if v is transformed to v'. So the critical solution, i.e. the one for which $dv/dr > 0$ at all distances, passes through a critical point where

$$v_c \equiv v(r_c) = \frac{a}{(1-B)^{1/2}} \quad \text{at} \quad r_c = \frac{GM_*}{2a^2} \tag{3.46}$$

provided that $B < 1$. So in this case the location of the critical point is the same as in the wind without the extra force, i.e. at the distance where $v(r_c) = v_{esc}(r_c)/2$. However the velocity at the critical point is no longer the sound speed a but it is increased by a factor $(1 - B)^{-1/2}$. The critical point is not the sonic point anymore! This is an important conclusion because it shows that *the critical point in an isothermal wind is not necessarily the sonic point*. The critical point is a mathematical point where the sum of all the terms which contain the factor dv/dr in the momentum equation, as well as the sum of all the terms that do not contain the velocity gradient, vanish.

3.3.2 The mass loss rate

The transformation of v into $v(1 - B)^{-1/2}$ allows an easy estimate of the mass loss rate from an isothermal wind with a radiative force $f = Bv(dv/dr)$ by means of the application of the mass continuity equation at the bottom of the isothermal region. The initial velocity $v_0' = v_0(1 - B)^{-1/2}$ is given by the solution of the velocity equation in an isothermal wind (Eq. 3.32). This gives a mass loss rate

$$\dot{M} \simeq \frac{4\pi \rho_0 a r_0^2}{\sqrt{(1-B)}} \left\{ \frac{v_{esc}(r_0)}{2a} \right\}^2 \exp\left\{ -\frac{v_{esc}^2(r_0)}{2a^2} + \frac{3}{2} \right\} \tag{3.47}$$

which was derived from Eqs. (3.16) and (3.23) with $v_0 \rightarrow v_0(1-B)^{-1/2}$. This mass loss rate is a factor $(1 - B)^{-1/2}$ larger than in the case without the vdv/dr force (Eq. 3.32), because the velocity $v(r_0)$ of the critical solution is larger by this factor. The velocity distribution is the same as in Eq. (3.24) with the correction factor $(1 - B)^{-1/2}$ to the velocity. With a velocity that is larger by a factor $(1 - B)^{-1/2}$ at every distance and a mass loss rate which is larger by the same factor, compared to the original case without the extra force, the density distribution in the wind is exactly the same as before. So in this case we find again that the density distribution in the subcritical region is approximately that of a hydrostatic atmosphere.

Let us consider now what happens in a wind in which the force is so large that $B > 1$. In that case the denominator of the momentum equation (3.45) is negative for all values of v and so the velocity will increase in the region $r_0 < r < r_c$, reach a maximum at r_c and decrease further outwards. Depending on the velocity v_0 the maximum velocity reached at r_c may be subsonic or supersonic. Since the resolution is no longer critical, neither the initial velocity nor the mass loss rate is fixed. This situation is physically unrealistic for radiation driven stellar winds, as will be shown in Chapter 8. In reality the radiation pressure is always due to a mixture of optically thick lines with $f_{\rm rad} \sim v dv/dr$ and optically thin lines with $f_{\rm rad} \sim r^{-2}$ with a ratio that varies through the wind. This results in a momentum equation which has a critical point.

3.3.3 Conclusions

The presence of a force which is proportional to $v\ dv/dr$, i.e. $f = Bv\ dv/dr$ in an isothermal wind, acts like a scaling factor of the velocity. The velocity law is the same as in a wind without such a force apart from this scaling factor which increases the velocity by a factor $(1-B)^{-1/2}$ at all distances. The mass loss rate is also increased by the same factor $(1-B)^{-1/2}$, but the density distribution is the same as in a wind with $B = 0$. The critical point of the momentum equation is located at the same distance $r_c = GM_*/2a^2$ as in a wind without such a force, but the velocity at the critical point is no longer the sound velocity a. In this case it is $a(1-B)^{-1/2}$. This shows that the critical point of the momentum equation is not necessarily the sonic point, but that it can be any velocity. The velocity at the critical point is simply the velocity for which the denominator of the momentum equation of $d \ln v/dr$ is zero.

3.4 Isothermal winds with general additional forces

In this section we consider the effects of forces of a more general nature on the structure of isothermal winds. The purpose of this section is to demonstrate how the critical solutions and the mass loss rates depend on the forces applied in the wind. We will consider forces which depend on r, v and dv/dr as $f(r,v) + g(r,v) \times v\ dv/dr$. At this point more complicated forces, such as those proportional to $(dv/dr)^\alpha$

which play a role in radiation driven winds, will be ignored for the purpose of clarity.

3.4.1 The momentum equation and the critical point

The momentum equation is

$$\frac{1}{v}\frac{dv}{dr}\left(v^2 - a^2\right) - \frac{2a^2}{r} + \frac{GM_*}{r^2} - f(r,v) - g(r,v)v\frac{dv}{dr} = 0 \tag{3.48}$$

which yields

$$\frac{1}{v}\frac{dv}{dr} = \left\{\frac{2a^2}{r} - \frac{GM_*}{r^2} + f(r,v)\right\} / \left\{v^2[1 - g(r,v)] - a^2\right\} \tag{3.49}$$

The critical point occurs where the velocity is

$$v_c^2\left\{1 - g(r_c, v_c)\right\} = a^2 \tag{3.50}$$

at a distance where

$$\frac{2a^2}{r_c} - \frac{GM_*}{r_c^2} + f(r_c, v_c) = 0 \tag{3.51}$$

Equation (3.50) shows that the critical velocity is not the isothermal sound velocity. Notice that these two conditions at the critical point depend on *local* quantities only and not on the forces elsewhere in the wind. However, a solution of the momentum equation that goes through the critical point does depend on the forces over the full subcritical region because this solution requires that $v = v_c$ is exactly reached at $r = r_c$.

The existence of solutions which start at small velocities ($v \ll a$) at the base of the isothermal region and reach high velocities ($v \gg a$) at large distances depends on the properties of the functions $g(r,v)$ and $f(r,v)$ and hence on the forces described by these functions. It can be shown easily from Eq. (3.49) that the conditions for a critical solution with $(dv/dr)_{r_0} > 0$ and $(dv/dr)_\infty \geq 0$ to exist are

$$g(r_0, v_0) > 1 - \frac{a^2}{v_0^2} \quad \text{and} \quad f(r_0, v_0) < \frac{GM_*}{r_0^2} - \frac{2a^2}{r_0} \tag{3.52}$$

at the base, and

$$g(\infty, v_\infty) \leq 1 - \frac{a^2}{v_\infty^2} \simeq 1 \quad \text{and} \quad f(\infty, v_\infty) \geq \frac{GM_*}{r^2} - \frac{2a^2}{r} = 0 \tag{3.53}$$

for $r \gg R_*$ and $v > a$ at large distances. This shows that at the bottom, r_0, of the isothermal region f must be smaller than some specific value and g must be larger. In § 3.2 we have shown that this is because the velocity gradient in the subcritical region decreases when the force increases. So too large a force f produces a negative velocity gradient in the subcritical region. At the critical point the forces must satisfy Eqs. (3.50) and (3.51). At large distances it is sufficient for the force to be outward directed because the terms $2a^2/r$ and GM_*/r^2 in Eq. (3.49) go to zero. The Eqs. (3.52) and (3.53) suggest that even inward directed forces are allowed at large distances because the isothermal gas will produce the pressure force of a^2/r everywhere. This will generally not be the case, however, because in real winds the temperature at large distances will decrease, so that a^2/r goes to zero.

In this discussion we have tacitly assumed that the wind has only one critical point. This is only a correct assumption if $f(r, v)$ and $g(r, v)$ vary in such a way that the velocity gradient does not become negative beyond the critical point. However, if $f(r, v)$ decreases so drastically that the numerator of Eq. (3.49) becomes negative at $r > r_c$, or $g(r, v)$ increases so drastically that the denominator becomes negative at $r > r_c$, the wind may have more than one critical point. We will see in Chapter 9 that multiple critical points occur in the winds of fast magnetic rotators.

3.4.2 Conclusions

We have shown that for isothermal winds with general outward directed forces of the type $f(r, v)$ and $g(r, v)v dv/dr$ the location r_c and the velocity $v(r_c)$ of the critical point depend on local conditions. However the velocity law in the subcritical region and hence the initial velocity v_0 at the bottom of the isothermal region depends on the variation of the forces throughout the subcritical region. Because the mass loss is fixed by r_0, ρ_0 and v_0, the mass loss rate also depends on the variation of the forces between r_0 and r_c. The mass loss rate does not depend on the variation of f and g above the critical point.

3.5 The analogy of rocket nozzles

Some of the properties of the critical solutions of the momentum equation of stellar winds can be understood in terms of an anology with the flow in a rocket nozzle. The analogy is in the following two properties. (a) The flow through a prespecified nozzle will only increase in velocity from subsonic to supersonic for a particular mass flow which passes through a critical point at the throat of the nozzle. This is analogous to the critical solution of stellar winds. (b) The application of an extra pressure in the subsonic part of the flow through the nozzle results in a decrease of the velocity gradient, whereas it results in an increase of the gradient if the force is applied to the supersonic part. This is analogous to the behaviour of the velocities of a stellar wind.

In this section the equations for nozzle flows will be derived and compared with those of isothermal winds. The concept of a critical solution can easily be understood in the case of the flow through a nozzle. The analogy with the stellar winds will then be used to explain some wind properties.

3.5.1 The momentum equation of a flow through a nozzle

Consider the flow through a horizontal pipe with a throat (Fig. 3.5). The throat is the place where the cross section A reaches a minimum value and $dA/d\ell = 0$. Assume that the inlet of the pipe is connected to a high pressure tank and that the flow can leave the outlet of the pipe freely, e.g. into a vacuum. In order to avoid complications of shocks at either end of the pipe, we assume that the inlet and the outlet are far removed from the throat. The flow is considered to be steady, smooth and one dimensional, so viscosity and the interaction with the inner surface of the pipe are neglected.

The cross section, A(ℓ) of the pipe is variable along its length, ℓ. The minimum cross section is reached at the position of the throat, ℓ_T. The equation of mass continuity through the flow is

$$\rho(\ell)v(\ell)A(\ell) = \rho_0 v_0 A_0 = F = \text{constant} \tag{3.54}$$

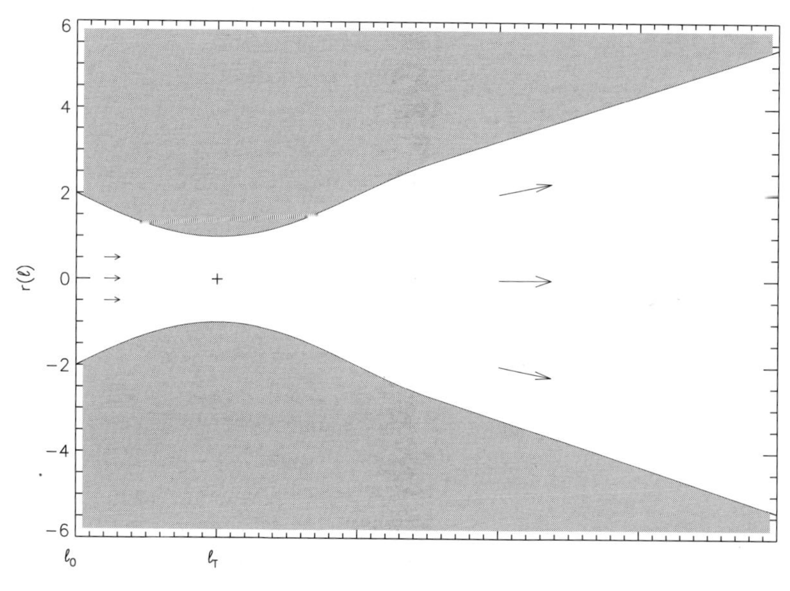

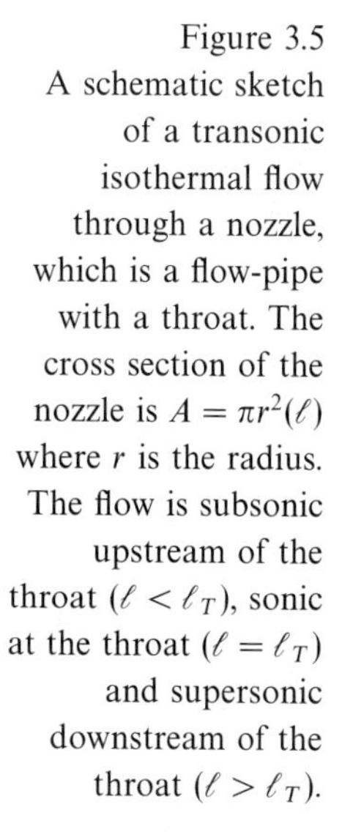

Figure 3.5 A schematic sketch of a transonic isothermal flow through a nozzle, which is a flow-pipe with a throat. The cross section of the nozzle is $A = \pi r^2(\ell)$ where r is the radius. The flow is subsonic upstream of the throat ($\ell < \ell_T$), sonic at the throat ($\ell = \ell_T$) and supersonic downstream of the throat ($\ell > \ell_T$).

The subscripts zero refer to the condition at $\ell = 0$, and $v(\ell)$ is the flow velocity. The equation of motion is

$$\frac{dv}{dt} = \frac{\partial v}{\partial t} + v\frac{dv}{d\ell} = v\frac{dv}{d\ell} = -\frac{1}{\rho}\frac{dp}{d\ell} \tag{3.55}$$

where the term $\partial v/\partial t$ is zero for a stationary flow. For an adiabatic flow of a perfect gas with $p \sim \rho^{\gamma}$ and $p/\rho = \gamma \mathcal{R}T/\mu = \gamma a^2$ the equation of motion is

$$\frac{vdv}{d\ell} = -\frac{\gamma a^2(\ell)}{\rho}\frac{d\rho}{d\ell} \tag{3.56}$$

where $\gamma = c_p/c_v$ is the adiabatic constant and $a^2(\ell)$ is the isothermal sound velocity. Let us first assume that the flow is kept isothermal by a device which keeps the pipe at a constant temperature throughout its length, so $a(\ell) = a$. This corresponds to $\gamma = 1$. Combining Eqs. (3.56) and (3.54) gives the momentum equation

$$\frac{1}{v}\frac{dv}{d\ell} = \frac{a^2}{A}\frac{dA}{d\ell}\left(v^2 - a^2\right)^{-1} \tag{3.57}$$

This equation has the same structure as the momentum equation (3.8) for isothermal winds, where the nozzle term $a^2 d\ ln\ A/d\ \ell$ is replaced by the terms $2a^2/r - GM_*/r^2$. [In fact, if the shape of the flow-pipe is such that its radius varies as

$$r(\ell) = \left(\ell/\sqrt{\pi}\right)\exp(L/\ell) \quad \text{with} \quad L = GM_*/2a^2 \tag{3.58}$$

then Eqs. (3.57) and (3.8) would be identical and so the velocity $v(\ell)$ through the flow would be exactly the same as $v(r)$ through the wind. The shape of such a nozzle is shown in Fig. (3.6). The length and the radius of the nozzle are both normalized to L. If L were equal to $GM_*/2a^2$, the velocity $v(\ell)$ would be exactly the same as that of an isothermal wind of a star with mass M_* and wind temperature equal to the temperature of the flow through the nozzle.]

Equation (3.57) shows that the transonic flow through the nozzle must reach $v = a$ at the throat where $dA/d\ell = 0$. It also shows that the nozzle must be converging, $dA/d\ell < 0$, in the subsonic part of the flow and diverging in the supersonic part in order to obtain a flow with $dv/d\ell > 0$ in both parts. In the region far away from the throat, where $A(\ell)$ is constant, the velocity is constant. So a constant width of the nozzle far downstream in the flow will result in a limiting constant velocity, which is analogous to a terminal velocity of a stellar wind. Whether the actual flow through the nozzle will be transonic depends of course on the lower boundary condition, i.e., the velocity v_0 at $\ell = 0$. For a given profile of the nozzle v_0 has to be exactly the right value. If v_0 is too small the flow will remain subsonic. If v_0 is too large the velocity will reach $v = a$ upstream from the throat, at $\ell < \ell_T$, where $dA/d\ell < 0$ and so $dv/d\ell \rightarrow -\infty$ at that point. The flow will stagnate, which results in a shock between the incoming flow and the low velocity region. So in this case there is no smooth transonic flow. The velocity v_0 necessary for a transonic flow can be found by solving Eq. (3.57), which yields

$$\frac{1}{v} \exp\left(\frac{v^2}{2a^2}\right) = A(\ell) \cdot C \tag{3.59}$$

The constant C is found by applying the condition at the throat $v(\ell_T) = a$. This yields

$$\frac{1}{M} \exp\left(\frac{M^2}{2} - \frac{1}{2}\right) = A(\ell)/A_T \tag{3.60}$$

where A_T is the cross section at the throat of the nozzle and the Mach number is $M(\ell) = v(\ell)/a$. The Mach number at $\ell = 0$ of the transonic solution is given by Eq. (3.60).

We now turn to the important question of the difference in the reaction of the velocity to a variation of A between the subsonic and the supersonic parts of the flow. To this purpose we rewrite Eq. (3.57) as

$$(M^2 - 1)\frac{1}{v}\frac{dv}{d\ell} = \frac{1}{A}\frac{dA}{d\ell} \tag{3.61}$$

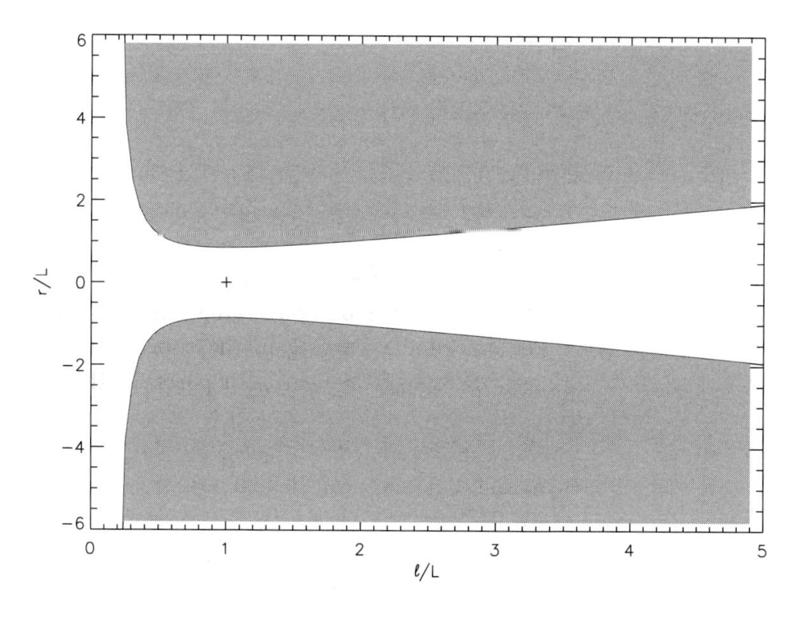

Figure 3.6 The shape of a scaled-down nozzle that produces the same momentum equation and velocity distribution as an isothermal stellar wind in which gas pressure and gravity provide the only forces.

When A decreases ($dA/d\ell < 0$) the velocity will decrease if $M > 1$ and increase if $M < 1$. This difference is due to the fact that the velocity does not vary proportional to density for a compressible gas in a flow pipe. This can be shown by writing the momentum equation (3.56) as

$$v^2\frac{dv}{v} = -a^2\frac{d\rho}{\rho} \quad \text{or} \quad -\frac{d\rho}{\rho} = M^2\frac{dv}{v} \tag{3.62}$$

which implies that

$$\rho \sim v^{-M^2} \quad \text{or} \quad \rho v \sim v^{1-M^2} \tag{3.63}$$

This equation shows that ρ decreases as v increases, except if $M = 0$. At the limit of $M = 0$, the density is independent of the velocity. In that case Eq. (3.61) shows that $dv/v = -dA/A$ and so $vA =$ constant. This result is well known for incompressible fluids for which $a = \partial p/\partial \rho \to \infty$ and so $M \to 0$: the water from a tap squirts out faster if you reduce the size of the exit.

If $M = 1$ the density decreases proportional to v^{-1} (Eq. 3.63), so $\rho v =$ constant. So in order to satisfy the continuity equation (3.54) A must be constant when $M = 1$. This is at the throat of the nozzle. In the subsonic region where $M < 1$ the density decreases slower than the velocity increases so $\rho v \sim v^{1-M^2}$ increases with increasing v. In that case the continuity equation can only be satisfied if A decreases along the flow. In the supersonic region where $M > 1$ the density decreases faster than the velocity increases, so ρv decreases with increasing velocity. Continuity then requires that A increases along the flow.

These arguments can also be expressed differently. If A decreases along a stationary flow, the continuity equation requires that ρv increases proportionally. For an accelerating flow this can only occur if ρv increases with v, which requires $d\ \ln\rho/d\ \ln v = M^2 < 1$. On the other hand, if A increases then ρv must decrease proportionally. For an accelerating flow this occurs only when ρ decreases faster than v increases, which is the case if $-d\ \ln\rho/d\ \ln v = M^2 < 1$. The throat of the nozzle can only be passed by an accelerating flow when the increase in velocity is compensated by the decrease in density so that ρv is constant. Notice that the crucial difference between the subsonic and the supersonic flows is the fact that $1/\rho$ increases slower than v in the subsonic region and faster than v in the supersonic region, with $d\ \ln\rho/d\ \ln v = -1$ at the sonic point.

So the crucial criterion for a transonic isothermal flow through a nozzle is $-d\ \ln\rho/d\ \ln v < 1$, $-d\ \ln\rho/d\ \ln v > 1$ and $-d\ \ln\rho/d\ \ln v = 1$ in the converging and the diverging part of the nozzle and at the throat respectively.

3.5.2 An isothermal flow with an extra force through a nozzle

Now suppose that the flow through the nozzle is isothermal again but that there is an additional force $f(\ell)$ in the downstream direction, i.e., outward. The momentum equation

$$v\frac{dv}{d\ell} = -\frac{a^2}{\rho}\frac{d\rho}{d\ell} + f(\ell) \tag{3.64}$$

can then be written as

$$\frac{d \ln \rho}{d\ell} = -M^2\frac{d \ln v}{d\ell} + \frac{f(\ell)}{a^2} \tag{3.65}$$

As before, the throat of the nozzle must be located where $d \ln\rho/d \ln v = -1$, because only in that case can the continuity equation with $dA/d\ell = 0$ at ℓ_T be satisfied. This implies that

$$M_T^2 - 1 = \frac{\{f(\ell)/a^2\}}{\{d \ln v/d\ell\}_T} \tag{3.66}$$

The combination of the momentum equation (3.65) with the continuity equation (3.54) yields the momentum equation in the familiar form

$$\frac{1}{v}\frac{dv}{d\ell} = \left\{\frac{a^2}{A}\frac{dA}{d\ell} + f(\ell)\right\}\{v^2 - a^2\}^{-1} \tag{3.67}$$

which has a critical point where $dA/d\,\ell = -Af(\ell)/a^2 \neq 0$ at the place where $v^2 = a^2$. If $f(\ell)$ is positive, the critical (= sonic) point now occurs in the converging part of the nozzle before the throat. So the sonic point of a rocket which is fired horizontally will occur at the throat of the nozzle, but in a rocket which is fired vertically upwards with gravity providing a force in the downstream direction, $f(\ell) = -g$, the sound velocity is reached above the throat.

3.5.3 The analogy to a stellar wind

In a stellar wind the nozzle-term $a^2\ d \ln A/d\ell$ is provided by the inward gravity $-GM_*/r^2$ and the outward gas-pressure force due to the spherical divergence of the flow, $2a^2/r$. The gravity provides the 'convergence of the nozzle' and the gas pressure provides the divergence. Close to the star where gravity dominates the nozzle is converging but at large distances the nozzle diverges because the gas pressure term dominates. In a wind without an extra force the sonic velocity is reached at the 'throat' where the two terms cancel one another. In a stellar wind with an extra outward directed force the sonic point occurs closer to the star. In fact the conditions for the location of the critical point in a wind (Eq. (3.51) with $g(r, v) = 0$) and in a nozzle (Eq. (3.67)) are very similar.

The expression (3.67) for a nozzle shows a characteristic which we have also encountered in the discussion of the stellar winds. If the force $f(\ell)$ is larger than the nozzle term $2a^2/r - GM_*/r^2$ in the subsonic part of the wind, the velocity will not become supersonic because its gradient is negative. The reason for this behaviour can be understood by noticing that the force $f(\ell)$ acts like an extra pseudo-divergence of the nozzle. If this pseudo-divergence is larger than the actual convergence of the nozzle $dA/d\ell$ in the subsonic part, there is no net throat anymore, but only a net diverging flowpipe in which the subsonic flow will decelerate.

3.5.4 The adjustment of the flow to the critical solution

We now turn to the important question of how a smooth transonic flow can be established through a nozzle if the initial condition v_0 is so critical to its existence. After all, rockets which are fired reach transonic flows in a very short time, despite the fact that during the ignition the flow certainly does not start transonic right away. The reason is that the flow can adjust itself to reach exactly $M = 1$ at the throat provided that the fuel injection rate is higher than some lower limit.

Consider a highly simplified picture of a rocket motor, consisting of a 'chamber' into which fuel is injected and burned, i.e. converted into gas at a constant temperature T. The rate of fuel injection and hence of the gas production rate F in the chamber can be regulated. For the sake of simplicity we assume that the flow through the nozzle is isothermal at temperature T, although this is not essential for the arguments. The mass flow F through the nozzle depends on the variable setting of the fuel injection tap. For any setting of the tap the flow has fixed values for F, T and $A(\ell)$. The density $\rho(\ell)$ and the velocity $v(\ell)$ are not fixed.

Suppose the rocket is ignited at a low injection rate. A low injection rate results in a small pressure in the chamber and a small velocity v_0 at ℓ_0. The velocity v_0 is determined by the pressure difference between the chamber pressure $p \sim \rho T$ and the ambient pressure. If v_0 is below the critical value, the flow will not reach the sound speed at the throat and so it will remain subsonic. When the injection tap is opened more, the gas production rate will increase so F and p increase, and the velocity v_0 increases. When the tap is opened sufficiently, v_0 will reach the critical value so that $M = 1$ is reached at the throat. The flow is transonic. When the tap is opened even further the flow will initially not be transonic, because the velocity v_0 will be larger than critical. This will result in a stalling of the flow near the throat because $dv/d\ell \rightarrow -\infty$ which in turn will result in a very rapid increase of the pressure in the subsonic part of the nozzle. This will decrease the pressure difference between the chamber and the nozzle so v_0 will decrease until its velocity reaches the critical value again. The flow then settles to a new stationary transonic situation with the same $v(\ell)$ as before but with a higher F, a higher density $\rho(\ell)$ and a higher pressure in the chamber. Increasing the fuel injection rate even further results in transonic flows with higher p, ρ and F but always with the same $v(\ell)$. So the flow will remain transonic for a whole range of values of F, provided that it is above some critical limit. (In reality the fuel injection rate is set to an optimum value such that the gas pressure at the downstream end of the nozzle is equal to the ambient gas pressure. In that case the rocket has its maximum efficiency because it minimizes shocks at its end.)

The crucial property of the rocket motor described here is its self-adjustment to the transonic flow by means of sending information (sound-waves) about pressure differences through the subsonic part of the flow back to the chamber. This regulates v_0 and ρ_0 in such a way that the gas production rate F in the chamber can leave the flow. So F is fixed by the fuel injection and the flow adjusts v so that it is transonic.

The analogy between a rocket motor and a stellar wind may now be used to understand qualitatively the adjustment of the wind to the transonic solution of the momentum equation. In an isothermal stellar wind the density ρ_0 is fixed by the lower boundary of the isothermal region (e.g. the bottom of a corona) and the velocity v_0 is the adjustable, which in turn determines the mass loss rate. If the initial velocity would be too large the stalling of the flow would result in a pressure inbalance which reduces v_0 until the flow is transonic again, analogous to the adjustment of the rocket flow. Contrary to a rocket flow, however, the transonic wind does not have a minimum density ρ_0 or a minimum mass loss rate, because the ambient pressure of the interstellar medium is negligible so an isothermal wind will always adjust itself to a transonic solution. If the wind has built up a circumstellar shell where it runs into the interstellar medium, the information about its ambient pressure cannot be propagated back to the star so it does not effect the flow nor the mass loss rate.

3.5.5 Conclusions

The momentum equation of a gas flow through a flowpipe with a nozzle is very similar to that of an isothermal wind: the convergence of the pipe acts like an inward directed force in the wind (e.g. the gravity) and the divergence of the pipe acts like an outward force, $-dp/dr$, in the wind. The isothermal flow through the flowpipe can only be transonic if the velocity reaches the sound speed at the location of the nozzle where the flowpipe is narrowest. The flow through the pipe can adjust itself to be transonic at the nozzle because soundwaves can travel upstream in the subsonic region and create a pressure gradient which can regulate the flow.

3.6 Conclusions about isothermal winds

The momentum equation that describes the variation of $d \ln v/dr$ in an isothermal wind has only one critical solution which goes from subsonic velocities in the lower layers to supersonic velocities further outward. The *location* r_c and the *velocity* v_c of the critical point are only determined by the local conditions at r_c. If the forces are only a function of r and v, the critical point is the sonic point. If the forces depend on dv/dr, as in the case of line driven winds, the critical point is no longer the sonic point.

The velocity between the bottom of the isothermal region, r_0, and the critical point is determined by the forces that act upon the gas between r_0 and r_c and by the boundary condition that $v(r_c) = v_c$. This fixes the value of v_0 at r_0. If the density at r_0 is given, the mass loss rate is fixed by $\dot{M} = 4\pi\rho_0 v_0 r_0^2$. The variation of the velocity above the critical point is determined by v_c and by the forces at $r > r_c$. The mass loss rate is *not* affected by forces above the critical point.

An extra outward force in the subcritical region changes the velocity structure of that region and increases the mass loss rate. An extra outward force in the region, $r > r_c$, changes the velocity law in the supercritical region but does not change the mass loss rate, because the information about this force cannot be transmitted down to the subcritical region where the mass loss is fixed.

In an isothermal wind with only gas pressure and gravity the density structure is very close to the hydrostatic one in the subcritical region. This is because the term $v\, dv/dr$ in the momentum equation is negligible compared to the pressure term $(1/\rho)dp/dr$ if $v \leq 0.5a$. This is no longer true in the region above the sonic point, where the density is determined by the variation of v. *So below the sonic point the structure of the wind is mainly determined by hydrostatic equilibrium, and above the sonic point it is mainly determined by the forces that produce the velocity increase.*

The energy of a transonic wind is negative at r_0 and positive at large distances. *This implies that a wind can only become transonic if energy is added to the gas, either in the form of heat or in the form of work done by a force.*

The momentum equation of an isothermal wind is similar to that of the flow from a rocket through a nozzle. The understanding of a flow though a nozzle helps to explain the physics of stellar winds, in particular the way in which the wind may adjust itself to the critical transonic flow.

3.7 Suggested reading

Parker, E.N. 1958 'Dynamics of the interplanetary gas and magnetic fields', *Ap. J.* **128**, 664
(The first supersonic gas driven wind theory)

Brandt, J.C. 1970 *Introduction to the solar wind* (San Francisco: Freeman)
(The basic solar wind theory and extensions)

4 Basic concepts: non-isothermal winds

The structure of a wind in which there are temperature gradients will be different from an isothermal wind for two reasons. (a) The presence of temperature gradients implies that extra pressure gradients enter into the momentum equation. These will change the velocity structure of the wind. (b) The variation of temperature with distance implies a variation of the sound speed and the Mach number of the flow. This affects the location of the critical point and hence the mass loss rate.

A useful way of discussing non-isothermal winds is to use the energy per unit mass, $e(r)$, which is the sum of the kinetic and gravitational energies, and the enthalpy

$$e(r) = \frac{v(r)^2}{2} - \frac{GM_*}{r} + \frac{\gamma}{\gamma - 1}\frac{\mathcal{R}T}{\mu} = \frac{v(r)^2}{2} - \frac{GM_*}{r} + \frac{5\mathcal{R}T}{2\mu} \tag{4.1}$$

The form of the latter expression is valid if the wind consists of a monatomic† ideal gas for which $\gamma = c_p/c_v = 5/3$ and the enthalpy is $(5/2)\ \mathcal{R}T/\mu$. In the lower boundary of the wind the energy is *negative*, because the photosphere is gravitationally bound to the star, $v(r_0) \ll v_{esc}(r_0)$ and $(5/2)\ \mathcal{R}T(r_0)/\mu \ll v_{esc}^2(r_0)/2$. If the wind is to escape from the stellar gravitational well, its energy must be *positive* at large distance r because $v^2(r) > v_{esc}^2(r)$. This implies that a stellar wind can only reach super-escape velocities if energy is added to it or work is performed on the gas.

In the previous chapter on isothermal winds the addition of energy was implicit in the assumption of $T(r)$ = constant, because the

† The term 'monatomic' = 'mono-atomic' implies that the gas consists of electrons, ions or atoms which have only three degrees of freedom (translation in three dimensions) so that $c_v = 3\mathcal{R}/2\mu$ and $\gamma = 5/3$. Ionized winds and neutral atomic winds are monatomic. Winds with variable ionization and molecular winds are not monatomic since $\gamma \neq 5/3$. The word monatomic does not imply that the wind consists of one type of particles only.

deposition of energy is necessary to compensate the cooling due to expansion of the gas. In fact, the energy added to the isothermal wind was used for three effects: to prevent the adiabatic cooling of the expanding gas, to lift the gas out of the potential well of the stellar gravity, and the remaining energy went into kinetic energy. In general the effect of energy deposition on a wind depends not only on the amount deposited but also on the place where it is deposited.

In this chapter we will discuss the effects that energy deposition in a non-isothermal wind has on the temperature, the velocity structure, and the mass loss rate. We first discuss the general case of winds with both energy and momentum deposition (outward forces) and describe the differences that these two types of deposition have on the structure of the wind (§ 4.1). We then illustrate the effect of energy input in the special case of the polytropic winds (§ 4.2). We conclude with a discussion of winds with multiple critical points that can occur in cases with energy and momentum deposition.

4.1 Winds with momentum and energy deposition

The addition of energy that is needed to drive a wind can be in the form of heat input (for instance by dissipation of waves from the star in the wind), or in the form of momentum input (for instance by radiation pressure). Newton's law $f = d\mathbf{p}/dt$ states that a force produces an increase of momentum. Therefore the presence of a outward force is usually called momentum deposition, in contrast to energy deposition.

Here we consider some general consequences of energy and momentum input into the wind. We will address such questions as: what fraction of the energy deposition goes into heating and what fraction goes into kinetic energy or potential energy? What is the effect of momentum or energy deposition applied at different locations in the winds? The discussion of these important aspects of stellar winds will help us understand the results of the specific mass loss mechanisms described in the subsequent chapters.

4.1.1 The energy of the wind

The momentum equation of a stationary spherical wind with an outward directed force $f = f(r)$ per unit mass is

$$v\frac{dv}{dr} + \frac{1}{\rho}\frac{dp}{dr} + \frac{GM_*}{r^2} = f \tag{4.2}$$

where f is in units of cm s^{-2}.

The effects of energy deposition on the wind can be expressed by means of the first law of thermodynamics $du = dQ - p\,d\rho^{-1}$ (Appendix A), where du is the change in the internal energy, dQ is the amount of heat added, and $pd\rho^{-1}$ is the work done by the gas; all are in units of energy per unit mass. The rate of change of internal energy can thus be written as

$$\frac{du}{dt} = \frac{dQ}{dt} - p\frac{d\rho^{-1}}{dt} \tag{4.3}$$

In a stationary wind where u, Q and ρ depend only on r, the time derivatives can be replaced by $v d/dr$, so this equation can be written in derivatives with respect to r. Also by using $du = da^2/(\gamma - 1)$ (Appendix A.33) and the perfect gas law, we find an expression for the gradient of the heat addition per unit mass, q in erg $\mathrm{g}^{-1}\mathrm{cm}^{-1}$

$$q \equiv \frac{dQ}{dr} = \frac{1}{\gamma - 1}\frac{\mathcal{R}}{\mu}\frac{dT}{dr} + p\frac{d\rho^{-1}}{dr} = \frac{1}{\gamma - 1}\frac{d(p/\rho)}{dr} + p\frac{d\rho^{-1}}{dr} \tag{4.4}$$

Adding the energy equation (4.4) and the momentum equation (4.2) and combining the two terms $\rho^{-1}dp/dr + p\,d\rho^{-1}/dr = d(p/\rho)/dr$ results in a new form of the energy equation

$$\frac{d}{dr}\left(\frac{v^2}{2} + \frac{\gamma}{\gamma - 1}\frac{\mathcal{R}T}{\mu} - \frac{GM_*}{r}\right) \equiv \frac{d\,e(r)}{dr} = f + q. \tag{4.5}$$

The energy equation in this form is called the *Bernouilli equation*. On integrating this, we find

$$\begin{aligned} e(r) &= \frac{v^2}{2} + \frac{\gamma}{\gamma - 1}\frac{\mathcal{R}T}{\mu} - \frac{GM_*}{r} \\ &= e_0 + \int_{r_0}^{r} f(r)dr + \int_{r_0}^{r} q(r)dr \\ &= e_0 + W(r) + Q(r) \end{aligned} \tag{4.6}$$

This equation shows that the total energy $e(r)$ per unit mass, which is the sum of the kinetic energy, the enthalpy and the potential energy, is equal to the initial energy e_0 at the lower boundary r_0 plus the energy added into the wind. The energy input is the result of heat deposition $Q(r)$ and the work $W(r)$ done by the force with

$$Q(r) = \int_{r_0}^{r} q(r)dr \tag{4.7}$$

and

$$W(r) = \int_{r_0}^{r} f(r)dr \tag{4.8}$$

Notice that $Q(r)$ is not the heat content but the heat *added* between

r_0 and r. The total energy added to the wind per unit mass between the lower boundary r_0 and infinity is

$$\Delta e = W(\infty) + Q(\infty) \tag{4.9}$$

This energy is used to increase the kinetic energy and the enthalpy of the wind and to lift it out of the potential well

$$\Delta e = \frac{v_\infty^2}{2} - \frac{v_0^2}{2} + \frac{\gamma}{\gamma - 1}\frac{\mathcal{R}}{\mu}(T_\infty - T_0) + \frac{GM_*}{r_0} \tag{4.10}$$

For a wind with $v_\infty \gg v_0$ and $T_\infty \ll T_0$ Eq. (4.10) provides an estimate of the amount of energy input per unit mass that is needed for the gas to reach its terminal velocity

$$\Delta e \simeq \frac{v_\infty^2}{2} + \frac{GM_*}{r_0} - \left(\frac{\gamma}{\gamma - 1}\right)\frac{\mathcal{R}T_0}{\mu} \tag{4.11}$$

or

$$\frac{v_\infty^2}{2} \simeq \left(\frac{\gamma}{\gamma - 1}\right)\frac{\mathcal{R}T_0}{\mu} - \frac{GM_*}{r_0} + \Delta e \tag{4.12}$$

The energy of the wind per unit mass at $r = \infty$ is equal to the initial energy at r_0 plus the energy that was added per unit mass.

If there is no energy input into the wind, neither in the form of heat input, $q = 0$, nor in the form of work done by a force, $f = 0$, then the wind cannot escape from the star because the initial energy at r_0 is negative for an atmosphere that is gravitationally bound to the star.

4.1.2 The momentum of the wind

The momentum equation (4.2) in a wind with energy and momentum deposition does not contain the thermal energy deposition $q(r)$. This does not mean, however, that the velocity in the wind is unaffected by heat deposition, because the pressure gradient in the momentum equation depends on the thermal structure and hence on the energy input. The effect of energy input on the velocity can be derived from the momentum equation by taking into account its effects on the pressure gradient. This will be studied here.

The pressure term $\rho^{-1}dp/dr$ in the momentum equation can be expressed in terms of p/ρ

$$\frac{1}{\rho}\frac{dp}{dr} = \frac{d(p/\rho)}{dr} - p\frac{d\rho^{-1}}{dr} = \frac{d(p/\rho)}{dr} + \frac{p}{\rho}\frac{d\ \ln\rho}{dr} \tag{4.13}$$

The density gradient can be eliminated by means of the mass continuity equation $d \ln\rho = -d \ln v - 2d \ln r$. Substitution of the isothermal sound speed $a(r)^2 = p/\rho = \mathcal{R}T/\mu$ for a perfect gas yields

$$\frac{1}{\rho}\frac{dp}{dr} = \frac{da^2}{dr} - \frac{2a^2}{r} - a^2\frac{d \ln v}{dr} \tag{4.14}$$

This results in the momentum equation in a general form that involves both the effects of a change in temperature and the external force

$$v\frac{dv}{dr} - \frac{a^2}{v}\frac{dv}{dr} + \frac{da^2}{dr} - \frac{2a^2}{r} + \frac{GM_*}{r^2} = f \tag{4.15}$$

or

$$\frac{1}{v}\frac{dv}{dr} = \left\{\frac{2a^2}{r} - \frac{da^2}{dr} - \frac{GM_*}{r^2} + f\right\}\{v^2 - a^2\}^{-1} \tag{4.16}$$

This equation contains the term $da^2/dr = (\mathcal{R}/\mu)dT/dr$. The energy equation (4.5) shows that in a wind with momentum and energy deposition

$$\frac{da^2}{dr} = \frac{\mathcal{R}}{\mu}\frac{dT}{dr} = \frac{\gamma - 1}{\gamma}\left\{f + q - \frac{GM_*}{r^2} - v\frac{dv}{dr}\right\} \tag{4.17}$$

Substituting (4.17) for da^2/dr in the momentum equation one finds after multiplication of the result by γ

$$\frac{1}{v}\frac{dv}{dr} = \left\{\frac{2c_s^2}{r} - \frac{GM_*}{r^2} + f - (\gamma - 1)q\right\}\{v^2 - c_s^2\}^{-1} \tag{4.18}$$

In this expression $c_s = \sqrt{\gamma a^2}$ is the adiabatic speed of sound.

This is the most general form of the momentum equation of a spherically symmetric stellar wind with energy input and momentum input. Notice that *the energy input, $q > 0$, produces an inward directed force $(\gamma - 1)q$ which counteracts the outward force f*. This is because the energy input heats the gas which reduces the outward temperature gradient and thus the outward force of the pressure gradient.

The two forms of the momentum equation (4.16) and (4.18) show a curious difference. The first one suggests that the critical point occurs where $v = a$, whereas the second one suggests that the critical point occurs where $v = c_s$. The difference is due to the fact that Eq. (4.16) still contains the temperature gradient in the numerator. This temperature gradient depends on the velocity gradient, as can easily be seen from Eq. (4.17). This extra vdv/dr term shifts the critical point from the distance where $v = a$ to the distance where $v = c_s$.

Equation (4.18) might seem to suggest that in the most general case of a wind with energy and momentum deposition the critical point is always the sonic point. This, however, is only true if both f and

q do not depend on the velocity gradient dv/dr. If the momentum or energy deposition depends on dv/dr, they will produce additional terms in the left hand side of Eq. (4.18) and move the critical point to another location. We have shown an example of this effect in § 3.3 where we considered an isothermal wind with a force proportional to vdv/dr. In that case the critical point occurs at $v > a$, whereas it is at $v = a$ in an isothermal wind with a force that depends only on r. In the later chapters it will be shown that for several kinds of wind models the momentum equation has critical points where $v \neq c_s$. For the present discussion we will assume that f and q depend only on r or on v but not on dv/dr, so that the critical point is the sonic point, where $v = c_s$.

4.1.3 The momentum equation in terms of the Mach number

We have derived expressions for the energy equation (4.6) and the momentum equation (4.15). These two expressions along with the boundary conditions fully describe the velocity and temperature structure in a wind with momentum and energy deposition. The interesting solution for a stellar wind is the one that passes through the critical point. This critical solution can only be reached for certain boundary conditions.

As the wind structure is described by two coupled equations (4.6) and (4.15) which can only be solved numerically, it is not easy to understand the consequences of the conditions posed by the critical point. Therefore we will express the momentum equation (4.15) in a closed form that is independent of c_s and which can thus be solved without the energy equation. The momentum equation can be expressed in terms of the Mach number defined by $M = v/c_s$, as was first done by Holzer and Axford (1970).

The momentum equation in terms of M and c_s is

$$\frac{(M^2-1)}{2M^2}\frac{dM^2}{dr} = \frac{2}{r} - \frac{(M^2-1)}{2c_s^2}\frac{dc_s^2}{dr} + \frac{f+(\gamma-1)q-(GM_*/r^2)}{c_s^2} \tag{4.19}$$

The terms c_s^2 and dc_s^2/dr can be expressed in terms of the energy equation

$$c_s^2 = \frac{\gamma \mathcal{R} T}{\mu} = (\gamma-1)\left\{e(r) + \frac{GM_*}{r} - \frac{v^2}{2}\right\} \tag{4.20}$$

Substitution of $v = Mc_s$ gives

$$c_s^2 = \frac{2(\gamma-1)}{(\gamma-1)M^2+2}\left\{e(r) + \frac{GM_*}{r}\right\} \tag{4.21}$$

and

$$\frac{dc_s^2}{dr} = (\gamma - 1)\left\{f + q - \frac{GM_*}{r^2} - \frac{1}{2}\frac{dv^2}{dr}\right\} \tag{4.22}$$

and so

$$\frac{dc_s^2}{dr} = \frac{2(\gamma - 1)}{(\gamma - 1)M^2 + 2}\left\{f + g - \frac{GM_*}{r^2} - \frac{c_s^2}{2}\frac{dM^2}{dr}\right\} \tag{4.23}$$

Substitution of Eqs. (4.21) and (4.23) into the momentum equation gives for $\gamma = 5/3$

$$\frac{(M^2 - 1)}{M^2}\frac{dM^2}{dr} = \frac{4(3 + M^2)}{3}\left\{\frac{[e(r)/r] + f - (3 + 5M^2)(q/12)}{e(r) + GM_*/r}\right\} \tag{4.24}$$

with $e(r) = e_0 + W(r) + Q(r)$. The denominator of the right hand side, $e(r) + GM_*/r$, is always positive because it is the sum of the kinetic energy and the enthalpy which are both positive.

The integral of $\{3(M^2 - 1)/4(3 + M^2)M^2\}dM^2/dr$ is $\ln(M^2 + 3) - (1/4)\ln(M^2)$ and so the momentum equation for $\gamma = 5/3$ can also be written in an integral form as

$$\left\{\ln(M^2 + 3) - \frac{\ln(M^2)}{4}\right\} - \left\{\ln(M_0^2 + 3) - \frac{\ln(M_0^2)}{4}\right\} = \int_{r_0}^{r}\left\{\frac{[e(r)/r] + f - (3 + 5M^2)(q/12)}{e(r) + GM_*/r}\right\} dr \tag{4.25}$$

where M_0 is the Mach number at the lower boundary. (We will use this form of the momentum equation later to discuss the solution in the case that the right hand side has multiple zero-points). If there is no energy input and the momentum input depends only on distance, then the right hand side depends only on distance. If it is a simple function of r, the momentum equation has an analytic solution.

The momentum equation (4.24) has a critical point r_c where $M = 1$. A continuous wind velocity requires that

$$\frac{e(r_c)}{r_c} + f(r_c) - \frac{2q(r_c)}{3} = 0 \tag{4.26}$$

at this critical point. Although this is a *local* condition it depends on the energy and momentum input throughout the subsonic region because of the term $e(r_c)/r_c$.

The momentum equation (4.24) expresses the variation of the Mach number in terms of the initial energy of the wind, the gravity, the force and the heat input in the wind. It is valid for any force or energy deposition, including energy loss, e.g. by radiative cooling or conduction, which can be considered as negative deposition. This form of the momentum equation is useful if the force and the energy input depend on the distance only. If the force or the energy deposition depends on the velocity, the equation becomes more complicated but it is still useful. If the force or the energy deposition depends on the velocity gradient, Eq. (4.24) becomes less useful because the right hand side will contain factors of the type dM/dr.

The momentum equation (4.24) can be solved numerically for any distribution of the force $f(r)$ and the heat input $q(r)$. In order to obtain a transonic solution, the functions $f(r)$ and $q(r)$ have to satisfy the critical point condition (4.26) where $M = 1$. The integration can be carried out both inward and outward from the critical point to find the velocity at the lower boundary and hence the mass loss rate. With $M(r)$ and $e(r)$ known, the temperature distribution follows easily from the energy equation (4.20) with $v^2 = M^2\gamma\mathscr{R}T/\mu$, so

$$T(r) = \frac{6}{5}\frac{\mu}{\mathscr{R}}\left\{e_0 + W(r) + Q(r) + \frac{GM_*}{r}\right\}/\{M^2(r) + 3\} \qquad (4.27)$$

for $\gamma = 5/3$. The velocity distribution is then found from $M(r)$ and $T(r)$, and the density distribution follows from the mass continuity equation and the velocity, if $\rho(r_0)$ is known.

4.1.4 Energy and momentum deposition at the critical point

For the Mach number to increase from $M < 1$ below the critical point, r_c, to $M > 1$ at $r > r_c$ the numerator of the momentum equation has to increase through the critical point from a negative value to a positive value. This means that energy has to be added to the gas right at the critical point. Let us consider the required momentum or heat deposition.

The condition at the critical point is

$$\frac{d}{dr}\left\{\frac{e(r)}{r} + f - (3 + 5M^2)\frac{q}{12}\right\}_{r_c} > 0. \qquad (4.28)$$

Since $de/dr = f + q$ and the energy at the critical point is $e(r_c) = \{(2/3)q - f\}r_c$ (Eq. 4.26), the transonic solution requires

$$\frac{2f}{r} + \frac{df}{dr} + \frac{1}{3}\frac{q}{r} - \frac{2}{3}\frac{dq}{dr} - \frac{5}{12}q\frac{dM^2}{dr} > 0 \qquad (4.29)$$

at the critical point.

If there is no heat input near the critical point, $q(r_c) = 0$ and $(dq/dr)_{r_c} = 0$, but only momentum input, then the force at the critical point must satisfy the condition

$$\frac{d}{dr}\left(r^2 f\right)_{r_c} > 0 \tag{4.30}$$

Hence a wind without energy input can only become transonic if there is an outward force which decreases less rapidly than r^{-2} near the critical point!

This has important implications for a stellar wind driven by continuum radiation pressure. If the absorption coefficient per unit mass κ is constant with height, the radiative force varies as $f \propto \kappa L_*/r^2$ so the wind cannot become transonic unless there is additional heat input. Alternatively a radiation driven wind without heat input can only be transonic if the absorption coefficient increases with distance through the critical point. We will show in Chapter (7) that dust driven winds indeed have this property.

If there is no momentum input but only heat input, the condition (4.29) requires that

$$\frac{r_c^{1/2}}{q(r_c)}\frac{d}{dr}\left\{\frac{q}{r^{1/2}}\right\}_{r_c} = \frac{d}{dr}\left\{\ln(qr^{-1/2})\right\}_{r_c} > \frac{5}{8}\left\{\frac{dM^2}{dr}\right\}_{r_c} > 0 \tag{4.31}$$

This requires that q increases slower than $r^{-1/2}$ through the critical point. In reality it must decrease even slower because $(dM^2/dr) > 0$ at r_c. The reason that q must increase more gradually than some limit is because heat input increases the temperature of the gas and therefore it reduces the outward decrease of the temperature and the gradient in the gas pressure. Thus heat input acts like an extra inward directed force and we have seen above that the net outward force should not decrease too fast through the critical point.

These considerations show that *a stellar wind can only go from subsonic to supersonic velocity if energy is added to it right at the sonic point*. This energy input can be in the form of work done by a force which varies less rapidly than r^{-2} or by heat input due to e.g. dissipation of mechanical energy, or by magnetic energy or conduction.

4.1.5 The effect of momentum and energy deposition on the velocity and on the mass loss rate

The effect of energy deposition on the velocity distribution of a stellar wind is described by the momentum equation (4.24). From this equation we see that energy deposition has two effects: (i) energy deposition changes dM^2/dr because of the negative term containing q which describes the decrease of the outward pressure gradient due to the heating. This is a local effect which depends on $q(r)$. (ii) Energy deposition changes the absolute value of the gradient dM^2/dr because of the term $e(r)$ in the denominator which describes the change in the temperature structure due to the integrated effect of energy deposition between r_0 and r.

The momentum equation shows the differences between the effect of momentum input and heat input on the *velocity gradient*, dv/dr or dM/dr.

(a) The effects of an outward force is opposite to the effects of heat input on the Mach number.

(b) Heat input increases dM^2/dr in the subsonic region and decreases dM^2/dr in the supersonic region.

(c) Momentum input decreases dM^2/dr in the subsonic region and increases dM^2/dr in the supersonic region.

For a given density and temperature at the lower boundary of the wind (r_0), the mass loss rate scales with $v(r_0)$ and hence with the Mach number $M(r_0)$. So the addition of momentum or heat to the wind has the following effects on the *mass loss rate* and *velocity*:

(a) Adding momentum in the subsonic part of the wind results in an increase of the mass loss rate and higher velocities with a smaller gradient dv/dr in the subsonic region.

(b) Adding heat in the subsonic part of the wind results in an increase of the mass loss rates and in higher velocities in the subsonic part of the wind.

(c) Adding momentum or energy in the supersonic part of a wind does not affect the mass loss rate, but it results in higher velocities in the supersonic regions.

These conclusions are generalizations of those derived earlier in § 3.4, where we discussed the effect of different forces on an isothermal wind. They can be understood most easily by realizing that the velocity and density structure in the subsonic part of the wind is essentially determined by hydrostatic equilibrium (see § 3.1.3). The addition of momentum by means of an outward force and the input of heat both work effectively as an increase of the scale height in the subsonic region. An increase of the subsonic scale height results in an increase of the density at the critical point and hence produces an increase of the mass loss rate.

4.2 Polytropic winds

To investigate the effects of a temperature gradient on the structure of the wind, we consider a very simple case of $T(r) \sim \rho(r)^{\Gamma-1}$ with Γ as a free parameter. Such a relation is called a 'polytropic' relation, and the winds are 'polytropic winds'. The value of the 'polytropic index' Γ, which determines the temperature structure, depends of course on the energy deposition in the wind: we will show below that $\Gamma = 5/3$ corresponds to a monatomic wind which has no energy deposition, and that $\Gamma = 1$ corresponds to the isothermal case for which the energy deposition was described by Eq. (3.18). We will discuss winds with $T(r)$ in between these two extremes, so $1 < \Gamma < 5/3$. For simplicity we will assume that Γ is constant throughout the wind. It should be kept in mind that small values of Γ correspond to large energy input, and large Γ to small energy input.

The temperature and pressure of a polytropic flow of an ideal gas can be written as

$$T = T_0 \left(\frac{\rho}{\rho_0} \right)^{\Gamma-1} \tag{4.32}$$

and

$$p = \frac{\mathscr{R}\rho T}{\mu} = p_0 \left(\frac{\rho T}{\rho_0 T_0} \right) = p_0 \left(\frac{\rho}{\rho_0} \right)^{\Gamma} \tag{4.33}$$

If Γ is chosen to be equal to the ratio of the specific heats $\Gamma = \gamma = c_p/c_v$, which is 5/3 for a monatomic gas such as in an ionized wind, then $p \sim \rho^{\gamma}$ corresponds to adiabatic expansion, i.e., without energy input or energy loss (see Appendix 2).

4.2.1 The momentum equation and the critical point

The conservation of mass and momentum is described by Eqs. (3.1) and (4.2). The term $\rho^{-1}dp/dr$ in the momentum equation can be eliminated by the equation of mass conservation and the ideal gas law

$$\frac{1}{\rho}\frac{dp}{dr} = \frac{p}{\rho}\frac{d\ \ln p}{dr} = \frac{\Gamma\mathscr{R}T}{\mu}\frac{d\ \ln\rho}{dr} = \frac{\Gamma\mathscr{R}T}{\mu}\left\{-\frac{1}{v}\frac{dv}{dr} - \frac{2}{r}\right\} \tag{4.34}$$

where we have used the polytropic relation $d\ \ln p = \Gamma d\ \ln\rho$. The momentum equation is

$$v\frac{dv}{dr} + \frac{\Gamma\mathscr{R}T}{\mu}\left\{-\frac{1}{v}\frac{dv}{dr} - \frac{2}{r}\right\} + \frac{GM_*}{r^2} = 0 \tag{4.35}$$

or

$$\frac{1}{v}\frac{dv}{dr} = \left\{\frac{2\Gamma a^2}{r} - \frac{GM_*}{r^2}\right\} / \left\{v^2 - \Gamma a^2\right\} \tag{4.36}$$

where a is the isothermal soundspeed which varies with r. Equation (4.36) has the same form as the momentum equation (3.8) of the isothermal wind. A solution that starts subsonic at the base of the wind and becomes supersonic further outward must pass through a critical point where the numerator is zero

$$r_c = \frac{GM_*}{2\Gamma a^2(r_c)} \tag{4.37}$$

and the denominator is zero

$$v^2(r_c) = \Gamma a^2(r_c) = \frac{GM_*}{2r_c} = \frac{v_{\rm esc}^2(r_c)}{4} \tag{4.38}$$

This condition for a transonic velocity law in a polytropic wind is the same as for an isothermal wind apart from the fact that a has been replaced by $\sqrt{\Gamma}a$. If $\Gamma = \gamma = c_p/c_v$ then $\sqrt{\Gamma}a$ is the adiabatic sound speed. If $\Gamma \neq \gamma$ the critical point is not the sonic point. This example corroborates our previous conclusion from isothermal winds that the velocity at the critical point is not necessarily either the isothermal or the adiabatic sound speed.

There is one important difference between the momentum equations (3.8) and (4.36) of isothermal and polytropic winds: the sound speed, a, depends on distance in a non-isothermal wind. In this case the momentum equation does not have an analytic solution. It can be solved by standard numerical integration techniques and we will discuss some of its properties below. We first discuss the energy of polytropic winds.

4.2.2 The energy of polytropic winds

The momentum equation (4.36) can be written in an alternative form by making use of the polytropic relation $p \sim (p/\rho)^{\Gamma/\Gamma-1}$, so

$$\frac{1}{\rho}\frac{dp}{dr} = \frac{p}{\rho}\frac{d\ln p}{dr} = \frac{\Gamma}{\Gamma-1}\frac{p}{\rho}\frac{d\ln(p/\rho)}{dr} = \frac{\Gamma}{\Gamma-1}\frac{d}{dr}(p/\rho) \tag{4.39}$$

for $\Gamma \neq 1$, which gives

$$v\frac{dv}{dr} + \frac{\Gamma}{\Gamma-1}\frac{d}{dr}\left(\frac{p}{\rho}\right) + \frac{GM_*}{r^2} =$$
$$\frac{d}{dr}\left\{\frac{v^2}{2} + \left(\frac{\Gamma}{\Gamma-1}\right)\frac{\mathcal{R}T}{\mu} - \frac{GM_*}{r}\right\} = 0 \tag{4.40}$$

This implies that a polytropic wind has a constant value of the parameter

$$e_\Gamma \equiv \frac{v^2}{2} + \left(\frac{\Gamma}{\Gamma - 1}\right) \frac{\mathscr{R}T}{\mu} - \frac{GM_*}{r} \tag{4.41}$$

for $\Gamma \neq 1$.

The constant e_Γ has the dimension of energy and it contains the kinetic energy, the potential energy and the term $\Gamma \mathscr{R} T / (\Gamma - 1)\mu$. The quantity e_Γ will be called 'the energy constant'. The physical meaning of e_Γ can be understood by considering the value of e_Γ as $r \to \infty$. At very large distance from the star the potential energy goes to zero and $T(r)$ also goes to zero because $T \sim \rho^{\Gamma-1}$ and $\rho(r \to \infty) = 0$. Thus the energy constant is simply the kinetic energy at $r \to \infty$

$$e_\Gamma = e(r \to \infty) = \frac{v_\infty^2}{2} \tag{4.42}$$

The energy constant must be positive for the wind to escape.

The range of allowed values of e_Γ can be derived from the condition that the velocity law of the wind should be transonic with a critical point in the range $r_0 < r_c < \infty$. Applying the expression (4.41) of e_Γ at the critical point, r_c, where the conditions (4.38) must be satisfied, one finds

$$e_\Gamma = \left(\frac{5 - 3\Gamma}{\Gamma - 1}\right) \cdot \frac{\Gamma a_c^2}{2} = \left(\frac{5 - 3\Gamma}{\Gamma - 1}\right) \cdot \frac{v_c^2}{2} = \left(\frac{5 - 3\Gamma}{\Gamma - 1}\right) \frac{GM_*}{4r_c} \tag{4.43}$$

This holds for $1 < \Gamma < 5/3$. The special case of $\Gamma = 5/3$ will be discussed below. Equation (4.43) can be used to set a limit to the value of e_Γ because the critical point must be located in the wind at $r_0 < r_c < \infty$, which yields

$$0 < e_\Gamma < \left(\frac{5 - 3\Gamma}{\Gamma - 1}\right) \frac{GM_*}{4r_0} \tag{4.44}$$

If e_Γ is negative the wind will not escape. If e_Γ is too large, the critical point will be inside the lower boundary of the polytropic layer.

We now consider the global energy balance and the energy input in the wind. Equation (4.41) at the lower boundary r_0 provides an expression for the terminal velocity as a function of the lower boundary conditions.

$$\frac{v_\infty^2}{2} = \frac{v_0^2}{2} + \left(\frac{\Gamma}{\Gamma - 1}\right) \frac{\mathscr{R}T_0}{\mu} - \frac{GM_*}{r_0} \tag{4.45}$$

The condition that $v_\infty > 0$ sets lower limits to the boundary conditions for a wind with a fixed value of Γ. Alternatively, if the lower boundary conditions are fixed the value of $\Gamma/(\Gamma - 1)$ has to be sufficiently large, i.e., Γ has to be sufficiently small, so that the wind can reach $v_\infty > 0$. Since the value of Γ is related to the energy input in the wind, because $T(r) \sim \rho(r)^{\Gamma-1}$, this requirement for Γ is in fact a requirement for a sufficiently large energy input.

A comparison between e_Γ and $e(r)$ shows that the energy per unit mass of a polytropic wind can be written as

$$e(r) = \frac{v^2}{2} - \frac{GM_*}{r} + \frac{5}{2}\frac{\mathcal{R}T}{\mu} = e_\Gamma - \frac{(5-3\Gamma)}{2(\Gamma-1)}\frac{\mathcal{R}T(r)}{\mu} \tag{4.46}$$

The total energy input into a polytropic wind is per unit mass

$$\Delta e = e(\infty) - e(r_0) = \frac{(5-3\Gamma)}{2(\Gamma-1)}\frac{\mathcal{R}T_0}{\mu} \tag{4.47}$$

The relation between the energy input in the wind and the temperature gradient can be found by differentiation of $e(r)$, which gives

$$\frac{d\, e(r)}{dr} = -\frac{(5-3\Gamma)}{2(\Gamma-1)}\frac{\mathcal{R}}{\mu}\frac{dT}{dr} \tag{4.48}$$

for $\Gamma \neq 1$.

If $\Gamma = 5/3$ there is no energy input into the wind. This agrees with the fact that a polytrope of $p \sim \rho^{5/3}$ corresponds to adiabatic expansion. In that case the energy of the wind is $e(r_0)$, which is negative for a gravitationally bound atmosphere, so the wind cannot escape the potential well. If $1 < \Gamma < 5/3$ the energy input is positive and the temperature decreases outwards. If $\Gamma = 1$ the wind is isothermal. If $\Gamma < 1$ the energy input produces an outward increase of the temperature.

The way in which Eq. (4.48) is written suggests that for a given value of Γ in the range of $1 < \Gamma < 5/3$ the temperature gradient will become more negative as the energy input de/dr becomes more positive. One might thus conclude that 'the more energy is put into a polytropic wind, the faster it will cool'. This is a wrong conclusion. What Eq. (4.48) really states is this: 'in a polytropic wind with a certain Γ, most of the energy input is needed in those layers where the temperature falls most steeply outwards'. We will show later that the values of Γ, dT/dr and de/dr are strictly related to one another, therefore it is physically not correct to assume one of these three is fixed and that the other two are free parameters related to one another by Eq. (4.48). The reason for the connection between Γ, dT/dr and de/dr is as follows: the pressure, temperature and velocity structure of a wind in which gas pressure and gravity provide the only forces is fully determined by the lower boundary conditions and the energy input de/dr. This pressure and temperature structure imply that dT/dr and $\Gamma \equiv d\ \ln p/d\ \ln \rho$ are fully determined by de/dr.

4.2.3 The velocity and temperature distribution

The velocity structure of a polytropic wind can be derived by solving the momentum equation (4.36) or its integral form (4.41). Equation (4.41) can be written as

$$e_\Gamma = \frac{v^2}{2} + \frac{\Gamma}{\Gamma - 1} a_c^2 \left(\frac{r}{r_c}\right)^{2-2\Gamma} \left(\frac{v}{v_c}\right)^{1-\Gamma} - \frac{v_{\rm esc}^2(r_c)}{2} \frac{r_c}{r} \tag{4.49}$$

where $\mathscr{R}T/\mu = a^2$ has been replaced by the polytropic relation $a^2 = a_c^2(T/T_c) = a_c^2(\rho/\rho_c)^{\Gamma-1}$ with ρ expressed in v and r by means of the mass continuity equation. The subscript c refers to the values at the critical point r_c. This equation reaches a convenient form if the dimensionless parameters

$$x = r/r_c \quad \text{and} \quad w = v/v_c \tag{4.50}$$

are introduced

$$\frac{v_c^2}{2} w^2 + \frac{1}{\Gamma - 1} v_c^2 x^{2-2\Gamma} w^{1-\Gamma} - \frac{2v_c^2}{x} = e_\Gamma \tag{4.51}$$

We have used the condition (4.38) at the critical point to express $v_{\rm esc}(r_c)$ and a_c in terms of v_c. The solution will only pass through the critical point if e_Γ and v_c^2 are related to one another by Eq. (4.43). Multiplying (4.51) by $w^{\Gamma-1}$ yields the equation

$$w^{\Gamma+1} - w^{\Gamma-1}\left(\frac{4}{x} + \frac{5 - 3\Gamma}{\Gamma - 1}\right) + \frac{2}{\Gamma - 1} x^{2-2\Gamma} = 0 \tag{4.52}$$

This equation describes the dependence of w on x in an implicit way. It can easily be solved numerically.

For $w \ll 1$ the first term can be neglected and so

$$w \simeq \left\{\frac{2}{\Gamma - 1} x^{3-2\Gamma} \Big/ \left(4 + \frac{5 - 3\Gamma}{\Gamma - 1} x\right)\right\}^{1/(\Gamma-1)} \tag{4.53}$$

far below the critical point.

At large distance the asymptotic value of w for $x \to \infty$ is $\sqrt{(5 - 3\Gamma)/(\Gamma - 1)}$ if $\Gamma > 1$ so

$$v_\infty^2 = \left(\frac{5 - 3\Gamma}{\Gamma - 1}\right) v_c^2 = \left(\frac{5 - 3\Gamma}{\Gamma - 1}\right) \frac{GM_*}{2r_c} \tag{4.54}$$

Figure (4.1) shows the structure of several polytropic winds, derived by solving the momentum equation (4.52) for three values of Γ, in between the isothermal case of $\Gamma = 1$ and the adiabatic case of $\Gamma = 5/3$. For $\Gamma = 5/3$ there is no solution because the wind cannot escape from the star. The upper part shows the velocity law in terms of the Mach number, which in this case is defined as $M(r) = v(r)/\sqrt{\Gamma}a(r)$. The Mach number increases from subsonic to supersonic velocities. The lower part of this figure shows the temperature distribution normalized to the critical point, which is derived from the velocity distribution by $T \sim \rho^{\Gamma-1} \sim v^{1-\Gamma} \cdot r^{2-2\Gamma}$ so

$$\frac{T(r)}{T(r_c)} = \left(\frac{v}{v_c}\right)^{1-\Gamma} \left(\frac{r}{r_c}\right)^{2-2\Gamma} \tag{4.55}$$

The temperature decreases faster with distance if there is less energy input, i.e. if Γ is larger. The middle part of the figure shows the velocity normalized to the critical point. Since the temperature decreases outwards, the velocity increases slower outwards than the Mach number.

Figure (4.1) shows that $v(r)$ increases if $1 < \Gamma < 3/2$ and decreases for $3/2 < \Gamma < 5/3$. For $\Gamma = 3/2$ the velocity is constant. This means that the energy input de/dr plus the loss of enthalpy of the wind, $(5\mathcal{R}/2\mu)dT/dr$, is all used for the increase of potential energy with no energy left for acceleration. The reason that this happens when

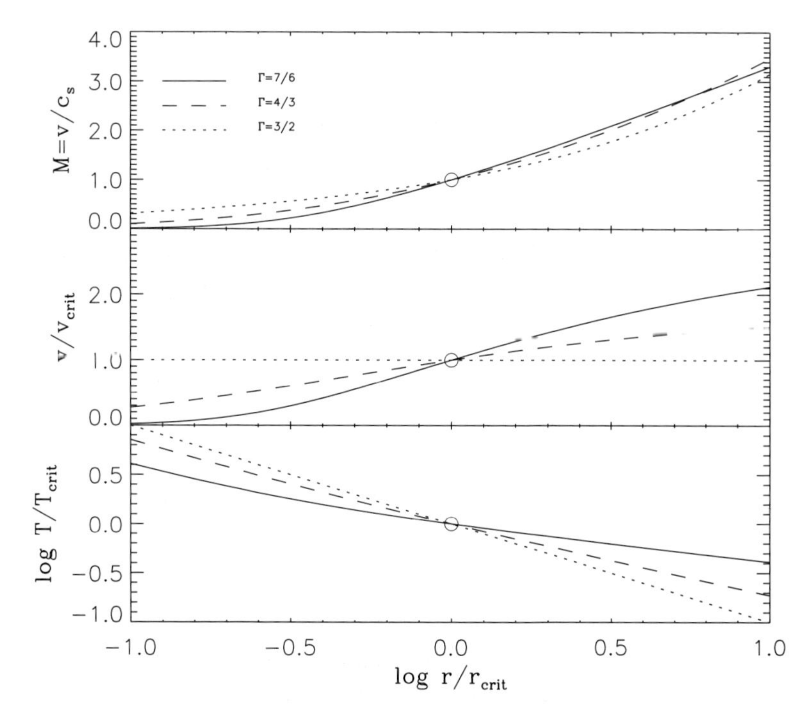

Figure 4.1 The velocity $w = v/v_c$ and temperature structure of polytropic winds with Γ=7/6, 4/3, 3/2 as a function of the normalized distance $x = r/r_c$. Notice that all velocity laws in terms of the Mach number are transonic, but the velocity in km s^{-1} is constant for $\Gamma = 3/2$ and decreases outward for $3/2 < \Gamma < 5/3$ (not shown).

$\Gamma = 3/2$ can be understood by considering the energy constant e_Γ [Eq. (4.41)]. In the case of a polytropic wind with a constant velocity $v^2/2 = v_\infty^2/2 = e_\Gamma$ the value of $\{\Gamma/(\Gamma - 1)\}\mathcal{R}T/\mu$ must be equal to GM_*/r at all distances, so T must be proportional to r^{-1} and $\rho \sim r^{1/(1-\Gamma)}$ because T and ρ are coupled by the polytropic relation $T \sim \rho^{\Gamma-1}$. However, for v = constant, mass conservation requires that $\rho \sim r^{-2}$. These two conditions for ρ can only be satisfied if $1/(1-\Gamma) = -2$ and $\Gamma = 3/2$.

The velocity laws shown in Fig. (4.1) clearly demonstrate that the acceleration depends strongly on Γ and hence on the energy input into the wind. The sensitivity of $v(r)$ to the energy input is in reality even stronger than suggested by these normalized velocity laws, because the normalization factors r_c and v_c depend on energy input. If the velocities and distances were shown in absolute values, i.e. not normalized to the critical point, the differences between the curves for different values of Γ would be much more significant.

Consider a set of wind models with the same initial values of v_0 and T_0 at the lower boundary r_0, but with different values of Γ and hence of e_Γ. Decreasing Γ from a value slightly smaller than 5/3 to about 1 affects the velocity law in three ways: the relation v/v_c versus r/r_c becomes steeper, as shown in Fig. (4.1); the velocity v_c at the critical point increases; the critical point r_c moves closer in to r_0 [Eq. (4.43)]. These three changes imply that the velocity at any r increases rapidly with e_Γ. At large distances these velocities approach the value of $v_\infty = \sqrt{2e_\Gamma}$ [Eq. (4.42)].

Figure (4.2) shows the potential energy, the kinetic energy, the enthalpy and the total energy of a polytropic stellar wind of $\Gamma = 7/6$ with lower boundary conditions: $T(r_0) = 4 \times 10^6$ K, $\mu = 0.60$ and $v_{esc}(r_0) = 600$ km s^{-1}. The total energy per gram increases from a negative value at r_0 to positive values at $r \gtrsim 1.5r_0$. This shows that a polytropic wind with $\Gamma < 5/3$ requires energy deposition at all distances. The energy input is largest close to the star where dT/dr is largest.

This section has shown that the velocity and temperature structure of polytropic winds depends strongly on the energy input. We now turn our attention to the effect of energy deposition on the mass loss rates of polytropic winds.

4.2.4 The mass loss rates of polytropic winds

The mass loss rate of a polytropic wind with a given density ρ_0 at its lower boundary, r_0, can be derived from the mass continuity equation if the velocity v_0 at r_0 is known. So we have to derive an expression for v_0 or $w_0 = w(x_0)$. The value of w_0 follows from the condition that the wind has to pass through the critical point.

The conditions at the critical point (Eq. 4.38) require that $\Gamma \mathscr{R} T_c/\mu = GM_*/2r_c$. The temperature of the polytropic wind is related to the density via $T \sim \rho^{\Gamma-1}$ and the density is related to velocity and distance via the mass continuity equation, so $T \sim v^{1-\Gamma} r^{2-2\Gamma}$. The temperature at the critical point can thus be written in terms of the lower boundary conditions

$$T_c = T_0 \, w_0^{\Gamma-1} \, x_0^{2\Gamma-2} \tag{4.56}$$

because $x = 1$ and $w = 1$ at the critical point. The conditions (4.38) at the critical point are now expressed in boundary values as

$$\frac{\Gamma \mathscr{R} T_0}{\mu} \cdot w_0^{\Gamma-1} \, x_0^{2\Gamma-2} = \frac{GM_*}{2r_0} \cdot x_0 \tag{4.57}$$

The last equation implies that

$$w_0^{\Gamma-1} \cdot x_0^{2\Gamma-3} = \left(\frac{GM_*}{2r_0}\right) \cdot \frac{\mu}{\Gamma \mathscr{R} T_0} \tag{4.58}$$

This equation describes the value of the product $w_0^{\Gamma-1} \cdot x_0^{2\Gamma-3}$ in terms of the lower boundary conditions. On the other hand, the relation between w and x is given by the momentum equation (4.52). Together they determine the values of x_0 and w_0. From x_0 the location of the critical point can be derived because $r_c = r_0/x_0$. The temperature at the critical point follows from Eq. (4.56) and so the velocity at the

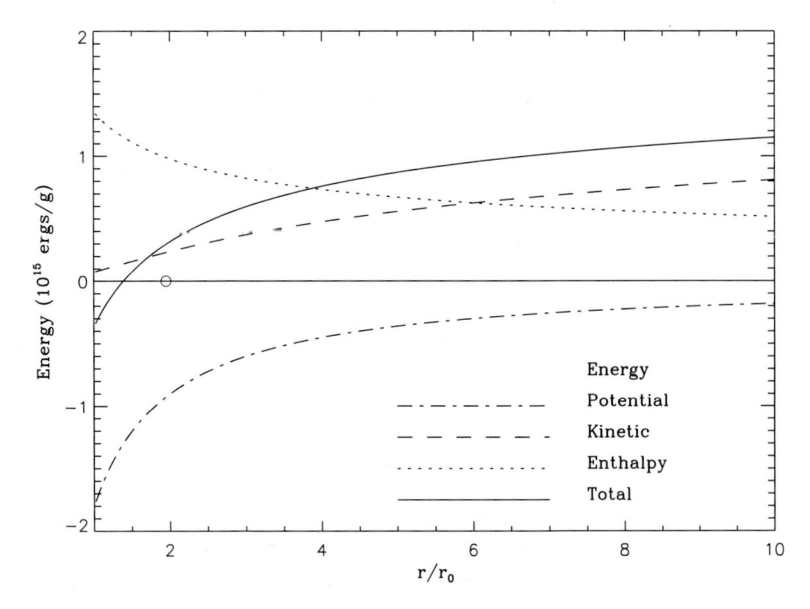

Figure 4.2 The energy of a polytropic wind with $\Gamma = 7/6$, $T(r_0) = 4 \times 10^6$ K, μ=0.60 and $v_{esc}(r_0) = 600$ km s^{-1}. The location of the critical point is indicated by a circle.

critical point, $v_c = \sqrt{\Gamma \mathcal{R} T_c/\mu}$ is also known. The velocity at the lower boundary is $v_0 = w_0 v_c$

$$v_0 = \sqrt{\frac{\Gamma \mathcal{R} T_0}{\mu}} \cdot w_0^{(\Gamma+1)/2} x_0^{\Gamma-1} \tag{4.59}$$

and so the mass loss rate is

$$\dot{M} = 4\pi r_0^2 \rho_0 \sqrt{\frac{\Gamma \mathcal{R} T_0}{\mu}} \cdot w_0^{(\Gamma+1)/2} x_0^{\Gamma-1} \tag{4.60}$$

In many cases, viz. when $v_{\rm esc}^2(r_0) \gg 4\Gamma \mathcal{R} T_0/\mu$, the values of x_0 and w_0 can be found explicitely from the approximate $w(x)$ relation (4.53) for $w \ll 1$. Substitution of Eq. (4.53) into the expression (4.58) for w_0 yields

$$x_0 \simeq \left\{ \frac{8\Gamma \mathcal{R} T_0/\mu}{v_{\rm esc}^2(r_0)} - 4\Gamma + 4 \right\} /(5 - 3\Gamma) \tag{4.61}$$

and thus the critical point is located approximately at

$$r_c \simeq \frac{(5 - 3\Gamma) r_0}{\{4 - 4\Gamma + 8\Gamma \mathcal{R} T_0/\mu v_{\rm esc}^2(r_0)\}} \tag{4.62}$$

Equation (4.61) shows that a realistic solution with $0 < x_0 < 1$ can only be reached if $(\Gamma - 1)/2\Gamma < \Gamma \mathcal{R} T_0/\mu v_{\rm esc}^2(r_0) < (\Gamma + 1)/8\Gamma$.

The value of w_0 is given by Eq. (4.58) which yields

$$w_0 = \left\{ \frac{v_{\rm esc}^2(r_0)}{4\Gamma \mathcal{R} T_0/\mu} \right\}^{1/(\Gamma-1)} x_0^{(3-2\Gamma)/(\Gamma-1)} \tag{4.63}$$

The substitution of these values of w_0 and x_0 into Eq. (4.60) gives an estimate of the mass loss rate for polytropic wind models with $v_{\rm esc}^2(r_0) \ll 4\Gamma a_0^2$

$$\dot{M} \simeq 4\pi r_0^2 \rho_0 \sqrt{\left(\frac{\Gamma \mathcal{R} T_0}{\mu}\right)} \left\{ \frac{v_{\rm esc}^2(r_0)}{4\Gamma \mathcal{R} T_0/\mu} \right\}^{(\Gamma+1)/2(\Gamma-1)}$$
$$\left[\left\{ \frac{8\Gamma \mathcal{R} T_0/\mu}{v_{\rm esc}^2(r_0)} - 4\Gamma + 4 \right\} /\{5 - 3\Gamma\} \right]^{(5-3\Gamma)/2(\Gamma-1)} \tag{4.64}$$

Table 4.1 gives the characteristics of polytropic winds derived from the numerical solutions of the equations for different values of Γ. The lower boundary of the polytropic region has a radius of $R_\odot$, an escape velocity of 600 km s^{-1}, a density of 10^{-14} g cm^{-3} and a temperature of 4 10^6 K. The last column gives the amount of energy input into the wind per gram. The last line gives the result for an isothermal wind. Smaller values of Γ correspond to larger energy input and higher terminal velocities. Notice that the mass loss rate is not very sensitive

Table 4.1 *Velocities and mass loss rates of polytropic winds*

Γ	v_0	v_c	v_∞	r_c/r_0	T_c (K)	$\dot{M}$	ΔE erg g^{-1}
16/12	283	290	502	1.069	$3.80\ 10^6$	$2.7\ 10^{-10}$	$1.00\ 10^{15}$
15/12	241	269	603	1.239	$3.50\ 10^6$	$2.3\ 10^{-10}$	$1.66\ 10^{15}$
14/12	215	262	785	1.312	$3.54\ 10^6$	$2.1\ 10^{-10}$	$2.99\ 10^{15}$
13/12	196	258	1186	1.342	$3.72\ 10^6$	$1.9\ 10^{-10}$	$6.98\ 10^{15}$
1.00	173	257	∞	1.353	$4.00\ 10^6$	$1.7\ 10^{-10}$	∞

1. Velocities in km s^{-1} and mass loss rates in $M_\odot/yr$
2. Lower boudary conditions: $r_0 = R_\odot$, $v_{esc}(r_0) = 600$ km s^{-1}, $T_0 = 4.10^6$K, $\rho_0 = 1.10^{-14}$ g cm^{-3} and $\mu = 0.5$

to the energy input, but that the terminal velocity increases rapidly with increasing energy input. This is because for a polytropic wind, most of the energy input occurs above the sonic point.

4.2.5 Conclusions about polytropic winds

In this section we have shown that polytropic winds in which gas pressure provides the only outward directed force require energy input throughout the wind. We have shown that the energy must be negative at the lower boundary and positive at large distances. The polytropic index $\Gamma = d\ln p/d\ln \rho$ determines the energy input. If Γ is equal to $\gamma = c_p/c_v = 5/3$ there is no energy input and the material cannot escape the potential well of the star. If Γ is between 5/3 and 3/2 the winds are transonic, but the increase in Mach number is slower than the decrease of the sound velocity, so the flow velocity decreases outwards. A polytropic wind with an increasing velocity requires $\Gamma < 3/2$ which corresponds to a total energy input per unit mass higher than $(5/2)\mathcal{R}T_0/\mu$.

Polytropic winds have an energy constant e_Γ [Eq. (4.41)] consisting of the kinetic and potential energy plus the term $(\{\Gamma/(\Gamma-1)\}\mathcal{R}T/\mu)$ which does not vary through the wind. The terminal velocity reached by the wind depends on e_Γ as $v_\infty^2/2 = e_\Gamma$.

The mass loss rates of polytropic winds can easily be estimated from the lower boundary conditions. For given lower boundary values r_0, T_0, ρ_0, the terminal velocity and the mass loss rate depend on the value of Γ and hence on the amount of energy input.

4.3 The critical point of momentum equations with multiple zero-points

In the examples discussed above the momentum and the energy deposition were assumed to be smooth functions of r or v, hence the numerator of the momentum equation is zero at only one distance in the wind. In several wind driving mechanisms this is not the case and the numerator is a complicated function with more than one zero point. The question is: which of these zero points corresponds to the critical point?

If the extra forces in a stellar wind, (i.e. all forces except the gradient in the gas pressure and gravity), depend only on the distance from the star, the momentum equation can be written in a general form

$$\frac{M^2-1}{M^2}\frac{dM^2}{dr} = n(r) \tag{4.65}$$

for isothermal winds and

$$\frac{3(M^2-1)}{4(M^2+3)M^2}\frac{dM^2}{dr} = n(r) \tag{4.66}$$

for non-isothermal winds with $\gamma = 5/3$, where

$$n(r) = \frac{e(r)/r + f(r) - (3+5M^2)q/12}{e(r) + GM_*/r} \tag{4.67}$$

see Eqs. (3.7) and (4.24). These equations can also be written in their integral form

$$\{M^2 - \ln M^2\} - \{M_0^2 - \ln M_0^2\} = \int_{r_0}^{r} n(r)dr \equiv N(r) \tag{4.68}$$

for isothermal winds and

$$\left\{\ln(M^2+3) - \frac{\ln M^2}{4}\right\} - \left\{\ln(M_0^2+3) - \frac{\ln M_0^2}{4}\right\} = N(r) \tag{4.69}$$

for non-isothermal winds. The left hand sides of Eqs. (4.68) and (4.69) reach a minimum at $M = 1$.

Suppose that $n(r)$ is a 'wavy' function with several zero points. The momentum equation shows that at these zero points either $dM^2/dr = 0$, which results in a maximum or a minimum of M^2, or $M^2 = 1$ which is a sonic point. Figure (4.3) shows an example of a function $n(r)$ which has seven zero points. The figure also shows the integral $N(r)$ of $n(r)$. $N(r)$ has seven extremes $(a, b, ..., g)$: three maxima and four minima. Notice that $n(r)$ is negative at r_0. This is required to obtain a velocity law that starts to increase at r_0, because $M(r_0) < 1$.

(i) To obtain a solution that is subsonic at the base r_0 and supersonic at a large distance, the sonic velocity, $M = 1$, must be reached in at least one point. It may be reached at more points, but the number of sonic points must be odd, 1, 3, 5, etc.

(ii) The zero points of n correspond to either sonic point(s), where $M^2 - 1 = 0$ or to extremes of M^2 where $dM^2/dr = 0$.

(iii) The sonic point(s) occur(s) at one or more of the zero points of n, which also are the extremes of N.

(iv) There must be at least one sonic point where M increases outwards from $M < 1$ to $M > 1$. This must be at a zero point of n where n increases from $n < 0$ to $n > 0$, because only in that case is the gradient of M^2 positive at both sides of the sonic point [Eq. (4.65)]. This sonic point therefore occurs at one of the *minima* of N (i.e. a, c, e, or g).

(v) For both isothermal and non-isothermal winds the left hand side of the integral form of the momentum equations (4.68) and (4.69) reaches an absolute minimum at $M = 1$. So in both cases the sonic point represents the absolute minimum of the left hand side of the integral form of the momentum equation. Suppose that the minimum a is a sonic point. Then points k and l would also have

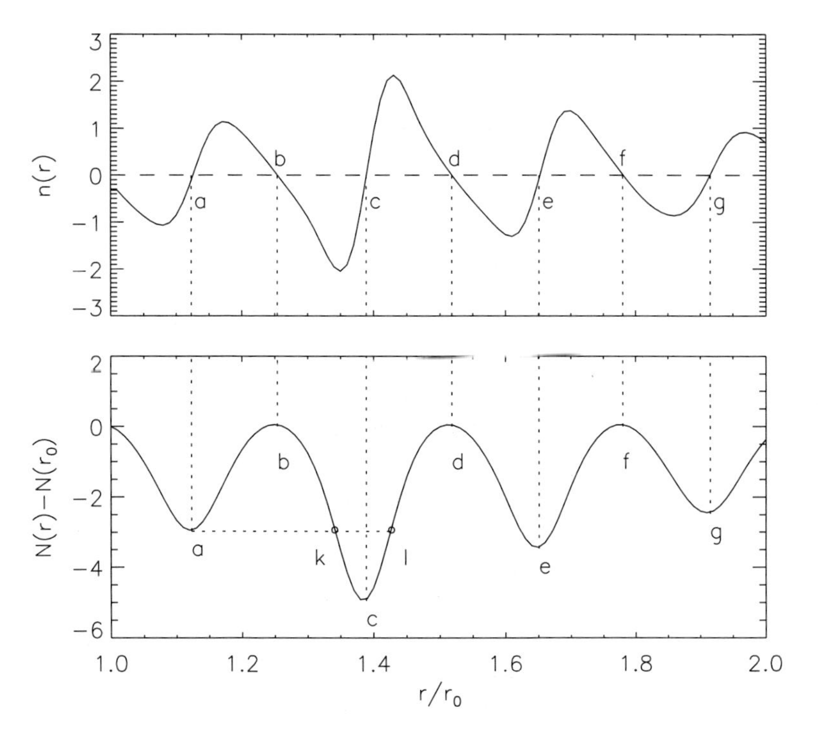

Figure 4.3 Upper panel: the right hand side $n(r)$ of a momentum equation with seven zero points. Lower panel: the integral $N(r)$ of $n(r)$ has seven maxima or minima. The sonic point occurs at the *absolute minimum* of $N(r)$, i.e. at c.

to be sonic points because the values of $N(a), N(k)$ and $N(l)$ are the same and so the left hand side of Eq. (4.65) would reach the same minimum value corresponding to $M = 1$ at all three points. However, we have shown before (iii) that the sonic points can only occur at the extremes of N. So a cannot be a sonic point. The same applies to the minima e and g. The sonic point can thus only occur at the *absolute minimum* of the function $N(r)$, which is at c.

We have thus shown that a transonic solution of momentum equation of the type described by Eq. (4.65) has a sonic point at the location where $N(r)$ has an absolute minimum. This also implies that the momentum equation will have only one sonic point, unless the function $N(r)$ happens to have more than one equal absolute minimum.

4.3.1 Jumping critical points

Let us now consider the structure of a wind that has a 'wavy' momentum equation but with a right hand side that has two almost equal minima. We have seen above that the critical point will be located where $N(r)$ has an absolute minimum. Now we ask how the structure of the wind would change if we add a small extra force at the right distance so that the absolute minimum of $N(r)$ occurs at a different distance. Intuitively we might expect that two winds in which the forces are almost identical will also have almost identical wind structures. But on the other hand, if we require that the solutions of the momentum equations go through the critical point and if the critical point of the two stars are at very different locations, the winds might be very different.

We can investigate this by calculating the wind structure for two stars with almost the same forces in the wind. Such a situation is sketched in Fig. (4.4) for an isothermal wind. The upper part of the figure shows the right hand side of the momentum equation, when it is written in the form of Eq. (4.65), with the right hand side only depending on the distance from the star and not on the velocity. The full line is for a star for which the absolute minimum of $N(r)$ occurs at $r \simeq 1.4\ r_0$. The function $N(r)$ is shown in the middle panel. Because $N(r)$ is the integral of $n(r)$ it has an integration constant which is set by the condition that the absolute minimum value of $N(r)$ must be unity, since the left hand term of the isothermal momentum equation has a minimum of 1 where $M = 1$. The resulting velocity law is shown in the lower panel by a full line. It was calculated from the numerical solution of the momentum equation, with the imposed condition that the wind should be transonic. This implies that the isothermal sound speed is reached at r_c, where $N(r)$ reaches its minimum, and that $M < 1$ below r_c and $M > 1$ above r_c.

The dashed line is for a star with almost identical forces but now the absolute minimum of $N(r)$ is at $r \simeq 1.9\ r_0$. The critical point is thus located at $r_c \simeq 1.9\ r_0$. The solution of the momentum equation, shown in the bottom figure, is very different from the previous one. Since the velocity must be transonic at r_c, the velocity is now subsonic below $1.9\ r_0$, whereas it was supersonic between 1.4 and $1.9\ r_0$ for the other star. The velocity laws above $1.9\ r_0$ and below $1.4\ r_0$ are almost the same for the two models. The similar velocity at the base of the wind implies that the mass loss rates of the two stars will be the same, if they also have the same density $\rho(r_0)$. So we see that the two stars will have the same mass loss rate, the same terminal velocity, but a very different velocity structure between the locations of their respective critical points. This is a consequence of the imposed condition that both winds are transonic.

The situation of two models with almost the same forces but with very different critical points and velocity laws might seem artificial. However, the situation does occur in several types of models of stellar

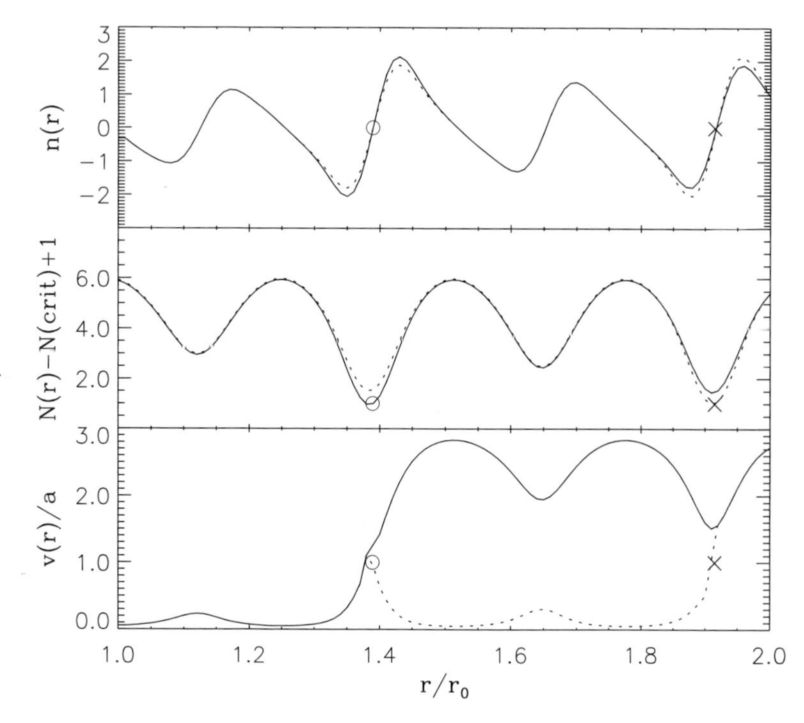

Figure 4.4 Upper panel: the right hand side $n(r)$ of the momentum equation for two isothermal winds with almost the same distribution of forces. Middle panel: the function $N(r)$ for the two models. Lower panel: the velocity laws of the two models. The critical points of the two models are indicated by a circle for the full lines and a cross for the dotted line.

winds. For instance in wave driven wind models with an exponential dissipation of the wave energy, the location of the critical point can jump from one distance region in the wind to another, if the dissipation length is smaller or larger than some critical value (Pijpers and Hearn, 1989). Another example is the radiation driven wind of stars with an effective temperature near 20 000 K. Near this temperature the degree of ionization in the wind, and hence the number densities of the ions that can drive the wind, is very sensitive to the optical depth in the wind. So small changes in the wind density or its optical depth can result in drastic changes in the driving force and in the wind structure (Pauldrach and Puls, 1990; Lamers and Pauldrach, 1991).

This situation of two very different wind structures resulting from almost the same distribution of forces in the wind suggests that winds might be 'bi-stable' if their momentum equation allows an easy jump from one type of solution to another. In such particular cases small local changes in the wind may result in large changes in its structure. Since real stellar winds will *never* be completely smooth and stationary at all distances, such jumps might occur continuously with different sections or sectors of the wind showing large temporal differences. The time-average structure of these bi-stable winds can only be derived by means of time-dependent hydrodynamical calculations.

4.4 Conclusions about non-isothermal winds

We have shown in this chapter that stellar winds require heat input or momentum input (outward directed forces) to be transonic. In the previous chapter where we discussed isothermal winds, this condition was hidden in the assumption of a constant temperature. In this section we discussed the effect of heat and momentum input explicitly.

The energy per unit mass of an atmosphere which is gravitationally bound must be negative and will be of the order of $e(r) \simeq -GM_*/R_*$ if $c_s \ll v_{\rm esc}$. The energy of a wind which escapes the potential well of the star is positive and of the order of $e(r) \simeq v_\infty^2/2$ at large distance. This implies that energy must be added to the gas to let it escape. The energy can be added in the form of heat input or work done by an outward directed force (momentum input).

Winds without an extra force (i.e. apart from the gravity and the gradient of the gas pressure) require heat input in the subsonic region to become supersonic. The heat input in terms of $q = dQ/dr$ must increase slower than $r^{1/2}$ through the critical point. If the heat input increases faster, e.g. q=constant, the resulting gradient in the gas pressure prevents the wind from becoming supersonic. A wind without heat input but with momentum input can only be transonic if the outward directed force decreases less rapidly than r^{-2} through the critical point.

The velocity in the subsonic region and the mass loss rate depend strongly on the heat or momentum input below the critical point, but does not depend on the input above the critical point. The critical point occurs where the Mach number $M \equiv v/\sqrt{\gamma}a=1$ and where the energy is $e(r_c) = \{-f + 2q(r_c)/3\}r_c$. The addition of heat or momentum above the critical point increases the velocity of the wind.

Polytropic winds with $T \propto \rho^{\Gamma-1}$ without extra forces can only be transonic if $\Gamma > 5/3$, which corresponds to heat input. The momentum equation has a critical point where $v_c^2 = \Gamma a^2 = v_{\rm esc}^2(r_c)/4$. The wind velocity increases outwards if $\Gamma > 3/2$, it is constant if $\Gamma = 3/2$ and decreases outwards if $5/3 < \Gamma < 3/2$. In all these cases the Mach number increases outwards so the wind is transonic. The mass loss rate is set by the conditions at the critical point and by the density at the bottom of the polytropic region. A simple estimate of the mass loss rate is given by Eq. (4.64).

If the addition of heat or momentum at various distances in the wind results in a momentum equation that is a 'wavy' function of distance, the numerator of the momentum equation will have multiple zero points. The critical point is located where the integral $N(r)$ of this numerator reaches an *absolute* minimum. If the integral has two almost equal zero points, the solution of the momentum equation may jump between the two corresponding solutions. This may lead to variable bi-stable winds.

4.5 Suggested reading

Holzer, T.E. & Axford, W. 1970 'The theory of stellar winds and related flows', *An. Rev. Astr. Ap.* **8**, 31
(The formulation of the wind equations for isothermal winds, winds with heat addition, conduction and galactic winds with the addition of mass)

5 Coronal winds

Coronal winds are stellar winds driven by gas pressure due to a high temperature of the gas. In the case of the sun a coronal temperature of about 2×10^6 K is reached in the outer layers of the solar atmosphere. The solar photosphere, where the visual radiation from the sun is emitted, has a temperature of about 6000 K. Above the photosphere the temperature rises with height to a few times 10^6 K. The temperature rise beyond the photosphere is due to the dissipation of mechanical energy or the reconnection of magnetic fields that originate in the convection zone below the photosphere. Other forces, such as those produced by Alfvén waves, may play a role in the coronal holes which are regions of lower temperatures and higher mass flux. However in this chapter on coronal winds, we will only consider the effects of gas pressure and heat conduction in the production of a stellar wind.

All non-degenerate stars with effective temperatures less than about 6500 K are expected to have a convection zone below their surface, so in principle chromospheres and coronae could exist around all cool stars. However, very luminous cool stars can also have winds driven by other mechanisms such as wave pressure or radiation pressure on dust grains. If these stars have a high mass loss rate, then the heating cannot compete with the cooling of the outflowing gas. Therefore cool stars with high mass loss rates do not have a corona. Only main sequence stars later than about F5V, and stars that have evolved off the main sequence and that have spectral types between about F5IV and about K1III are expected to have coronae and winds that are driven by gas pressure.

Coronal winds have much lower mass loss rates than the winds driven by other mechanisms discussed in this book. The mass loss rate of the sun is only 10^{-14} $M_\odot$ yr^{-1}. Yet coronal winds are included in this book because the sun has a coronal wind and studies of the solar wind have played a crucial role in understanding stellar winds in general.

The solar coronal wind theory was first develloped by Parker (1958). He derived the equations and showed that the high velocity of the solar wind observed near the earth requires a critical transonic solution to the momentum equation. His suggestion that the solar wind follows the critical solution was initially controversial. However, the measurements of the solar wind velocity and density near the earth and the improvement in the theory led to acceptance of the basic idea.

The basic mechanism of coronal winds, as we discuss it in this chapter, is quite similar to that of the isothermal winds or the non-isothermal winds with only gravity and gas pressure, as discussed in Chapters 3 and 4. The main difference is the inclusion of thermal conduction which can transport energy from the high temperature regions to the low temperature regions. So wind models with conduction will have a flatter temperature distribution than models without conduction. In this chapter we discuss the basic properties of coronal winds. For a more extended discussion we refer to the book 'Introduction to the Solar Wind' by Brandt (1970).

5.1 The energy of a coronal wind with heat conduction

In a high temperature gas thermal conduction can transport energy from hotter to cooler regions. The conductive heat flux, in erg cm^{-2} s^{-1}, is

$$F_c = -\kappa_c \frac{dT}{dr} \tag{5.1}$$

where κ_c is the coefficient for thermal conductivity in erg cm^{-1} s^{-1} K^{-1}, which is a function of temperature, $\kappa_c = \kappa_0 T^{5/2}$, with $\kappa_0 = 1.0 \times 10^{-6}$ erg cm^{-1} s^{-1} K$^{-7/2}$ (Allen, 1973, p 50). Notice that κ_c depends on a high power of T, therefore conduction is most effective at high temperatures. The minus sign indicates that the conductive flux is positive if T decreases outwards.

The total amount of heat conducted through a sphere of radius r per second, i.e. the conductive luminosity, is

$$L_c = 4\pi r^2 F_c = -4\pi r^2 \kappa_0 T^{5/2} \frac{dT}{dr} = -\frac{8\pi\kappa_0}{7} r^2 \frac{dT^{7/2}}{dr} \tag{5.2}$$

If $T \sim r^{-2/7}$ then the conductive luminosity is constant. If T decreases outwards faster than $r^{-2/7}$ then the conductive luminosity decreases with distance, whereas L_c increases with distance if T decreases slower than $r^{-2/7}$. (The conductive luminosity can also be negative, of course, if $dT/dr > 0$. This occurs in the transition region between the chromosphere and the corona.)

The total energy flow per second through a sphere of radius r of a stellar wind driven by gas pressure with thermal conduction but without other sources of heat or momentum input is

$$\dot{M}\left\{\frac{v^2}{2}+\frac{\Gamma}{\Gamma-1}\frac{\mathcal{R}T}{\mu}-\frac{GM_*}{r}\right\}-\frac{8\pi\kappa_0}{7}r^2\frac{dT^{7/2}}{dr}=\text{constant} \quad (5.3)$$

This is the Bernouilli equation for a wind with thermal conduction. The factor $\Gamma/(\Gamma-1)$ in the enthalpy term is 5/2 for a wind without additional energy input, i.e. for $\Gamma = 5/3$. In case of polytropic energy input, Γ can be smaller than 5/3 (see § 4.2). The constant of the energy equation can be found by comparing the energy flow at the lower boundary r_0 of the wind to that at $r \to \infty$, where $T \to 0$. If the conduction also goes to zero at large distances, i.e. $dT^{7/2}/dr \to 0$, we find the expression for the terminal velocity of the wind

$$\frac{v_\infty^2}{2}=\frac{v_0^2}{2}+\frac{\Gamma}{\Gamma-1}\frac{\mathcal{R}T_0}{\mu}-\frac{GM_*}{r_0}-\frac{8\pi\kappa_0}{7\dot{M}}\left(r^2\frac{dT^{7/2}}{dr}\right)_{r_0} \quad (5.4)$$

If $\Gamma/(\Gamma-1) = 5/2$ the second right hand term is the enthalpy. In that case the first three right hand terms together are the energy per unit mass, $e(r)$, at the bottom of the wind. This equation states that the kinetic energy per gram at $r \to \infty$ is equal to the initial energy, $e(r_0)$, plus the amount of thermal energy that was conducted into the wind per unit mass at r_0. This is because the conductive luminosity at infinity goes to zero, so the conductive luminosity has been deposited into the gas.

The energy input into the wind by the decrease of the conductive luminosity per unit mass and per unit distance is

$$q(r)=\frac{d\,e(r)}{dr}=+\frac{8\pi\kappa_0}{7\dot{M}}\frac{d}{dr}\left(r^2\frac{dT^{7/2}}{dr}\right) \quad (5.5)$$

This equation shows that $q(r) < 0$, i.e. energy is removed from the gas in the layers where the temperature decreases slower than $r^{-2/7}$, and it is deposited into the gas in the layers where T decreases faster than $r^{-2/7}$. So *conduction can redistribute energy over the stellar wind.*

A stellar wind can only become supersonic if its energy is sufficiently large to escape the potential well of the star. This means that for a coronal wind either the energy must already be positive at the lower boundary of the wind, or the energy at the lower boundary is still negative but the thermal conduction at the lower boundary is sufficienty large to enable the wind to escape.

5.2 The momentum equation

The driving force of a coronal wind is gas pressure. Therefore the momentum equation is the same as Eq. (4.16) with f=0.

$$(v^2 - a^2)\frac{1}{v}\frac{dv}{dr} = \frac{2a^2}{r} - \frac{da^2}{dr} - \frac{GM_*}{r^2} \tag{5.6}$$

This equation has the characteristic form that we have encountered in the two previous chapters.

It is convenient to write the momentum equation and the energy equation in a dimensionless form by normalizing T and v. Parker (1958) introduced the notation

$$\tau = T(r)/T_c \tag{5.7}$$

and

$$\Psi = \frac{v^2\mu}{\mathcal{R}T_c} \tag{5.8}$$

where T_c and v_c are the temperature and the velocity at the critical point r_c of the coronal wind. We also introduce the normalized inverse distance scale†

$$z = \frac{r_c}{r} \tag{5.9}$$

The momentum equation and the energy equation expressed in these nondimensional units are

$$\frac{1}{\Psi}\frac{d\Psi}{dz} = 2\left(A - \frac{2\tau}{z} - \frac{d\tau}{dz}\right) / (\Psi - \tau) \tag{5.10}$$

and

$$\frac{\Psi}{2} + \frac{\Gamma}{\Gamma - 1}\tau - Az + B\frac{d\tau^{7/2}}{dz} = \frac{\Psi(0)}{2} \tag{5.11}$$

with

$$A = \frac{GM_*\mu}{r_c\mathcal{R}T_c} \tag{5.12}$$

and

$$B = \frac{8\pi\kappa_0\mu}{7\dot{M}\mathcal{R}}r_cT_c^{5/2} \tag{5.13}$$

† Parker used the distance scale $\lambda = GM_*\mu/\mathcal{R}T_cr$ instead of z, and $\Psi = v^2\mu/\mathcal{R}T_0$.

The critical point of the momentum equation occurs at $z_c = 1$ where both the numerator and the denominator of Eq. (5.10) vanish. So

$$\Psi(z_c) = \tau(z_c) = 1 \tag{5.14}$$

and

$$\left(\frac{d\tau}{dz}\right)_c = A - 2 \tag{5.15}$$

The value of $d\tau/dz$ at $z_c = 1$ can also be derived from the energy equation (5.11) because $d\tau^{7/2}/dz = (7/2)\tau^{5/2}d\tau/dz$. Applying this at the critical point we find a condition for the wind velocity at $r \to \infty$

$$\Psi(0) = 6 - 2A + 7B(A - 2) \tag{5.16}$$

So the condition for the temperature gradient at the critical point sets the value of $\Psi(0)$ which describes the ratio between the terminal velocity and the critical velocity.

The slope of the velocity law at the critical point $z_c = 1$ is found by applying de l'Hopitals rule to the momentum equation (Appendix 3) which gives the quadratic expression of $d\Psi/dz$

$$\left(\frac{d\Psi}{dz}\right)_c^2 + (2 - A)\left(\frac{d\Psi}{dz}\right)_c + 4A - 12 + 2\left(\frac{d^2\tau}{dz^2}\right)_c = 0 \tag{5.17}$$

The term $(d^2\tau/dz^2)_c$ can be expressed in Ψ by differentiation of the energy equation (5.11) at the critical point. This gives

$$\left(\frac{d^2\tau}{dz^2}\right)_c = -\frac{1}{7B}\left(\frac{d\Psi}{dz}\right)_c - C \tag{5.18}$$

with

$$C = \frac{5}{2}(A - 2)^2 + \left(\frac{2A - 4\Gamma}{7B(\Gamma - 1)}\right) \tag{5.19}$$

This gives a quadratic equation for the velocity gradient at the critical point

$$\left(\frac{d\Psi}{dz}\right)_c^2 + \left(2 - A - \frac{2}{7B}\right)\left(\frac{d\Psi}{dz}\right)_c + (4A - 12 - 2C) = 0 \tag{5.20}$$

This equation has zero, one or two solutions, depending on the values of A and B. If it has two solutions the critical point will have the topology of a 'tilted X' of the type $(d\Psi/dz)_c = a \pm b$. In analogy to the case of the isothermal wind driven by gas pressure, we adopt the solution that goes through the singularity with increasing velocity: $(dv/dr)_c > 0$ or $(d\Psi/dz)_c < 0$. This sets the value of $(d\Psi/dz)_c$.

5.3 The calculation of coronal wind models

The stucture of a coronal wind can be calculated by means of a numerical solution of the two equations (5.10) and (5.11), starting at the critical point where $z = 1$, $\Psi = 1$, $\tau = 1$, $d\tau/dz = A - 2$ and $d\Psi/dz$ is given by Eq. (5.20). The parameters A, B and $\Psi(0)$ should be adjusted so that the solution is transonic and fits the boundary conditions at the lower and upper boundary of the corona.

Notice two important differences between the momentum and energy equations for a wind with conduction and without conduction.

(a) In the isothermal models and the polytropic models that we discussed in the previous chapters, the density did not enter into these equations, because both the energy input and the momentum were simply proportional to the density. So the density dropped out of the equations. As a consequence, the solution of the momentum and energy equations gave the velocity law and the temperature, but not the density. The density was a free scaling factor determined by the density at the lower boundary.

(b) In the models with conduction, the mass loss rate enters into the equations because the heat transport and heat deposit per unit volume is independent of the density. This means that the energy deposition *per unit mass* is inversely proportional to the density. So the energy equation depends on ρ^{-1}. We see this in the factor B, which is proportional to $\dot{M}^{-1}$, in the energy equation (5.11).

Because of this effect the numerical solution of the equations is more complicated, as it depends on two parameters, $\tau(z_0) \sim T_0$ and $\Psi(z_0) \sim v_0^2$, which have to be adjusted to the boundary conditions. The mass loss rate then follows from v_0 with the lower boundary values r_0 and ρ_0.

5.4 A model for the solar corona

The equations for a coronal model with heat conduction were first solved numerically in this form by Noble and Scarfe (1963). They searched for a transonic solution with conduction but without heat input ($\Gamma = 5/3$) that satisfied the condition of the observed solar wind at the distance of the earth, where the wind velocity is 350 km s^{-1}, the temparature is 2.7 10^5 K, and the density is 6.75 particles cm^{-3} (atoms and electrons), with $\mu = 0.62$. This corresponds to a mass loss rate of 1.1 10^{-14} $M_\odot$ yr^{-1}. The result is shown in Fig. (5.1) in terms of $\tau(z)$ and $\Psi(z)$. The critical point where $z = 1$, $\tau = 1$ and $\Psi = 1$ is indicated. We see that τ increases with increasing z and Ψ decreases rapidly. The solution corresponds to $A = 2.343$ and $B = 20.92$.

The scaling of the solution to the temperature and velocity at the outer boundary r=1 AU = 215 $R_\odot$ requires that the the critical point is located at 5.23 $R_\odot$. The temperature and velocity at r_c are $T_c = 1.16\ 10^6$ K and $v_c = 125$ km s^{-1}. The inner boundary of this coronal model is at 1.175 $R_\odot$, so $z_0 = 4.451$. The temperature at the inner boundary is 2.0 10^6 K so $\tau_0 = 1.72$ and the density at the inner boundary is 1.1 10^7 particles cm^{-3} or 1.1 10^{-17} g cm^{-3}.

The temperature and velocity, derived from $\tau(z)$ and $\Psi(z)$ with the scaling to the critical point, is shown in Fig. (5.2). The temperature decreases outwards and the velocity increases outwards.

In the wind model discussed here we have assumed that the temperature structure is the same for electrons and protons. However, since coronal winds typically have small mass loss rates and low densities, the interaction between these two types of particles may be insufficient for equipartition of energy. We can derive a criterion for the equipartition of energy. The typical timescale for the energy exchange between protons and electrons is

$$\tau_{\rm exch} \simeq 10^1 \frac{T_e^{3/2}}{n_e}, \tag{5.21}$$

in s, when T_e is in K and n_e is in cm^{-3} (Brandt, 1970). The typical

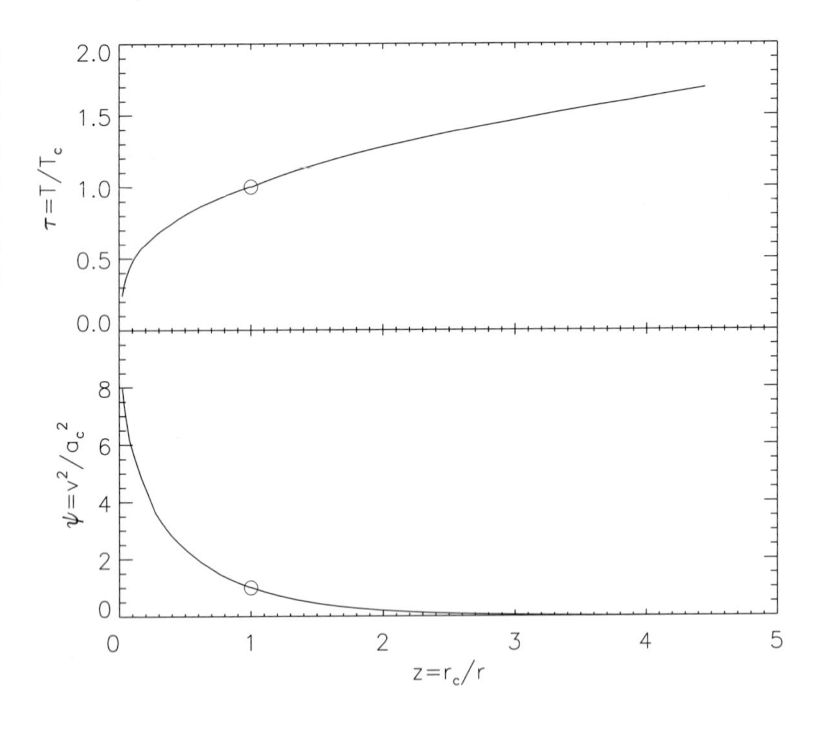

Figure 5.1 The solution of the momentum energy equations for the solar corona with conduction. (Adapted from Noble and Scarfe, 1963.)

timescale for the expansion of the wind is

$$\tau_{\rm exp} \equiv \frac{-\rho}{d\rho/dt} = -v\left(\frac{d\ \ln\rho}{dr}\right)^{-1} \tag{5.22}$$

For a mean velocity of $\langle v \rangle$ the density varies as $\rho \sim r^{-2}$, hence $\tau_{\rm exp} \simeq 0.5r/\langle v \rangle$. If the expansion timescale is shorter than $\tau_{\rm exch}$, there is not enough time for the exchange of energy between the protons and the electrons and the proton temperature will be different from the electron temperature.

Models for the solar corona with different temperature structures for protons and electrons (two fluid models) have been calculated by Hartle and Sturrock (1968). They found that the electron temperature of the solar wind at the distance of the earth is more than a factor ten higher than the proton temperature.

5.5 Mass loss rates and terminal velocities of coronal winds

Mass loss by means of coronal winds is due to the gas pressure at high temperature. Therefore we can expect that the mass loss rate will be of the same order of magnitude as the mass loss rate of an isothermal or a polytropic stellar wind with the same lower boundary conditions.

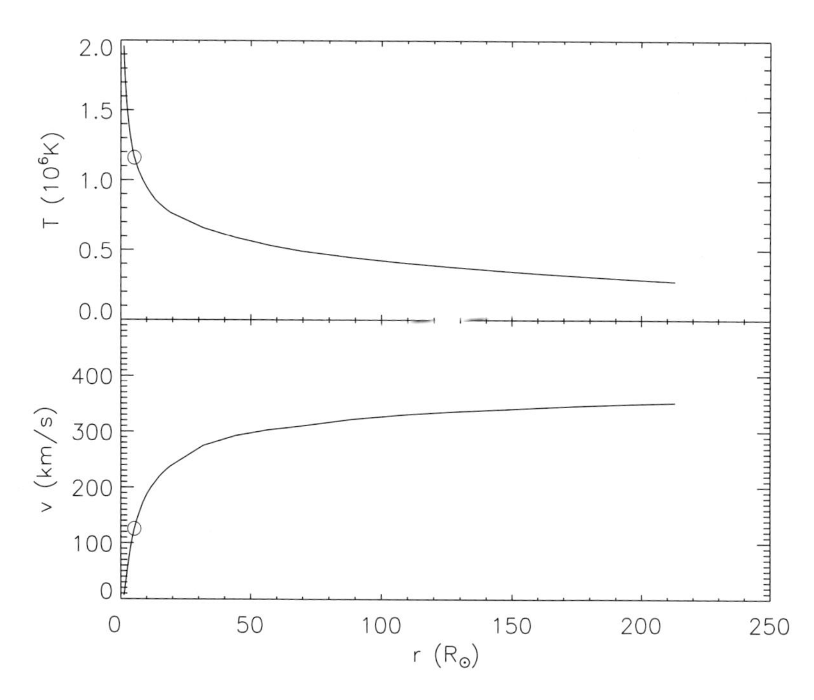

Figure 5.2 The temperature and velocity structure of the solar coronal model of Noble and Scarfe (1963).

Table 5.1 *Comparison between coronal and polytropic wind models*

Γ	$r_c/R_\odot$	T_c K	v_c km s^{-1}	v_∞ km s^{-1}	$\dot{M}$ $M_\odot$yr^{-1}
1	3.55	2.00 10^6	164	∞	9.6 10^{-14}
1.025	3.91	1.77 10^6	156	2741	2.0 10^{-14}
1.050	4.39	1.54 10^6	147	1794	1.7 10^{-14}
1.075	5.06	1.30 10^6	137	1335	1.3 10^{-14}
1.100	6.08	1.06 10^6	125	1032	8.9 10^{-15}
1.125	7.79	0.81 10^6	110	798	5.3 10^{-15}
1.150	11.27	0.55 10^6	92	590	2.3 10^{-15}
1.175	22.19	0.27 10^6	66	380	4.0 10^{-16}
Coronal model	5.23	1.16 10^6	125	>350	1.07 10^{-14}

We can test this by comparing the structure of the solar coronal wind model discussed above with that of isothermal or polytropic winds for different values of Γ, but with the same temperature of 2×10^6 K and density of 1.1×10^{-17} g cm^{-3} at the lower boundary of r_0=1.175 $R_\odot$ as the coronal model. This comparison is shown in Table (5.1).

We see that the coronal model is quite similar to a polytropic model of $\Gamma \simeq 1.08$ in terms of the conditions at the critical point and the mass loss rate. This indicates that the heat input by conduction below the critical point is very similar in the two models. The major difference between the two models is in the terminal velocity, which is much higher in the polytropic model than observed in the solar wind at the distance of the earth. This is due to fact that the polytropic model has energy input at all distances, even far beyond the critical point, whereas the energy input in the coronal model beyond the critical point is very small.

5.6 Conclusions

We have shown that a coronal wind that starts at a high temperature (typically a few times 10^6 K) can be transonic because heat is conducted away from the hottest layers at the bottom of the corona and deposited at the cooler layers. The deposition or removal of energy by conduction depends on the second derivative of $T(r)$ (Eq. 5.5). Energy is deposited in the layers where the temperature decreases faster than $r^{-2/7}$ and removed from the layers where it decreases slower than $r^{-2/7}$.

A coronal wind is driven by gas pressure at high temperature, similar to that of isothermal winds and polytropic winds discussed in the previous two chapters. Therefore the structure of coronal winds and the mass loss rates are quite similar to that of isothermal or polytropic winds. A comparison between the solar wind model and polytropic wind models in Table (5.1) confirms this. The mass loss rate of the coronal wind is an order of magnitude smaller than that of an isothermal wind with the same lower boundary conditions. This is because an isothermal wind model requires a much higher energy input. The mass loss rate of the solar wind agrees very well with that of a polytropic wind of $\Gamma = 1.08$, i.e. with a slowly outward decreasing temperature in the subsonic region.

The terminal velocity of coronal winds is much smaller than predicted for isothermal or polytropic wind because the energy input in the supersonic region quickly goes to zero in the solar wind, whereas it remains finite in a polytropic model. One can estimate the terminal velocity of coronal winds by calculating the velocity at the critical point of the polytropic model and multiplying it by about a factor three.

The calculation of coronal models for stars requires knowledge of the heating of the chromosphere and the corona. The mechanisms that have been proposed include dissipation of sound waves, which are generated in the convection zone below the photosphere, and of hydromagnetic waves. Stellar coronal models heated by the dissipation of sound waves have been calculated, e.g. by Hammer (1982). These models depend on the stellar parameters, the energy flux of the acoustic waves at the top of the photosphere and the dissipation length, which is the characteristic distance over which the wave energy is dissipated into heat.

5.7 Suggested reading

Brandt, J.C. 1970 *Introduction to the Solar Wind* (San Francisco: Freeman)
(This book describes the basic solar wind theory and extensions)

Parker, E.N. 1971 'Present developments in theory of the solar wind' in *Solar Wind*, NASA-SP **308**, p. 161
(A review of the solar wind theory and its problems)

Cram, L.E. 1985 'Investigations of solar and stellar atmospheric heating' in *Relations between chromospheric-coronal heating and mass loss in stars*, eds. R. Stalio & J. Zirker (Trieste: Osservatorio Astronomico) p. 93
(A historical review of the search for coronal heating mechanisms)

6 Sound wave driven winds

In stars with a convection zone just below the photosphere, the convective motions might create acoustic waves which propagate outwards through the photosphere. These sound-waves produce an extra pressure, i.e. 'wave pressure' in the atmosphere. This pressure will depend on the density and on the amplitudes of the waves. The gradient of the wave pressure results in a force that can drive a stellar wind. If a stellar wind is driven by acoustic wave pressure it is called a 'sound wave driven wind'.

In this chapter we will first explain the concept of wave pressure by studying the motion of a particle in the presence of an oscillating force. This simple case, first developed by Landau and Lifshitz (1959) shows that oscillations may result in a net force in the direction of the oscillations. In § 6.1 we discuss the motions of particles in an oscillatory field, such as in a sound wave, and we show that this produces a 'wave pressure'. In § 6.2 we introduce the concepts of the 'wave action density' and the 'acoustic wave luminosity'. These are useful concepts for describing sound wave driven winds. The pressure due to acoustic waves is described in § 6.3. Section (6.4) descibes sound wave driven wind assuming no dissipation of acoustic energy. This results in estimates of the both the mass loss rate and the wind velocity. In § 6.5 we discuss sound wave driven winds with dissipation of the acoustic energy. In the last section we review the properties of sound wave driven winds.

6.1 A particle in an oscillatory field

Consider the one dimensional motion of a particle of mass m due to a steady force $f(r)$, where r is the coordinate along the trajectory of the particle. This motion is described by Newton's law

$$md^2r/dt^2 = f(r) = -m\, d\varphi(r)/dr \tag{6.1}$$

where $\varphi(r)$ is the potential associated with the force $f(r)$. Suppose that the particle is also subjected to an additional, rapid oscillatory force, $g(r)\cos\omega t$, with an oscillation period $\tau = 2\pi/\omega$ that is small compared to the typical timescale of the motion due to the force $f(r)$. We assume that the amplitude of the oscillatory force, $g(r)$, depends only on r. The equation of motion is

$$md^2r/dt^2 = f(r) + g(r)\cos\omega t \tag{6.2}$$

The particle will move with a motion that consists of a smooth component corresponding to the outward motion of the fluid, and an oscillatory component of frequency $\omega/2\pi$, associated with the wave. Therefore we can express $r(t)$ as

$$r(t) = \bar{r}(t) + \delta(r,t) \tag{6.3}$$

with

$$\delta(r,t) = \xi(r)\cos\omega t \tag{6.4}$$

Taking the time average over one period, the second term on the right side of Eq. (6.3) vanishes and so $\bar{r}(t)$ represents the period-averaged motion. We assume that the amplitude ξ of the oscillations around the time-averaged motion is small compared to the distance traveled during one period: $\xi(r) \ll (d\bar{r}(t)/dt)2\pi/\omega$. This will be the case if $g(r) \ll f(r)$. We can then expand the equation of motion in terms which are linear in $\delta(r)$. The equation of motion is

$$md^2\bar{r}(t)/dt^2 + md^2\delta/dt^2 = f(\bar{r}) + \delta(\bar{r})\frac{df}{d\bar{r}} + g(\bar{r})\cos\omega t + \delta(r)\frac{dg}{d\bar{r}}\cos\omega t \tag{6.5}$$

Substitution of expression (6.4) for $\delta(r)$ into (6.5) gives

$$md^2\bar{r}/dt^2 - m\omega^2\xi(\bar{r})\cos\omega t = f(\bar{r}) + \left\{\xi(\bar{r})\frac{df}{d\bar{r}} + g(\bar{r})\right\}\cos\omega t + \xi(\bar{r})\left(\frac{dg}{d\bar{r}}\right)\cos^2\omega t \tag{6.6}$$

We can solve this equation to derive the mean motion of $\bar{r}(t)$ by taking the time-average of each term over a period $2\pi/\omega$. The terms containing $\cos\omega t$ will vanish since the time-average $\langle\cos\omega t\rangle = 0$, but the last term will not vanish since $\langle\cos^2\omega t\rangle = 1/2$. The mean motion is thus described by

$$md^2\bar{r}/dt^2 = f(\bar{r}) + \frac{1}{2}\xi(\bar{r})\frac{dg}{d\bar{r}} \tag{6.7}$$

A comparison with Eq. (6.1) shows that the oscillatory force introduced an extra term in the equation of motion, that depends on the amplitude of the oscillations and the gradient of the amplitude of the oscillatory force. We can find the amplitude $\xi(\bar{r})$ of the oscillations by equating the terms containing the factor $\cos \omega t$ in Eq. (6.6)

$$-m\omega^2 \xi(\bar{r}) \cos \omega t = \left\{ \xi(\bar{r}) \frac{df}{d\bar{r}} + g(\bar{r}) \right\} \cos \omega t \tag{6.8}$$

For small amplitude $\xi(\bar{r})$ or a small gradient df/dr, we can ignore the first right hand term. In that case the amplitude of the oscillatory motions is

$$\xi(\bar{r}) = -g(\bar{r})/m\omega^2 \tag{6.9}$$

The minus sign indicates that the oscillatory motion of the particle is out of phase with the oscillating force by half a period. Equation (6.9) shows that the local amplitude of the motion is proportional to the local amplitude of the oscillatory force: if the amplitude of the force is constant, the amplitude of the motions is constant. We can now substitute Eq. (6.9) into Eq. (6.7) to find the equation for the mean motion of the particle

$$\begin{aligned} md^2\bar{r}/dt^2 &= f(\bar{r}) - (m\omega^2/2) \cdot \xi(\bar{r}) d\xi/dr \\ &= -\frac{d}{dr} \left\{ m\varphi(\bar{r}) + \frac{m\omega^2}{4} \xi^2(\bar{r}) \right\} \end{aligned} \tag{6.10}$$

The oscillatory force has added the second term to the potential that describes the mean motion. This second term is simply the mean kinetic energy of the oscillations, as can be seen from

$$\frac{m}{2} \langle v_{\text{osc}}^2 \rangle = \frac{m}{2} \langle \left(\frac{d\delta}{dt} \right)^2 \rangle = \frac{m}{2} \langle (-\xi \omega \sin \omega t)^2 \rangle = \frac{m}{4} \omega^2 \xi^2 \tag{6.11}$$

where $\langle\rangle$ indicates the average over an oscillation period. So the equation for the mean motion of the particle is

$$md^2(\bar{r})/dt^2 = -\frac{d}{dr} \left\{ m\varphi(r) + \frac{m}{2} \langle v_{\text{osc}}^2 \rangle \right\} \tag{6.12}$$

This equation shows that a particle can be accelerated by an oscillatory force if the kinetic energy of the oscillations decreases with distance. Multiplying both sides of the equation by dr/dt and then integrating over t, we can express Eq. (6.12) in terms of energy conservation

$$\frac{1}{2} mv^2 + \frac{1}{2} m \langle v_{\text{osc}}^2 \rangle + m\varphi(r) = \text{ constant} \tag{6.13}$$

This implies that the mean kinetic energy of the oscillations can be transferred to the kinetic energy of the flow if the amplitude of the forced oscillations decreases with distance.

This simple case of a particle moving by a longitudinal oscillating force, i.e. a force acting along the path, shows the basic concept of 'wave force'. Let us look at the effects which produce this force. Take the case of an oscillating force whose amplitude, $g(\bar{r})$ in Eq. (6.2), decreases with distance r. Due to this force the particle oscillates longitudinally with an amplitude proportional to $g(r)$ but in opposite phase [Eq. (6.9)]. If the particle is in the $+r$ segment of its oscillation around its mean motion ($\delta > 0$ in Eq. 6.3), there will be a restoring force that is negative, i.e. in the $-r$ direction, and vice-versa. However, since the amplitude of the oscillatory force decreases with distance r, the negative force in the phases when $\delta > 0$ is smaller in the absolute value than at the other half of the oscillation when $\delta < 0$. As a result, there will be a net positive force (i.e. in the $+r$ direction), that arises because the oscillations have amplitudes that decrease with r. Similarly, an oscillatory force with a increasing amplitude would produce a negative net force.

Since the force depends on the gradient of the amplitude of the oscillatory force, it is convenient to express the effect in terms of a *'wave pressure'* with the net force depending on the gradient of the wave pressure. If both sides of Eq. (6.12) are multiplied by the number density $n(r)$ of the particles, then we see that the wave pressure is equal to the kinetic energy density of the oscillations

$$P_w = \frac{1}{2}\rho\langle v_{\rm osc}^2\rangle \tag{6.14}$$

6.2 The acoustic wave luminosity

The example of a particle in an oscillatory field, discussed above, has shown that oscillations can produce pressure gradients which result in motion. In this section we will discuss the effects of waves on the medium through which they propagate. We will discuss the energy density of waves and we will define a quantity called '*acoustic wave luminosity*' which is useful for describing the dynamics of wave driven stellar winds.

Consider acoustic waves where the density variation can be described as $\delta\rho\exp(i\omega t - \mathbf{k}\cdot\mathbf{r})$ where $\delta\rho$ is the density amplitude, $\omega/2\pi$ is the frequency of the waves and $\mathbf{k}$ is the wave vector with $|\mathbf{k}| = 2\pi/\lambda$, which describes the direction of propagation of the waves. The wavelength is λ. If $\mathbf{k}$ and $\mathbf{r}$ are parallel, the wave is traveling in the $+r$ direction.

We have used the mass continuity equation, $\dot{M} = 4\pi r^2 \rho v$, to express ρ in terms of the mass loss rate. We see that $L_{\rm ac}$ is of the order of the kinetic energy loss of the wind if $v \simeq a$ and $\delta v \simeq a$. If there is no dissipation of acoustic waves then $L_{\rm ac}$ is constant.†

The definition of $L_{\rm ac}$ by Eq. (6.19) is for monochromatic waves. If there is a spectrum of waves the acoustic luminosity is

$$L_{\rm ac} = 4\pi r^2 v_g \frac{\int \{F(\omega)\omega E(\omega)/(\omega - kv)\}\, d\omega}{\int F(\omega) d\omega} \tag{6.21}$$

where $F(\omega)$ describes the distribution of the energies at the bottom of the wind over frequencies. In the case of a monochromatic spectrum, $F(\omega)$ is a delta-function and Eq. (6.21) reduces to (6.19).

6.3 The pressure due to acoustic waves

Since the pressure produced by acoustic waves is not isotopic, the pressure has to be described as a tensor (Landau and Lifschitz, 1959). For monochromatic adiabatic sound waves, which travel radially outwards the pressure tensor is

$$P_w = \begin{pmatrix} P_{rr} & 0 & 0 \\ 0 & P_{\theta\theta} & 0 \\ 0 & 0 & P_{\phi\phi} \end{pmatrix} = E \begin{pmatrix} a_1 & 0 & 0 \\ 0 & a_2 & 0 \\ 0 & 0 & a_3 \end{pmatrix} \tag{6.22}$$

(Bretherton, 1970) where r, ϕ, and θ are the polar coordinates and $a_1 = (\gamma + 1)/2$ and $a_2 = a_3 = (\gamma - 1)/2$. In these expressions $\gamma = c_p/c_v$ is the ratio of the specific heats which is 5/3 for an ideal gas. For isothermal sound waves the diagonal elements can be found by setting $\gamma = 1$, so $a_1 = 1$, and $a_2 = a_3 = 0$.

The pressure gradient due to monochromatic radial waves is the radial component of the divergence of the stress tensor

$$(\nabla \cdot P_w)_r = \frac{dP_{rr}}{dr} + \frac{2}{r}\left\{P_{rr} - \frac{1}{2}\left(P_{\theta\theta} + P_{\phi\phi}\right)\right\} \tag{6.23}$$

† If the definition of $L(\omega)$ is modified by replacing v_g with the Alfvén speed and $E(\omega)$ is the energy density of the Alfvén waves, then $L(\omega)$ is constant for Alfvén waves in a stellar wind. (See Chapter 10)

which is

$$\frac{dP_w}{dr} = (\nabla \cdot P_w)_r = \frac{(\gamma + 1)}{2}\frac{dE}{dr} + \frac{2E}{r} \tag{6.24}$$

for acoustic waves. For isothermal waves $\gamma = 1$.

The Eqs. (6.23) to (6.24) are only valid for sinusoidal waves of low amplitude, i.e. δv less than the speed of sound, and with wavelengths smaller than the pressure or density scale height of the gas.

6.4 Sound wave driven winds without dissipation

To explain the basic physical aspects of sound wave driven winds, we will discuss the models of isothermal sound wave driven winds in which there is no dissipation of acoustic energy. Such a wind is driven by the wave pressure of the sound waves and by the gas pressure.

6.4.1 The energy equation

The momentum equation of a stellar wind driven by wave pressure from acoustic waves is

$$v\frac{dv}{dr} + \frac{1}{\rho}\frac{dP_g}{dr} + \frac{1}{\rho}\frac{dP_w}{dr} + \frac{GM_*}{r^2} = 0 \tag{6.25}$$

which contains separate terms for the gradients of the gas pressure and the wave pressure. For an isothermal wind model without dissipation we can derive an analytic expression that shows the effects of wave pressure.

The gradient of the wave pressure for outward traveling waves is given by Eq. (6.24), with $\gamma = 1$ for isothermal waves. The energy E can be expressed in terms of the acoustic wave luminosity, L_{ac} given by Eq. (6.21), which is constant through the wind if there is no dissipation. Using these two equations one finds that the force due to the wave pressure can be written as

$$\frac{1}{\rho}\frac{dP_w}{dr} = \frac{1}{\rho}\left(\frac{dE}{dr} + \frac{2E}{r}\right) = \frac{1}{\rho r^2}\frac{d}{dr}\left(r^2 E\right) = \frac{L_{ac}}{4\pi r^2 \rho}\frac{d}{dr}\left\{\frac{a}{(v+a)^2}\right\} \tag{6.26}$$

Substitution of the mass continuity equation $4\pi r^2 \rho = \dot{M}/v$ and introduction of the Mach number $M = v/a$ gives

$$\frac{1}{\rho}\frac{dP_w}{dr} = \frac{L_{ac}}{\dot{M}}M\frac{d}{dr}\left\{(M+1)^{-2}\right\} = \frac{L_{ac}}{\dot{M}}\frac{d}{dr}\left\{\frac{(2M+1)}{(M+1)^2}\right\} \tag{6.27}$$

This is a perfect differential, since L_{ac} and $\dot{M}$ are constant through the wind. The other three terms of Eq. (6.25) can also be written as differentials of $v^2/2$, $a^2 \ln \rho$, and $-GM_*/r$ respectively. So the momentum equation reduces to

$$\frac{d}{dr}\left\{\frac{v^2}{2} + a^2 \ln \rho + \frac{L_{ac}}{\dot{M}}\frac{(2M+1)}{(M+1)^2} - \frac{GM_*}{r}\right\} = 0 \tag{6.28}$$

or

$$\frac{v^2}{2} + a^2 \ln \rho + \frac{L_{ac}}{\dot{M}}\frac{(2M+1)}{(M+1)^2} - \frac{GM_*}{r} = \text{constant} \tag{6.29}$$

This is the energy equation for an isothermal wind driven by the pressure of sound waves.

In this case we cannot derive the value of the constant by applying the energy equation to the outer boundary, $r = \infty$, because the terminal velocity of the isothermal wind will go to infinity, as in the case of isothermal winds driven by gas pressure only, discussed in Chapter 3. Therefore we derive the constant by applying this equation at the lower boundary of the wind. This gives, after division by a^2,

$$\begin{aligned}&\frac{M^2 - M_0^2}{2} + 2 \ln\left(\frac{r_0}{r}\right) + \ln\left(\frac{M_0}{M}\right) - \frac{GM_*}{a^2 r_0}\left(\frac{r_0}{r} - 1\right)\\ &+ \frac{L_{ac}}{a^2 \dot{M}}\left\{\frac{2M+1}{(M+1)^2} - \frac{2M_0+1}{(M_0+1)^2}\right\} = 0\end{aligned} \tag{6.30}$$

where we have used the mass continuity equation to express ρ in v and r. This is the Bernouilli equation for an isothermal wave driven wind without dissipation of acoustic energy.

In models where the flow velocity is very small at the bottom of the wind, we may set $M_0 \approx 0$ and Eq. (6.30) reduces to

$$\begin{aligned}&\frac{M^2}{2} - \ln\left(\frac{M}{M_0}\right) - 2 \ln\left(\frac{r}{r_0}\right)\\ &-\frac{L_{ac}}{a^2 \dot{M}}\frac{M^2}{(M+1)^2} - \frac{GM_*}{a^2 r_0}\left(\frac{r_0}{r} - 1\right) \simeq 0\end{aligned} \tag{6.31}$$

The first term is the kinetic energy gained by the gas in units of a^2. This term is positive. The second and third terms together represent the work per gram due to the gradient of the gas pressure. They are both negative. The fourth term is the work per gram due to the gradient of the wave pressure. This term is also negative. The last term is the increase of the potential energy of the gas. This term is positive. So we see that the wind has gained kinetic and potential energy due to the work done by the gas pressure and the wave pressure.

At the bottom of the wind the total energy input due to the waves, i.e. the energy density of the waves plus the capacity to do work is about $L_{ac}/\dot{M}$ per gram. When the flow is accelerated to a Mach number M, the remaining energy is $(L_{ac}/\dot{M})(2M+1)/(M+1)^2$. So the difference between the originally available wave energy and the actual wave energy is the amount of wave energy that is transferred into wind due to the work done by the wave pressure.

6.4.2 The momentum equation and the critical point

Let us now consider the momentum equation (6.25) and derive the conditions for the critical point. We can express each term in Mach numbers. The first term is simply $a^2 M dM/dr$. For the second term we use the equation of continuity and find

$$\frac{1}{\rho}\frac{dP_g}{dr} = -\frac{a^2}{M}\frac{dM}{dr} - \frac{2a^2}{r} \tag{6.32}$$

(see also Eq. 3.6). The third term is given in Eq. (6.27). Writing the differential we find

$$\frac{1}{\rho}\frac{dP_w}{dr} = -\frac{L_{ac}}{\dot{M}}\frac{2M}{(M+1)^3}\frac{dM}{dr} \tag{6.33}$$

Substitution of these expressions in the momentum equation (6.25) and dividing by a^2 gives

$$\left\{M - \frac{1}{M} - \frac{L_{ac}}{\dot{M}a^2}\frac{2M}{(M+1)^3}\right\}\frac{dM}{dr} = \frac{2}{r} - \frac{GM_*}{a^2r^2} \tag{6.34}$$

This can be written in the familiar form (see § 3.4.1)

$$\frac{1}{M}\frac{dM}{dr} = \left\{\frac{2}{r} - \frac{GM_*}{a^2r^2}\right\} \Big/ \left\{M^2 - 1 - \frac{2L_{ac}}{\dot{M}a^2}\frac{M^2}{(M+1)^3}\right\} \tag{6.35}$$

This expression shows that the velocity law of a wind driven by acoustic waves without dissipation has a critical point. The Mach number increases through the critical point if both the numerator and the denominator vanish. The numerator is zero at the critical point so

$$r_c = \frac{GM_*}{2a^2} \tag{6.36}$$

This means that the critical point is at the same location as in a wind without wave pressure. The denominator is also zero at the critical point, so

$$M_c^2 - \frac{2L_{ac}}{\dot{M}a^2}\frac{M_c^2}{(M_c+1)^3} = 1 \tag{6.37}$$

or

$$\frac{(M_c+1)^4(M_c-1)}{M_c^2} = \frac{2L_{ac}}{\dot{M}a^2} = \frac{A_0}{M_0} \tag{6.38}$$

with

$$A_0 \equiv \frac{L_{\rm ac}}{2\pi r_0^2 \rho_0 a^3} = \left(\frac{v_0 + a}{a}\right)^2 \left(\frac{\delta\ v}{a}\right)_0^2 \tag{6.39}$$

We have expressed $\dot{M} = 4\pi r_0^2 \rho_0 v_0$ and $L_{\rm ac}$ in terms of its definition (6.20) for monochromatic sound waves. If the flow velocity v_0 at the lower boundary is much smaller than a, then A_0 is simply the square of the Mach number of the amplitude of the waves: $A_0 \simeq (\delta\ v/a)_0^2$. So A_0 depends only on the lower boundary conditions.

Equation (6.38) shows that $M_c \rightarrow 1$ in the limit of very small acoustic wave luminosity as expected, because in that case the critical point condition is the same as for an isothermal wind without sound waves.

6.4.3 The mass loss rate of sound wave driven winds

The mass loss rate can be derived from the density and the radius at the lower boundary of the wind, if the velocity v_0 or the Mach number M_0 is known.

Equation (6.38) describes the relation between M_0 and M_c, derived from the condition at the critical point. The upper part of Fig. (6.2) shows this relation for several values of $A_0 \simeq (\delta\ v/a)_0^2$.

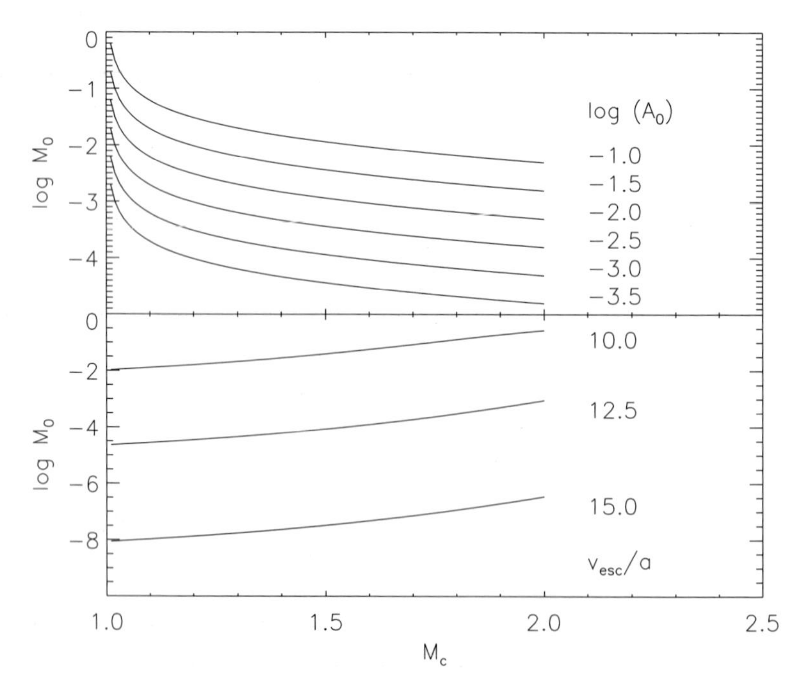

Figure 6.2 The relation between the Mach number at the critical point and at the lower boundary as a function of the constant A_0 and of $v_{\rm esc}/a$. The upper curves show the relation derived from the critical point condition. The lower curves show the relation derived from the energy equation.

We can also derive a relation between M_c and M_0 from the energy equation (6.31) by applying it to the critical point and substituting $GM_*/r_0 = v_{esc}^2/2$. This gives

$$(M_c^2 - M_0^2) - 2 \ln\left(\frac{M_c}{M_0}\right) + f(M_c, M_0) = -B_0 \qquad (6.40)$$

with

$$f(M_c, M_0) \equiv \frac{(M_c - 1)(M_c + 1)^4}{M_c^2} \left\{ \frac{-M_c^2}{(M_c + 1)^2} + \frac{M_0^2}{(M_0 + 1)^2} \right\} \qquad (6.41)$$

and the constant B_0 depending only on the lower boundary conditions, in particular on (v_{esc}/a),

$$B_0 \equiv + \left(\frac{v_{esc}}{2a}\right)^2 - 8 \ln\left(\frac{v_{esc}}{2a}\right) - 4 \qquad (6.42)$$

The constant is positive for a strongly gravitationally bound atmosphere of $v_{esc}/a > 7.69$ and negative for a loosely bound atmosphere of $1.28 < v_{esc}/a < 7.69$.

Equation (6.40) describes the relation between M_c and M_0 derived from the energy equation. The lower part of Fig. (6.2) shows relation (6.40) for several values of (v_{esc}/a). If the initial wind velocity is highly subsonic, so $M_0 \ll 1$, then $f(M_c, M_0) \simeq -(M_c - 1)(M_c + 1)^2$. In that case we can approximate

$$M_0 \simeq M_c e^{-(1+B_0+M_c^3-M_c)/2} \qquad (6.43)$$

This equation is equivalent to the expression (3.23) for isothermal winds without waves, where $M_c = 1$. For any set of lower boundary conditions given by A_0 and B_0 we can derive two sets of M_c versus M_0 relations. The crossing point of these two relations gives the value of M_0 from which the mass loss rate can be derived with the mass continuity equation.

We have calculated the initial velocities of isothermal wave driven winds, for a range of model parameters by simultaneously solving the two equations (6.38) and (6.40) for the initial velocity v_0 or the initial Mach number M_0. The two parameters are: (1) the amplitude of the sound waves in Mach number at the lower boundary, $(\delta v/a)_0$, and (2) the ratio v_{esc}/a. The first parameter determines the acoustic wave luminosity that goes into the wind at the lower boundary and thus the wave pressure or the amount of work that can be done by the waves. The second parameter determines how strongly the lower boundary of the wind is bound to the star by gravity.

The result is shown in Fig. (6.3) where we have plotted the initial Mach number M_0 as a function of $(\delta v/a)_0$ for different values of v_{esc}/a.

The full line refers to wind models for which the assumption of no dissipation is valid. The dotted line shows the relation in the range of parameters for which this assumption is not justified (see below). The figure shows that for any value of v_{esc}/a, the value of M_0 decreases very rapidly with decreasing $(\delta v/a)_0$, in fact $M_0 \sim (\delta v/a)_0^2$, until it reaches some lower limit and then remains constant. This lower limit is reached when $M_c \rightarrow 1$ because in that case the velocity at the critical point is exactly the same as for a wind without wave pressure and so M_0 approaches the value given by Eq. (3.23). This value depends very critically on v_{esc}/a.

The mass loss rate is $4\pi r_0^2 \rho_0 a M_0$. The data in Fig. (6.3) show that the mass loss rates of cold wave driven winds ($v_{esc}/a > 10$) will be very small unless the initial amplitude of the sound waves is substantial, $(\delta v/a)_0 > 0.1$. In that case, however, the assumption of no dissipation is no longer valid.

6.4.4 No dissipation?

In the calculations of M_0 and the mass loss rates above we have assumed that there is no dissipation of acoustic waves. This assumption is valid for sound waves with an amplitude less than about half the sound speed, so $(\delta v/a) < 0.5$. If the amplitude of the waves is larger, part of the energy of the waves will be transferred into thermal energy. This will result in an outward decrease of the acoustic wave luminosity.

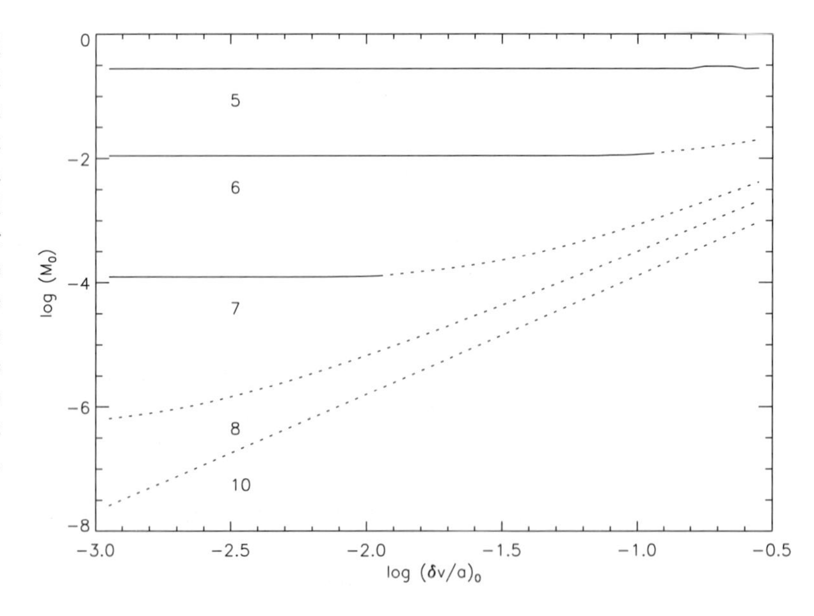

Figure 6.3 The initial velocity of isothermal sound wave driven winds without dissipation, as a function of the initial wave amplitude, for various values of the parameter v_{esc}/a. The dotted line refers to models for which the assumption of no dissipation is not valid.

If dissipation occurs the assumption of constant L_{ac} that we have used above is no longer valid.

We can easily derive the condition for which the assumption of no dissipation is justified. A constant value of L_{ac} (Eq. 6.20) in an isothermal wind implies that the wave amplitude increases outwards as

$$\frac{\delta v}{a} = \left(\frac{2L_{ac}}{\dot{M}a^2}\right)^{0.5} \left(\frac{v}{a}\right)^{0.5} \frac{a}{v+a} \sim \frac{\sqrt{v/a}}{(v+a)} \tag{6.44}$$

This shows that in the subsonic part of the wind, where $v < a$ but where v increases almost exponentially, the amplitude of the waves will rapidly increase outwards, roughly as $\sqrt{v} \simeq 1/\sqrt{\rho r^2}$. We can compare the wave amplitude at the lower boundary with that at the critical point

$$\left(\frac{\delta v}{a}\right)_c = \left(\frac{\delta v}{a}\right)_0 \sqrt{\frac{M_c}{M_0}} \left(\frac{M_0+1}{M_c+1}\right) \simeq \left(\frac{\delta v}{a}\right)_0 \sqrt{\frac{4}{M_0}} \tag{6.45}$$

For the latter approximation we have assumed that $M_0 \ll 1$ and $M_c \simeq 1$.

The assumption of no dissipation is valid in the subcritical region of the wind if $(\delta v/a)_c < 0.5$, which implies that $(\delta v/a)_0 < \sqrt{M_0}$. This is shown in Fig. (6.3) by the full lines. We see that the full lines coincide approximately with the region where the value of M_0 approaches that of an isothermal wind without waves. This shows that *the influence of the sound waves without dissipation on the mass loss rates of isothermal winds is only marginal.*

6.5 Sound wave driven winds with dissipation

If the velocity amplitudes of sound waves becomes larger than about half the sound speed, microscopic effects will result in a dissipation of the acoustic energy into heat. The details of this dissipation have been described in the studies of the acoustic heating of the solar chromosphere. In general, one finds that the dissipation is a steep function of the velocity amplitude of the waves.

In the subsonic part of the wind the amplitude will increase approximately as $(\delta v/a) \sim \sqrt{v/a^3}$, Eq. (6.44). So when v increases or a decreases outwards the Mach number of the amplitude rises rapidly. We have shown in chapters 3 and 4 that the density structure of the subsonic part of the wind is approximately hydrostatic, so the flow velocity increases exponentially outwards with the pressure scale height $\mathcal{H}$ as $v \sim e^{+(r-r_0)/\mathcal{H}}$. This implies that the amplitude of the sound waves is expected to rise very rapidly in the subsonic region and dissipation is likely to occur. Dissipation keeps the Mach number limited to a value close to one, because the dissipation increases very rapidly for higher $M > 1$. So, since dissipation will result in an almost constant Mach number ($M \simeq 1$) of the wave amplitude, one can expect that dissipation will decrease L_{ac} exponentially with approximately the pressure scale height in the subsonic region.

The dissipation of the acoustic waves has two major effects on the dynamical and thermal structure of the wave driven wind. First, as the acoustic wave luminosity decreases outwards, the wave pressure will decrease outward faster than if there were no dissipation. This means that the force produced by the gradient of the wave pressure will be affected. Second, the dissipation of acoustic energy into thermal energy provides heat-input into the wind.

Let us first consider the change in the wave pressure force due to the dissipation. The force, f_w, depends on the gradient of the wave pressure, but the wave pressure itself is proportional to the acoustic wave luminosity so $f_w = -(1/\rho)(dP_w/dr) \sim -(1/\rho)\, d(L_{ac}/dr)$. In the lower layers where the dissipation occurs and where L_{ac} has not yet diminished considerably, the gradient in the wave pressure and the outward force will be *larger* due to the dissipation. However, further outward where the acoustic wave luminosity has decreased considerably due to the dissipation, the force due to the wave pressure is *smaller* compared to models without dissipation, because L_{ac} is small. So dissipation of acoustic waves has the effect of a larger force in the subsonic region and a smaller force in the supersonic region. We have seen in Chapters 3 and 4 that an increase of the force in the subsonic region increases the mass loss rate.

Now we consider the effect of dissipation on the thermal structure of the wind and the resulting effect on the mass loss rate. Dissipation of wave energy in the lower layers of the wind provides heat input. We have seen in Chapter 4 that heat input in the subsonic part of the wind results in an increase in the mass loss rate.

So both effects of dissipation of wave luminosity in the subsonic part of the wind result in a increase of the mass loss rate compared to a non-dissipative wind. However, the dissipation in the lower layers of the wind also means that there is little or no acoustic wave luminosity left in the higher layers. This means that the force due to the wave pressure will become negligible after a few pressure scale heights. So in that case one has to invoke another mechanism to provide the acceleration of the wind in these higher layers: for instance an isothermal wind with sufficient gas pressure at large distance or radiation pressure on dust grains.

Sound wave driven wind models with dissipation have been calculated by several authors, e.g. Pijpers and Hearn (1989) and Koninx and Pijpers (1992). They conclude that acoustic wave pressure can only play a role in cool stars of very low gravity, such as stars on the asymptotic giant branch (AGB), if the winds are nearly isothermal or if radiation pressure on dust grains provides the acceleration of the supersonic region. Moreover these models require a fine tuning of the input parameters L_{ac} and the dissipation length to explain the observed mass loss rates of cool stars. This implies that the acoustic wave pressure is probably not the dominant driving mechanism for the winds of cool stars.

6.6 Conclusions

The convection zones of cool stars generate acoustic waves in their photospheres. These waves propagate outwards and carry wave energy into the wind. Sound waves produce a wave pressure in the atmospheres of cool stars and the gradient of this wave pressure results in an outward directed force which can drive a stellar wind. The crucial quantity for sound wave driven winds is the 'acoustic wave luminosity' defined in § 6.2. This quantity describes the rate at which energy flows with the group velocity into the wind.

The momentum equation of isothermal wave driven winds has a critical point. The requirement that the wind velocity should increase through the critical point determines the velocity at the lower boundary. This initial velocity together with the density at the lower boundary determines the mass loss rate of wave driven winds. The velocity at the lower boundary is plotted in Fig. (6.3) for a grid of models. For small values of $(\delta v/a)_0$ the initial velocity and hence the mass loss rates are similar to those of isothermal winds driven without wave pressure. For higher initial wave amplitudes the assumption of no dissipation is no longer valid. This shows that sound waves have a negligible effect on the mass loss rates of stars, unless dissipation is taken into account.

Sound waves only effect the mass loss rates of very low gravity stars if there is strong dissipation in the subsonic region. In that case the dissipation affects both the thermal structure of the winds and the force that drives the wind. Models with a large acoustic wave luminosity and strong dissipation in the subsonic region might be able to accelerate the wind in the lower wind layers, however these models fail to accelerate the wind beyond the critical point. To solve this problem one has to invoke another mechanism that can lift the material above the critical point out of the potential well. This could be done by radiation pressure on dust grains that forms in the winds of cool stars with high mass loss rates. Therefore sound wave driven wind models are hybrid models with two different mechanisms co-operating in the subcritical and the supercritical region.

Model calculations in the literature suggest that sound wave driven winds with radiation pressure on dust might generate substantial mass loss from cool stars with a very low gravity, such as stars on the asymptotic giant branch. However, to explain the observed mass loss rates and wind velocities of AGB stars requires a fine tuning of the input parameters which makes these models implausible. Moreover the existence of a correlation between the mass loss rates and the pulsation periods of AGB stars (shown in Fig. 2.26) suggests that pulsation is more important than acoustic waves for driving the winds of AGB stars.

6.7 Suggested reading

Pijpers, F.P. & Hearn, A.G. 1989, 'A model for a stellar wind driven by linear acoustic waves', *A & A* **209**, 198
(The theory of sound wave driven winds)

Koninx, J.P.M. & Pijpers, F.P. 1992 'The applicability of the theory of linearized sound wave driven winds', *A & A* **165**, 183
(About the limits of the wave driven wind theory)

7 Dust driven winds

The outer atmospheres of luminous cool giant stars and early-type stars can be driven outward by the strong radiation fields from the stellar photospheres. In the case of the cool stars, radiative driving occurs because of absorption of photons by dust grains that can form in the outer atmospheres. The grains can absorb radiation over a broad range of wavelengths, so the outflows of the cool stars are said to be 'continuum driven' winds. In the case of hot early-type stars the winds are driven by the scattering of radiation by line opacity, so their outflows are called 'line driven' winds.

The essential difference between continuum driven and line driven winds is the role of the Doppler shift between a parcel of outflowing matter and the photosphere. In the acceleration of a stellar wind to terminal velocity, the stellar light incident on the parcel of the wind is increasingly redshifted up to the final value of $\Delta\lambda = \lambda v_\infty/c$. For a cool star with a continuum driven wind, this redshift corresponds to a few Å, which is such a narrow band that within it neither the continuum opacity nor the incident radiation field changes significantly. So the Doppler shifting is not important in continuum driven winds. In the case of line driven winds both the line opacity and the radiation field in the lines change significantly over the Doppler shifts associated with the winds. In general, the velocity field in a line driven wind increases more steeply with radius. In spite of these differences, both continuum and line driving can produce large mass loss rates, of order $10^{-5}\ M_\odot\ \mathrm{yr}^{-1}$. However, the terminal speeds are quite different. Dust driven winds are slow with speeds of about 10 to 30 $\mathrm{km\,s^{-1}}$, while line driven winds usually have speeds larger than $10^3\ \mathrm{km\,s^{-1}}$. Part of this difference can be traced to the differences in the escape speeds of the stars involved. The cool star wind speeds are comparable to the escape speed at several stellar radii above the surface, where the transsonic point occurs, while the hot stars achieve a speed above the escape speeds (of ≈ 800 km s^{-1}), within a few tenths of a stellar radius. Line driven winds will be discussed in the next chapter.

Some major features of continuum driven winds have been introduced in Chapter 3. There we defined Γ to be the ratio of radiative acceleration to gravity, (Eq. 3.36). In the case of radiation forces on dust, this ratio is

$$\Gamma_d = \frac{k_{rp} L_*}{4\pi c G M_*} \tag{7.1}$$

where k_{rp} is the 'radiation pressure mean opacity' which will be defined in § 7.3.2. We found in Chapter 3 that:

- Winds can occur if Γ_d increases outwards and becomes greater than unity, beyond some radius.
- The sonic point occurs just prior to the radial distance at which Γ_d becomes greater than unity.
- The closer the transonic point is to the surface of the star, the larger the mass loss rate.
- The greater the momentum addition in the supersonic portion of the flow, the faster the terminal velocity.

Dust driving of winds occurs in the very restricted portion of the HR diagram that contains the cool, luminous red supergiants and asymptotic giant branch (AGB) stars. The effective temperatures of such stars range from 2000 to 3000 K. The luminosities are $\geq 10^5$ $L_\odot$ for the supergiants, and $\geq 10^4$ $L_\odot$ for the AGB stars. The low stellar temperatures allow grain formation and grain growth to occur in the upper atmospheres of these stars. The large luminosity to mass ratios allow the outward acceleration to exceed gravity as is required for radiation driven winds.

Section (7.1) provides an overview of the physical processes that must occur in dust driven winds, and we consider whether the name 'dust driven' is appropriate since stellar pulsation often plays a crucial role in the mass loss. In § 7.2, we derive some fundamental limits on continuum driven winds, from a general consideration of the momentum transfer requirements. The specific properties of dust opacities that play a role in winds are described in § 7.3. The absorptive properties of grains are used in § 7.4 to derive temperatures of 'test grains' and the radial distance in the wind at which radiative driving can begin. Coupled gas and dust momentum equations are derived in § 7.5. In § 7.6, we discuss the importance of stellar pulsation in determining the mass loss rate that can occur from the star. To illustrate the basic wind driving process we discuss a single grain size model in § 7.7, and derive the velocity distribution in terms of Γ_d. We find in § 7.8 that the magnitude of Γ_d depends on the mass loss rate because of (a) the drift effect, and (b) the reddening effect. The drift effect allows us to make an estimate of the minimal mass loss rate for a dust driven wind, while the reddening effect leads to a maximal mass loss rate. The conclusions are in § 7.9.

7.1 The physical processes of dust driven winds

7.1.1 The necessary conditions for driving winds with dust

The special conditions needed for a dust driven wind can be derived by considering a test grain in the outer atmosphere of the star. We assume that a test grain is present at some arbitrary radius in the outer atmosphere, and we determine whether or not it can survive against the tendency for a hot grain to sublimate. If it can, we then determine whether the grain is sufficiently coupled to the gas to drive out a dust/gas wind mixture.

The survivability of a test grain is most easily determined by deriving its radiative equilibrium temperature. If this exceeds the 'condensation temperature' then the grain will be so hot that it will sublimate rather than grow. The radiative equilibrium temperature of a grain depends on the balance between the radiative heating by light from the star, and the radiative cooling by the thermal emission of the grain itself. The temperature derived by this balance is determined by the grain's opacity, its distance from the star, and the effective temperature of the star. Similar properties determine the theoretical temperatures of planets in the solar system. As with the inner planets, test grains placed close to the star, where the radiative heating rate is rapid, tend to be very hot. As a result, there is some minimal distance at which the test grain can be cooler than the condensation temperature, independent of the density of the ambient gas. An additional requirement, that the grain temperature be low at a radius where the density is sufficient for grain growth, means that only stars in the coolest section of the HR diagram can have dust driven winds.

When solids condense from the atmospheric gas, the opacity blocking the flow of radiation increases by several orders of magnitude. The absorption of the stellar photon has two main effects. The grain acquires the energy of the photon, $h\nu$, and thus the grain heats up. The grain also receives the momentum of the photon, $h\nu/c$. Since the photons are from the star the incident momentum is approximately radial, while the re-emission of the photon energy is approximately isotropic, the absorption/re-emission process gives rise to a net acceleration of the grain in the outward direction.

Another essential feature of the dust driven wind process is the coupling of the dust grains to the gas. As a grain gains momentum from the absorption or scattering of photons and is driven outward, it will collide with gas molecules and produce a drag force on them. Gilman (1972) was the first to fully explain the 'momentum coupling' of the grains with the gas molecules. The driving of the outflow of the gas is produced by the drift of the grains through the gas. The process is like the motion of a helium balloon; as it rises in the air, collisions transmit some of its upward momentum to the gas. Another useful analogy is the drag force on a parachute falling through the atmosphere, although in this case the direction of the drift is inverted, i.e. a grain drifts outward relative to the gas.

The momentum coupling between the grains and the gas sets a lower limit on the mass loss rate that can be driven. If the mass loss rate were (somehow) steadily decreased, the coupling between the grains and gas would also decrease until the rate of momentum transfer to the gas was insufficient to lift the gaseous material out of the star's potential well. The minimal mass loss rate that can be driven by dust is of order $10^{-7}\ M_\odot\ \mathrm{yr}^{-1}$. The grain/gas coupling can also set a limit to the speed of a dust driven wind. Kwok (1975) argued in one of the earliest papers on dust driven winds, that there is a maximal drift speed of a grain through the gas. If the drift speed is too large the collisions between the grains and gas particles will be energetic enough to destroy the grains. Since drift speeds tend to increase inversely with the square root of the gas density, there is a radius beyond which the collisions exceed this limiting energy, so there will be no further increase in the wind speed.

The transition to supersonic flow does not usually occur very close to the star. Instead, models typically have a dust free extended photospheric region which lies below the dust condensation radius. The dust grains give rise to a large increase in the opacity, and this rather abruptly leads to an acceleration from subsonic to supersonic speeds. The properties of dust opacity determine both where this transition occurs and the radiative acceleration that subsequently results. So we must have a basic understanding of dust grain opacities to be able to derive properties of the dust driven winds.

7.1.2 The types of grains that drive the winds

The most difficult aspect of the dust driven wind theory concerns the processes of grain formation and growth. Once a cluster of about ten atoms or so forms, any condensible molecule that collides with the cluster is likely to accrete onto the grain (Gail and Sedlmayr, 1987). A distribution in grain sizes will arise from the competing processes of growth by accretion and erosion by 'sputtering', which is the loss of material caused by collisions with the ambient gas. Although grain nucleation and growth are crucial for the occurrence of dust driven winds, these processes lie beyond our goal of understanding the basics of wind driving. In this section we will discuss only the properties of grains needed to determine their temperatures, formation radii and radiative accelerations.

The grains are generally thought to have a range of sizes with an average radius of about 0.05 μm to 0.1 μm. The composition of the grains depends on that of the gas phase material. Gilman (1969) proposed a simple picture to explain infrared features seen in cool stars with dust envelopes. The spectra indicate that dust grains usually belong to one of two types, as determined by the carbon to oxygen number ratio, C/O. The 'silicate grains' are formed around *oxygen-rich* stars, for which $C/O < 1$, whereas the 'carbonaceous grains' are formed around *carbon-rich* stars, for which $C/O > 1$.

For the oxygen-rich stars the grains are thought to have a composition such as Mg_2SiO_4 or $MgSiO_3$ (Dorschner and Henning, 1995). For the carbon-rich stars the grains tend to be amorphous (poorly structured) carbon, along with other abundant compounds that can condense into the solid phase such as SiC, MgS, or Fe. The dichotomy between silicate and carbonaceous winds arises because in the gas phase carbon and oxygen combine to form the very stable molecule CO. This leaves only the remainder of the more abundant element to take part in the nucleation and the grain accretion process.

Since the cosmic abundance of oxygen is greater than that of carbon, stars with nearly normal atmospheric abundances, such as the M supergiants, have silicate grains in their winds. Carbon rich atmospheres can be produced by a dredge-up process that results from thermal pulses occuring while the stars are on the asymptotic giant branch (Iben & Renzini, 1983; and Chapter 13). The carbon rich dust driven winds occur from stars that had initial masses of about 1.5 to 2.0 $M_\odot$. Although this is a rather restricted class of stellar objects, Jura & Kleinmann (1990) have estimated that the C stars are responsible for perhaps as much as half of the stellar material that is lost by cool stars into the interstellar medium.

7.1.3 The role of stellar pulsation in the origination of the wind

While the temperature and radiation field determine whether grains form, and at what distance from the star, it is the density at the dust formation radius that determines the mass loss rate of the wind. In regards to the existence of the wind, the density determines whether there is sufficient momentum coupling between the grains and the gas. The density of the gas decreases with increasing distance from the star. If the value of the density is so low at the condensation radius that the grain is driven through the gas with too few collisions, no wind can be generated. There must be a sufficient *drag* by the grains in order that the gas experiences a net outward acceleration. Observations indicate that dust driven winds have large mass loss rates. The need for a sufficiently high density to provide the large mass loss rates has led to the conclusion that most dust driven winds are 'hybrid' winds. The stars can have the observed large mass loss rates only if there is a mechanism to increase drastically the atmospheric density scale heights relative to those of gas pressure supported hydrostatic atmospheres. Most cool stars seem to achieve an extended atmosphere structure through the outward push provided by shocks that originate in the pulsations of the underlying star. Examples are the Mira variables, OH/IR stars and related types of long period variables. So it is not surprising that the mass loss rate of pulsating stars depends on the pulsation period (see § 2.8.2).

Given the importance of pulsation in extending the atmosphere, it is perhaps questionable whether the mass loss should be called 'dust driven'. Most of the energy required to drive the matter to escape speed is provided by the pulsational levitation. Bowen (1988) addressed this issue of the importance of grains by calculating models with and without grains. The presence of the grains increased the mass loss rates by several orders of magnitude. The dust acceleration thus plays a crucial role in actually driving the matter from the star, and so 'dust-driven' is an appropriate name for the winds for many luminous and cool stars.

The mass loss rate of a dust driven wind is determined by the conditions in the subsonic region, similar to the wind models we have discussed in previous chapters. This implies that $\dot{M}$ is very sensitive to the scale height in the subsonic region, and this is strongly affected by the levitation produced by stellar pulsation. In this chapter, however, we concentrate on the effects of dust on the wind, rather than on the effects of pulsation. Therefore, we will treat the mass loss as a free parameter in most of this chapter. We return to the effects of pulsation on $\dot{M}$ in § 7.6. The use of $\dot{M}$ as an independent parameter has been defended from an observational point of view by Habing *et al.* (1994), who note that stars with nearly the same mass, luminosity, and T_{eff}, can have quite different winds. Also, the M supergiants and AGB stars can have about the same effective temperature, but have very different luminosities, yet have about the same mass loss rate.

With the mass loss rate taken to be a free parameter, there is little need to discuss the subsonic portion of the wind. Instead we assume that the stellar pulsation provides an extended atmosphere with a significant density at the dust condensation radius. The combination of this radius, the local density, and the sound speed determine the mass loss rate by way of the mass conservation equation.

Before considering the details of grain properties and dust driven winds, we discuss the more general problem of continuum driven winds, and the properties of these winds that can be derived from momentum constraints.

7.2 The momentum of a continuum driven wind

7.2.1 The single scattering limit

The momentum of a radiation driven wind originates in the momentum of the stellar photons incident on the wind. As each photon has a momentum of $h\nu/c$, the total momentum of the light that leaves the photosphere per second is L_*/c. The efficiency of the coupling is measured by the ratio of the final momentum per second of the wind $\dot{M}v_\infty$ to the photospheric radiation momentum:

$$\eta_{\text{mom}} = \frac{\dot{M}v_\infty}{L_*/c} \tag{7.2}$$

It has been thought for many years that the maximal value of η_{mom} is near unity, so a plausible upper limit on the wind momentum results from equating the wind and radiation momentum fluxes. This equality yields the 'single scattering maximum mass loss rate',

$$\dot{M}_{\text{max},1} = \frac{L_*}{v_\infty c} \tag{7.3}$$

Assuming a typical luminosity of an asymptotic giant branch star of about $3 \times 10^4\ L_\odot$, and a terminal wind speed of $v_\infty = 30\ \text{km s}^{-1}$, in Eq. (7.3), we get a mass loss rate of $2 \times 10^{-5}\ M_\odot\ \text{yr}^{-1}$. This is within the range of values derived from observations of asymptotic giant branch stars, so the equation provides a useful first estimate of the mass loss rates.

If we define the 'wind kinetic energy luminosity' as $L_w = \frac{1}{2}\dot{M}v_\infty{}^2$ (erg s^{-1}), then Eq. (7.3) shows that even this 'maximum' mass loss rate corresponds to a wind for which L_w is a small fraction of the radiative luminosity; $L_w = \frac{1}{2}(v_\infty/c)L_*$. For this reason the temperatures in winds are often derived assuming that the winds are in radiative equilibrium, which corresponds to assuming that the only significant flux of energy is in the radiation field.

The phrase 'single scattering', used in regards to Eq. (7.3), simply means that each photon has interacted with wind material just once on its way from the star to infinity. However, it is not necessary that it be a scattering interaction. Both isotropic scattering and true absorption lead to the same transfer of momentum to the particle interacting with the radiation field (see problem 7.1).

The energy associated with the photons that are scattered or absorbed is not lost. The photons which are scattered will be only slightly redshifted as they are reflected to a different direction. These photons can be scattered again and deposit further momentum in the wind. The photons which are absorbed, heat the dust, and the energy is re-emitted as thermal radiation. The reprocessed photons can also interact with the wind material and deposit more momentum. The multiple interactions of the radiation with the grains require us to consider the momentum transfer process in more detail than we did in deriving Eq. (7.3).

7.2.2 General momentum considerations

Let us assume that the dust grains are fully formed at a certain distance, r_d, from the star and that the dust opacity increases over a short distance, such that $\Gamma_{\rm d}$ is a constant larger than unity from that radius outwards. We have shown in § 3.2.2 that such a wind will reach its sonic point in the region of the rapid increase in $\Gamma_{\rm d}$. So we can assume that the condensation radius, the dust formation radius and the sonic radius are all about equal, $r_c \simeq r_d \simeq r_s$.

The momentum equation of a dust driven wind is

$$v\frac{dv}{dr} + \frac{1}{\rho}\frac{dp}{dr} + \frac{GM_*}{r^2} = g_{\rm rad} = \frac{GM_*}{r^2}\Gamma_{\rm d} \tag{7.4}$$

see Eq. (3.33).

We can derive the efficiency of the radiation field in ejecting mass from a star, by integrating Eq. (7.4) over the mass of the outflowing gas ($dm = 4\pi r^2 \rho dr$) from the photosphere, at R_*, through the sonic radius, $r_s = r_d$, to infinity.

Integrating the momentum equation over dm gives

$$\int_{R_*}^{\infty} 4\pi r^2 \rho\, v \frac{dv}{dr} dr + \int_{R_*}^{r_s} \left[\frac{1}{\rho}\frac{dp}{dr} + \frac{GM_*}{r^2} \right] dm$$
$$+ \int_{r_s}^{\infty} \frac{1}{\rho}\frac{dp}{dr} dm + \int_{r_s}^{\infty} \frac{GM_*}{r^2}(1-\Gamma_{\rm d})\,\rho\, 4\pi r^2 dr = 0 \tag{7.5}$$

The first integral gives the rate at which momentum is ejected from the star, $\dot{M}v_\infty$, because we assume that the velocity at the photosphere is negligible compared to v_∞; the second integral has a negligible value because the integrand is the hydrostatic equilibrium balance condition that holds very well in the subsonic region of the wind; the third integral is also negligible because beyond the sonic point the gas pressure gradient is very small compared to the radiation pressure gradient. So taking only the first and last term of Eq. (7.5) as important, gives

$$\int_{R_*}^{\infty} \dot{M}(dv/dr)dr = \dot{M}v_\infty = 4\pi GM_*(\Gamma_{\rm d}-1)\int_{r_s}^{\infty} \rho dr \tag{7.6}$$

We define the optical depth of the wind as

$$\tau_W = \int_{r_s}^{\infty} k_{\rm rp}\rho\, dr \tag{7.7}$$

which is the optical depth of the supersonic wind. So Eq. (7.6) reduces to

$$\dot{M}v_\infty = \frac{L_*}{c}\left(\frac{\Gamma_{\rm d}-1}{\Gamma_{\rm d}}\right)\tau_W \tag{7.8}$$

This shows that the final wind momentum is determined not only by the single scattering limit, L_*/c, but also by $\Gamma_{\rm d}$ and the optical depth of the supersonic wind. It is often the case for dust driven winds that radiative acceleration greatly exceeds gravity in the supersonic portion of the flow; $\Gamma_{\rm d} \gg 1$. Accounting for this leads to the simpler momentum flux expression

$$\dot{M}v_\infty = \frac{L_*}{c}\tau_W \tag{7.9}$$

This states that the wind momentum flux is given by the single scattering limit, Eq. (7.3), multiplied by the wind optical depth.

7.2.3 An upper limit on the momentum transfer

Equation (7.9) for the wind momentum appears to pose a problem because the wind momentum is bounded only by the wind optical depth, which could in principle be arbitrarily large. However, if we consider the energy of the wind, we will find that τ_W is restricted, and so is $\dot{M}v_\infty$. At maximum, all of the radiative *energy* of the star is transferred to the wind, so that no more of the star's radiation penetrates through the wind to the observer. With this extreme limit we have

$$\tfrac{1}{2}\dot{M}{v_\infty}^2 < L_* \tag{7.10}$$

This implies that

$$\dot{M} < \frac{2L_*}{{v_\infty}^2} \quad \text{and} \quad \dot{M}v_\infty < \frac{2L_*}{v_\infty} \tag{7.11}$$

Combining this with Eq. (7.9), we find that

$$\tau_W < \frac{2\ c}{v_\infty} \tag{7.12}$$

We know from observations that this limit on τ_W is never reached for dust driven winds, so Eq. (7.12) is a severe upper limit. With a more detailed consideration of the energy transfer, Ivezic and Elitzur (1995) have derived a limit of $\tau_W < c/v_\infty$, (see also problem 7.2). This results in an upper limit for τ_W of

$$\tau_W \ < \ \frac{c}{v_\infty} = \sqrt{\frac{\frac{1}{2}\dot{M}\ c^2}{L_*}} = 27\sqrt{\frac{\dot{M}/10^{-5}}{L_*/10^5}} \tag{7.13}$$

where v_∞ was eliminated using Eq. (7.10), and where $\dot{M}$ is in $M_\odot$ yr^{-1}, and L_* is in $L_\odot$.

From Eq. (7.13), we conclude that the wind momentum flux, Eq. (7.9), can be proportional to τ_W to values significantly larger than unity. Therefore, continuum driven winds can have mass loss rates well above the single scatter limit given in Eq. (7.3).

When we consider the maximal mass loss rates for dust driven winds in § 7.8, we will find that there are other limits on the magnitude of τ_W, which arise because of the restricted range in temperatures over which grains can exist.

7.3 Dust grain opacities

In this section we present some basic definitions regarding the opacity of grains, and we show some results of detailed calculations of mean opacities, and describe some simplifications that are often applicable to dust driven wind models.

In dust driven winds, the transonic point occurs at a radius just beyond the dust condensation region. The dust condensation radius, r_c, can be derived by considering the distance at which the temperature of a test particle is equal to the condensation temperature, T_c, of dust grains. Once the grains can form they block some of the flow of radiation from the star. Thus there are two aspects of grain opacity that are of special interest to us: (a) the absorptive grain opacity, κ_d, which determines the temperature and condensation radius and (b) the radiation pressure mean opacity k_{rp}, which determines the radiative acceleration of the grains.† A significant difference between absorption and radiation pressure mean opacities is that *the scattering opacity does not affect the grain temperature determination but scattering does contribute to the radiative acceleration of grains.*

Let us assume that each of the grains is a sphere of uniform density, ρ_d. Since the atmosphere is either oxygen-rich or carbon-rich, let us also assume that the grains in a given star are all of one composition, either silicate or carbonaceous material. The grains are not all likely to have the same radius, so let us consider initially that there is a size distribution to the grain radii, a. Let $n(a)da$ be the number of grains per cm^3 with radius in the range a to a + da. A well known dust size distribution is that of Mathis, Rumpl & Nordsieck (1977; MRN), which was derived for interstellar grains

$$n(a)da = Ka^{-3.5}da, \tag{7.14}$$

where K is a scale factor and a ranges from $a_{min} = .005\ \mu m$ (= 50 Å) to $a_{max} = .25\mu m$ (=2500 Å).

In the portion of the wind where grain growth is occurring there would be a need to account for changes in the size distribution function versus the distance from the star. Given the MRN range in sizes, the value for K is determined by the composition of the gas and the condition that all of the least abundant element in the grain composition is condensed. For example, for interstellar material with solar abundances, the maximal value of K for silicate material is $K_{Si} = n_H 10^{-15.11} \mu m^{2.5}$, where n_H is the number of hydrogen nuclei per cm^3 (Draine, 1981).

7.3.1 Grain cross sections

The cross section, C [in cm^2], for the interaction of a grain with radiation, is traditionally written as a product of the geometrical cross section πa^2, and the 'efficiency', Q, which depends on the grain radius and the wavelength, λ, of the radiation. If we account for both

† Following Mihalas (1978) we use the symbol κ for pure absorption and k for extinction by absorption plus scattering.

absorption and scattering of light, then the total cross section is

$$
\begin{aligned}
C^{\mathrm{T}} &= C^{\mathrm{A}} + C^{\mathrm{S}} \\
&= \pi \mathrm{a}^2 \left\{ Q^{\mathrm{A}}(\mathrm{a}, \lambda) + Q^{\mathrm{S}}(\mathrm{a}, \lambda) \right\} \\
&= \pi \mathrm{a}^2 Q^{\mathrm{T}}
\end{aligned}
\tag{7.15}
$$

where Q^{A},Q^{S}, and Q^{T} are the absorptive, scattering, and total extinction efficiencies. These can be calculated by using Mie theory as developed for astronomical applications by van de Hulst (1957).

The efficiency factors have simple asymptotic limits for interactions with single grains. If the wavelength of the incident radiation is much larger than the particle size, then the Rayleigh limit applies, and Q^{A} is approximated by a power law in wavelength, $Q^{\mathrm{A}} \propto \mathrm{a} \times \lambda^{-p}$. In general the exponent, p, depends on the particular grain composition, but at sufficiently long wavelength, p approaches 2 for all compositions. From figures shown in the next section we will see that for the wavelength region of most interest to us (λ = 1 to 30 μm), the exponent p is about 1. The absorptive efficiency, Q^{A}, is of special interest to us because at the IR wavelengths, which are most important for absorbing stellar radiation and driving the flow, the scattering efficiency decreases strongly with wavelength, as $Q^{\mathrm{S}} \sim \lambda^{-4}$, and so $Q^{\mathrm{S}} \ll Q^{\mathrm{A}}$ in the IR.

If the wavelength of the incident light is much smaller than the particle radius, the geometrical optics limit applies and Q^{T} approaches two, with Q^{A} and Q^{S} each having values of approximately unity. The short wavelength (or UV) limit for Q is not especially important for the dust driven winds of cool stars. However, the short wavelength opacities are needed for studies involving UV sources with dust in their envelopes, such as novae shells, Wolf-Rayet stars of the carbon rich WC class, and massive protostars.

To find the opacity per gram of gas we must integrate the cross section per grain over the distribution in grain sizes. The mass absorption and scattering opacities, κ_λ and σ_λ (in cm^2 g^{-1}), are related to the grain cross sections and the number density by the expressions:

$$
\kappa_\lambda \rho = \int_{\mathrm{a}_{\mathrm{min}}}^{\mathrm{a}_{\mathrm{max}}} n(\mathrm{a})\ \pi \mathrm{a}^2 Q^{\mathrm{A}}(\mathrm{a}, \lambda) d\mathrm{a}
\tag{7.16}
$$

$$
\sigma_\lambda\ \rho = \int_{\mathrm{a}_{\mathrm{min}}}^{\mathrm{a}_{\mathrm{max}}} n(\mathrm{a})\ \pi \mathrm{a}^2 Q^{\mathrm{S}}(\mathrm{a}, \lambda) d\mathrm{a}
\tag{7.17}
$$

Figure (7.1) shows the absorption and scattering opacities per gram of gas as a function of wavelength for the MRN grain size distribution. In both cases the abundances of the heavy elements are assumed to be solar. If the abundances of the elements in the star exceed the solar value the opacities must be increased in proportion to the abundances.

In Fig. (7.1) one can see the 2200 Å bump associated with graphites in the interstellar extinction. For circumstellar dust the bump is much weaker and is shifted to longer wavelengths and appears to be due to amorphous carbon, (Hecht *et al.* 1984, Martin & Rogers, 1987). In the IR portion of Fig. (7.1b), the silicate bumps at 9.7 and 18 μm are noticeable. There are differences in the appearances of these features in different stars, and the presence of one of these emission features in the IR continuum is often used to determine whether C-rich dust or O-rich dust is the wind driving material (see § 2.5).

7.3.2 Mean opacities

Averages of the grain opacities over wavelength occur in the radiative equilibrium equation, and in acceleration terms in momentum equations. The 'Planck mean absorption efficiency' for a grain of radius a is given by

$$Q_{\rm P}^{\rm A}(\mathrm{a}, T) = \frac{\int_0^\infty Q^{\rm A}(\mathrm{a}, \lambda) B_\lambda(T) d\lambda}{\int_0^\infty B_\lambda(T) d\lambda} \tag{7.18}$$

A plot of $Q^{\rm A}(\mathrm{a}, \lambda)$ versus λ, for a grain of radius $\mathrm{a} = 0.05\ \mu$m, is shown as a function of λ in Fig. (7.2, left) for three types of grains that might occur in dust driven winds: silicate, graphite, or amorphous carbon grains. The Planck mean opacities, $Q_{\rm P}^{\rm A}$, for each of these grains as a function of temperature is shown in Fig. (7.2, right). The bump in the silicate distribution arises from the 9.7 micron absorption feature that is seen in Fig. (7.2, left).

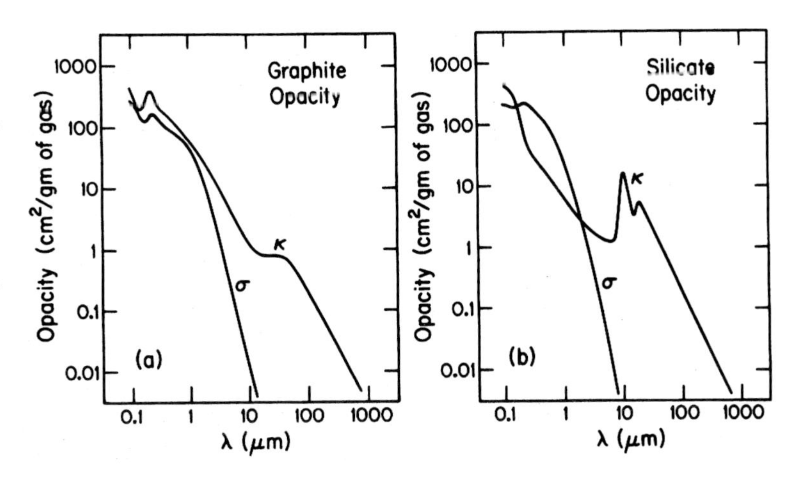

Figure 7.1 The absorptive κ_λ and scattering σ_λ opacity of grains in units of cm^2 g^{-1} as a function of wavelength. Here it is assumed that the grains follow the MRN distribution of sizes. Results are shown for both silicate and graphite grains. (From Wolfire and Cassinelli, 1986)

The Planck mean opacity appears in the radiative equilibrium equation for deriving the temperature of a grain, and for finding the radial distance from the star at which the grains first form. The Planck mean opacity for a mixture of grains is given by

$$\kappa_P \rho = \int_{a_{min}}^{a_{max}} n(a, r)\, \pi a^2 Q_P^A(a, T) da \tag{7.19}$$

Even though the cross section efficiency itself does not depend on T, both Q_P^A and κ_P have a temperature dependence because they involve an integral over the Planck function $B_\lambda(T)$.

The radiation pressure mean opacity, k_{rp}, that is needed in the momentum equation is somewhat more complicated than the Planck mean because it generally involves the scattering of the grains and this scattering can be very non-isotropic. Dust scattering tends to be strongly peaked in the forward direction, i.e. along the direction of the incident radiation. A convenient way to account for the angular re-distribution of the radiation was introduced by Whitworth (1975), who approximated the scattering phase function by an isotropic part plus a forward throwing δ function. For a spherically symmetric wind we can define directions relative to the outward radial direction. Let θ be the angle between the scattered ray and the radius vector from the center of the star, and $\mu = \cos\theta$. Let μ' be the cosine of the angle between the radius and the incident light. Then the Whitworth

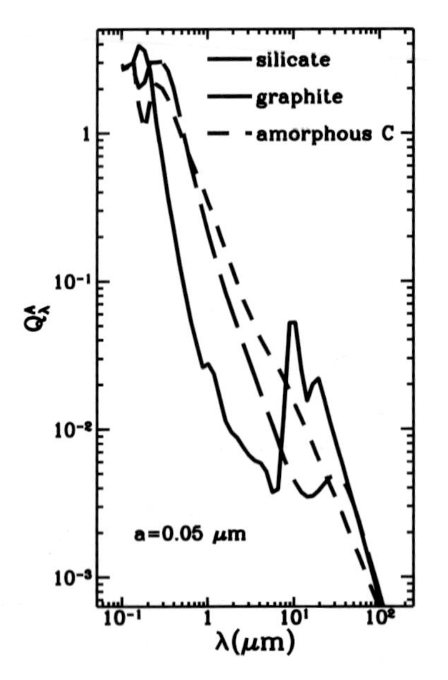

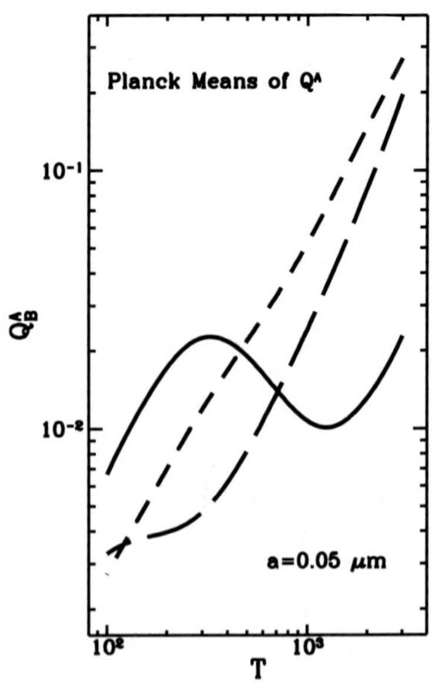

Figure 7.2 On the left is shown the absorptive efficiency as a function of wavelength for a grain or radius a=0.05 μm, for the three types of grains indicated. On the right is shown the Planck mean of the grains as a function of the temperature of the grains. (Courtesy of John Mathis)

scattering phase function is

$$p(\mu, \mu') = (1 - g) + 2g\ \delta(\mu - \mu'), \tag{7.20}$$

This function is normalized so that the integral of $p(\mu, \mu')$ over all values of μ', is unity. In this equation, g is the mean cosine of the scattering angle, and it can be derived from Mie scattering theory for spherical grains. For isotropic scattering $g = 0$, and $p(\mu, \mu') = 1$. For forward scattering, $g = 1$ and $p(\mu, \mu')$ is a delta function in the forward direction; while for complete backward scattering $g = -1$. Figure (7.3) shows the wavelength dependence of g_λ, for both silicates and amorphous carbon grains, which have been averaged over the MRN distribution of grain sizes. Note that the anisotropy of the dust scattering is most severe at ultraviolet wavelengths. The scattering becomes nearly isotropic at infrared wavelengths beyond about 2 μm.

Photons scattered in the forward direction can be considered as not interacting at all with the grain as far as radiative accelerations are concerned. Therefore, for hydrodynamical calculations one should use a 'radiation pressure mean efficiency', $Q_{\rm rp}$, that is defined by:

$$Q_{\rm rp}(a)\ \mathscr{F} = \int_0^\infty \{Q^{\rm A}({\rm a}, \lambda) + (1 - g_\lambda)\, Q^{\rm S}({\rm a}, \lambda)\}\ \mathscr{F}_\lambda d\lambda \tag{7.21}$$

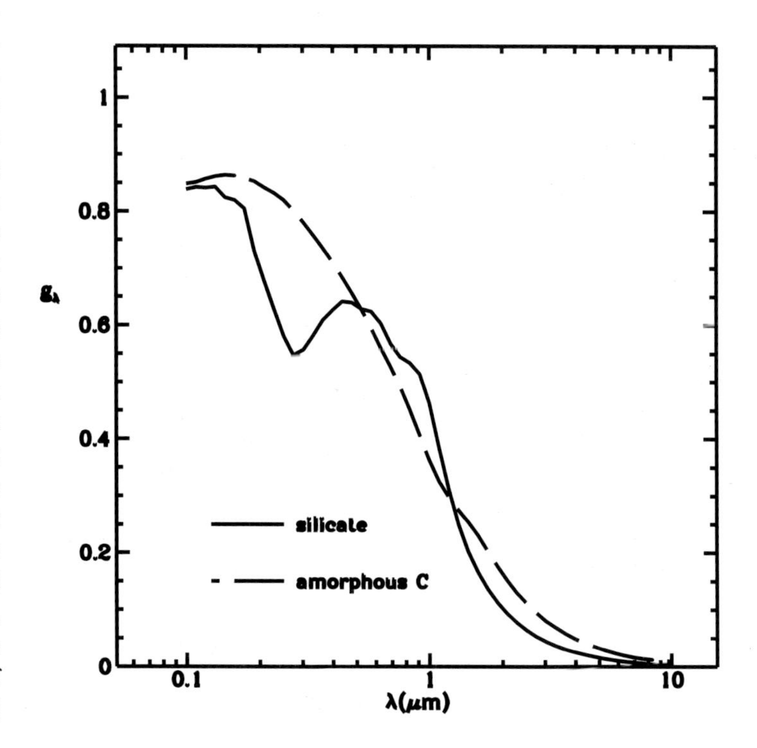

Figure 7.3 The mean cosine of the scattering angle, g_λ, as a function of wavelength for both silicate and amorphous carbon material. These results were calculated assuming the particles have an MRN grain size distribution. At short wavelengths the grain scattering is highly peaked in the forward direction. At long wavelengths scattering follows a Rayleigh phase function, with light scattered equally in the forward and backward directions, thus g_λ approaches zero. (Courtesy of John Mathis)

or the associated 'radiation pressure mean opacity', k_{rp}

$$k_{\mathrm{rp}}\,\mathscr{F} = \int_0^\infty \{\kappa_\lambda + (1 - g_\lambda)\,\sigma_\lambda\,\}\,\mathscr{F}_\lambda d\lambda \tag{7.22}$$

where $\mathscr{F}_\lambda$ is the monochromatic flux, which is related to the monochromatic luminosity at r by

$$\mathscr{F}_\lambda = \frac{L_\lambda}{4\pi r^2} \tag{7.23}$$

Note that for the case of an isotropic scatterer, for which $g_\lambda = 0$, the radiation pressure mean opacity, k_{rp} is the same as the flux mean opacity, k_{F}. For gaseous atmospheres and winds from hot stars which involve electron scattering, one can assume that $k_{\mathrm{rp}} = k_{\mathrm{F}}$, but this is not generally true for dust driven winds.

To find $\mathscr{F}_\lambda$ one must in general solve the radiation transfer equation through the dust shell envelope. However, there are several approximations that are useful. For optically thin winds we can relate $\mathscr{F}_\lambda$ to the intensity at the photosphere of the star, $I_\lambda{}^*$, by the relation $\mathscr{F}_\lambda = \pi I_\lambda{}^*(R_*/r)^2$. If we also assume that the star is radiating as a blackbody at temperature T_*, then $I_\lambda{}^* = B_\lambda(T_*)$, and the radiation pressure mean may be approximated by the blackbody mean, k_{B}, where

$$k_{\mathrm{B}} B = \int_0^\infty [\kappa_\lambda + (1 - g_\lambda)\sigma_\lambda\,]\,B_\lambda d\lambda \tag{7.24}$$

with $B = \sigma T_*{}^4/\pi$ and k_{B} in $\mathrm{cm}^2\ \mathrm{g}^{-1}$.

On the other hand, if the dust shell is very optically thick, we can use as a radiation pressure mean opacity, a generalization of the Rosseland mean opacity, which is a well known transmissivity mean opacity used in stellar interior and gray stellar atmosphere theory (Mihalas, 1978). It is defined in terms of the local temperature via an integral over the gradient of the Planck function, as:

$$\frac{1}{k_{\mathrm{Ros}}}\frac{dB}{dT} = \int_0^\infty \left\{\frac{1}{\kappa_\lambda + (1 - g_\lambda)\sigma_\lambda}\right\}\frac{dB_\lambda}{dT}d\lambda \tag{7.25}$$

As has been the case for the other mean opacities, the Rosseland mean can be expressed in terms of a Rosseland extinction efficiency, Q_{Ros}. For example, for a wind with particles of just one size, say $\mathsf{a} = 0.05$ μm, $k_{\mathrm{Ros}}\rho \equiv n_{\mathrm{d}}\pi\mathsf{a}^2 Q_{\mathrm{Ros}}$, where n_{d} is the number of grains per cm^{-3}. Figure (7.4) shows Q_{Ros} as a function of grain temperature for both graphite and silicate grains, with $\mathsf{a} = 0.05$ μm. The Rosseland mean is of interest only for the range of temperatures that can occur in thick shells, i.e. with values ranging from the sublimation temperature of about 1500 K down to the outer boundary temperature which might be of order 100 K. We will discuss thick winds in § 7.8.

7.4 The temperature of grains and the wind inner boundary

7.4.1 Radiative equilibrium

The temperature of a grain is determined by a balance of the heating rate and the cooling rate. If the energy added per second exceeds that lost per second, the grain heats up until the balance is achieved. Grains could in principle be heated either because of their collisions with fast gas particles or because of the absorption of direct stellar radiation or diffuse ambient radiation. The grains could also be cooled by either collisional energy transfer or by emitting thermal radiation. We will assume that the radiative processes are dominant. In such a case, the balance that determines the temperature of the grain is the 'radiative equilibrium' condition. (Radiative equilibrium is also the energy balance condition that determines the temperature stratification

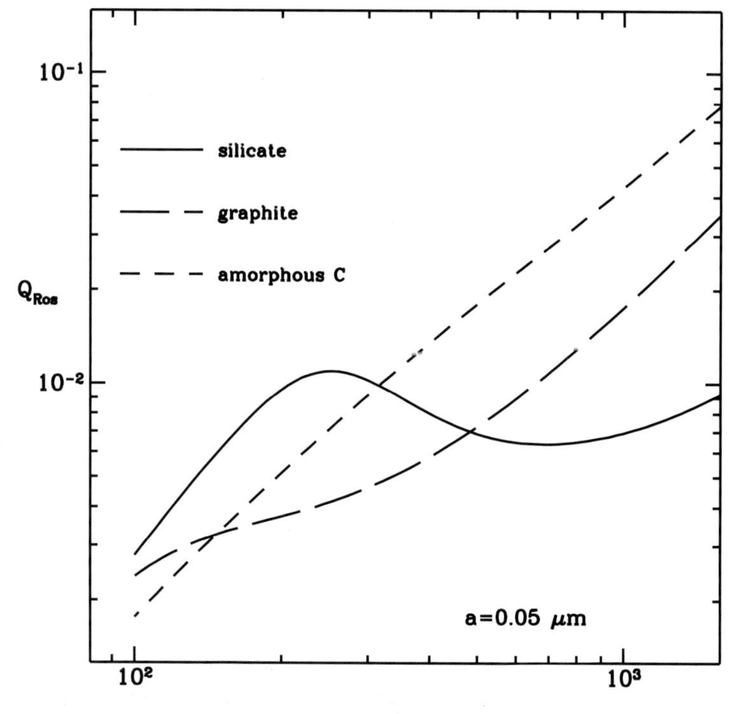

Figure 7.4 The Rosseland extinction efficiency coefficient $Q_{\rm Ros}$ as a function of grain temperature for grains with radius a $= 0.05\ \mu$m. Results are shown for both silicate and graphite material, and mean opacity as a function of the temperature in an optically thick dust shell. (Courtesy of John Mathis)

in stellar photospheres and in the radiative zones of stellar interiors.) The radiative equilibrium condition for determining the temperature, T_d, of a grain is

$$\int_0^\infty \kappa_\lambda B_\lambda(T_d)\, d\lambda = \int_0^\infty \kappa_\lambda J_\lambda \, d\lambda \tag{7.26}$$

The left hand side of this equation is the radiative cooling of a grain because the product, $\kappa_\lambda B_\lambda$, is the thermal emissivity at λ of a grain at temperature T_d. The right hand side is the radiative heating owing to the opacity κ_λ and the ambient radiation field. The quantity J_λ is the monochromatic 'mean intensity' of the radiation field incident on the grain, and it is the angle averaged mean of the intensity, I_λ

$$J_\lambda = \frac{1}{4\pi} \int_0^{4\pi} I_\lambda \, d\Omega \tag{7.27}$$

where $d\Omega$ is the differential of the solid angle. Given the intensity, I_λ, at the location of any grain, Eq. (7.26) determines the grain temperature, no matter whether the region is optically thick or thin.

For the optically thin case in which the intensity incident on the grain is direct star light with a uniform intensity, $I_\lambda^* = B_\lambda(T_*)$, the mean intensity is given by

$$J_\lambda = W(r) B_\lambda(T_*) \tag{7.28}$$

where $W(r)$ is the 'geometrical dilution factor' (see Eq. 2.8)

$$W(r) = \frac{1}{2}\left\{1 - \sqrt{1 - (R_*/r)^2}\right\} \tag{7.29}$$

which reduces to $W(r) = (R_*/2r)^2$ for $r \gg R_*$.

The radiative equilibrium condition for a dust grain of radius a is now obtained by expressing the absorptive opacity, κ_λ, in terms of the cross section efficiencies, and we find

$$\int_0^\infty \pi \mathrm{a}^2 Q^{\mathrm{A}}(\mathrm{a}, T_d) B_\lambda(T_d)\, d\lambda = \int_0^\infty \pi \mathrm{a}^2 Q^{\mathrm{A}}(\mathrm{a}, T_*) B_\lambda(T_*) W(r)\, d\lambda \tag{7.30}$$

One can assume that all grains attain a local steady state so this equation holds for each grain size individually. Note that $\pi \mathrm{a}^2$ cancels from both sides of Eq. (7.30), but there remains a dependence of the equilibrium temperature on grain size because of $Q^{\mathrm{A}}(\mathrm{a}, T)$. From this equation we can derive for the optically thin case the temperature of grains as a function of distance from the star.

7.4.2 The grain temperature

Both the heating and cooling integrals in Eq. (7.30) are related to the Planck mean efficiency, $Q_{\rm P}^{\rm A}$(a,T), given in Eq. (7.18). Therefore, the temperature of a dust grain as a function of the radial distance, r, is given by

$$T_d{}^4 Q_{\rm P}^{\rm A}({\rm a}, T_d) = T_*{}^4 W(r) Q_{\rm P}^{\rm A}({\rm a}, T_*) \qquad (7.31)$$

This equation shows that at $r > 2R_*$, where we can use $W(r) \simeq (R_*/2r)^2$, the dust temperature varies approximately as $r^{-0.5}$. More explicitly, the temperature distribution is given by

$$T_d(r) \simeq T_* \left\{ \frac{R_*}{2r} \right\}^{1/2} \left\{ \frac{Q_{\rm P}^{\rm A}({\rm a}, T_*)}{Q_{\rm P}^{\rm A}({\rm a}, T_d)} \right\}^{1/4} \qquad (7.32)$$

The Planck mean of the cross section efficiencies $Q_{\rm P}^{\rm A}$ are given in Fig. (7.2). If r is small, say just slightly greater than the stellar radius, where $W(r) \simeq 1/2$, then Eq. (7.31) indicates that the grains would have a temperature nearly as high as the stellar temperature. However, such hot grains would be destroyed by sublimation. Therefore, by setting the grain temperature on the left hand side of Eq. (7.31) to the condensation temperature, T_c, we can use that equation to derive the innermost radial distance, r_c , at which a grain can survive

$$W(r_c) = \frac{Q_{\rm P}^{\rm A}(T_c) T_c{}^4}{Q_{\rm P}^{\rm A}(T_*) T_*{}^4} \qquad (7.33)$$

A solution of this provides the inner boundary radius of the dust region, r_c, for the optically thin case. Assuming $r_c \gg R_*$, this solution for r_c is

$$r_c \approx \frac{R_*}{2} \left(\frac{T_*}{T_c} \right)^2 \sqrt{\frac{Q_{\rm P}^{\rm A}(T_*)}{Q_{\rm P}^{\rm A}(T_c)}} \qquad (7.34)$$

If the opacity can be approximated by a power-law of wavelength, $Q^A \sim \lambda^{-p}$, the ratio between the two Planck-mean opacities in Eq. (7.32) can be expressed in a simple analytic expression

$$\left\{ \frac{Q_{\rm P}^{\rm A}(T_*)}{Q_{\rm P}^{\rm A}(T_c)} \right\} = \frac{\int_0^\infty \lambda^{-5-p} \{\exp(hc/\lambda k T_*) - 1\}^{-1} d\lambda}{\int_0^\infty \lambda^{-5-p} \{\exp(hc/\lambda k T_c) - 1\}^{-1} d\lambda} \cdot \frac{T_c{}^4}{T_*{}^4} = \frac{T_*{}^p}{T_c{}^p} \qquad (7.35)$$

The last equality is derived by substituting $z = hc/\lambda k T_*$ and $y = hc/\lambda k T_c$, which makes the two integrals equal, apart from a factor $(T_*/T_c)^{p+4}$. This results in a dust temperature distribution of

$$T_d(r) = T_* \times W(r)^{\frac{1}{(4+p)}} \simeq T_* \left(\frac{2r}{R_*} \right)^{-\frac{2}{4+p}} \qquad (7.36)$$

Table 7.1 *Condensation radii*

	silicate T_c=1500[b]	graphite T_c=1500	amorphous carbon[a] T_c=1500
T_*	r_c/R_*	r_c/R_*	r_c/R_*
3000	2.99	4.03	3.42
2500	1.85	2.34	2.12
2000	1.15	1.29	1.24

[a] the EDOH form of amorphous carbon grains, Hanner (1988)
[b] refractory nuclei of grains can form at temperatures of about 1500 K or higher and the rest of the silicate grain SiO etc. will follow because they have a stable grain on which to deposit their heat of formation

and a solution for the radius at which a grain has the temperature T_d,

$$r_c \simeq \left(\frac{R_*}{2}\right)\left(\frac{T_d}{T_*}\right)^{-\frac{(4+p)}{2}} \tag{7.37}$$

These equations give the proportionality $T_d(r) \sim r^{-2/5}$ for $p = 1$.

7.4.3 The radial distance for grain condensation

Table (7.1) gives several solutions of Eq. (7.33) for the inner boundary radius, r_c, for both silicate and graphite grains. These have been derived using the efficiencies given in Fig. (7.2). These solutions assume that the only contribution to the radiation field is the direct star light. However, if there is a significant contribution by the diffuse light from the dust driven wind, the values for r_c will be increased. This table shows that the distance where the grains condense is a very steep function of the stellar temperature. Since the mass loss rate depends critically on the condensation distance we can also expect the mass loss rate to be a very steep function of T_*.

Detailed studies show that grain growth can be very efficient near this condensation radius because the temperature no longer impedes the growth of grains, and the gas density is high. Models show that the grain growth just beyond the condensation radius rather quickly leads to a transonic flow. For Table (7.1), we have chosen a condensation temperature of 1500 K, because it is roughly at the middle of the range of temperatures at which different types of grains condense (Whittet, 1992).

7.5 Combined dust and gas flow

Dust driven winds are 'multi-fluid' flows. This means there is a separate momentum equation for grains of each size, and another equation for the gas. These grain and gas equations are coupled because the collisions between grains and atoms produce a drag force. The grains have the large opacity that causes them to be driven outward by radiation, but the drag force due to the atoms, which are not radiatively driven, acts to decelerate their motion. The gas is driven outward only because of the presence of the drag force in the momentum equation.

7.5.1 The dust momentum equation

To keep our discussion focused on the basic wind driving process, we assume for the remainder of this chapter that the grains are all of one size, with radius a. Let $u(r)$ be the speed of a grain at a radial distance r, let n_d be the number of grains per cm^3 and m_d be the mass of each grain. Then the dust momentum equation for a steady state radial flow is

$$u\frac{du}{dr} = -\frac{GM_*}{r^2} + \frac{\pi a^2 Q_{rp} L_*}{4\pi r^2 c\ m_d} - \frac{f_{Drag}}{m_d} \tag{7.38}$$

where Q_{rp} is the radiation pressure mean efficiency from Eq. (7.21). We will also ignore scattering, so there is no difference between the radiation pressure and flux mean opacities, thus

$$Q_{rp}\ \mathscr{F} = \int_0^\infty Q^A(a, \lambda)\mathscr{F}_\lambda d\lambda \equiv\ Q_F^A \mathscr{F} \tag{7.39}$$

and f_{Drag} is the drag force per grain, i.e. the momentum per second transfered by the grain to the gas. So $n_d f_{Drag}/\rho$ is the force per gram. The outward radiation force is much greater than the inward pull of gravity on a dust grain, so gravity will be ignored. The drag force depends on the relative speed or *drift speed*, w_{dr}, of the grain through the gas,

$$w_{dr} = u - v \tag{7.40}$$

where v is the flow speed of the gas, and $u > v$. There are two limiting expressions for the drag force depending on the value of w_{dr} relative to the thermal speed of the gas particles

$$a_{th} = \sqrt{\frac{2kT}{\mu' m_H}} \tag{7.41}$$

with $\mu' m_H$ being the mean mass of all the atoms and molecules in the gas phase. For a solar abundance atmospheric mixture with helium and molecular hydrogen, one would derive $\mu' \approx 2.4$. However, Bowen (1988) argues that the pulsational shocks that extend the subsonic density structure dissociate H_2, so we use $\mu' = 1.3$ in calculations of thermal speeds.

If the drift speed is much larger than the thermal speed of the gas particles, the drag force is given by the product of the relative dynamic pressure $\rho\, w_{dr}^2$ times the area of the dust grain; or $f_{Drag} = \pi a^2 \rho\, w_{dr}^2$, where ρ is the density of the gas. If, on the other hand, the drift speed is small compared to the thermal speed of the gas, then $f_{Drag} = \pi a^2 \rho\, a_{th} w_{dr}$. These two limiting cases are combined in the expression

$$f_{Drag} = \pi a^2 \rho\, w_{dr} \sqrt{w_{dr}^2 + a_{th}^2} \tag{7.42}$$

A major simplification of dust driven wind theory was introduced by Gilman (1972), who showed that a grain should rather quickly reach 'terminal speed', and all of the momentum that a dust grain receives from the radiation field will be transmitted to the gas. We can picture this as being analogous to a helium filled balloon moving upwards in the earth's atmosphere, and dragging gas along as it goes up. In the case of the grains, the upward driving is caused by the radiation field. The grains seldom collide with other grains and they collide with only a small fraction of the gas particles. It is the subsequent collisions among the gas particles that allows the outward momentum to be transmitted to the gas as a whole.

The grain will move at terminal speed if the momentum equation is dominated by the two opposing forces of the drag force and the force of radiation. The balance of these two terms in the momentum equation leads to the result

$$\frac{\pi a^2 Q_{rp} L_*}{4\pi r^2 c} = \pi a^2 \rho\, w_{dr} \sqrt{w_{dr}^2 + a_{th}^2} \tag{7.43}$$

This equation provides an expression for the drift speed. Squaring both sides of (7.43) we find

$$w_{dr}^4 + w_{dr}^2 a_{th}^2 - \left(\frac{Q_{rp} L_*}{4\pi r^2 \rho\, c}\right)^2 = 0 \tag{7.44}$$

In the limiting case where $a_{th} \ll w_{dr}$ is small, we get the result

$$w_{dr} = \sqrt{\frac{Q_{rp} L_*}{4\pi r^2 \rho\, c}} \tag{7.45}$$

This shows that the drift speed is inversely proportional to the square root of the gas density. The inverse dependence on density is reasonable, because for example, we should expect parachutes going at terminal speed to move faster at altitudes where the atmospheric density is lower.

Because of the inverse density dependence of the drift speed, the gas-grain collisions also become increasingly energetic with increasing distance from the star. It is possible that sputtering of the grains could eventually occur. In principle, processes that change the size distribution can be treated by using multi-fluid momentum equations along with equations to account for the change in grain sizes with radial distance.

The drift of the grains through the gas is an important aspect of dust driven winds. The drift speed depends on density, so it introduces a dependence of the wind driving on the mass loss rate. Eliminating the gas density from Eq. (7.45) using conservation of mass, we get

$$w_{\rm dr} = \sqrt{\frac{Q_{\rm rp} L_* v}{\dot{M}\; c}} \tag{7.46}$$

This expression for the drift speed explicitly shows the dependence on $\dot{M}$ and the gas flow speed v. Later we will find that the drift speed affects the outward acceleration of a wind, and as a result, it leads to a mimimal value for the mass loss rate for dust driven winds.

7.5.2 The gas momentum equation

The grains enter the gas momentum equation because of the outward drag force that arises from their drift through the gas. Of course, in the case of the gas, the drag force is in the outward direction because the gas is being pushed by the faster moving grains. The gas momentum equation can therefore be written as

$$v\frac{dv}{dr} + \frac{1}{\rho}\frac{dp}{dr} + \frac{GM_*}{r^2} = n_{\rm d}\frac{f_{\rm Drag}}{\rho} \tag{7.47}$$

where $n_{\rm d}$ is the number of grains per cm^3, and thus the term on the right hand side of this equation is the momentum transfered per second to a unit mass of the gas. As we did in deriving in the dust momentum equation, let us assume that the grains are moving at terminal speed, hence we can use (7.42) and (7.43) to eliminate $f_{\rm Drag}$ from (7.47), and obtain

$$v\frac{dv}{dr} + \frac{1}{\rho}\frac{dp}{dr} + \frac{GM_*}{r^2} - \frac{n_{\rm d}}{\rho}\frac{\pi a^2 Q_{\rm rp} L_*}{4\pi r^2 c} = 0 \tag{7.48}$$

where $n_{\rm d}/\rho$ is the number of dust grains per unit mass. Since the last two terms in this equation vary as r^{-2}, it is convenient to combine them to provide the form of the wind momentum equation

$$v\frac{dv}{dr}+\frac{1}{\rho}\frac{dp}{dr}+\frac{GM_*}{r^2}(1-\Gamma_d)=0 \tag{7.49}$$

with

$$\Gamma_d=\frac{k_{rp}L_*}{4\pi cGM_*}=\frac{n_d}{\rho}\frac{\pi a^2 Q_{rp}L_*}{4\pi cGM_*} \tag{7.50}$$

Here, we have introduced k_{rp}, the opacity (in cm^2 g^{-1}) which is related to the cross section of the grains for radiation pressure by $k_{rp}\rho = n_d \pi a^2 Q_{rp}$.

Even though the gas is not directly driven by radiation forces, the gas momentum equation takes the same form as a radiatively driven gaseous wind (see § 3.2). This is because of the 'momentum coupling' of the grains and the gas. The grains move at terminal speed, and this implies that all of the radiative driving of the grains is directly transmitted to the gas. Also, since Eq. (7.49) has the form of the Parker momentum equation, it has a critical point, where the speed of the gas reaches the sound speed.

The transonic point occurs at a radius just before the point where Γ_d becomes larger than unity. The location of the critical point requires that the grains be able to form and grow efficiently, but once they do, the gas is quickly brought to supersonic speed. Then the gas density drops rather rapidly with radius and this terminates the dust growth. This provides the justification for the common assumption that grain sizes are conserved, and that the mass loss rates of grain material and gaseous material are separately conserved throughout the entire wind that lies beyond the grain formation region.

7.6 The mass loss rate of a dust driven wind

The gas momentum equation (7.49), can be used to derive the subsonic and supersonic structure of a dust driven wind. Gail & Sedlmayr (1987) have developed self-consistent models of the outflows from carbon stars, in which the process of grain growth and the transistion from subsonic to supersonic flow is followed in some detail. The formation of the grains leads to an outward increase in the ratio, Γ_d, and the critical point of the flow occurs at the sonic point, where v equals the sound speed, a_s. The density and radius at that point determine the mass loss rate. So, in the self-consistent models, one specifies just the basic stellar parameters, L_*, M_*, T_{eff}, and the composition. The mass loss rate follows from the solution. Although the mathematics of the problem is well defined, observations indicate that actual stars behave differently. This is because in stars with significant dust driven winds, the structure of the underlying atmosphere is not determined by the effective temperature and surface gravity, but rather by the effects of pulsation that extend the atmosphere. The enhancement of the density scale height can increase by several orders of magnitude the density at the dust condensation radius. Therefore, the mass loss of a star is determined by conditions other than those that determine the structure of a classical hydrostatic atmosphere. In this section we briefly describe the effects of pulsation as have been derived by detailed modelling by Bowen (1988), and we discuss the properties at the base of the supersonic flow.

7.6.1 The effect of pulsation on the density structure of the subsonic region

In a classical stellar atmosphere, the density at the condensation radius, r_c, would be extremely small because that radius would be many density scale heights above the photosphere. This would imply very small mass loss rates. (See for instance the model for $T_* = 3000$ K, and $R_* = 1000 R_\odot$ in Table 3.1.) However, the asymptotic giant branch stars of interest here have a pulsational instability in their envelopes that causes them to be Mira variables and related long period variable stars. The pulsation provides the stars with a mechanism for 'levitating' the material in their atmospheres. The combination of the pulsation and dust formation can produce very large mass loss rates, as high as $10^{-5}\ M_\odot\ \mathrm{yr}^{-1}$ or more. Such rates are large enough to affect the future evolution of these stars. Also slight changes in the stellar properties occurring during the asymptotic giant branch evolution of the stars can give rise to major differences in the total amount of mass driven off by the star. The most extreme of the pulsating AGB stars are the OH/IR sources which have optically thick dust shells in which OH masering, pumped by dust IR emission, occurs (see § 2.8).

Mira variables execute large amplitude radial pulsations, and these pulsations lead to the propagation of strong outwardly moving shocks, as is illustrated in Fig. (7.5). This figure shows, as a function of time, the radial location of the material originating at various heights, or Lagrangian zones, in the atmosphere. The convergence of these time-lines illustrates the compression that is brought about by the shocks that propagate outward through the atmosphere. The material is shock heated, and the shock motion plus post-shock pressure gradients push the cooling gas outward. It moves along roughly ballistic trajectories until the infalling material encounters the next shock. The pattern is of the drift speed, $w_{\rm dr}$, and then combine (7.65) with the expression for $\Gamma_{\rm d}$ from Eq. (7.1), we obtain

$$\Gamma_{\rm d} = \Gamma_0 \frac{v}{(w_{\rm dr} + v)} \tag{7.66}$$

where Γ_0 is the value of $\Gamma_{\rm d}$ at the condensation radius.

Equation (7.66) clearly shows the effect of drift speed in reducing the outward acceleration of the wind. For low density winds the drift speed may be large enough that it leads to a significant decrease in $\Gamma_{\rm d}$.

We can derive a *lower limit on the mass loss rate* by using the condition that the wind acceleration requires that $\Gamma_{\rm d}$ be larger than unity. For this purpose, let us make the following assumptions: (a) at the low densities associated with the minimal mass loss rate, the gas will not be accelerated to speeds much above its initial speed, $v_c = a_s(r_c)$, which is the sonic speed at the condensation radius, (b) the drift speed will be much larger than the gas speed, so we can approximate $w_{\rm dr} + v \simeq w_{\rm dr}$ in Eq. (7.66). Finally we use for $w_{\rm dr}$, Eq. (7.46) which shows an inverse dependence on the mass loss rate. Therefore, if we start with a wind model and reduce the mass loss rate, we would find that the drift speed would increase, and most importantly, the value of $\Gamma_{\rm d}$ would decrease. Eventually, when the mass loss rate is reduced to the point at which $\Gamma_{\rm d} < 1$, the wind would stop, because it would no longer be driven out by the drag from the radiatively accelerated grains.

Setting $\Gamma_{\rm d}$ equal to unity, we can derive from Eq. (7.66), and the assumptions above, the minimal mass loss rate

$$\dot{M}_{\rm min} = \frac{(4\pi c G M_*)^2}{v_c\ c} \left(\frac{m_d}{\delta^0_{\rm dg} \pi {\rm a}^2} \right)^2 \frac{L_*}{Q_{\rm F}} \tag{7.67}$$

To derive a numerical value for this minimal mass loss rate, let us use typical values, $Q_{\rm F} = 0.04$, $v_c = 2$ km s^{-1}, $\delta^0_{\rm dg} = 1/200$, ${\rm a} = 0.05$ μm. Expressing the stellar mass in solar units we obtain for the minimal mass loss rate for dust driven winds

$$\dot{M}_{\rm min} = 6.6 \times 10^{-8} \left(\frac{M_*}{M_\odot} \right)^2 \left(\frac{L_*}{10^4 L_\odot} \right)^{-1} \quad M_\odot\ {\rm yr}^{-1} \tag{7.68}$$

So we see that a typical value for the minimal mass loss rate for dust driven winds is of order 10^{-7} $M_\odot$ yr^{-1}. When compared with mass loss rates of other stars across the HR diagram this is actually a rather large mass loss rate. The pulsation of the underlying star therefore must supply a significant density at the dust formation zone for a combined dust/gas driven wind to occur.

In the previous section we found that there is a minimal luminosity to mass ratio for dust driven winds; now we find that there is a minimal mass loss rate for a coupled dust/gas wind to exist.

What happens if the star has properties that would lead to an $\dot{M}$ that is below that limit? For example, what if the pulsation delivers too low a density at the condensation radius to drive a coupled dust/gas wind? If that occurs, we can expect that a wind would not be driven, but interesting results would occur, nevertheless. This is because grains that form in the outer regions of the pulsationally supported extended atmosphere would still be accelerated outward. Since they do not interact sufficiently with the gas, they will be driven away from the atmosphere in the form of 'free particle ejection'. So the star will have a 'wind' that consists purely of grains. This can lead to a stripping of metals from the outer atmopheres of cool giant stars. Mathis and Lamers (1992) have applied this idea to explain abundance anomalies in post-AGB stars which show deficiencies in elements that can condense onto grains. Waters *et al.* (1992) have suggested the same mechanism for the last phase of the accretion of pre-main sequence stars to explain the occurrence of young A-type stars with low metalicity.

7.8.2 The reddening effect and the maximal mass loss rate

Now let us consider the opposite extreme in which the dust/gas flow is dense and very optically thick. For the large mass loss rates that occur for example among the OH/IR sources, the shell around the star produced by the dust/gas wind must certainly be very optically thick in the visible and near IR region, because we receive essentially no direct stellar radiation. The stellar radiation is absorbed deep in the shell and is re-radiated at longer wavelengths with an emissivity determined by the temperature of the grains. If the grain opacities decrease monotonically with increasing wavelength, as is the case for carbonaceous grains, (Fig. 7.2a), the reddening of the radiation field will lead to a decrease in the outward radiative acceleration. This is the reddening effect of mass loss and it is important only if the dust driven wind is optically thick to the direct stellar radiation.

Detailed numerical models of the radiation transfer in optically thick shells have been carried out by Rowan-Robinson & Harris (1983) and numerous other authors. We can make some analytic approximations to the radiation transfer that will be sufficient for discussing an upper limit on the mass loss rate.

Let us assume the 'cocoon approximation' (Davidson and Harwit,

1967), that is often used in star formation theory (e.g. Wolfire and Cassinelli, 1987). In this approximation, all of the direct star light is absorbed at the base of the shell. Since the grains at that location all have the condensation temperature, T_c, the stellar luminosity is fully converted to infrared radiation characterized by the Planck function, $B_\lambda(T_c)$. Thus we can consider the circumstellar shell of dust and gas to be made up of the *interface zone*, which has an Rosseland optical depth of about unity, that absorbs the direct stellar light, and an *outer envelope*, in which the diffusion of the dust emitted light occurs. We assume that the energy is carried through the outer envelope by radiation alone. Thus, as in a classical stellar atmosphere, the radiative luminosity through the shell is constant, and the temperature of the grains is determined by the radiative equilibrium equation (7.26). The decrease in the grain temperature with increasing distance from the star is the essential feature of the reddening effect. Figure (7.8) shows this model schematically.

In an optically thick and geometrically extended dust shell, the temperature will decrease with distance from the star for two reasons: diffusion and dilution. Firstly, the outward diffusion of the radiative flux requires that there be a negative temperature gradient with ra-

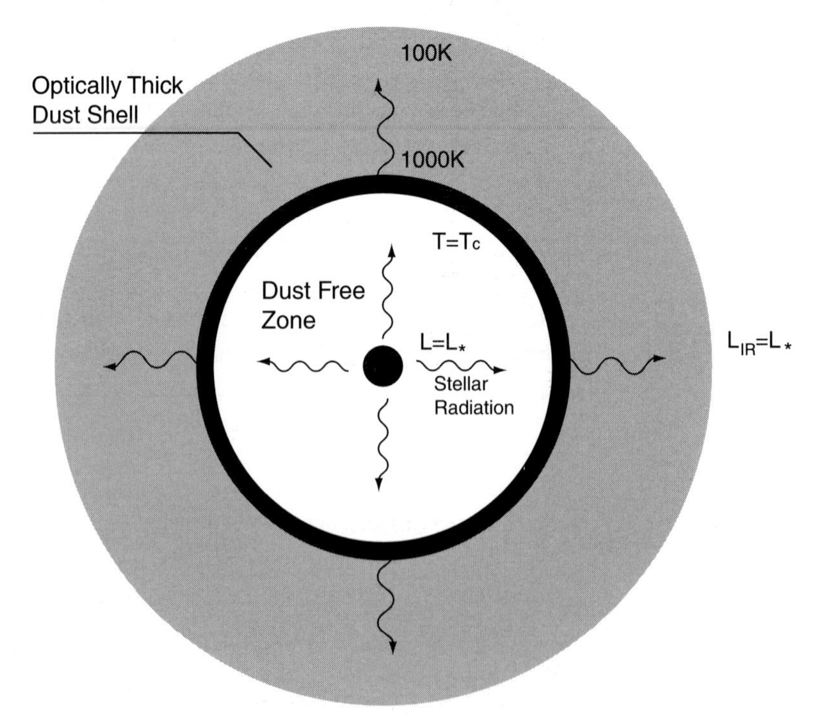

Figure 7.8 The cocoon model for optically thick dust shells. The star is surrounded by an optically thick dust shell whose inner radius is the condensation radius, r_c. The direct starlight is absorbed and re-emitted by the dust at the 'interface zone' (black) near the inner radius. The re-emitted infrared radiation that drives the dusty wind diffuses outward through the 'outer envelope'.

dius, as for example is the case for stellar interiors and atmospheres. Secondly, in a geometrically extended shell, there is a decrease in the mean intensity caused by the dilution of the radiation field. The effects of geometrical dilution occur for both optically thin and optically thick shells, because in both cases the flux, integrated over all frequencies, decreases as r^{-2}. In an optically thin shell case we can use the geometrical dilution factor $W(r)$, defined by Eq. (7.29), to describe the effect of dilution on the mean intensity, $J(r)$. However, in an optically thick shell discussed here, the dilution of the radiation can be found only by a solution of the radiative transfer equation. Since our main interest is in discussing the effect of reddening on the mass loss rate, we choose to neglect the dilution effect and consider just the decrease in the temperature associated with radiative diffusion. This approach is equivalent to assuming that the geometrical thickness of the shell is small compared to its inner radius, in which case we can use the plane parallel approximation for radiative diffusion (Mihalas, 1978). The temperature decrease then only depends on the optical depth through the shell.

For radiative transfer through an optically thick outer envelope the appropriate flux mean opacity is the Rosseland mean opacity, k_{Ros}, which we defined in Eq. (7.25). It is a 'transmissivity' average opacity, which means that the opacity is weighted most heavily by the spectral regions where the diffusive flux is the greatest. Radiative diffusion theory relates the integrated Planck function, $B = \sigma T^4/\pi$, to the radiative flux, $\mathscr{F}$. In the case of a plane parallel optically thick shell that we are assuming here, the radiative flux could be written as $\mathscr{F} = \sigma {T_{\text{eff}}}^4$, where T_{eff} is the effective temperature of the shell, and is approximately equal to the temperature at Rosseland mean radial optical depth, $\tau_{\text{Ros}} \approx 2/3$. However, to avoid confusion with the T_{eff} of the central star, let us call the effective temperature of the shell the 'surface temperature', T_{surf}. Thus the temperature distribution in the thick shell can described by the well known $T(\tau)$ relation from plane parallel stellar atmospheres theory

$$T^4 = \frac{3}{4} T_{\text{surf}}^4 \left(\tau_{\text{Ros}} + \frac{2}{3} \right) \tag{7.69}$$

The essence of the reddening effect is that the opacity and the radiation driving force decrease as the temperature decreases. For this discussion, let us consider the Rosseland mean of the amorphous carbon dust grains plotted in Fig. (7.4), which can be fitted by the expression $Q_{\text{Ros}} \propto T^{4/3}$. Converting this to a Rosseland mean opacity using $a = .05\ \mu\text{m}$, $\delta_{\text{dg}}^0 = 1/200$ and $\rho_g = 2.25\ \text{gm cm}^{-3}$ as the density of an amorphous carbon grain, we find

$$k_{\rm Ros}(T) = 24. \left(\frac{T}{1500K}\right)^{4/3} \quad {\rm cm}^2\ {\rm gm}^{-1} \tag{7.70}$$

In a optically thick dust shell, this Rosseland mean opacity is also the radiation pressure mean opacity, thus $\Gamma_{\rm d}$, of Eq. (7.50), also varies as $T^{4/3}$. Therefore, as the temperature decreases, $\Gamma_{\rm d}$ and the radiation acceleration also decrease. At a radius where the reddening effect has decreased $\Gamma_{\rm d}$ to unity, the net outward acceleration of the shell ceases. The layers above that are decelerated by gravity, but the material in the layers may still escape if the velocity is larger than the local escape speed. It is plausible that a shell operating at maximal mass driving efficiency will become optically thin at about the same radius that radiative driving ceases to be dominant. This is because the flow would have positive acceleration throughout the thick region, and there would be no pile up of material in the outer part of the optically thick shell.

We can use the conjecture that $\Gamma_{\rm d} \geq 1$ throughout the thick region of the shell, to derive an upper limit on the mass loss rate owing to the reddening effect. Given that the temperature at the base of the dust shell, $T(\tau_{\rm Ros}({\rm tot}))$ is 1500 K, and also given the value of $\Gamma_{\rm d}$ at the base of the shell, we can use Eq. (7.70) to find the temperature $T_{\rm surf}$ at which $\Gamma_{\rm d}$ is reduced to unity. Then, with both $T(\tau_{\rm Ros}({\rm tot}))$ and $T_{\rm surf}$ known, we can find $\tau_{\rm Ros}({\rm tot})$ by using Eq. (7.69). For example, if $\Gamma_{\rm d}$= 2 at the base of the dust shell where $T = 1500$ K, then $T_{\rm surf} = 890$ K where $\Gamma_{\rm d}$= 1, and we find that this range in temperatures corresponds to $\tau_{\rm Ros}({\rm tot}) = 10$. A lower value for $T_{\rm surf}$ would occur at the top of the same optical depth range if we had accounted for geometical dilution effects.

An optical depth of about 10 is a reasonable upper limit for a dust driven wind. Primarily this is because the radius of the base of the dust shell, r_c, is affected by the total optical depth of the shell and large values of r_c are not compatible with dust driven winds. Recall that r_c is determined by the condition that $T_d = T_c(\simeq 1500$ K), where the temperature of the test grain, T_d, is determined by the mean intensity J_λ that appears in the radiative equilibrium equation (7.26). In the case of an optically thick shell, $J_\lambda(r_c)$ has a contribution from both the direct radiation from the star, and the diffuse radiation from the overlying dust shell. As seen at the location of a grain at the inner boundary of the shell, the star subtends a solid angle equal to the $4\pi W$, and the shell subtends the much larger solid angle $4\pi(1 - W)$. For a very optically thick shell, the diffuse radiation originates where the temperature is only slightly lower than the dust condensation temperature. This leads to a significant increase in $J_\lambda(r_c)$, and thus causes r_c to increase with $\tau_{\rm Ros}$(tot). If the increase in the condensation radius is by more than a factor of two or so, then we need to question whether the wind can be considered to be dust driven as opposed to pulsationally driven. In a pulsationally driven wind the dust formation would just be a phenomenon occuring in the outflow, but having little effect on either the mass loss rate or velocity of the flow.

Another effect of the total optical depth is to change the degree by which $\Gamma_{\rm d}$ is reduced by the reddening effect. In the example above of a plane parallel dust shell of amorphous carbon grains, $\Gamma_{\rm d}$ decreased by a factor of two. A larger change in $\Gamma_{\rm d}$ occurs if we account for the steeper $T(\tau)$ relation associated with spatially extended dust shells. In summary, for dust shells with $\tau_{\rm Ros}({\rm tot}) \simeq 10$, (a) there can be significant changes in $\Gamma_{\rm d}$ and (b) the condensation radius can remain near enough to the star that levitation effects can supply matter to the inner boundary of the dust shell.

Now, let us assume a value for the upper limit on the radial Rosseland mean optical depth, $\tau_{\rm Ros}$(tot), and use it to find a maximum mass loss rate of a dust driven wind. The total radial optical depth of the shell is given by

$$\tau_{\rm Ros}({\rm tot}) \equiv \tau_{\rm Ros}(r_c) = \int_{r_c}^{r_{\rm max}} k_{\rm Ros}\ \rho\ dr \tag{7.71}$$

We can eliminate ρ from this equation by using the mass conservation equation. If we also assume the velocity law $v(r)^2 = v_\infty{}^2(1 - r_c/r)$, we can eliminate dr/r^2, in terms of dv, and we find

$$\tau_{\rm Ros}({\rm tot}) = \frac{\dot{M}}{4\pi} \int_{r_c}^{\infty} k_{\rm Ros} \frac{dr}{v\ r^2} \simeq \langle k_{\rm Ros} \rangle \frac{\dot{M}}{2\pi v_\infty r_c} \tag{7.72}$$

where $\langle k_{\rm Ros} \rangle$ is the average value of the Rosseland mean opacity through the shell; for this we use $\langle k_{\rm Ros} \rangle = 18\ {\rm cm}^2\ {\rm gm}^{-1}$, based on the discussion following Eq. (7.70). Assuming the typical values of $v_\infty =$ 10 km s^{-1}, $r_c = 3.0\ R_*$, and $L = 10^{4.5} L_\odot$, we find that Eq. (7.72) leads to an upper limit for the mass loss rate of a dust driven wind of

$$\dot{M}_{\rm max} \approx 1.1 \times 10^{-5} \left\{ \frac{\tau_{\rm Ros}({\rm tot})}{10} \right\} \left\{ \frac{L_*}{10^{4.5} L_\odot} \right\}^{0.5} \left\{ \frac{T_{\rm eff}}{2500K} \right\}^{-2} M_\odot\ {\rm yr}^{-1} \tag{7.73}$$

This agrees fairly well with the observed mass loss rates of Miras and OH/IR stars, discussed in § 2.8.2, if $\tau_{\rm Ros}({\rm tot}) \simeq 10$. In particular the saturation of the mass loss rates for OH/IR stars in Fig. (2.26) supports values of a few times $10^{-5}\ M_\odot\ {\rm yr}^{-1}$.

Observational estimates of the optical depth through the shells of a few stars have led to mass loss rates that are much larger than that given in Eq. (7.73) (Ivezic and Elitzur, 1995). A possible explanation is that stars have episodes of large mass loss rates by dust driven winds followed by periods of reduced mass loss activity. Large total optical depths could thus be accumulated after many episodes. Also, Bowen and Willson (1991) discuss the 'superwinds' that could be produced primarily by the large amplitude pulsations of stars at the top of the asymptotic giant branch. Such models suggest that dust does not always determine the stellar mass loss properties of AGB stars. However, for most of the AGB stars, the dust driven wind theory appears to be able to explain the basic observational properties.

7.9 Conclusions

The largest mass loss rates across the HR diagram are produced by radiation forces. Winds driven by continuous opacity have been considered in this chapter, with special consideration to the dust accelerated winds of very cool stars. Dust driven winds can occur only if both the temperature and density in the outer layers of the stellar atmosphere allow for the condensation of grains. Dust condensation begins to occur at a temperature of about 1500 K for silicate grains in O-rich atmospheres and for the amorphous carbon grains in C-rich atmospheres. The temperature of the grains is determined by the equilibrium between the heating by the absorption of stellar photons and the cooling by thermal emission. Therefore the grain temperature depends on the opacity of the grain material and on the ambient radiation field. It will roughly decrease with distance from the star as $r^{-2/5}$ if the dust shell is optically thin.

Assuming that the radiation is primarily direct light from the star, the distance where the grains form, i.e. the condensation radius, can be found by setting the expected temperature of the dust grains equal to the condensation temperature. The condensation radius of cool stars is typically between 1.1 and 2.6 R_* for stars with $2200 < T_{\rm eff} < 3000$ K. Once the grains are formed they are quickly accelerated to the sound speed and beyond because of their large opacity, so the gas is also expected to quickly reach the sound speed. In dust driven winds *the sonic point and critical point both occur at about the condensation radius.* This is because $\Gamma_{\rm d}$ quickly grows to values larger than unity. Hence *the mass loss rate is determined by the density at the condensation radius.*

If the atmosphere below the condensation radius were static, the mass loss rates of dust driven winds would be very small because the density in the subsonic region would decrease approximately with the hydrostatic pressure scale height (see Chapter 3). So the density at the sonic point would be extremely small, and the mass loss rate would also be so. However, the luminous cool stars have pulsations that levitate the atmosphere, i.e. the pulsations increase the scale height in the subsonic portion of the atmosphere, and thus increase the density at the sonic point. Given the radius of the sonic point from the condensation considerations, *the pulsations directly determine the mass loss rate.*

The theory of dust formation and growth is a difficult one, and is still not fully developed. Therefore, we have chosen to avoid that aspect of the problem by assuming that, once the grains form, there is rapid transition to a supersonic flow in which all of the condensible material is in grains with equal radii. Our discussion of the wind has therefore been concerned with only the supersonic region in which grain formation is complete and the gas pressure gradients are unimportant.

We have shown that dust driven winds can only occur within a range of mass loss rates. The lower limit is set by the drift effect, which requires sufficient coupling between the gas and the dust. We find that this lower limit is about 10^{-7} $M_\odot$ yr^{-1}, for a typical AGB star (Eq. 7.68). The upper limit is set by the reddening effect and requires that the re-emitted radiation is not shifted too far in the infrared where the opacity decreases strongly and the radiation pressure gradient would drop below some critical value. This upper limit is about 10^{-5} $M_\odot$ yr^{-1} for a typical AGB star (Eq. 7.73). These limits agree with the observations.

7.10 Suggested reading

Habing, H.J. 1996, 'Circumstellar Envelopes and Asymptotic Giant Branch Stars' *A & A Reviews* **7**, 97

(A broad overview regarding both the theory and observations of winds from cool stars.)

Iben, I, & Renzini, A. 1983 'Asymptotic Giant Branch Evolution and Beyond', *Ann. Rev. Astr. Ap.* **21**, 271

(A review of the structure and evolution of the stars that produce dust driven winds.)

Lafon, J.-P., & Berruyer, N. 1991, 'Mass Loss Mechanisms in Evolved Stars', *A & A Reviews* **2**, 249

(A broad review of models of cool star winds.)

Netzer, N., & Elitzur, M. 1993, 'The Dynamics of Stellar outflows Dominated by Interaction of Dust and Radiation ', *Ap .J* **410**, 701
(A paper on the techniques and results of dust driven wind models.)
Whittet, D.C.B. 1992, *Dust in the Galactic Environment* (Bristol: Institut. Phys. Publ).
(This book has a thorough discussion of the physical properties of dust grains.)

8 Line driven winds

The winds of luminous hot stars are driven by absorption in spectral lines and they are called *line driven winds.*

Hot stars emit the bulk of their radiation in the ultraviolet where the outer atmospheres of these stars have many absorption lines. The opacity in absorption lines is much larger than the opacity in the continuum. The opacity of one strong line, say the C IV resonance line at 1550 Å, can easily be a factor of 10^6 larger than the opacity for electron scattering.

The large radiation force on ions due to their spectral lines would not be efficient in driving a stellar wind if it were not for the Doppler effect. In a static atmosphere with strong line-absorption, the radiation from the photosphere of the star will be absorbed or scattered in the lower layers of the atmosphere. The outer layers will not receive direct radiation from the photosphere at the wavelength of the line, and so the radiative acceleration in the outer layers of the atmosphere due to the spectral lines is strongly diminished. However, if the outer atmosphere is moving outward, there is a velocity gradient in the atmosphere allowing the atoms in the atmosphere to see the radiation from the photosphere as redshifted. This is because in the frame comoving with the gas the photosphere is receding. As a result the atoms in the outer atmosphere can absorb radiation from the photosphere which is not attenuated by the layers in between the photosphere and the outer atmosphere. The Doppler shift thus allows the atoms to absorb undiminished continuum photons in their line transitions. This makes the radiative acceleration due to spectral lines in the atmospheres of hot luminous stars very efficient for driving a stellar wind.

The radiative acceleration caused by spectral lines drives the winds of many different types of stars: main-sequence stars, giants and supergiants of types O, B and A (possibly also F-type supergiants);

central stars of planetary nebulae and white dwarfs. Winds from quasars may also be radiation driven.

We will first describe some ingredients for line driven winds: momentum transfer and Coulomb coupling. We then introduce in §§ 8.3 and 8.4 the very important simplification of the radiative transfer in spectral lines formed in moving atmospheres: the Sobolev approximation. With this simplification we can describe the radiative acceleration due to spectral lines. In § 8.6 we describe the ions and the lines that contribute to the line radiation pressure in the winds of hot stars and we derive a simple parametrization in terms of a force multiplier. The solution of the resulting momentum equation in § 8.7 gives expressions for the mass loss rate, the velocity law and the terminal velocity in the limit that the star is a point source. In § 8.9 we improve the theory by taking into account the finite size of the star. This changes the predictions for the mass loss rate and the terminal velocity significantly. We conclude this chapter with discussions of the effects of multiple scattering, (§ 8.10), wind blanketing, (§ 8.11), stellar rotation, (§ 8.12), and the stability of line driven winds, (§ 8.13). We will show in a simple analysis that small density or velocity perturbations will quickly grow into shocks, an important aspect of the most recent studies of hot star winds.

8.1 Physical processes in line driven winds

8.1.1 Momentum and energy transfer by absorption and scattering

Let us consider the transfer of momentum and energy from radiation to the gas if an atom absorbs a photon and re-emits it again. For simplicity we assume that the atom is moving in the radial direction at velocity v_r and that it absorbs a photon going in the same direction (see Fig. 8.1).

After the absorption the atom has gained the momentum of the photon and its velocity is v_r' with

$$mv_r' = mv_r + \frac{h\nu}{c} \tag{8.1}$$

So there is an increase of velocity by $\Delta v_r = h\nu/mc$. If the atom subsequently emits a photon of frequency ν' at an angle α with respect to the radial direction, the new velocity v_r' of the atom in the radial direction will be

$$mv_r'' = mv_r' - \frac{h\nu'}{c}\cos\alpha \tag{8.2}$$

Suppose the atom can absorb and emit photons only at a frequency ν_0

in its rest frame. Then for an outside observer, the absorbed photon has a frequency $\nu_0(1 + v_r/c)$ and the emitted photon has a frequency $\nu_0(1 + v_r'/c)$. Therefore the velocity of the atom after absorption and re-emission is

$$v_r'' = v_r + \frac{h\nu_0}{mc}\left(1 + \frac{v_r}{c}\right) - \frac{h\nu_0}{mc}\left(1 + \frac{v_r'}{c}\right)\cos\alpha \tag{8.3}$$

$$= v_r + \frac{h\nu_0}{mc}\left(1 + \frac{v_r}{c}\right)(1 - \cos\alpha) - \frac{1}{c}\left(\frac{h\nu_0}{mc}\right)^2\left(1 + \frac{v_r}{c}\right)\cos\alpha$$

For $v \ll c$ and $h\nu_0 \ll mc$ we find that the radial velocity component of the atom has increased by

$$v_r'' - v_r = \frac{h\nu_0}{mc}(1 - \cos\alpha) \tag{8.4}$$

Forward scattering ($\cos\alpha = 1$) does not increase the momentum of the atom, and backward scattering ($\cos\alpha = -1$) increases the momentum by $2h\nu_0/c$. Since the re-emission of photons by the atom will be in a random direction in the rest frame of the atom, which is also random in the rest frame of the observer if $v \ll c$, the mean transfer of momentum after scattering of photons coming in the radial direction is the integral of Δmv over a sphere

$$\langle \Delta mv \rangle = \frac{h\nu_0}{c}\frac{1}{4\pi}\int_{-\pi/2}^{\pi/2}(1 - \cos\alpha)\, 2\pi \sin\alpha d\alpha = \frac{h\nu_0}{c} \tag{8.5}$$

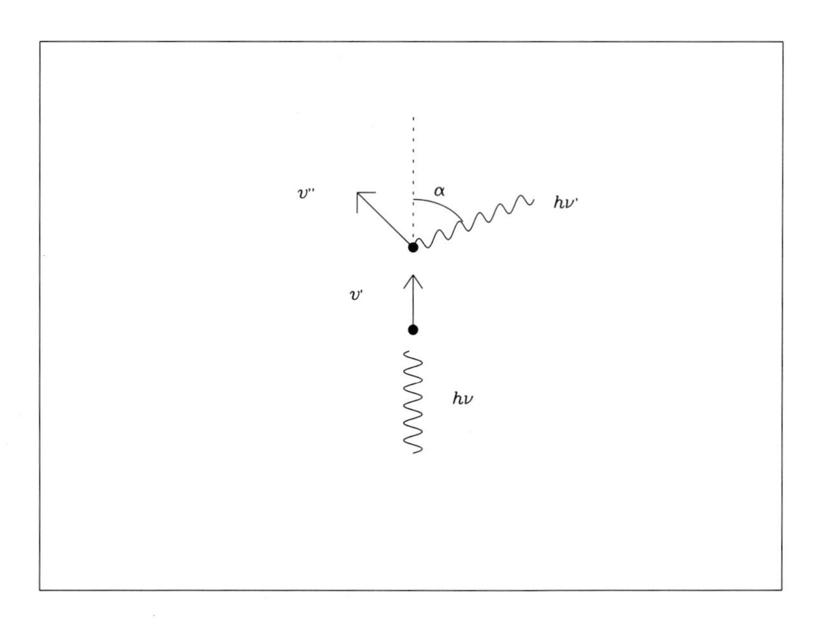

Figure 8.1 The transfer of momentum by absorption and emission of a photon. Since the absorbed photons come from the direction of the star but the re-emission is random in the frame of the atom, the momentum transfer for scattering is about the same as for absorption.

As a consequence, the momentum increase by isotropic scattering is the same as for the case of pure absorption of the photon.

If the absorbed photons were coming from all directions as in the case of an isotropic diffuse radiation field, the net effect on the momentum would also be zero. Hence, isotropic diffuse radiation can be ignored in the acceleration of the wind, as long as $v \ll c$. If the diffuse radiation due to continuum emission and continuum scattering is not isotropic, it cannot be ignored.

The radiative acceleration in the winds of hot stars is provided mainly by the absorption and re-emission of UV photons in the resonance lines of ions of abundant elements such as C, N and O and Fe-group elements in the Lyman continuum. The resonance lines of N IV for instance are at a wavelength of 765 Å. This means that a N^{+++} ion that absorbs such a photon increases its velocity by $h\nu_0/mc = 37$ cm s^{-1}. To accelerate a single N^{+++} ion to the terminal velocity of the wind, say 2000 km s^{-1}, thus requires 5×10^6 absorptions. However, the wind is a plasma and the N^{+++} ions interact frequently with other electrons and other ions, mainly H^+ and He^+ or He^{++}. Due to this interaction the momentum gained by the N^{+++} ions is shared by all constituents of the wind. This means that the absorbing ions need many more absorptions to drive a stellar wind.

Suppose that the ions which provide the dominant radiative acceleration in the winds of hot stars constitute about 10^{-5} of all ions by number and 10^{-4} by mass. We will see in § 8.6 that this is a reasonable assumption because the majority of the ions, H and He, contribute very little to the radiative acceleration because they are fully ionized (H^+, He^{++}) or, in the case of He^+, the strongest lines are in the extreme UV where the stellar flux is small. Suppose that a typical absorption by an ion increases its velocity by 20 cm s^{-1}. This momentum has to be shared by the rest of the gas so the effective velocity increase per absorption is only 2×10^{-3} cm s^{-1}. To accelerate the wind to $v_\infty = 2000$ km s^{-1} requires 10^{11} absorptions per absorbing ion. The terminal velocity of a wind is reached within a few stellar radii, so the time it takes to accelerate the gas to v_∞ is of the order of $3R_*/v_\infty \simeq 10^4$ s if $R_* = 10R_\odot$. Therefore the ions that provide the radiative acceleration have to absorb about 10^7 photons s^{-1}. This implies that only transitions to levels with short lifetimes ($\tau \lesssim 10^{-7}$ s), i.e. transitions with oscillator strengths $f \gtrsim 0.01$, will contribute effectively to the radiative acceleration. Lines with smaller oscillator strengths will also contribute to the radiative acceleration if their number is very large. We will see later that this is indeed the case in the winds of hot stars.

Photons transfer not only momentum to the wind but also energy. The radiation provides the kinetic energy of the wind; the potential energy to lift the gas out of the potential well of the star and the thermal energy of the wind. We will see in § 8.3 that all these forms of energy are less than 10^{-2} of the luminosity of the star. As a result the luminosity of the star decreases very little by the radiative acceleration of the wind.

8.1.2 Momentum transfer by Coulomb coupling

In the line driven winds of hot stars the transfer of momentum from photons to the gas occurs via specific absorbing ions: mainly those of C, N, O, Ne, Si, P, S and Fe-group elements (see § 8.6). The momentum gained by these ions must be shared by the surrounding field particles — protons, helium ions and electrons — to obtain a steady outflow of the whole plasma. This requires that the absorbing ions are slowed down efficiently by interactions with the field particles. This is called *Coulomb coupling* because the interactions are the result of the electric charge of the ions. We will derive a condition for the Coulomb coupling in the winds to be efficient. This process is analogous to the motion of a dense crowd of people in a market place. If a group of heavy-set policemen going after a criminal were forcing their way through the inert crowd by pushing and shoving, soon the whole crowd would be on the move. However, if the density of the crowd was small, the policemen would just run through the crowd and cause little motion.

The transfer of the momentum from the absorbing ions to the field particles is efficient if the characteristic time, t_s, for slowing down these ions by interactions is small compared to the time, t_d, it takes the absorbing ions to gain a large drift velocity with respect to the field particles. This was first shown by Lucy and Solomon (1970) and improved by Lamers and Morton (1976). Hence the condition for Coulomb coupling is

$$t_s < t_d \tag{8.6}$$

The slowing down time, t_s, for moving charged particles with mass A (in units of m_H) and charge Z (in units of the electron charge) due to interaction with H^+, He^{++} and electrons is approximately

$$t_s = 0.305 \frac{A}{Z^2} \frac{T_e^{3/2}}{n_e(1 - 0.022 \ln n_e)} \tag{8.7}$$

in seconds (Spitzer, 1962, p. 135) where n_e is the electron density. The term $(1-0.022 \ln n_e)$ is about 0.5 in stellar winds with $10^8 \leq n_e \leq 10^{12}$.

We define the drift time, t_d, as the time required to increase the velocity of the absorbing ions due to momentum gain from photons by an amount equal to the thermal velocity, $v_{\rm th}$, of the field particles. The most probable thermal velocity is

$$v_{\rm th} = \left(\frac{2k}{m_H}\right)^{1/2} \left(\frac{T_e}{A_f}\right)^{1/2} \tag{8.8}$$

We adopt an atomic mass of $A_f \simeq 1$ for the field particles, i.e. for protons. The drift time is

$$t_d = \frac{v_{\rm th}}{g_i} \tag{8.9}$$

where g_i is the acceleration of the absorbing ions.

The acceleration of the absorbing ions with mass $A \times m_H$, if there is no interactions with field particles, is determined by the momentum transfer from photons to the ions

$$\frac{d(mv)}{dt} = Am_H\, g_i = \frac{\pi e^2}{m_e c} f \frac{\mathscr{F}_{\nu_0}}{c} \tag{8.10}$$

where $(\pi e^2/m_e c)f$ is the cross section for absorption. The flux $\mathscr{F}_{\nu_0}$ at distance r from the star at the frequency of the line ν_0 is

$$\mathscr{F}_{\nu_0} = \mathscr{F}^*_{\nu_0} \cdot \left(\frac{R_*}{r}\right)^2 \tag{8.11}$$

where $\mathscr{F}^*_{\nu_0} = L^*_{\nu_0}/4\pi R_*^2$ is the flux at the stellar surface.

The condition (8.6) for efficient Coulomb coupling can now be written as

$$\frac{L^*_{\nu_0} T_e}{4\pi r^2 n_e} < \frac{Z^2 c}{0.61} \sqrt{2km_H} \left(\frac{\pi e^2}{m_e c} f\right)^{-1} A_f^{-1/2} = 3.6 \times 10^{-6} \tag{8.12}$$

(in cgs units) where the constant is for $A_f = 1$, $f = 0.1$ and $Z = 3$. Notice that this condition is independent of the mass of the absorbing ion, Am_H, because both the slowing down time and the drift time are proportional to A.

Assume that $L^*_{\nu_0}$ is the peak monochromatic luminosity of a star that radiates like a blackbody with a temperature of $T_{\rm eff}$ and the temperature in the wind is $T_e \simeq 0.5\, T_{\rm eff}$. The peak of the Planck curve is at a frequency $\nu_m = 5.83 \times 10^{10}\ T_{\rm eff}$ and the peak flux is $\mathscr{F}_{\nu_m} = 5.97 \times 10^{-16}\ T_{\rm eff}{}^3$ (erg cm^{-2} s^{-1} K^{-3}). So $L^*_{\nu_0} T_e = 5.26 \times 10^{-12} L_*$ (in cgs units). The electron density for a fully ionized wind can be written as $n_e = 5.2 \times 10^{23} \rho$ (cm^{-3}), and expressed in the mass loss rate via the mass continuity equation. This gives the simple condition for Coulomb coupling

$$\frac{L_* v}{\dot{M}} < 5.9 \times 10^{16} \tag{8.13}$$

if L_* is in $L_\odot$, v in km s^{-1} and $\dot{M}$ in $M_\odot$ yr^{-1}. Notice that this condition is independent of distance because both the radiative acceleration and the density drop approximately as r^{-2} if v is about constant. For a typical O star with $L_* \simeq 10^5 L_\odot$, $\dot{M} \simeq 10^{-5}$ $M_\odot$ yr^{-1} and $v \simeq 2000$ km s^{-1} this condition is easily satisfied. So we can safely assume Coulomb coupling for line driven winds of O stars, but it may not be valid for line driven winds of late-B and A main sequence stars which have much lower mass loss rates. Stellar wind models for which the Coulomb coupling is insufficient to distribute the momentum over all components of the gas have been calculated by Springmann and Pauldrach (1993) and Babel (1995).

8.2 The energy and momentum of line driven winds

Before going into detail in the theory of line driven winds we will describe two simple examples that show some of the basic physical effects of these winds.

Suppose a luminous star has a stellar wind that is driven by the radiative acceleration on *only one absorption line.* In this example we assume that the wind is optically very thick in that spectral line. Let ν_0 be the rest frequency of the line and $\mathscr{F}^*_\nu$ be the flux (in erg cm^{-2} s^{-1} Hz^{-1}) at frequency ν leaving the stellar photosphere. For simplicity we assume that the photosphere emits a continuum without photospheric absorption lines, and that the radiation leaves the photosphere radially (as if it were coming straight from a point in the center of the star). The velocity of the stellar wind is almost zero at the photosphere and increases with distance, reaching a terminal velocity v_∞ at $r \to \infty$.

If the *wind is optically very thick* for the line transition, each photon leaving the star at the right wavelength will be absorbed in the wind. All photons with frequency ν_0 will be absorbed in the wind layers just above the photosphere where the velocity is still very small. All photons with frequency $\nu_0(1 + v_\infty/c)$ will be absorbed in the wind by the outer layers which have the terminal velocity v_∞. The difference in the location at which the absorption occurs is due to the Doppler effect. The ions in the wind see the star receding so the radiation from the star is redshifted in the frame of the ions in the wind. (We ignore the thermal motions of the ions and the intrinsic width of the atomic absorption profile and assume the profile is a δ-function.) Since the absorbing ions can only absorb photons at frequency ν_0 *in their rest frame*, these photons must have been emitted by the photosphere at a frequency $\nu_0(1+v/c)$ in the frame of the star. Since the wind is optically thick in the line, all the photospheric radiation in the frequency range between ν_0 and $\nu_0(1+v_\infty/c)$ will be absorbed by the wind layers with a velocity between $v = 0$ and $v = v_\infty$, i.e. by the complete wind (see Fig. 8.2).

The total photospheric radiation energy absorbed per second by the wind is

$$L_{\rm abs} = \int_{\nu_0}^{\nu_0(1+v_\infty/c)} 4\pi R_*^2 \mathscr{F}_\nu d\nu \qquad (8.14)$$

The wind has not only absorbed the energy of these photons but also their momentum: $L_{\rm abs}/c$. If the wind is driven by the radiation pressure of this single line, then the rate at which momentum is carried away in the wind, $\dot{M}v_\infty$, must be equal to the momentum transferred

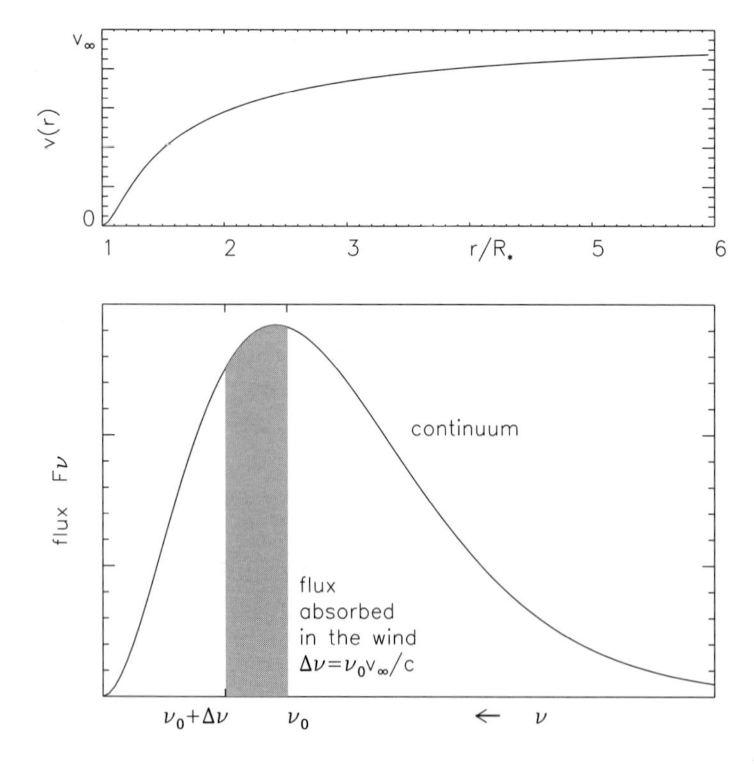

Figure 8.2 The upper figure shows the velocity law $v(r)$ in a stellar wind with a terminal velocity v_∞. The lower figure shows the fraction of the stellar flux absorbed in the wind by one strong absorption line with rest frequency ν_0 near the maximum of the energy curve.

from the photons to the gas in the wind by way of the spectral line opacity

$$\dot{M}v_\infty = \frac{1}{c}\int_{\nu_0}^{\nu_0(1+v_\infty/c)} 4\pi R_*^2 \mathscr{F}_\nu^* \, d\nu \simeq \frac{4\pi R_*^2}{c}\mathscr{F}_{\nu_0}^* \cdot \nu_0 \frac{v_\infty}{c} \tag{8.15}$$

For the last equality we have assumed that the continuum spectrum near the frequency ν_0 is about constant over the frequency interval of the integral. For a typical strong UV line formed in the wind of a hot star at say 1500Å, the absorption is only 15Å wide.

We see that Eq. (8.15) provides an estimate for the mass loss rate of a wind driven by one strong line. Since both sides of the equation contain v_∞, the velocity drops out of the expression, which makes $\dot{M}$ independent of the velocity! The physical reason for this is the following. If the terminal velocity of the wind were increased, say by a factor of two, then photons over a twice as large a frequency band could be absorbed in the wind. This means that the amount of momentum transferred from the radiation to the gas would also increase by a factor of two. But since the momentum of the wind is proportional to v_∞ which is also increased by a factor of two, the amount of momentum transferred to the wind *per unit mass* remains the same.

It should be realized that we have ignored the role of the gravity in this simple analysis. In a later section (8.7) we will show that the inclusion of the gravity restricts the value of v_∞ that can be reached for a given mass loss rate.

To derive a quantitative estimate for the mass loss rate produced by one optically thick line let us assume that the continuum spectrum of the photosphere is a blackbody spectrum with an effective temperature $T_{\rm eff}$, and that the line is situated at the peak of the Planck curve. It turns out that $\nu_m \mathscr{F}_{\nu_m} \simeq 0.62\,\sigma T_{\rm eff}{}^4$, where $\sigma = 5.67 \times 10^{-5}$erg cm^{-2} s^{-1} K^{-4} is the Stefan-Boltzmann constant. Substituting this in Eq. (8.15) we find an estimate for the mass loss rate for one optically thick line

$$\dot{M} \simeq 0.62 L/c^2 \simeq L/c^2 \tag{8.16}$$

For a star of luminosity $10^5 L_\odot$ this implies a mass loss rate of $6.8 \times 10^{-9}\ M_\odot\ {\rm yr}^{-1}$.†

If the wind has $N_{\rm eff}$ optically thick absorption lines the mass loss rate would be

$$\dot{M} \simeq N_{\rm eff} L/c^2 \tag{8.17}$$

† The mass loss driven by one optically thick line happens to be about the same as the mass lost in the stellar *interior* due to nuclear fusion: $\dot{M} = L/c^2$.

where

$$N_{\mathrm{eff}} \equiv \frac{\sum_{i=1}^{N} \int_{\nu_i}^{\nu_i(1+v_\infty/c)} \nu \mathscr{F}_\nu \, d\nu}{\int_0^\infty \mathscr{F}_\nu d\nu} \tag{8.18}$$

and the summation is over all lines for which the wind is optically thick.

In this estimate we have assumed that the absorption bands due to the Doppler shift of the lines in the wind do not overlap. Therefore we call N_{eff} the 'effective number' of optically thick lines, with $N_{\mathrm{eff}} < N$, where N is the real number of absorption lines. This equation does not require that the lines are at the peak of the spectral energy distribution of the star. We see that the mass loss rate increases proportionally to the effective number of optically thick lines in the wind. An observed mass loss of about $10^{-6}\ M_\odot\ \mathrm{yr}^{-1}$ for a star of $10^5 L_\odot$ would require about 150 strong lines in the wind.

There is, however, an upper limit to the mass loss rate driven by line radiation. This is the limit when there are so very many strong spectral lines that *all* the photons leaving the star are absorbed or scattered in the wind. In that case the momentum of the wind will be equal to the momentum of the radiation. So

$$\dot{M}_{\mathrm{max}} v_\infty = L/c \tag{8.19}$$

This is the 'single scattering upper limit' for the wind momentum that is based on the assumptions that only the first scattering of a photon contributes to the radiation pressure, because scattered photons are more or less isotropic and do not contribute significantly to the momentum transfer anymore (see § 8.1.1). In general this is a fairly strict upper limit for radiation driven winds, and most stars abide by this limit. (In § 7.2 we have shown that multiple scattering can increase the momentum of a continuum driven wind up to $\dot{M} v_\infty = (L/c)\tau_W$, where τ_W is the optical depth of the wind if the opacity is distributed evenly over frequency. In § 8.10 we discuss multiple scattering in a line driven wind.)

Let us consider the consequences of Eqs. (8.16) to (8.18). Define the efficiency for the transfer of radiative momentum into wind momentum as

$$\eta_{\mathrm{mom}} = \dot{M} v_\infty/(L/c) \simeq N_{\mathrm{eff}}\ v_\infty/c \tag{8.20}$$

with $0 \leq \eta_{\mathrm{mom}} \leq 1$. If the wind is driven by only one optically thick line, then $\dot{M} \simeq L/c^2$, the maximum efficiency of momentum transfer, i.e $\eta_{\mathrm{mom}} = 1$, requires a wind velocity of $v_\infty \simeq c$. This is because one line can absorb all the radiation from the star only if its absorption

Table 8.1 *Five typical early type stars*

Parameter	ζ Pup	ϵ Ori	P Cyg	τ Sco	WR1
Type	O4 f	B0 Ia	B1 Ia$^+$	B0 V	WN5
$L_*/L_\odot$	7.9×10^5	4.6×10^5	7.2×10^5	3.2×10^4	1.0×10^5
T_{eff} (K)	42 400	28 000	19 300	30 000	40 000
$R_*/R_\odot$	17	33	76	6.5	2.2
$M_*/M_\odot$	59	42	30	15	9
g (cm s^{-2})	5.6×10^3	1.1×10^3	1.4×10^2	9.7×10^3	5.1×10^4
$\dot{M}$ ($M_\odot$ yr^{-1})	2.4×10^{-6}	4.0×10^{-6}	1.5×10^{-5}	7.0×10^{-9}	6×10^{-5}
v_∞ (km s^{-1})	2200	1500	210	2000	2000
N_{eff}	45	129	309	3.2	8800
η_{mom}	2.3×10^{-1}	6.4×10^{-1}	2.2×10^{-1}	2.2×10^{-2}	5.9×10^1
η_{kin}	1.2×10^{-3}	1.6×10^{-3}	7.6×10^{-5}	7.2×10^{-5}	2.0×10^{-1}
η_{pot}	3.3×10^{-4}	3.5×10^{-4}	2.6×10^{-4}	1.6×10^{-5}	7.8×10^{-2}
η_{th}	3.8×10^{-7}	7.0×10^{-7}	1.2×10^{-6}	1.9×10^{-8}	6.8×10^{-5}

extends over the whole spectral range. On the other hand, if the wind is driven by N_{eff} optically thick lines, the maximum velocity that can be reached is $v_\infty \simeq c/N_{eff}$. This is easily understood because the N_{eff} lines have to share the radiation over the full width of the spectrum.

This argument can also be turned around since we usually know v_∞ of a star from observations. The effective number of optically thick lines driving the wind is at most $N_{eff}(\max) \simeq c/v_\infty$, even if the real number of absorbing lines is much larger. If the spectrum is crowded with absorption lines formed in the wind, the frequency ranges where radiation from the star can be absorbed by lines in the wind starts to overlap, which reduces the effective number to at most $N_{eff}(\max)$.

In Table (8.1) we give information about the wind of four typical luminous hot stars. The star P Cygni is a hypergiant with a very low effective gravity and a very high mass loss rate. The stars ζ Pup and ϵ Ori are normal hot supergiants. The star τ Sco is a main sequence star of high gravity and small mass loss rate. The Wolf-Rayet star WR1 has a very high mass loss rate for its luminosity. The table shows that the mass loss rates of hot stars increase rapidly with increasing luminosity, and that the efficiency η_{mom} for converting the momentum of the radiation into the momentum of the wind is of the order unity for early type supergiants but higher than one for WR stars. We also compare the energy loss per second by the star due to the wind to the energy loss by radiation. The energy loss of the wind has three components: the potential energy needed to lift the wind out of the potential well, the kinetic energy when it reached its terminal velocity and the enthalpy, H, of the wind, i.e. the kinetic energy of the gas plus the capacity to do work, (Appendix 2).

$$-\frac{dE_{\rm wind}}{dt} = -\frac{dE_{\rm pot}}{dt} + \frac{dE_{\rm kin}}{dt} + \frac{dH}{dt} = \frac{\dot{M}GM_*}{R_*} + \frac{\dot{M}v_\infty^2}{2} + \frac{5}{2}\frac{\dot{M}\mathscr{R}T_{\rm w}}{\mu} \quad (8.21)$$

where $\mathscr{R}$ is the gas constant and μ is the mean mass per particle in units of m_H. We define the wind efficiency factors

$$\eta_{\rm pot} = \dot{M}GM_*/R_*L \quad (8.22)$$

$$\eta_{\rm kin} = \dot{M}v_\infty^2/2L \quad (8.23)$$

and

$$\eta_{\rm th} = 5\dot{M}\mathscr{R}T_{\rm w}/2\mu L \quad (8.24)$$

with $T_{\rm w} \simeq 0.5\,T_{\rm eff}$ and $\mu \simeq 0.6$ for an ionized wind.

Table (8.1) shows that the energy loss by the wind is very small compared with the luminosity of the star. The fact that $\eta_{\rm mom}$ is of order unity whereas $\eta_{\rm kin}$, $\eta_{\rm pot}$ and $\eta_{\rm th}$ are several orders of magnitude smaller shows that the winds of luminous stars are driven by the transfer of momentum from radiation to the gas, rather than by the transfer of energy.

8.3 The absorption of photons in an expanding atmosphere

In this section we will describe the radiation pressure due to spectral lines in the winds in more detail. We start with a simple consideration of the atomic processes involved and gradually work our way to the general descriptions in macroscopic terms. We will allow for the finite width of the spectral line opacity in this section.

Consider an ion of atomic weight m_i that has an electronic transition between levels ℓ (lower) and u (upper) with energy $h\nu_0$ and wavelength λ_0. The absorption coefficient κ_ν in cm^2 g^{-1} in this line is

$$\kappa_\nu \rho = \frac{\pi e^2}{m_e c} f_\ell n_\ell \left\{1 - \frac{n_u}{n_\ell}\frac{g_\ell}{g_u}\right\} \phi(\Delta\nu) = \kappa_\ell \rho\, \phi(\Delta\nu) \quad (8.25)$$

with

$$\kappa_\ell = \frac{\pi e^2}{m_e c} f_\ell \frac{n_\ell}{\rho} \left\{1 - \frac{n_u}{n_\ell}\frac{g_\ell}{g_u}\right\} \quad (8.26)$$

(κ_ℓ in cm^2 g^{-1} and $\kappa_\ell\rho$ in cm^{-1}) with n_ℓ and n_u the number density in cm^{-3} of the ion in levels ℓ and u respectively, g_ℓ and g_u are the corresponding statistical weights and f_ℓ is the oscillator strength of the transition. The term in brackets contains the correction for stimulated emission. The constant $\pi e^2/m_e c$, where m_e is the mass of the electron, has a value of 0.02654 cm^2 s^{-1}. The profile function $\phi(\Delta\nu)$ with $\Delta\nu = \nu - \nu_0$ is centered around $\Delta\nu = 0$ and normalized to

$$\int_{-\infty}^{\infty} \phi(\Delta\nu) d\Delta\nu = 1 \tag{8.27}$$

with $\phi(\Delta\nu)$ in Hz^{-1} or s. The profile function describes the width of the transition due to the intrinsic width, the damping broadening and the thermal or small scale turbulent motions ('micro-turbulence') in the gas. The density in stellar winds is so low ($n \lesssim 10^{11}$ atoms/cm^3) that damping can be neglected and the width of ϕ is determined by the thermal and turbulent motions only. In that case ϕ is a Gaussian function

$$\phi(\Delta\nu)\, d\Delta\nu = \frac{1}{\sqrt{\pi}} \frac{1}{\Delta\nu_G} e^{-(\Delta\nu/\Delta\nu_G)^2} d\Delta\nu \tag{8.28}$$

with a Gaussian width of

$$\Delta\nu_G = \frac{\nu_0}{c} \sqrt{\frac{2}{3}\left(\langle v_{\rm th}^2\rangle + \langle v_{\rm turb}^2\rangle\right)} \tag{8.29}$$

where $\langle v_{\rm th}^2\rangle = 3kT/m_i$ and $\langle v_{\rm turb}^2\rangle$ are the averages of the squared velocities due to thermal and turbulent motions, respectively. For practical purposes we can assume that the profile function extends to about $1.5\Delta\nu_G$ on either side of ν_0, because the absorption coefficient at $\Delta\nu = 1.5\Delta\nu_G$ is only 10 percent of that in the center of the line.

We now consider the absorption of photons from the star due to this line transition by gas in an outflowing wind. Suppose the wind has a monotonically increasing velocity law $v(r)$. This produces a Doppler shift of the absorption coefficient in the wind. Let us look at the fate of a photon of frequency ν_p emitted by the photosphere along a line z that makes an angle θ' with respect to the radial direction (Fig. 8.3), with $\mu' = \cos\theta'$. The photon can be absorbed by the line transition if somewhere along its path z it encounters ions which have a velocity in the z-direction such that the Doppler shift brings the photon within the line width of the transition in the frame of the ion:

$$\nu_0 - 1.5\Delta\nu_G \leq \nu_p(1 - v_z/c) \leq \nu_0 + 1.5\Delta\nu_G \tag{8.30}$$

(If the line-width was infinitely small, ν_p should be equal to $\nu_0/(1 - v_z/c)$.) The location in the wind where this happens is called the *'line interaction region'*.

Photons emitted at a frequency $\nu_p < \nu_0 - 1.5\Delta\nu_G$ will not have an interaction region with that transition along their path through the wind: they have too low a frequency to be absorbed by the transition anywhere in the receding wind. Photons with $\nu_p \geq (\nu_0 + 1.5\Delta\nu_G)/(1 - v_\infty/c) \simeq (\nu_0 + 1.5\Delta\nu_G)(1 + v_\infty/c)$ cannot be absorbed either: they have too high a frequency to be absorbed anywhere in the wind. Consequently only photospheric photons with

$$(\nu_0 - 1.5\Delta\nu_G) \leq \nu_p \leq (\nu_0 + 1.5\Delta\nu_G)\,(1 + v_\infty/c) \tag{8.31}$$

can be absorbed in the wind by that line.

The geometry of the model is shown in Fig. (8.3). The line of sight to the observer is defined by the impact parameter p. At a distance r from the center of the star the line of sight makes an angle θ with the radial vector. The angles θ and θ' are related by

$$r\,\sin\theta = R_*\,\sin\theta' = p \tag{8.32}$$

The velocity v_z along the line of sight is

$$v_z(r) = \cos\theta\; v(r) = \frac{z}{r} v(r) = \frac{\sqrt{r^2 - p^2}}{r}\, v(r) \tag{8.33}$$

Consider a photon of frequency ν_p emitted by the photosphere (where the subscript p refers to the photosphere) in the direction θ'. The photon can be absorbed in the wind by a transition with a rest frequency ν_0, in a frequency range $\nu_0 + \Delta\nu$, with $-1.5\Delta\nu_G < \Delta\nu < +1.5\Delta\nu_G$ within the profile function. The absorption occurs at a radial distance r from the stellar center which is described by the condition

$$\nu_0 + \Delta\nu = \nu_p \left\{1 - \cos\theta \frac{v(r)}{c}\right\} = \nu_p \left\{1 - \frac{z}{r}\frac{v(r)}{c}\right\} \tag{8.34}$$

or

$$v_z = v(r)\cos\theta = (c/\nu_p)\,\{\nu_p - (\nu_0 + \Delta\nu)\} \tag{8.35}$$

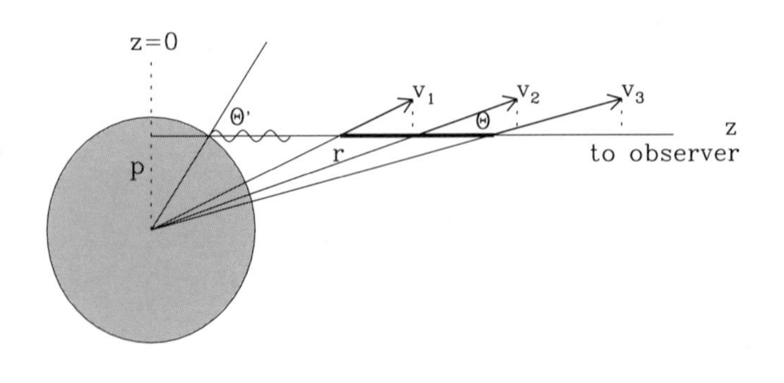

Figure 8.3 A photon of frequency ν_p is emitted from the photosphere at an angle θ' in the line of sight, z, to the observer. This photon can be absorbed in the wind by a line with central frequency $\nu_0 < \nu_p$ and a Doppler width of $\Delta\nu_G$ in the *interaction region*, indicated by a thick horizontal line and described by Eq. (8.36).

This means that the photon can be absorbed or scattered in the *interaction region* of its path, which is given by

$$\frac{\nu_p - \nu_0 - 1.5\Delta\nu_G}{\nu_p} \leq \frac{z}{r}\frac{v(r)}{c} \leq \frac{\nu_p - \nu_0 + 1.5\Delta\nu_G}{\nu_p} \tag{8.36}$$

For a given velocity law $v(r)$ the distance r where the interaction occurs can be calculated from this equation. This is shown in Fig. (8.3).

Figure (8.4) shows the interaction region in another way. The upper figures shows the wind velocity component v_z along the path z of the photon, emitted at frequence ν_p. The middle figure shows the line absorption coefficient κ_ν of a line with central frequency ν_0 for the photon ν_p along its path. The point z_S is the interaction point for the line center where $v_z = c(1 - \nu_0/\nu_p)$ (Eq. 8.30). The coefficient is non-zero in the interaction region, whose limits are indicated by the outer dashed lines. The lower figure shows the optical depth $\tau_\ell(z)$ along the line of sight. It increases from zero to its maximum value in the line interaction region.

8.4 The Sobolev approximation

We have seen in the previous section that a photon emitted from the photosphere can be absorbed by a line transition in the wind only in a narrow interaction region as illustrated in Fig. (8.4). The geometrical width of this region depends on both the velocity gradient in the wind and the frequency width of the profile function.

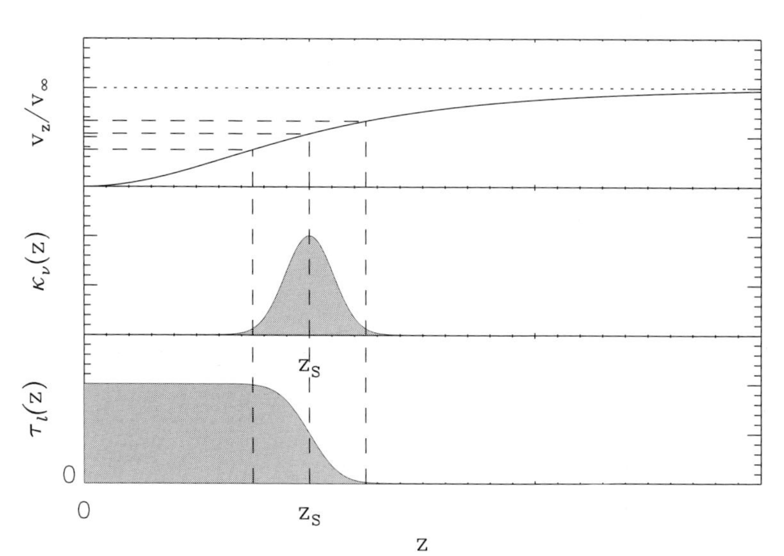

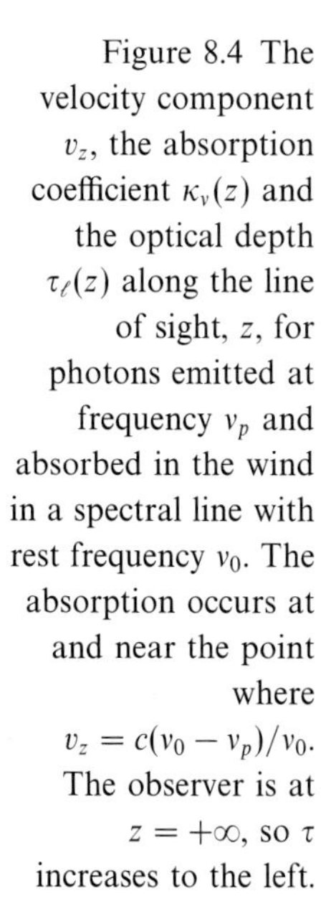

Figure 8.4 The velocity component v_z, the absorption coefficient $\kappa_\nu(z)$ and the optical depth $\tau_\ell(z)$ along the line of sight, z, for photons emitted at frequency ν_p and absorbed in the wind in a spectral line with rest frequency ν_0. The absorption occurs at and near the point where $v_z = c(\nu_0 - \nu_p)/\nu_0$. The observer is at $z = +\infty$, so τ increases to the left.

A narrow absorption profile will give a narrow interaction region. A steep velocity law with a large gradient dv/dr will also give a narrow interaction region. If the interaction region is *very narrow* it is possible to greatly simplify the radiative transfer problem in stellar winds.

Sobolev (1960) derived the equations for radiative transfer in the limit that the interaction region is infinitely narrow. This limit is reached when the width of the profile function, $\phi(\Delta\nu)$, of the line absorption is so small that ϕ can be treated as a delta-function. This is called *the Sobolev approximation.* In that case the optical depth, which normally contains an integration of the density over distance, depends only on the *local* condition at the point where the absorption occurs. This simplifies the solution of the radiative transfer equation in moving atmospheres enormously. In the Sobolev approximation, the interaction region where photons with a given frequency and coming from a given direction can interact with the ions in the wind by line absorption, reduces to a point along the path of the photon. This is called *the Sobolev point.*

The location of the Sobolev point for photons emitted by the photosphere at frequency ν_p along a ray of direction z that are absorbed in a line of frequency ν_0 is given by the conditions

$$\nu_0 = \nu_p \left(1 - \frac{v_z(r_S)}{c}\right) \tag{8.37}$$

or

$$v_z(r_S) = \frac{z}{r} v(r_S) = c \left(1 - \frac{\nu_0}{\nu_p}\right) \tag{8.38}$$

(see Eq. (8.33)), where r_S is the distance of the Sobolev point from the stellar center, and $v_z(r_S)$ is the component of the wind velocity at r_S in the direction z.

In the Sobolev approximation the interaction between the radiation and the ions in the wind is reduced to a 'local' process that can be described by the conditions at the Sobolev point only and does not require knowledge about the optical depth of the wind above and below the Sobolev point. This is shown in Fig. (8.5) which is the equivalent of Fig. (8.4) in the Sobolev approximation. This figure shows that the line absorption coefficient $\kappa_\nu(\Delta\nu)$ is a delta-function and that the optical depth of the wind for photons of frequency ν_p from the photosphere is a step-function with $\tau_\nu = 0$ up to the Sobolev point and constant beyond the Sobolev point.

8.4.1 The Sobolev optical depth

The optical depth for a photon of frequency ν_p along a line of direction z is defined as

$$\tau_{\nu_p}(z_1) = \int_{z_1}^{\infty} \kappa_{\nu_p}(z)\rho(z)dz \tag{8.39}$$

where κ_{ν_p} is the absorption coefficient in cm^2 g^{-1} given by Eq. (8.25). So

$$\tau_{\nu_p}(z_1) = \frac{\pi e^2}{m_e c} f_\ell \int_{z_1}^{\infty} n_\ell(z) \left\{1 - \frac{n_u}{n_\ell}\frac{g_\ell}{g_u}\right\} \phi(\Delta\nu)dz \tag{8.40}$$

(The optical depth is dimensionless.) The ion densities for the lower and upper level of the line transition, n_ℓ and n_u, will vary along the path z. The frequency $\Delta\nu$ in the profile function depends on z through the Doppler relation, Eq. (8.34)

$$\Delta\nu(z) = \nu_p \left\{1 - \frac{v_z}{c}\right\} - \nu_0 \tag{8.41}$$

In the Sobolev approximation $\phi(\Delta\nu)$ is a delta-function and Eq. (8.40) can be written as

$$\begin{aligned}
\tau_{\nu_p}(z_1) &= \frac{\pi e^2}{m_e c} f_\ell \int_{\Delta\nu(z_1)}^{\Delta\nu(z=\infty)} n_\ell(z) \left\{1 - \frac{n_u}{n_\ell}\frac{g_\ell}{g_u}\right\} \frac{dz}{d(\Delta\nu)} \phi(\Delta\nu)d(\Delta\nu) \\
&= \frac{\pi e^2}{m_e c} f_\ell n_\ell(r_S) \left\{1 - \frac{n_u}{n_\ell}\frac{g_\ell}{g_u}\right\}_{r_S} \left(\frac{dz}{d\Delta\nu}\right)_{r_S} \\
&= (\kappa_\ell \rho)_{r_S} \left(\frac{dz}{d\Delta\nu}\right)_{r_S}
\end{aligned} \tag{8.42}$$

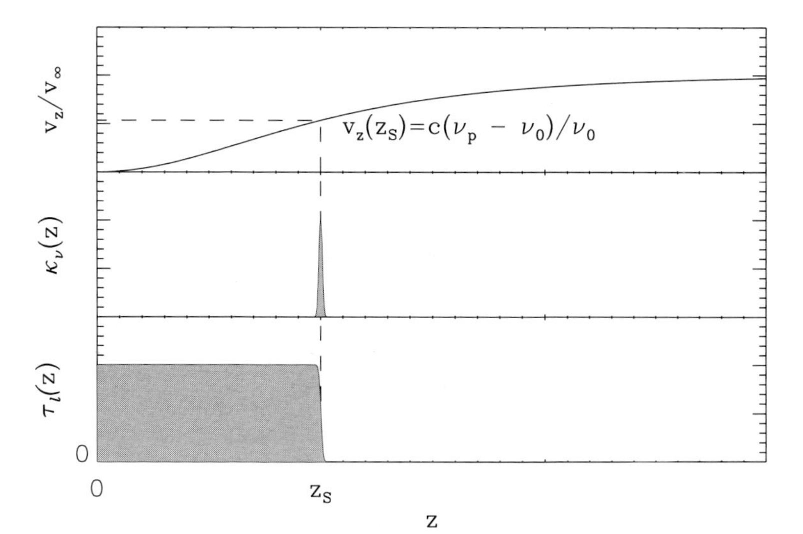

Figure 8.5 The same figure as the previous one but now for the Sobolev approximation of an extremely narrow absorption profile. The absorption occurs at one spot: *the Sobolev point* and the optical depth is a step function. The observer is at $z = +\infty$, so τ increases to the left.

where κ_ℓ, ρ and $dz/d(\Delta\nu)$ are to be evaluated at the Sobolev point given by condition (8.38). This expression is called the *Sobolev optical depth.* It is a step-function that has a value 0 if z_1 is in front of the Sobolev point, and the value of (8.42) beyond the Sobolev point (see Fig. 8.5).

We now derive an expression for the term $(dz/d\Delta\nu)_{r_S}$ in Eq. (8.42). The relation between z and $\Delta\nu$ is given by Eq. (8.34) so

$$\Delta\nu = \nu_p - \nu_0 - \left(\frac{\nu_p}{c}\right)\frac{z}{r}\,v(r) \tag{8.43}$$

Differentiation of $\Delta\nu$ with respect to z with $r = \sqrt{z^2+p^2}$ and $\cos\theta \equiv \mu = z/r$ gives

$$\left(\frac{dz}{d\Delta\nu}\right)_{rs} = \frac{c\,/\,\nu_p}{\left\{(1-\mu^2)\frac{v(r)}{r} + \mu^2\frac{dv(r)}{dr}\right\}_{rs}} \tag{8.44}$$

where μ, r and $v(r)$ should be evaluated at r_S. This provides the final expressions for the *Sobolev optical depth* in terms of the rest frequency ν_0 of the absorption line

$$\begin{aligned}\tau^S_{\nu_0} &= \frac{(\kappa_\ell\rho)_{rs}\,(c/\nu_0)}{\left\{(1-\mu^2)\frac{v(r)}{r} + \mu^2\frac{dv(r)}{dr}\right\}_{rs}}\\ &= \frac{(\kappa_\ell\rho)_{rs}\,\lambda_0\,(r/v)_{rs}}{1\,+\,\sigma\mu^2}\\ &= \frac{\pi e^2}{m_e c}f_\ell\lambda_0 n_\ell(r_S)\left\{1-\frac{n_u}{n_\ell}\frac{g_\ell}{g_u}\right\}_{rs}\frac{(r/v)_{rs}}{1+\sigma\mu^2}\end{aligned} \tag{8.45}$$

where σ is defined by

$$\sigma \equiv \frac{r}{v}\frac{dv}{dr} - 1 \equiv \frac{d\ln v}{d\ln r} - 1 \tag{8.46}$$

We have changed ν_p to ν_0 since $\nu_p \simeq \nu_0$ unless the wind velocity is a substantial fraction of the speed of light. For a β-type velocity law of $v(r) \simeq v_\infty(1 - R_*/r)^\beta$ we find that

$$\sigma = -\frac{r\,-\,(1+\beta)R_*}{r\,-\,R_*} \tag{8.47}$$

which shows that $\sigma > 0$ if $r < (1+\beta)R_*$ and $\sigma < 0$ if $r > (1+\beta)R_*$.

The Sobolev optical depth in the *radial* direction ($\mu = 1$) is

$$\tau^S_{\nu_0}(\mu=1) = (\kappa_\ell\rho)_{rs}\,\lambda_0\left(\frac{dv}{dr}\right)^{-1} \tag{8.48}$$

and in the *tangential* direction ($\mu = 0$) it is

$$\tau^S_{\nu_0}(\mu = 0) \; = \; (\kappa_\ell \rho)_{r_S} \, \lambda_0 \left(\frac{v}{r}\right)^{-1} \tag{8.49}$$

The difference between these two expressions arises because in the radial direction the gradient in the Doppler shift is due to the gradient in the velocity law, whereas in the tangential direction it is due to the spherical divergence of the wind.

8.4.2 Comparison with the optical depth in a static atmosphere

Let us consider the physical implications of Eq. (8.45) for the optical depth in the Sobolev approximation. (For simplicity here we ignore the stimulated emission.) The optical depth in a static atmosphere is

$$\tau_{\nu_0}(\text{static}) = \frac{\pi e^2}{m_e c} f_\ell \int \phi(\Delta\nu) n_\ell dr \tag{8.50}$$

The Sobolev optical depth in the radial direction is

$$\tau_{\nu_0}(\text{Sobolev}) = \frac{\pi e^2}{m_e c} f_\ell \left(\frac{c}{\nu_p}\right) n_\ell(r_S) \left(\frac{dr}{dv}\right)_{r_S} \tag{8.51}$$

We can easily understand the difference between the two expressions if we adopt a simple rectangular absorption profile for $\phi(\Delta\nu)$ with a full width 2Δ and height $\phi(\Delta\nu) = (2\Delta)^{-1}$ to ensure that the integral of ϕ over frequency is unity. In a moving atmosphere with a velocity gradient the value of $\phi(\Delta\nu) = 0$ except in the region where $\nu_0 - \Delta \leq \nu_p(1 + v/c) \leq \nu_0 + \Delta$, i.e. in the layer where $(c/\nu_p)\ (\nu_0 - \nu_p - \Delta) \leq v \leq (c/\nu_p)\ (\nu_0 - \nu_p + \Delta)$. The velocity interval where $\phi(\Delta\nu)$ is positive has a width of $\Delta v = 2\Delta c/\nu_p$. The path where $\phi(\Delta\nu) \neq 0$ has a length $\Delta r = \Delta v(dr/dv)$, and so the integral of Eq. (8.50) should be replaced for a moving atmosphere by

$$\begin{aligned} \int \phi(\Delta\nu) n_\ell dr \simeq \phi(\Delta\nu) n_\ell \Delta r &= (2\Delta)^{-1} n_\ell (2\Delta)(c/\nu_p) dr/dv \\ &= (c/\nu_p) n_\ell dr/dv \end{aligned} \tag{8.52}$$

where n_ℓ and dr/dv refer to the point where the velocity is $v = (c/\nu_p)(\nu_0 - \nu_p)$. Comparing this with Eq. (8.51), we see that Eq. (8.52) explains the difference between the expressions for τ_ν in a static atmosphere and in a wind, for a line of sight in the radial direction, $\mu = 1$, if the path length has been defined as the thickness of the interacting region.

8.4.3 The line interaction region

In the Sobolev approximation the profile function for line absorption is assumed to be infinitely narrow. In reality $\phi(\Delta\nu)$ is not a delta-function because the Gaussian velocity, v_G, in the wind due to thermal and turbulent motions is not zero. This means that the interaction region for a photon is in reality not a single point, but that the interaction can occur over a certain length along the photon path, as shown in Fig. (8.3). That length depends on v_G and on the velocity gradient in the wind.

We define the *Sobolev length*, $\mathscr{L}_s(r,u)$, at point r in the wind as the length over which the velocity component of the wind in the direction ℓ, that crosses the radial vector at an angle $\theta = \arccos\mu$, changes by an amount v_G. So

$$\mathscr{L}_s(r,\mu) = \frac{v_G}{d\mu v(r)/d\ell} = \frac{v_G}{(1-\mu^2)v/r + \mu^2 dv/dr} \tag{8.53}$$

We can consider $\mathscr{L}_s$ as a vector originating at r in the direction ℓ, described by μ, with a length $|\mathscr{L}_s|$. At any point r in the wind the ensemble of vectors $\mathscr{L}_s$ defines a three-dimensional surface. The volume within this surface is called the '*line interaction region*', because it is the region around a point r within which a photon, emitted at r by an atomic transition, can be absorbed by that same transition. The surface of the line interaction region is the locus where the gas is moving away from r with a velocity v_G.

If we assume that the absorption in the line can occur in a frequency range within $1.5\Delta\nu_G$ from the central frequency ν_0, the absorption of a photon with frequency $\nu_p = \nu_0\ (1 + v_\ell(r)/c)$ can occur within 1.5 Sobolev-lengths from r.

Figure (8.6) shows the shapes of the line interaction regions at several distances in a wind with a β-type velocity law

$$v(r) = v_\infty(1 - R_*/r)^\beta \tag{8.54}$$

with $\beta = 0.5$ and 1. The Gaussian velocity is $v_G = 0.015\ v_\infty$. Three-dimensional Sobolev regions are found by rotating the depicted shapes around the radial direction. Close to the star where $dv/dr \gg v/r$ the Sobolev region is flat and extended in the lateral direction. Far from the star where $dv/dr \ll v/r$, because the wind acceleration is much slower than the spherical expansion, the Sobolev region is extended in the radial direction.

An easy way to visualize the shapes of the line interaction regions is to consider the motions of the gas surrounding an ion that moves radially outwards with the wind. For an ion moving with the velocity of the wind, the gas in front of it (i.e. in the direction away from the star) is seen to be moving away because the gas at larger r has a higher outflow velocity. The gas behind it (i.e. in the direction of the star) also is seen to be moving away, because the wind closer to the star has a smaller outflow velocity. The gas on either side of it is also moving away, because of the spherical expansion. This was shown in Fig. (2.8). If the ion is close to the star where the velocity gradient dv/dr is large but v/r is small, the receding motions towards the front and towards the rear will be much larger than to the side. A larger differential velocity implies a shorter Sobolev length, thus the line interaction region will be 'flat'. If the ion is very far in the wind where the velocity has reached its terminal value, v_∞, the receding motions to the side will be larger than toward the front and toward the rear. So the line interaction region will be 'long'.

8.4.4 The condition for the validity of the Sobolev approximation

The Sobolev approximation in its 'pure form', as described in the beginning of this section, assumes an infinitely narrow profile function, so that $\phi(\Delta v)$ can be treated as a delta-function, and the optical depth

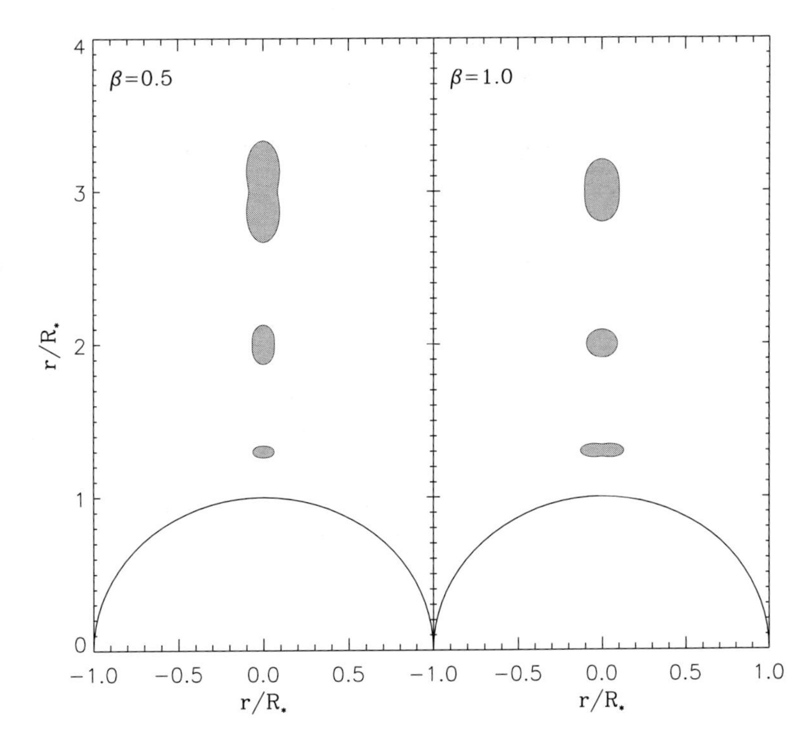

Figure 8.6 The shape and size of the *line interaction regions* at 1.3, 2.0 and 3.0 R_* in a wind with a β-type velocity law and a mean thermal or turbulent velocity of 0.015 v_∞. The left hand figure shows the regions for a wind with β=0.5 and the right hand figure for $\beta = 1$.

τ is a step function. In that case the optical depth τ_S, Eq. (8.43), depends only on the conditions at the Sobolev point. Now that we have introduced the concepts of the Sobolev length and the Sobolev region we can relax the condition for the width of the profile function in the Sobolev approximation.

In the derivation of the Sobolev optical depth we used the delta-function of $\phi(\Delta\nu)$ to write τ_S in the form of Eqs. (8.45) and (8.51). However, we would have obtained the same result if we had 'only' assumed that n_ℓ, n_u and $dz/d\Delta\nu$ were constant over the length of the integration where $\phi(\Delta\nu) \neq 0$. We have shown above that this is the case within three Sobolev lengths from the Sobolev point for a certain frequency and direction. Therefore, the condition for the Sobolev approximation to be valid is that *the changes in the particle densities and in the velocity gradient of the wind over a length of about* $3\mathscr{L}_s$ *should be negligible.*

8.5 The radiative acceleration due to spectral lines

The radiative acceleration, $g_{\rm rad}$, due to the absorption of stellar radiation by a spectral line at some distance in the wind depends on the line opacity at that distance and on the flux, $\mathscr{F}_\nu$, of the incoming radiation. This flux depends on the intensity emitted at the photosphere, and on the absorption in between the photosphere and the distance where we want to calculate the radiative pressure gradient. In general this is a complicated radiative transfer problem. However, if the Sobolev approximation can be adopted, the problem reduces greatly because in the Sobolev approximation a photon emitted by the photosphere can interact with an absorbing ion only at one point, viz. the Sobolev point r_S. This means that the photons from the photosphere which interact with the gas at some point, r, in the wind to provide the radiative acceleration, have had no interaction with the wind before they reached r. They only interact in the Sobolev region immediately surrounding the point r. Thus the flux of radiation that produces the radiative acceleration depends only on the intensity emitted by the photosphere and the *local* conditions at r. The local intensity can be described in terms of a quantity called the *penetration probability*, which describes the fraction of the incoming radiation that reaches a point r through the Sobolev region.

Similarly, the radiation emitted or scattered from that point in the wind can be expressed in terms of an *escape probability*. We will first derive expressions for these quantities and then use them to calculate the radiative acceleration due to one spectral line.

8.5.1 The penetration probability

Consider the intensity of radiation coming from the photosphere that reaches some point r in the wind. The radiation comes in at an angle θ with respect to the radial vector at point r, with $\mu = \cos\theta$. This angle cannot be larger than a certain limit θ_* which describes the angular extension subtended by the photospheric disk as seen from point r

$$\cos\theta_* \equiv \mu_* = \sqrt{1-(R_*/r)^2} \tag{8.55}$$

So the intensity reaching r is zero if $\mu < \mu_*$.

Suppose that the star radiates like a homogeneous disk, i.e. without limb darkening, then the intensity from the photosphere is $I^*_\nu(\mu) \equiv I^*_\nu$. The radiation that reaches point r is

$$I_{\nu_p}(\mu) = I^*_{\nu_p}\, e^{-\tau_{\nu_p}(\mu)} \tag{8.56}$$

where $\tau_\nu(\mu)$ is the optical depth due to the absorption line in the wind. This optical depth is

$$\begin{aligned}\tau_{\nu_p}(\mu) &= \int_{\rm phot}^{r} \kappa_{\nu_p}\rho d\ell \\ &= \kappa_\ell\rho\left(\frac{d\ell}{d\Delta\nu}\right)\int_{\Delta\nu(\rm phot)}^{\Delta\nu(r)} \phi(\Delta\nu)d(\Delta\nu) \\ &= \tau_{\nu_0}\int_{\Delta\nu(\rm phot)}^{\Delta\nu(r)} \phi(\Delta\nu)d\Delta\nu\end{aligned} \tag{8.57}$$

where the integral is along the path of the photons, ℓ, from the photosphere to point r in the wind. We have expressed the absorption coefficient in terms of Eq. (8.25) with $\Delta\nu = \nu_p - \nu_0\ (1 + \mu v(r)/c)$ (Eq. 8.43).

Define the integral of Eq. (8.57) as

$$\Phi(\Delta\nu_\mu) = \int_{\Delta\nu(\rm phot)}^{\Delta\nu(r)} \phi(\Delta\nu)d\Delta\nu \tag{8.58}$$

with

$$\Delta\nu_\mu = \nu_p - \nu_0(1 + \mu v(r)/c) \tag{8.59}$$

† The quantity Φ describes the integral of the absorption profile of the line with rest frequency ν_0 for a photon emitted with frequency ν_p by the photosphere. Since $\phi(\Delta\nu)$ is normalized it follows that $\Phi(\infty) = 1$ and $\Phi(-\infty) = 0$. The definition of Φ is illustrated in Fig. (8.7).

We will assume that there is no *photospheric* absorption line near ν_0,

† In Eq. (8.43) we expressed $\Delta\nu$ relative to ν_p, whereas in Eq. (8.59) it is expressed with respect to ν_0. The two expressions are quantitatively very similar because $\Delta\nu \ll \nu_0$.

consequently the intensity emitted by the photosphere is constant in the frequency interval between $\nu_0 - 1.5\Delta\nu_G$ and $(\nu_0 + 1.5\Delta\nu_G)(1 + v_\infty/c)$, which is the frequency range in which photospheric radiation can be absorbed by the line, so $I^*_{\nu_p} \simeq I^*_{\nu_0}$ where $I^*_{\nu_0}$ is the continuum intensity.

We want to express the mean intensity, J_ν, of the radiation at any point r in the wind in terms of the photospheric intensity $I^*_{\nu_o}$. The intensity of radiation emitted at the photosphere arriving at r with $\mu = \cos\theta$ is

$$I_{\nu_p}(\mu) = I^*_{\nu_o}\, e^{-\tau_{\nu_o}\Phi(\Delta\nu_\mu)} \tag{8.60}$$

The mean intensity J_{ν_p} per steradian of radiation of frequency ν coming from the photosphere and reaching r is the integral of $I_{\nu_p}(\mu)$ over the solid angle of the photospheric disk as seen from a distance r, divided by 4π,

$$J_{\nu_p}(r) = \frac{1}{2}\int_{\mu_*}^{1} I_{\nu_p}(\mu)d\mu = \frac{1}{2}I^*_{\nu_0}\int_{\mu_*}^{1} e^{-\tau_{\nu_o}\Phi(\Delta\nu_\mu)}d\mu \tag{8.61}$$

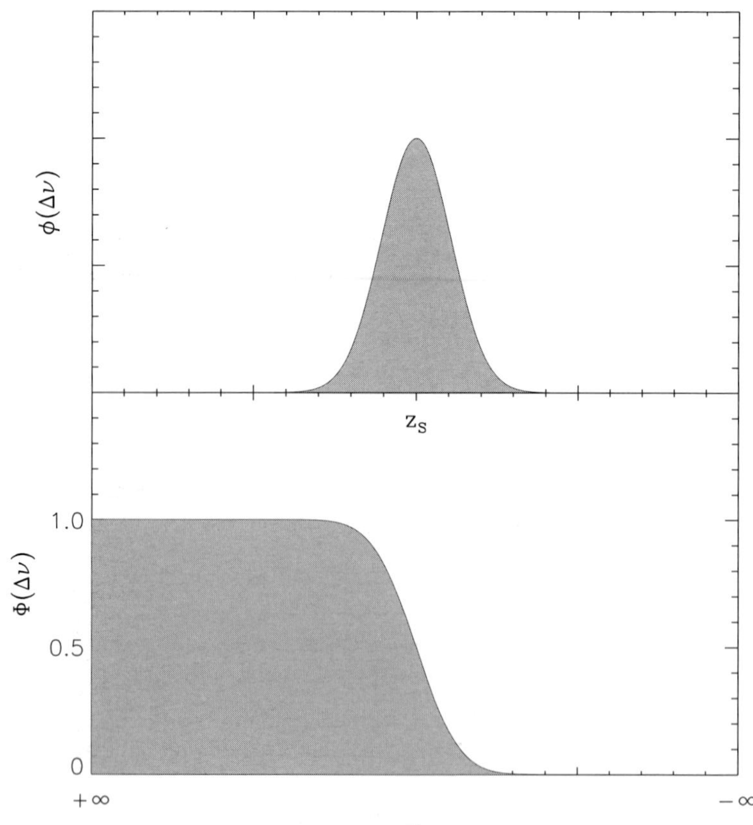

Figure 8.7 The schematic representation of the absorption profile $\phi(\Delta\nu)$ and its integral $\Phi(\Delta\nu)$.

The mean intensity averaged over the line profile is the integral of $J_{\nu_p}(r)\phi(\Delta\nu_\mu)$ over the full line profile

$$\bar{J}(r) = \frac{1}{2} I^*_{\nu 0} \int_{\mu_*}^{1} \int_{\Delta\nu_\mu=-\infty}^{\Delta\nu_\mu=\infty} \phi(\Delta\nu_\mu)\, e^{-\tau_{\nu_0}\Phi(\Delta\nu_\mu)}\, d\Delta\nu_\mu\, d\mu \tag{8.62}$$

The definition of $\Phi(\Delta\nu_\mu)$, given in Eq.(8.58), shows that we can write $\phi(\Delta\nu_\mu)d\Delta\nu_\mu = d\Phi(\Delta\nu_\mu)$, which reduces the integral to

$$\begin{aligned}\bar{J}(r) &= \frac{1}{2} I^*_{\nu_o} \int_{\mu_*}^{1} \int_{0}^{1} e^{-\tau_{\nu_o}\Phi(\Delta\nu_\mu)}\, d\Phi(\Delta\nu_\mu)\, d\mu \\ &= \frac{1}{2} I^*_{\nu_o} \int_{\mu_*}^{1} \frac{1-e^{-\tau_{\nu_o}}}{\tau_{\nu_o}}\, d\mu \\ &= \beta_c(r) I^*_{\nu_o}\end{aligned} \tag{8.63}$$

The *penetration probability* at point r is

$$\beta_c(r) = \frac{1}{2} \int_{\mu_*}^{1} \frac{1-e^{-\tau_{\nu_o}}}{\tau_{\nu_o}}\, d\mu \tag{8.64}$$

where the subscript c stands for the radiation from the 'core', i.e. the photosphere. The dependence of τ_{ν_o} on μ is given in Eq. (8.45). The penetration probability describes the amount of photospheric radiation that reaches a point r in terms of the ratio $\bar{J}(r)/I^*_{\nu_o}$. Let us consider this expression. If the Sobolev optical depth τ_{ν_o} at point r is very large the penetration probability approaches 0. If τ_{ν_o} is very small, $\tau_{\nu_o} \ll 1$, the integrand approaches 1 and $\beta_c \simeq (1-\mu_*)/2$. This is the geometrical dilution factor $W(r)$, (Eq. 2.8),

$$\begin{aligned}W(r) &= \frac{1}{2}\{1-\sqrt{1-(R_*/r)^2}\} \\ &\simeq (R_*/2r)^2 \quad \text{if} \quad r \gg R_*\end{aligned} \tag{8.65}$$

8.5.2 The escape probability

The *escape probability* for photons leaving the Sobolev region around r is analogous to the penetration probability because it is equally difficult (or easy) for a photon of frequency ν_p to reach point r from a far distance at an angle μ as it is for a photon of frequency ν_p to escape the Sobolev region around r at the same angle. So without proof we can write the expression for the *escape probability*

$$\beta(r) = \frac{1}{2} \int_{-1}^{1} \frac{1-e^{-\tau_{\nu_o}}}{\tau_{\nu_o}}\, d\mu = \int_{0}^{1} \frac{1-e^{-\tau_{\nu_o}}}{\tau_{\nu_o}}\, d\mu \tag{8.66}$$

In this case we have to integrate over *all* angles, and not just over μ_* to 1, because the photons can escape in all directions, not just in the direction towards the photosphere.

8.5.3 The radiative acceleration in the point source limit

Now that we have an expression for the intensity of the stellar radiation which penetrates into the Sobolev region around some point r at an angle μ, we can calculate the flux and derive an expression for the radiative acceleration at r due to a spectral line.

In this section and in the rest of this chapter until § 8.9 we only consider the acceleration caused by radiation that comes directly from the stellar photosphere. There is of course also diffuse line radiation in the wind which comes from multiple scatterings. This radiation is nearly isotropic and contributes little to the net acceleration of the wind. Continuum opacity due to bound-free transitions is usually much weaker than line opacity and it is ignored here. The radiative acceleration due to electron scattering will be taken into account, as it is important for the winds of hot stars.

The radiation force on one gram of gas is equal to the momentum of the absorbed radiation per second. This is equal to the radial component of the intensity multiplied by the absorption coefficient and divided by c. The radiative acceleration due to an absorption line is

$$g_{\rm rad} = \frac{2\pi}{c} \int_{\mu_*}^{1} \int_{-\infty}^{+\infty} \kappa_{\nu_p}(\Delta\nu) I_{\nu_p}(\mu) d(\Delta\nu)\mu d\mu \tag{8.67}$$

with the absorption coefficient per gram κ_{ν_p} given by Eq. (8.25), $\Delta\nu$ given by Eq. (8.34), and $I_{\nu_p}(\mu)$ given by Eq. (8.60). This is a difficult integral because of the coupling between $\Delta\nu$ and μ. It can be simplified greatly if we assume for simplicity that all the radiation from the photosphere that is absorbed at some point r in the wind is radially streaming from the star. This is called *the point source limit* because it is equivalent to the assumption that the star is a point source. This is of course a reasonable approximation far from the star, but close to the star it is a poor approximation because radiation may come in at large angle to the radial direction. The point source limit was introduced in the fundamental paper by Castor, Abbott and Klein (1975) with the first line driven wind models.

The reason why we accept this rather inaccurate approximation is that it results in a simple expression for the radiation acceleration due to spectral lines. With this simple expression we can solve the momentum equation of the line driven winds and derive the basic properties of these winds: the mass loss rate, velocity law and terminal velocity. These simple models are in agreement with more accurate models within about a factor of two in $\dot{M}$ and v_∞. In § 8.9 we find that the effects of a finite sized star can be treated as correction factors to the results of the point source model.

The expression for g_{rad} in Eq. (8.67) can be written in terms of the penetration probability analogous to Eq. (8.63)

$$\begin{aligned} g_{\text{rad}} &= \frac{2\pi}{c}\kappa_\ell I^*_{\nu_o}\int_{\mu_*}^{1}\int_{-\infty}^{\infty}\phi(\Delta\nu)\,e^{-\tau_{\nu_o}\Phi(\Delta\nu)}\,d(\Delta\nu)\mu d\mu \\ &= \frac{2\pi}{c}\kappa_\ell I^*_{\nu_o}\int_{\mu_*}^{1}\int_{0}^{1} e^{-\tau_{\nu_o}\Phi(\Delta\nu)}\,d\Phi(\Delta\nu)\mu d\mu \\ &= \frac{2\pi}{c}\kappa_\ell I^*_{\nu_o}\int_{\mu_*}^{1}\frac{1-e^{-\tau_{\nu_o}}}{\tau_{\nu_o}}\mu d\mu \end{aligned} \tag{8.68}$$

In the point source limit where the photons are only streaming radially from the star, we can ignore the factor μ in the integrand because $\mu \simeq 1$, and we find

$$g_{\text{rad}} \simeq \frac{2\pi}{c}\kappa_\ell I^*_{\nu_o}\int_{\mu_*}^{1}\frac{1-e^{-\tau_{\nu_o}}}{\tau_{\nu_o}}d\mu = \frac{4\pi}{c}\kappa_\ell I^*_{\nu_o}\beta_c(r) = \frac{4\pi}{c}\kappa_\ell\bar{J}(r) \tag{8.69}$$

[see Eqs. (8.63) and (8.64)]. The penetration probability for radially streaming photons is

$$\beta_c(r) = \frac{1}{2}\int_{\mu_*}^{1}\frac{1-e^{-\tau_{\nu_o}}}{\tau_{\nu_o}}d\mu \simeq \frac{(1-\mu_*)}{2}\,\frac{1-e^{-\tau_{\nu_o}(\mu=1)}}{\tau_{\nu_o}(\mu=1)} \tag{8.70}$$

with

$$\tau_{\nu_o}(\mu=1) = \kappa_\ell\rho\frac{c}{\nu_o}\left(\frac{dr}{dv}\right) \tag{8.71}$$

The term $(1-\mu_*)/2$ is the geometrical dilution factor $W(r)$, which reduces to $(R_*/2r)^2$ in the point source limit when $R_* \ll r$.

We can easily understand the result of Eq. (8.69) because, if the radiation is only radial then the flux is related to the mean intensity by

$$\mathscr{F}_{\nu_o}(r) = 4\pi\bar{J}(r) \tag{8.72}$$

and the expression for the radiative acceleration is

$$g_{\text{rad}} = \frac{\mathscr{F}_{\nu_o}\kappa_\ell}{c} \tag{8.73}$$

which is the standard form for g_{rad}.

The radiative acceleration due to one line in the point-source limit, where $\mu_* \to 1$ and $\int_{\mu_*}^1 d\mu = 1 - \mu_* \simeq 0.5(R_*/r)^2$ in Eq. (8.68) is

$$g_{\rm rad} = \frac{\pi I^*_{\nu_o}}{c} \left(\frac{R_*}{r}\right)^2 \kappa_\ell \left\{ \frac{1 - e^{-\tau_{\nu_o}(\mu=1)}}{\tau_{\nu_o}(\mu = 1)} \right\} \tag{8.74}$$

with $\tau_{\nu_0}(\mu = 1)$ given by Eq. (8.48).

Let us consider this expression and try to understand the physics behind it. For this we rearrange the terms of Eq. (8.74) as follows.

$$g_{\rm rad} = \left\{ \pi I^*_{\nu_o} \left(\frac{R_*}{r}\right)^2 \right\} \left\{ \frac{\kappa_\ell \rho}{\tau_{\nu_o}(\mu = 1)} \right\} \left\{ 1 - e^{-\tau_{\nu_o}(\mu=1)} \right\} \frac{1}{c\rho} \tag{8.75}$$

The first term in brackets is the monochromatic flux $\mathscr{F}^*_\nu$ from the star. The second term in brackets is

$$\frac{\kappa_\ell \rho}{\tau_{\nu_o}(\mu = 1)} = \frac{\nu_o}{c} \frac{dv}{dr} \tag{8.76}$$

which is the width of the frequency band that can be absorbed [see Eq. (8.48)]. The third term, $1 - e^{-\tau_{\nu o}}$, is the probability that absorption occurs within 1 cm^3. Hence the product of the first three terms is the amount of energy absorbed by the cm^3 of gas. The factor $1/c$ transforms this energy into absorbed momentum, which is the force per cm^3. The factor $1/\rho$ transforms the force into an acceleration. Equation (8.75) for $g_{\rm rad}$ of one line in the point source limit can also be written as

$$g_{\rm rad} = \frac{\mathscr{F}_{\nu_o} \nu_o}{c} \left\{ 1 - e^{-\tau_{\nu_o}(\mu=1)} \right\} \frac{dv}{dr} \frac{1}{c\rho} \tag{8.77}$$

Now consider the expression for the acceleration in the limits of small and large optical depth.

8.5.4 The radiative acceleration due to an optically thin line

For small optical depth $e^{-\tau_{\nu o}} \simeq 1 - \tau_{\nu_o}$ Eq. (8.74) predicts a radiative acceleration

$$g_{\rm rad}(\text{opt. thin line}) \simeq \frac{\pi I^*_{\nu_o}}{c} \left(\frac{R_*}{r}\right)^2 \kappa_\ell \tag{8.78}$$

In this optically thin case $g_{\rm rad}$ is proportional to $(R_*/r)^2$, which describes the decrease of the flux with distance from the star. Note that $g_{\rm rad}$ is proportional to the number of absorbing ions per gram, $\kappa_\ell \sim n_\ell$, because every ion receives the full radiative acceleration by the unattenuated radiation, and the larger the number of absorbing ions per cm^3 the larger the acceleration. Note also that this expression for the radiative acceleration is independent of the velocity gradient.

8.5.5 The radiative acceleration due to an optically thick line

For a line with a large optical depth, $\tau_{\nu_o} \gg 1$, the radiative acceleration is

$$g_{\rm rad}(\text{opt. thick line}) \simeq \frac{\pi I^*_{\nu_o}}{c}\left(\frac{R_*}{r}\right)^2 \frac{\nu_o}{c}\frac{1}{\rho}\frac{dv}{dr} \tag{8.79}$$

In this optically thick case the radiative acceleration does not depend on the number of absorbing particles, but instead it depends on the velocity gradient in the wind. The independence from the number of absorbers is easy to understand because all the radiation that enters into a cm^3 of gas in the wind will be absorbed, independent of n_ℓ, if the photons are in the frequency range that can be absorbed by the line. Because the profile function $\phi(\Delta\nu)$ is so narrow in the Sobolev approximation that it is a δ-function, the width $\Delta\nu$ of the frequency range that can be absorbed is proportional to the velocity difference within that cm^3, so $\Delta\nu = \left(\nu_o/c\right) dv/dr$.

8.5.6 The radiative acceleration due to an ensemble of lines

The radiative acceleration in a stellar wind is produced by a large ensemble of lines, with a wide range of optical depth. The total line radiative acceleration, g_L, is the result of the summation of the contributions by all individual lines, ℓ, with rest frequency ν_ℓ

$$g_L(r) = \frac{2\pi}{c}\sum_\ell \kappa_\ell \int_{\mu_*}^1 I_{\nu_\ell} \frac{1-e^{-\tau_{\nu_\ell}}}{\tau_{\nu_\ell}} \mu d\mu \tag{8.80}$$

In this expression I_{ν_ℓ} is the intensity at the line frequency ν_ℓ of the radiation from the photosphere that reaches point r under an angle $\arccos\mu$. If there are many absorption lines, the intensity from the photosphere may be attenuated by the absorption in other lines between the photosphere and r. If this effect can be neglected then $I_{\nu_\ell}(r,\mu) \simeq I^*_{\nu_\ell}(\mu)$. If the photospheric spectrum is 'smooth', i.e. without sharp and deep absorption lines or discontinuities, and if the photosphere radiates like a homogeneous disk without limb darkening, $I^*_{\nu_\ell}(\mu) = I^*_{\nu_\ell}$ for $\mu_* \le \mu \le 1$, then the radiative acceleration due to lines is

$$g_L(r) = \frac{2\pi}{c}\sum_\ell \kappa_\ell I^*_{\nu_\ell} \int_{\mu_*}^1 \frac{1-e^{-\tau_{\nu_\ell}}}{\tau_{\nu_\ell}} \cdot \mu d\mu \tag{8.81}$$

We will describe the calculations of this total radiative acceleration in the next section.

8.6 A realistic estimate of the radiative force due to lines

The radiative force due to spectral lines in stellar winds can be calculated by summing the contributions described by Eq. (8.80), of all possible lines. This requires the computation of the degree of ionization and excitation for large numbers of energy levels for many different elements. For each possible transition from these levels one also has to calculate the Sobolev optical depth. This is a difficult job that has been done by Abbott (1982), by Pauldrach *et al.* (1986) and by Shimada *et al.* (1994).

The number of spectral lines that has to be taken into account can be reduced drastically if we realize that the densities in the winds are low ($\rho \lesssim 10^{-12}$ g cm^{-3}). In this case collisional excitation can be neglected and only lines from the ground level, from other low-excitation levels and from metastable levels will contribute to the radiative force. With these restrictions the number of possible lines is reduced to about 10^5.

The distribution of the lines over the frequencies is not homogeneous. There are frequency ranges where only a few lines contribute to the radiative acceleration (for instance the range of $\lambda \gtrsim 1200$ Å in the winds of hot stars), and ranges which are crowded with lines (for instance at $300 \lesssim \lambda \lesssim 600$ Å in the winds of hot stars or $1000 \lesssim \lambda \lesssim 3000$ Å in the winds of B and A type stars). This means that many of the absorption lines may overlap in frequency space. It also means that the assumption, that the radiation from the photosphere reaching the Sobolev point of a particular line at a particular distance is not affected by the presence of other lines, is not strictly valid.

This effect of photons being absorbed by different lines at different locations in the wind is called *multiple scattering*. We will ignore it in our discussion of the line driven wind models. But it has been included in some of the sophisticated calculations of the wind models and its effect will be described in § 8.10.

Once the line force is tabulated for a wide range of conditions, the force can be parameterized in a simple way, such that the momentum equation can be solved and the mass loss rate and velocity law of radiation-driven winds can be computed. We first describe the calculation of the radiative acceleration.

8.6.1 The calculation of the radiative acceleration

The radiative acceleration due to spectral lines depends on their Sobolev optical depth. The expression for the optical depth of the lines, Eq. (8.45) contains a part that depends on the line, $f\lambda_0 n_\ell$, and a part that depends on the wind structure, through ρ and dv/dr. This second part is the same for all lines that contribute to the radiative acceleration. Therefore it is convenient to define an optical depth scale for the wind that depends only on the structure of the wind, i.e. proportional to $\rho(dr/dv)$. The optical depth of the lines can then be expressed in terms of this reference optical depth.

Castor, Abbott and Klein (1975), hereafter called 'CAK', introduced the dimensionless optical depth parameter

$$t \equiv \sigma_e^{\rm ref} v_{\rm th} \rho (dr/dv) \tag{8.82}$$

where $\sigma_e^{\rm ref}$ is some reference value for the electron scattering opacity. CAK used $\sigma_e^{\rm ref} = 0.325$ cm^2 g^{-1} (Abbott, 1982). The velocity $v_{\rm th}$ is the mean thermal velocity of the *protons* in a wind with a temperature equal to the effective temperature of the star

$$v_{\rm th} = \sqrt{\frac{2k_B T_{\rm eff}}{m_H}} \tag{8.83}$$

where k_B is the Boltzman constant.†

The total radiative acceleration g_L due to all spectral lines can be expressed in terms of the radiative acceleration due to electron scattering for the reference value of $\sigma_e^{\rm ref}$, times a multiplication factor $M(t)$ which is called the '*force multiplier*'

$$g_L \equiv g_e^{\rm ref} M(t) \tag{8.84}$$

with

$$g_e^{\rm ref} \equiv \frac{\sigma_e^{\rm ref} \mathscr{F}}{c} = \frac{\sigma_e^{\rm ref} L_*}{4\pi r^2 c} \tag{8.85}$$

The radiative acceleration, and hence the value of $M(t)$, depends on the chemical composition and on the ionization and excitation in the wind. The ionization depends on the stellar radiation, i.e. on $T_{\rm eff}$ and on the parameter n_e/W, where n_e is the electron density and $W(r)$ is the geometrical dilution factor. This is because the photo-ionization rate depends on the flux of the star at distance r, which is proportional to $W(r)$, whereas the recombination rate depends on the electron density.

The values of log $M(t)$ are plotted in Fig. (8.8) for different values of $T_{\rm eff}$ as a function of log t, based on calculations by Abbott (1982) and Shimada *et al.* (1994). Notice that these values can very well

† There is some confusion in the literature about the definition of $v_{\rm th}$ and $\sigma_e^{\rm ref}$. The formulae described here are the ones used by CAK and by Pauldrach *et al.* (1986).

be approximated, to an accuracy of about 10 percent, by a simple power law. The dependence on n_e/W, which is not shown here, can also be approximated by a power law. This means that $M(t)$ can be approximated by a simple function of the type

$$M(t) = k\, t^{-\alpha}\, (10^{-11} n_e/W)^{\delta} \tag{8.86}$$

The quantities k, α and δ are called the '*force multiplier parameters*'. They are listed in Table (8.2). The values of α are approximately 0.45 to 0.65. This means that the radiative acceleration is due to a mixture of optically thick lines and optically thin lines. The values of δ are small and of the order of 10^{-1}.

The radiative acceleration due to lines can thus be written as

$$g_L = \frac{\sigma_e^{\mathrm{ref}}}{4\pi c} \frac{L_*}{r^2}\, k\, t^{-\alpha}\, (10^{-11} n_e/W)^{\delta} \tag{8.87}$$

with $\sigma_e^{\mathrm{ref}} = 0.325\ \mathrm{cm}^2\ \mathrm{g}^{-1}$ and k, α and δ listed in Table (8.2). Figure (8.9) shows the variation of the line acceleration as a function of distance in the wind for a wind with a velocity law of $\beta = 0.70$ and force multipliers $\alpha = 0.6$ and $\delta = 0.0$. Notice that g_L increases very steeply above the photosphere, reaches a maximum at $r \simeq 1.1R_*$, and decreases outwards approximately as r^{-2}. This is because

$$g_L \sim r^{-2} \left\{ \rho \frac{dr}{dv} \right\}^{-\alpha} \sim r^{2(\alpha-1)} \left\{ v \frac{dv}{dr} \right\}^{+\alpha} \tag{8.88}$$

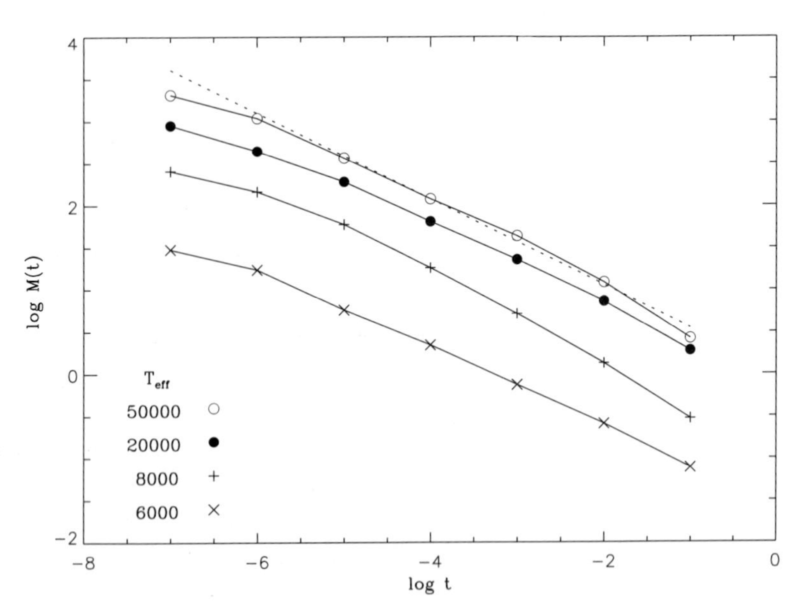

Figure 8.8 The force multiplier $M(t)$ as a function of t for different stellar models. Notice the power law dependence. The dotted line shows the power-law fit for $T_{\mathrm{eff}} = 50\,000$ K, given in the table. (Data by Abbott for 6000 and 8000 K, and Shimada *et al.* for 20 000 and 50 000 K.)

Table 8.2 *The force multiplier parameters*

$T_{\rm eff}$ K	log g	k	α	δ	reference
6000	0.5	0.064	0.465	0.160	A
8000	1.0	0.114	0.542	0.020	A
10000	1.5	0.866	0.454	0.058	S
15000	2.0	0.922	0.446	0.134	S
20000	2.5	0.709	0.470	0.089	S
30000	3.5	0.375	0.522	0.099	S
40000	4.0	0.483	0.526	0.061	S
50000	4.5	0.917	0.510	0.040	S
20000		0.320	0.565	0.020	P
30000		0.170	0.590	0.090	P
40000		0.124	0.640	0.070	P
50000		0.124	0.640	0.070	P

These values are for $\sigma_e^{\rm ref} = 0.325$ cm^2 g^{-1}.
A = Abbott (1982), modified to $(n_e/W)^{\rm ref} = 10^{11}$ cm^{-3}.
S = Shimada *et al.* (1994)
P = Pauldrach *et al.* (1986): non LTE calculations.

The values of the force multipliers listed in Table (8.2) and described by Eq. (8.87) are for solar abundances. Computations of the radiative acceleration for other abundances suggests that the force multiplier depends on the metallicity Z (i.e. the mass fraction of all species other than H and He) as $Z^{1.0}$, thus

$$M(t) = M(t)_{\odot}(Z/Z_{\odot})^{1.0} \tag{8.89}$$

with $Z_{\odot} = 0.017$ (Abbott, 1982; Shimada *et al.*, 1994). So the radiation pressure for stars in the Small Magellanic Cloud, which has a ten times lower metallicity than our Galaxy, is ten times smaller than for Galactic stars of the same mass, radius and temperature.

8.6.2 The lines that drive the winds

Now that we have described the results of the calculations of radiative acceleration, let us consider these in more detail.

Figure (8.10) shows the amount of stellar flux used by the lines in the wind for the radiative acceleration.

First of all, notice that a considerable fraction of the flux is used for the radiative acceleration. This explains why the values of $\eta_{\rm mom}$ which is the efficiency for transferring radiative momentum into the wind, Eq. (8.20), is between 0.1 and 1 for supergiants in Table (8.1). Secondly, notice that the wavelength bands where most of the radiative acceleration occurs gradually shift to longer wavelengths as $T_{\rm eff}$ decreases. There are two reasons for this effect: (i) as $T_{\rm eff}$ decreases the wavelength where most of the energy is emitted by the star moves to longer wavelength, (ii) as $T_{\rm eff}$ decreases the degree of ionization in the wind decreases and the wavelengths of the strong resonance lines move to longer wavelengths.

Figure (8.11) shows the contribution to the radiative acceleration by lines from different groups of ions.

At the highest temperature, $T_{\rm eff} = 50\,000$ K, the largest contribution comes from the lines of Ne to Ca, i.e. mainly Si, S, P. At temperatures in the range of $25\,000 \leq T_{\rm eff} \leq 40\,000$ K the dominant contribution is by C, N and O, i.e. mainly N IV and O IV. However, more

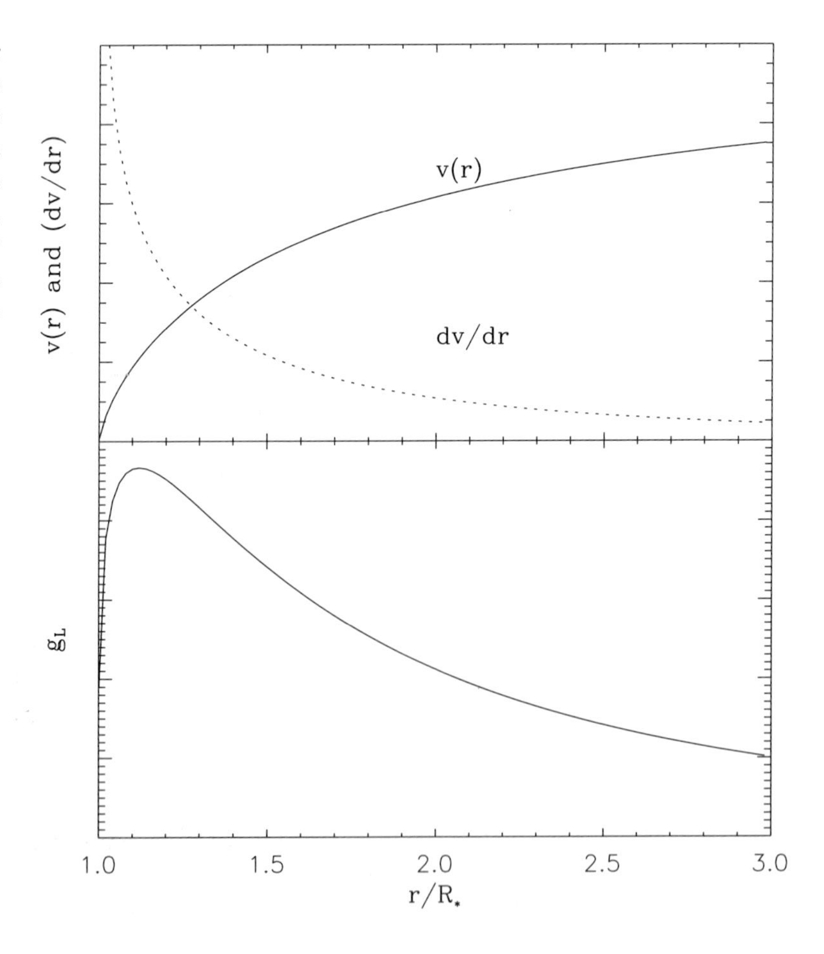

Figure 8.9 The upper figure shows the velocity and the velocity gradient of a $\beta = 0.7$ law. The lower figure shows g_L for this velocity law and for $\alpha = 0.6$, $\delta = 0$.

recent calculations with largely extended line lists indicate that the contribution by Fe-group elements is larger than in Abbott's calculations (Pauldrach, 1987). Between 6000 K and 25 000 K the Fe-group elements provide the dominant contribution, mainly in the form of doubly ionized metals near 20 000 K and as singly ionized metals at

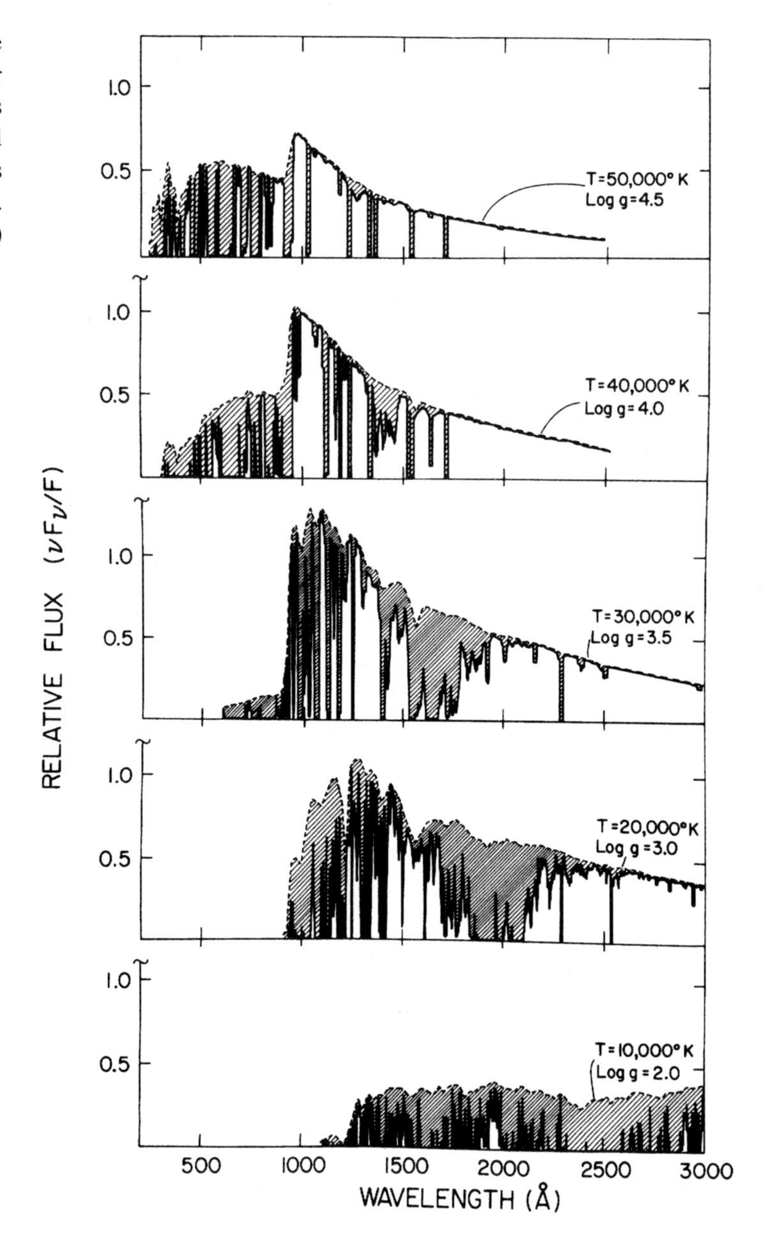

Figure 8.10 The fraction of the stellar radiation that is scattered or absorbed in the winds of stars of different $T_{\rm eff}$. (From Abbott, 1982)

10 000 K or cooler. Hydrogen and helium contribute very little to g_L except for winds of stars with $T_{\rm eff} \leq 6000$ K, where the Balmer lines of H become important.

8.7 Line driven winds in the point source limit

We have shown in the previous section that the radiative acceleration due to spectral lines can be approximated by a simple formula, Eq. (8.87). In this section we will derive radiation driven wind models based on this approximation for g_L and under the assumption that the photosphere can be treated as a point-source, which means that its radiation is only in the radial direction. In § 8.9 we will describe the corrections due to the finite extent of the photosphere.

The momentum equation of a radiation driven wind is

$$\frac{vdv}{dr} = -\frac{GM_*}{r^2} + \frac{1}{\rho}\frac{dp}{dr} + g_e + g_L \tag{8.90}$$

where g_e is the radiative acceleration due to the continuum opacity by electron scattering and g_L that due to the lines. We write

$$g_e(r) = \frac{\sigma_e(r)L_*}{4\pi r^2 c} = \frac{GM_*}{r^2}\Gamma_e(r) \tag{8.91}$$

with

$$\Gamma_e(r) = \frac{\sigma_e(r)L_*}{4\pi cGM_*} \tag{8.92}$$

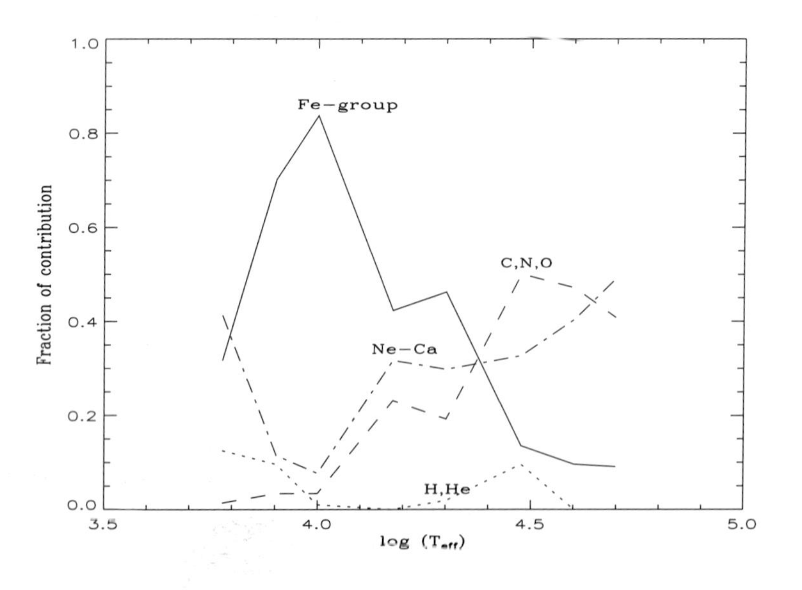

Figure 8.11 The contribution to the radiative acceleration by lines of different ions as a function of temperature. For $T_{\rm eff} \geq 25\,000\ K$ the lines of C, N, O and Ne-Ca dominate the radiative acceleration. For cooler stars the Fe-group elements produce the largest contribution to the line acceleration. H and He hardly contribute. (Data from Abbott, 1982)

The opacity for electron scattering, σ_e in cm^2 g^{-1}, depends on the chemical composition and the degree of ionization of the wind

$$\sigma_e = \sigma_T \frac{n_e}{\rho} = 0.401\,\{I_{\rm H}X + I_{\rm He}(Y/4) + I_{\rm Z}(Z/14)\} \tag{8.93}$$

where $\sigma_T = 6.6524\,10^{-25}$ cm^2 is the Thomson cross section of electrons, and n_e is the number of electrons per cm^3. The quantities X, Y, Z are the mass fractions of H, He and heavier atoms respectively, and $I_{\rm H}$, $I_{\rm He}$, $I_{\rm Z}$ are the number of electrons per ion of H, He or heavier elements. We adopted a mean mass of 14 m_H for an average heavy atom. For early type population I stars the electron scattering opacity is $0.28 < \sigma_e < 0.35$ cm^2 g^{-1}.

If the degree of ionization is constant in the wind, then $\sigma_e(r)$ and $\Gamma_e(r)$ are both constant. In the model discussed here we will assume that Γ_e is a constant. In that case the gravitational acceleration and g_e can be combined into one term $G\,M_{\rm eff}/r^2$, with

$$M_{\rm eff} = M_*(1 - \Gamma_e) \tag{8.94}$$

We will suppose for simplicity that the wind is isothermal and behaves like a perfect gas [see Eqs. (3.5) and (3.6)], thus dp/dr can be expressed in terms of v and the isothermal sound velocity a, which gives $\rho^{-1}dp/dr = (-2a^2/v)dv/dr - 2a^2/r$. The equation of motion then reads

$$\frac{vdv}{dr} = -\frac{GM_*(1-\Gamma_e)}{r^2} + \frac{a^2}{v}\frac{dv}{dr} + \frac{2a^2}{r} + g_L \tag{8.95}$$

After substitution of Eq. (8.87) for g_L and multiplication of Eq. (8.95) by r^2 and rearranging the terms, we obtain the momentum equation for a line driven wind

$$\left(1 - \frac{a^2}{v^2}\right) r^2 \frac{vdv}{dr} = -GM_*(1-\Gamma_e) + 2a^2 r + C(r^2 v dv/dr)^\alpha \tag{8.96}$$

The constant C is defined by

$$C = \frac{\sigma_e^{\rm ref} L_* k}{4\pi c} \left\{\frac{\sigma_e^{\rm ref} v_{\rm th} \dot{M}}{4\pi}\right\}^{-\alpha} \left\{\frac{10^{-11} n_e}{W}\right\}^{\delta} \tag{8.97}$$

In fact n_e/W is not strictly constant in the wind, since $n_e \sim \rho \sim r^{-2}v^{-1}$ and $W \sim r^{-2}$. However, since δ is very small we can assume that the last factor of Eq. (8.97) is about constant.

The momentum equation (8.96) is a nonlinear differential equation. However, it has the character of a critical equation that we have encountered in all earlier chapters. We are looking for a solution with a monotonically increasing velocity law, so the right-hand side of this equation should be zero at the sonic point where $v = a$ and the left-hand side vanishes. The solution of this equation is considerably more difficult than that of previous momentum equations because of its nonlinear character due to the $(r^2 v dv/dr)^\alpha$ term.

8.7.1 The solution with no gas pressure gradient

A first estimate of the solution can be obtained easily if we completely ignore the force due to the gradient of the gas pressure in the momentum equation. This approximation is justified in the supersonic region where the radiative acceleration dominates the acceleration due to the gradient of the gas pressure. The solution of the momentum equation in this first estimate differs only slightly from the more accurate solution that will be described later. This was first shown by Kudritzki (1988).

If we ignore the gradient of the gas pressure in Eq. (8.90) the two terms containing a^2 disappear from Eq. (8.96) and the momentum equation reads

$$r^2 \frac{v dv}{dr} - C \left(r^2 \frac{v dv}{dr} \right)^\alpha = -GM_*(1 - \Gamma_e) = \text{constant} \quad (8.98)$$

This condition can only be fulfilled at every distance r if

$$r^2 \frac{v dv}{dr} \equiv D = \text{constant} \quad (8.99)$$

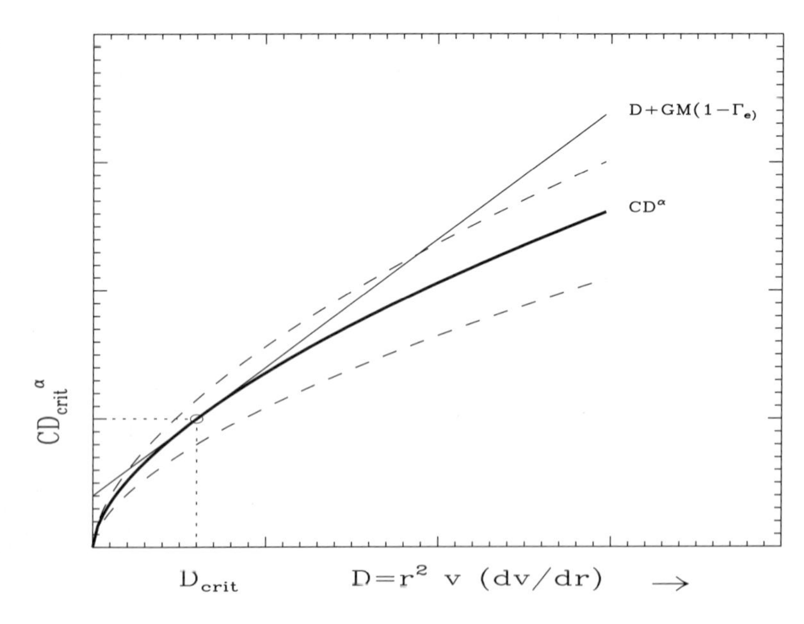

Figure 8.12 The simple solution of the momentum equation for line driven winds derived by ignoring the gas pressure. The figure shows the function $D + GM_*(1 - \Gamma_e)$ (thin straight line) and the function CD^α for three different values of C (dashed and thick curved lines). The thick line shows the critical solution. (After Kudritzki *et al.*, 1989)

thus Eq. (8.98) reduces to the very simple form

$$CD^{\alpha} = D + GM_*(1 - \Gamma_e) \tag{8.100}$$

The left hand side and the right hand side of this equation are plotted in Fig. (8.12) for three values of C. The figure shows that there are either none, or one or two solutions, depending on the value of C. Since C depends on the mass loss rate, Eq. (8.97), and we want the mass loss rate uniquely determined by the momentum equation, we can only accept the value of C for which there is only one solution of Eq. (8.100). This is the case if the straight line of the right hand side of Eq. (8.100) grazes the curved line of the left hand side. For this solution the mass loss rate and the velocity law are uniquely defined and they are self-consistent. If we bring the left hand side of Eq. (8.100) to the right hand side, we require that the resulting equation has a single solution at its minimum (see Fig. 8.12). This minimum is found by differentiating the equation and by finding its zero point. This gives

$$C = \frac{D^{1-\alpha}}{\alpha} \tag{8.101}$$

Substituting this into Eq. (8.100) gives an expression for the velocity law

$$r^2 \frac{v dv}{dr} = D = \frac{\alpha}{1-\alpha} \, GM_*(1 - \Gamma_e) \tag{8.102}$$

Realizing that the left hand side of this equation is $-0.5\, d(v^2)/d(1/r)$ we can easily integrate this expression from the photosphere, where $v(R_*)$ is negligible, outward. This gives

$$v(r) = \left\{ \frac{\alpha}{1-\alpha} \, 2GM_*(1 - \Gamma_e) \left(\frac{1}{R_*} - \frac{1}{r} \right) \right\}^{1/2} = v_\infty \sqrt{\left(1 - \frac{R_*}{r}\right)} \tag{8.103}$$

Notice that this is a β-type velocity law with $\beta = 1/2$, and with a terminal velocity

$$v_\infty = \sqrt{\frac{\alpha}{1-\alpha} \frac{2GM_*(1-\Gamma_e)}{R_*}} = \sqrt{\frac{\alpha}{1-\alpha}} \, v_{\rm esc} \tag{8.104}$$

So the terminal velocity of a radiation driven wind scales with the escape velocity at the photosphere and with $\sqrt{\alpha/(1-\alpha)}$. Notice that v_∞ depends only on α and not on k.

The mass loss rate for this simple model follows from the substitution of C, Eq. (8.101) into (8.97). This gives

$$\dot{M} = \frac{4\pi}{\sigma_e^{\rm ref} v_{\rm th}} \left(\frac{\sigma_e^{\rm ref}}{4\pi} \right)^{\frac{1}{\alpha}} \left(\frac{1-\alpha}{\alpha} \right)^{\frac{1-\alpha}{\alpha}} (k\alpha)^{\frac{1}{\alpha}} \left\{ \frac{10^{-11} n_e}{W} \right\}^{\frac{\delta}{\alpha}} \left(\frac{L_*}{c} \right)^{\frac{1}{\alpha}} \{GM_*(1 - \Gamma_e)\}^{\frac{\alpha-1}{\alpha}} \tag{8.105}$$

The mass loss rate and the terminal velocity derived in this section, by ignoring the effects of gas pressure, are very similar to the ones derived from the full solution of the momentum equation that we will discuss next. So obviously the gas pressure does not play a significant role in the line driven winds of hot stars.

8.7.2 The full analytical solution of the momentum equation

Here we consider the solution of an isothermal radiation driven wind with the gradient of the gas pressure taken into account. We will derive the full solution, following Castor, Abbott and Klein (1975) and Cassinelli (1979).

The momentum equation (8.96) can be written as

$$\left(1-\frac{a^2}{v^2}\right) r^2 vv' + \left\{GM_*(1-\Gamma_e) - 2a^2 r\right\} - C(r^2 vv')^\alpha = 0 \quad (8.106)$$

with $v' = dv/dr$. We will call the point where $GM_*(1-\Gamma_e) = 2a^2 r$ the *Parker point*, r_p, because it is the critical point in the momentum equation for the solar wind, first discussed by Parker (1958). If this equation is solved by numerical integration either from inside at the photosphere or from outside at $r = \infty$, there are four kinds of solutions. These are shown schematically in Fig. (8.13):

(a) Solutions that start subsonic at $r = R_*$ and remain subsonic. These solutions fail at and beyond the Parker point because all terms in Eq. (8.106) are negative. These solutions end at the forbidden region B.
(b) Solutions that start subsonic, go smoothly through the sonic point, become supersonic and fail at the Parker point where the solution of Eq. (8.106) requires $r^2 vv' = 0$, because otherwise the first and third term of Eq. (8.106) do not exactly cancel.
(c) One solution that starts subsonic, grazes the forbidden region A (see below) and fails at the Parker point where $r^2 vv' = 0$.
(d) Solutions that start subsonic at $r = R_*$ and reach supersonic velocities too close to the star where $GM_*(1-\Gamma)-2a^2$ is still large. These solutions fail somewhere between the sonic point and the Parker point when $dF/dv' = 0$ (see below). This occurs at the edge of the forbidden region A in Fig. (8.13).
(e) Solutions that start supersonic at $r = \infty$ with too high a velocity and enter the forbidden region A.
(f) One solution that starts supersonic at $r = \infty$, grazes the forbidden region A and fails at the sonic point.

(g) Solutions that start supersonic at large distances, $r = \infty$, go smoothly through the Parker point, and fail at the sonic point, where the velocity gradient becomes infinite, because the second and third term of Eq. (8.106) do not cancel at the sonic point.
(h) Solutions that start supersonic at large distances and enter the forbidden region at the sonic point.

We see that only *one* solution has the desired property of smoothly going from subsonic to supersonic velocities. This is a combination of solutions c and f which connect at the point where they graze the forbidden region A. This is a unique solution. The critical point of this solution is not the sonic point, nor the Parker point, but the point in between where the two solutions (curves c and f) meet each other with the same velocity gradient.

We can express this condition in mathematical terms and derive a formula for the resulting mass loss rate. The momentum equation (8.106) can be written as

$$F(r, v, v') = 0 \tag{8.107}$$

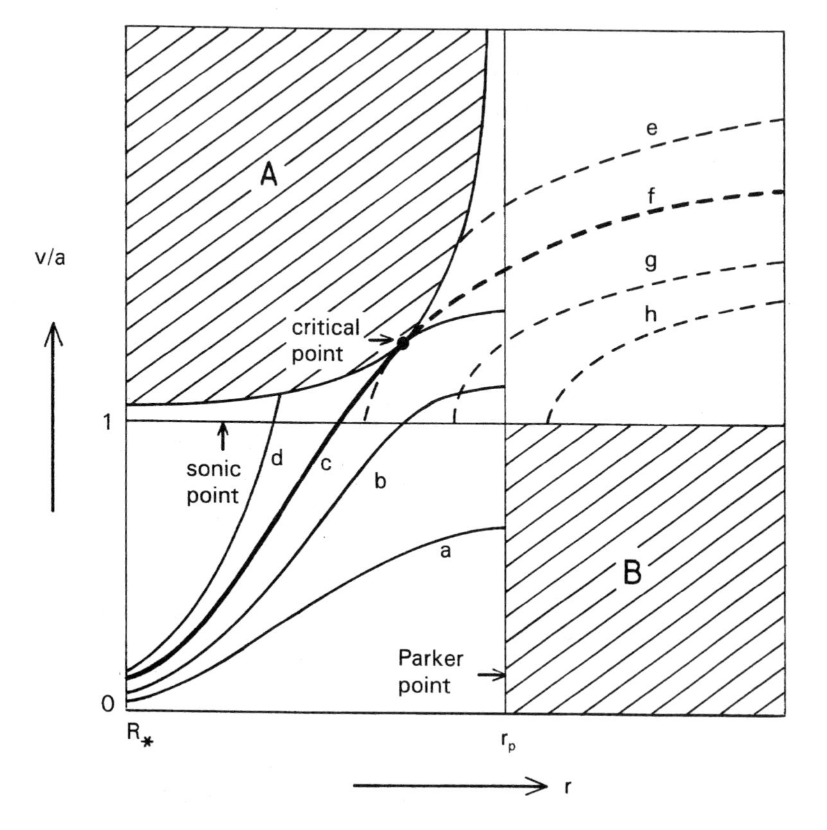

Figure 8.13 The topology of the solutions of the momentum equation (see text). The thick line gives the transonic solution. (After Cassinelli, 1979)

The condition that the solution 'grazes the forbidden region A', in other words that there is a point where the momentum equation in the form of Eq. (8.100) has only one solution, is given by the *singularity condition* that $\partial F/\partial v' = 0$ at the singular point. The condition that at this point the velocity gradient is continuous, where we switch from solution c to solution f in Fig. (8.13), requires the *regularity condition* that $v'' = d^2v/dr^2$ is defined at the singular point. We derive an expression for v''.

The momentum equation requires that $dF/dr = 0$ all along the solution curve. So

$$\begin{aligned}\frac{dF}{dr} &= \frac{\partial F}{\partial r} + \frac{\partial F}{\partial v}\frac{dv}{dr} + \frac{\partial F}{\partial v'}\frac{dv'}{dr} \\ &= \frac{\partial F}{\partial r} + v'\frac{\partial F}{\partial v} + v''\frac{\partial F}{\partial v'} = 0\end{aligned} \quad (8.108)$$

Solving this expression for v'' we find

$$v'' = -\frac{\left(\dfrac{\partial F}{\partial r} + v'\dfrac{\partial F}{\partial v}\right)}{\dfrac{\partial F}{\partial v'}} \quad (8.109)$$

We want v'' to be defined at the critical point where $\partial F/\partial v' = 0$. This means that the numerator of Eq. (8.109) should be zero at the critical point. The momentum equation and the *two conditions at the critical point* r_c, are

$$F = 0 \qquad \text{momentum equation} \quad (8.110)$$

$$\left(\frac{\partial F}{\partial v'}\right)_c = 0 \qquad \text{singularity condition} \quad (8.111)$$

$$\left(\frac{\partial F}{\partial r}\right)_c + \left(v'\frac{\partial F}{\partial v}\right)_c = 0 \quad \text{regularity condition} \quad (8.112)$$

These three equations can be solved to give r_c, $v_c \equiv v(r_c)$ and $v_c' \equiv v'(r_c)$. Since the solution of these three equations depends on the constant C in Eq. (8.96) which depends on the mass loss rate $\dot{M}$, the results give an expression for $\dot{M}$.

Equations (8.110) to (8.112) give, respectively,

$$\left(1 - \frac{a^2}{v^2}\right) r^2vv' + GM_*(1-\Gamma_e) - 2a^2r - C(r^2vv')^\alpha = 0 \quad (8.113)$$

$$\left(1 - \frac{a^2}{v_c^2}\right) r_c^2v_c - \frac{\alpha}{v_c'}C(r_c^2v_cv_c')^\alpha = 0 \quad (8.114)$$

$$\left(1 - \frac{a^2}{v_c^2}\right) 2r_c v_c v_c' - 2a^2 - \frac{2\alpha}{r_c} C(r_c^2 v_c v_c')^\alpha$$
$$+\frac{v_c'}{v_c}\left\{\frac{2a^2}{v_c^2} r_c^2 v_c v_c' + \left(1 - \frac{a^2}{v_c^2}\right) r_c^2 v_c v_c' - \alpha C(r_c^2 v_c v_c')^\alpha\right\} = 0 \quad (8.115)$$

Equation (8.113) is valid at all distances because it is the momentum equation of the wind. Equations (8.114) and (8.115) are only valid at the critical point. Equation (8.114) gives an expression for the radiative acceleration at the critical point

$$C(r_c^2 v_c v_c')^\alpha = \left(1 - \frac{a^2}{v_c^2}\right) \frac{r_c^2 v_c v_c'}{\alpha} \quad (8.116)$$

Substituting this into Eq. (8.113) gives an expression for $r_c^2 v_c v_c'$

$$r_c^2 v_c v_c' = \left(\frac{\alpha}{1-\alpha)}\right) \{GM_*(1-\Gamma_e) - 2a^2 r_c\} \left\{1 - \frac{a^2}{v_c^2}\right\}^{-1} \quad (8.117)$$

Substitution of Eq. (8.116) into (8.115) gives a very simple expression for the velocity gradient at the critical point

$$v_c' = v_c / r_c \quad (8.118)$$

With this velocity gradient we find an expression for v_c from Eq. (8.117)

$$v_c^2 = a^2 + \left(\frac{\alpha}{1-\alpha}\right) \left\{\frac{GM_*(1-\Gamma_e)}{r_c} - 2a^2\right\} \quad (8.119)$$

Note that $v_c > a$.

Now we try to find an expression for C, which will give us the mass loss rate. From Eqs. (8.116) and (8.119) we find

$$C = \left(\frac{1}{1-\alpha}\right) \frac{1}{r_c^\alpha v_c^{2\alpha}} \{GM_*(1-\Gamma_e) - 2a^2 r_c\} \quad (8.120)$$

Combining this with the definition of C, Eqs. (8.97) and (8.119), gives the mass loss rate

$$\dot{M} = \left(\frac{4\pi}{\sigma_e^{\text{ref}} v_{\text{th}}}\right) \left(\frac{\sigma_e^{\text{ref}} k}{4\pi c}\right)^{1/\alpha} (1-\alpha)^{1/\alpha} L_*^{1/\alpha} \left\{\frac{10^{-11} n_e}{W}\right\}^{\delta/\alpha}$$
$$\{GM_*(1-\Gamma_e) - 2a^2 r_c\}^{-1/\alpha}$$
$$\left[\left(\frac{\alpha}{1-\alpha}\right) \{GM_*(1-\Gamma_e) - a^2 r_c\} + a^2 r_c\right] \quad (8.121)$$

This expression, as well as the one for the velocity at the critical point, Eq. (8.119), contains the location of the critical point r_c. The way in which r_c is found is described in the next section.

8.7.3 The mass loss rate and the terminal velocity

Suppose we want to calculate the mass loss rate and the terminal velocity of the wind of a star with given stellar parameters L_*, R_*, M_* and Γ_e. Then one assumes a value of r_c, say about $2R_*$. This gives an estimate of $\dot{M}$, Eq. (8.121), v_c, Eq. (8.119) and v_c', Eq. (8.118). The momentum equation (8.106) is then numerically integrated inwards and outwards. This gives the velocity and the density structure below and above the critical point.

With the density structure known, one can calculate the optical depth in the wind for continuum electron scattering, τ_e. An optical depth of $\tau_e = 2/3$ should be reached at the pre-specified radius R_*. This is because the radiation pressure by lines can only become important in the wind when the continuum optical depth is less than 2/3. In the layers where $\tau_e \geq 2/3$ the radiation is almost isotropic which reduces the radiation pressure due to lines considerably. So the assumed expression for the line acceleration g_L is valid for $\tau_e < 2/3$. This means that the lower boundary of the momentum equation, R_*, is the layer where $\tau_e = 2/3$,

$$\int_{R_*}^{\infty} \rho(r)\sigma_e dr = 2/3 \tag{8.122}$$

If Eq. (8.122) is not satisfied, the value of r_c should be adjusted and the momentum equation solved again. This process is iterated until Eq. (8.122) is satisfied. It turns out that the mass loss rate is not very sensitive to the location of r_c, so the final solution can be reached in a few iterations.

We will show below (Eq. 8.128) that $r_c \simeq 1.5R_*$. This means that in most cases we can ignore the terms of order $2a^2 r_c$ compared to terms of order $GM_*(1-\Gamma_e)$ because the escape velocity at the photosphere is usually much larger than the sound speed. This simplifies the results enormously. In that case we find from Eqs. (8.119) and (8.121)

$$v_c^2 \simeq \left(\frac{\alpha}{1-\alpha}\right) \frac{GM_*(1-\Gamma_e)}{r_c} = \left(\frac{\alpha}{1-\alpha}\right) \frac{R_*}{2r_c} v_{\rm esc}^2 \tag{8.123}$$

and

$$\dot{M} \simeq \frac{4\pi}{\sigma_e^{\rm ref} v_{\rm th}} \left(\frac{\sigma_e^{\rm ref}}{4\pi}\right)^{\frac{1}{\alpha}} (k\alpha)^{\frac{1}{\alpha}} \left(\frac{1-\alpha}{\alpha}\right)^{\frac{1-\alpha}{\alpha}} \left(\frac{L_*}{c}\right)^{\frac{1}{\alpha}} \{GM_*(1-\Gamma_e)\}^{\frac{\alpha-1}{\alpha}} \left\{\frac{10^{-11} n_e}{W}\right\}^{\frac{\delta}{\alpha}} \tag{8.124}$$

This expression for $\dot{M}$ is exactly the same as Eq. (8.105) derived in

§ 8.7.1 which was found by ignoring the force due to the gas pressure in the momentum equation right from the beginning.

Note that this mass loss rate is proportional to $L_*^{1/\alpha} \cdot M_{\mathrm{eff}}^{(\alpha-1)/\alpha}$. Since α is of the order of 0.52 the line driven wind theory predicts that $\dot{M} \sim L_*^{1.9} M_{\mathrm{eff}}^{-0.9}$. The luminosity of luminous stars scales approximately as $M^{2.5}$, so we expect that $\dot{M} \sim L_*^{1.5}$. This agrees well with the observations.

The velocity law of the stellar wind can be found by integrating the momentum equation (8.106) from the critical point inward and outward. In the region where $v > a$, and where $GM_*(1-\Gamma_e) > 2a^2 r$, which is the case throughout most of the subcritical region, $r < r_c$, and in the supercritical region, $r > r_c$, up to large distances, we can ignore the terms of the order a^2/v^2 and the term $2a^2 r$ compared to $GM_*(1-\Gamma_e)$ in the equation of motion (8.106). This implies that the value of $r^2 v v'$ must be constant throughout that region, at the same value as at the critical point (Eq. 8.117), so

$$r^2 \frac{v dv}{dr} \simeq \left(\frac{\alpha}{1-\alpha}\right) GM_*(1-\Gamma_e) \tag{8.125}$$

The solution of this differential equation is

$$v^2(r) = v_c^2 + 2\left(\frac{\alpha}{1-\alpha}\right) GM_*(1-\Gamma_e)\left(\frac{1}{r_c} - \frac{1}{r}\right) \tag{8.126}$$

Note that this is very similar to Eq. (8.103), apart from the term v_c^2. Substituting v_c from Eq. (8.119) and ignoring the term of order a^2/v_{esc}^2, we find

$$v^2(r) = \left(\frac{\alpha}{1-\alpha}\right) v_{\mathrm{esc}}^2 \left\{\frac{3}{2}\frac{R_*}{r_c} - \frac{R_*}{r}\right\} \tag{8.127}$$

This velocity law is valid over the distance range where $v(r) > a$. Let us assume that the velocity at the lower boundary of the wind is approximately $v(R_*) \simeq a$. Substitution of this value into Eq. (8.127) provides an estimate for the location of the critical point

$$r_c = \frac{3}{2} R_* \left\{1 + \frac{1-\alpha}{\alpha}\frac{a^2}{v_{\mathrm{esc}}{}^2}\right\}^{-1} \tag{8.128}$$

The term is brackets is very close to unity, and we see that the critical point of a line driven wind is located at $r_c \simeq 1.5 R_*$. This is at a much larger distance than the sonic point! Substitution of Eq. (8.128) into Eq. (8.127) gives the final expression for the velocity law

$$v^2(r) = a^2 + \left(\frac{\alpha}{1-\alpha}\right) v_{\mathrm{esc}}{}^2 \left\{1 - \frac{R_*}{r}\right\} \tag{8.129}$$

This is approximately

$$v(r) \simeq v_\infty \sqrt{1 - \frac{R_*}{r}} \tag{8.130}$$

with

$$v_\infty \simeq \sqrt{\frac{\alpha}{1-\alpha}}\, v_{\rm esc} \tag{8.131}$$

The velocity at the critical point $r \simeq 1.5R_*$ of line driven winds is approximately

$$v_c \simeq \frac{v_\infty}{\sqrt{3}} = 0.577 v_\infty \tag{8.132}$$

This is much higher than the sound speed and it is of the order of the escape velocity at the photosphere.

So including the gas pressure into the momentum equation of line driven winds results in a solution that has a critical point at typically 1.5 R_*. The solution through the critical point gives a unique value for the mass loss rate and the terminal velocity that is almost identical to the solution of the momentum equation without gas pressure.

8.8 A physical explanation of the line driven wind models

In the preceeding section we have derived expressions for the mass loss rate, the velocity law and the terminal velocity of isothermal line driven winds in the point source limit. Let us try to understand these results in physical terms.

For this purpose we ignore the effects of gas pressure in the wind, as we did in § 8.7.1. This is allowed for two reasons: firstly, the force due to the gradient in the gas pressure is negligible compared to the radiative force and to the gravity and, secondly, the sound velocity is negligible compared to the flow velocity. The latter means that we can set the factor $1 - (a^2/v^2)$ in the momentum equation (8.96) to unity. This is not trivial and requires an explanation. In the discussion of isothermal wind models in previous chapters we have argued that the factor $v^2 - a^2$ in the momentum equation is crucial because it determines the velocity at the critical point of the momentum equation, $v_c = a$, and hence the mass loss rate. In the case of line driven winds, however, the critical point of the momentum equation is not the sonic point, but it is at the much higher velocity of $v_c \simeq v_\infty/\sqrt{3}$ where $(dv/dr)_c = v_c/r_c$, Eq. (8.118). This means that the sound speed is much lower than the critical velocity and the sonic point is far below the critical point. Therefore we can assume $a/v \simeq 0$ throughout isothermal line driven winds.

The momentum equation then reduces to that of Eq. (8.98) or (8.100). The solution requires that $r^2 v dv/dr$ is constant. This is because the velocity law depends on the radiative acceleration, but the radiative acceleration in turn depends on the velocity gradient to the power α. There is only one form of velocity law that is consistent with this mutual coupling between $v(r)$ and g_L. That is the one with a constant value of $r^2 v dv/dr$. This is the case for *any* value of α, also for $\alpha = 1$ (i.e. force due to optically thick lines) or $\alpha = 0$ (i.e. force due to optically thin lines) as can be seen immediately from Eq. (8.98).

If $r^2 v dv/dr$ is constant, the radiative acceleration is proportional to the flux and varies as r^{-2}, but the gravity also varies as r^{-2}. The resulting momentum equation is then of the type

$$v \frac{dv}{dr} = \frac{\text{constant}}{r^2} \equiv \frac{D'}{r^2} \tag{8.133}$$

which has a solution of a β-type velocity law with $\beta = 1/2$ and a terminal velocity of $v_\infty = \sqrt{2D'/R_*} = \sqrt{\alpha/(1-\alpha)} v_{\rm esc}$.

The terminal velocity does not depend on the mass loss rate. This is surprising because the radiative acceleration g_L depends on the density in the wind and hence on the mass loss rate. The expressions for g_L and t [Eqs. (8.87) and (8.82)] show that $g_L \propto \dot{M}^{-\alpha}$. Thus we might have expected that v_∞ would also depend on $\dot{M}^{-\alpha}$. This is not the case, however, because the mass loss rate is not a free parameter in line driven wind models.

The mass loss rate of line driven winds is determined by the fact that the radiative force must be able to accelerate the wind from $v(R_*) \simeq 0$ to $v(r) > v_{\rm esc}(r)$. This can be done for only one value of the mass loss rate. This is because the radiative acceleration g_L depends on the density and hence on the mass loss rate. The value of $\dot{M}$ is set by the condition that the velocity law should be continuous and smooth. At the critical point of the momentum equation the velocity law should smoothly pass from one solution into another, Eq. (8.111). This requires one particular value for the radiative acceleration, and hence one particular value for the mass loss rate [Eqs. (8.116) and (8.120)].

There is an important difference between the line driven wind models and the models discussed in the previous sections (coronal models, wave driven models and dust driven models). In the previous models the forces that drive the wind are proportional to the density, and hence the acceleration is *independent of density*. Therefore the solution of the momentum equation gives a velocity law that is independent of the mass loss rate. For a given velocity law the mass loss rate is set by the density at the lower boundary of the wind. However, in the line driven wind model the radiative acceleration depends on the density as $\rho^{-\alpha}$. Therefore there is only one value of the mass loss rate that gives an acceptable solution of the momentum equation. Thus, the important difference between the line driven wind models and the other models is: in a line driven wind both the velocity law and the mass loss rate are due to the same mechanism; whereas in other models the velocity law is due to one mechanism but the density is determined by another mechanism that sets the value at the lower boundary, and thus determines $\dot{M}$.

The coupling between the mass loss rate and the velocity law disappears if one considers line driven wind models for which the mass loss rate is set by a mechanism other than line radiation pressure, but the acceleration is done by line radiation. Such models have been proposed to explain the high velocities of winds from quasars. In those models the velocity law is found by solving the momentum equation *above the critical point*. The mass loss rate is then a free parameter for the supercritical region. If the adopted mass loss rate is smaller than the value set by the critical point, the terminal velocity of the wind will be higher than for a pure line driven wind model. This is because the radiative acceleration depends on $\dot{M}^{-\alpha}$, so a decrease of $\dot{M}$ results in an increase of v_∞. This can explain the very high velocities reached in quasar winds, although they are driven by almost the same lines as a typical O-star wind.

8.9 The correction for the finite size of the star

In the previous section we have described the models of line driven winds with the assumption that the radiation from the wind is only in the radial direction, i.e. the star is considered to be a point source. This is a good approximation far from the star, but it is inaccurate close to the star where the radiation has a significant amount of non-radial momentum. Therefore the line radiation force in the point source model is overestimated close to the star. We will show below that this results in an overestimate of the mass loss rate and an underestimate of the terminal velocity. We will derive a correction factor for the radiation pressure that takes into account the finite size of the star. This is called the *finite disk correction factor*.

8.9.1 The acceleration by radiation from a finite disk

Assume that the star radiates as a uniform disk. The intensity of the photospheric radiation at a point at distance r from the center of the star under an angle $\theta = \arccos\, \mu$ with the radial direction is

$$I_{\nu_0}(r,\mu) = I^*_{\nu_0} \quad \text{if} \quad \mu_* \leq \mu \leq 1 \tag{8.134}$$

$$I_{\nu_0}(r,\mu) = 0 \quad \text{if} \quad 0 \leq \mu \leq \mu_* \tag{8.135}$$

with $\mu_*^2 = 1 - (R_*/r)^2$ (see Fig. 8.14).

For this photospheric radiation the radiative acceleration produced by spectral lines is given by Eq. (8.68).

$$g_{\text{rad}} = \frac{2\pi}{c} \kappa_\ell I^*_{\nu_0} \int_{\mu_*}^{1} \frac{1 - e^{-\tau_{\nu_0}}}{\tau_{\nu_0}} \mu d\mu \tag{8.136}$$

This expression for g_{rad} differs from the expression (8.69) in the point source approximation by the extra μ factor under the integral. (In the point source approximation we have simply set its value to $\mu = 1$.) We will show that the difference between Eq. (8.136) and (8.69) can be expressed in terms of a correction factor.

The Sobolev optical depth τ_{ν_0} depends on distance and on μ (Eq. 8.45)

$$\tau_{\nu_0} = \kappa_\ell \rho \frac{c}{\nu_0} \frac{r/v}{1 + \sigma\mu^2} \tag{8.137}$$

Eq. (8.43) with $\sigma = (d\ln v / d\ln r) - 1$. We want to express τ_{ν_0} in terms of the optical depth parameters t, Eq. (8.82), with

$$t = \sigma_e^{\text{ref}} v_{\text{th}} \rho \frac{dr}{dv} = \sigma_e^{\text{ref}} v_{\text{th}} \rho \frac{r}{v} \frac{1}{1+\sigma} \tag{8.138}$$

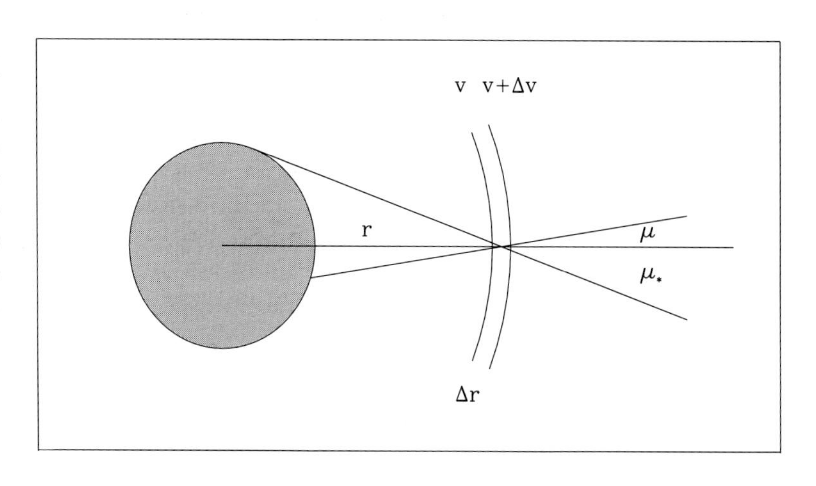

Figure 8.14 Schematic drawing of the finite disk effect on radiative acceleration. The radiation from the star comes in a cone with angle $\Theta_* = \arccos\, \mu_*$.

Elimination of $\rho r/v$ from Eq. (8.137) by means of Eq. (8.138) gives

$$\tau_{\nu_0} = \frac{\kappa_\ell}{\sigma_e^{\rm ref} v_{\rm th}} \frac{c}{\nu_0} \frac{1+\sigma}{1+\sigma\mu^2} t = \eta_\ell \frac{1+\sigma}{1+\sigma\mu^2} t \tag{8.139}$$

where

$$\eta_\ell = \frac{\kappa_\ell}{\sigma_e^{\rm ref} v_{\rm th}} \frac{c}{\nu_0}. \tag{8.140}$$

Substitution of this expression for τ_{ν_0} into Eq. (8.136) gives the acceleration due to a single line as

$$g_\ell = \frac{2\pi}{c} \kappa_\ell I^*_{\nu_0} \int_{\mu_*}^1 \left(\frac{1+\sigma\mu^2}{1+\sigma}\right) \frac{1}{\eta_\ell t} \left\{1 - e^{-\{(1+\sigma)/(1+\sigma\mu^2)\}\eta_\ell t}\right\} \mu d\mu \tag{8.141}$$

The radiative acceleration due to an ensemble of lines is

$$g_L = \frac{2\pi}{c} \int_{\mu_*}^1 \sum_\ell \kappa_\ell I^*_{\nu_0} \left(\frac{1+\sigma\mu^2}{1+\sigma}\right) \frac{1}{\eta_\ell t} \left\{1 - e^{-\{(1+\sigma)/(1+\sigma\mu^2)\}\eta_\ell t}\right\} \mu d\mu \tag{8.142}$$

where the summation is over all contributing lines.

The radiative acceleration can be expressed in terms of the force multiplier $M(t) = g_L c/\sigma_e^{\rm ref} \mathscr{F} = g_L/g_e^{\rm ref}$ (Eq. 8.84). In the case of a finite disk star, the force multiplier is $M'(t) = g_L/g_e^{\rm ref}$ with g_L from Eq. (8.142).

$$M'(t) = \frac{2\pi}{\sigma_e^{\rm ref} \mathscr{F}} \int_{\mu_*}^1 \sum_\ell \kappa_\ell I^*_{\nu_0} \left(\frac{1+\sigma\mu^2}{1+\sigma}\right) \frac{1}{\eta_i t} \left\{1 - e^{-\{(1+\sigma)/(1+\sigma\mu^2)\}\eta_\ell t}\right\} \mu d\mu \tag{8.143}$$

Let us compare this expression for the force multiplier associated with a finite disk with the one in the point source limit. If $\mu \to 1$ the term in square brackets becomes independent of μ and can be taken from the integral. This reduces $M'(t)$ to $M(t)$ in the point source limit

$$M(t) = \frac{1-\mu_*^2}{2} \frac{2\pi}{\sigma_e^{\rm ref} \mathscr{F}} \sum_\ell \kappa_\ell I^*_{\nu_0} \frac{1}{\eta_\ell t} \left\{1 - e^{-\eta_\ell t}\right\} \tag{8.144}$$

There are two differences between $M(t)$ and $M'(t)$: (1) the term $(1-\mu_*^2)/2$ is replaced by the integral over $\mu d\mu$ and (2) the term t has been replaced by $t(1+\sigma)/(1+\sigma\mu^2)$. This means that we can write the expression for $M'(t)$ as

$$M'(t) = \frac{2}{1-\mu_*^2} \int_{\mu_*}^1 M\left(t \frac{1+\sigma}{1+\sigma\mu^2}\right) \mu d\mu \tag{8.145}$$

We have shown in § 8.6.1 that M can be expressed as $M(t) = kt^{-\alpha}(10^{-11}n_e/W)^{\delta}$, therefore

$$M'(t) = kt^{-\alpha}\left\{\frac{10^{-11}n_e}{W}\right\}^{\delta} \frac{2}{1-\mu_*^2}\int_{\mu_*}^{1}\left(\frac{1+\sigma}{1+\sigma\mu^2}\right)^{-\alpha}\mu d\mu$$
$$\equiv M(t)D_f \qquad (8.146)$$

The *finite disk correction factor* D_f (where D stands for disk) is

$$D_f = \frac{2}{1-\mu_*^2}\int_{\mu_*}^{1}\left(\frac{1+\sigma}{1+\sigma\mu^2}\right)^{-\alpha}\mu d\mu$$
$$= \frac{(1+\sigma)^{\alpha+1}-(1+\sigma\mu_*^2)^{\alpha+1}}{(1-\mu_*^2)(\alpha+1)\sigma(1+\sigma)^{\alpha}} \qquad (8.147)$$

So we see that for a finite disk we only have to apply a correction factor D_f to the expression for $M(t)$.

The definition of $\sigma = (rdv/vdr) - 1$, Eq. (8.45), shows that $\sigma < 0$ if $dv/dr < v/r$, $\sigma = 0$ if $dv/dr = v/r$ and $\sigma > 0$ if $dv/dr > v/r$. This implies that $D_f < 1$ close to the star where $dv/dr > v/r$, $D_f = 1$ where $dv/dr = v/r$ and $D_f > 1$ where $dv/dr < v/r$.

Figure (8.15) shows the finite disk correction factor D_f as a function of distance from the star for a β-type velocity law of Eq. (8.54) with β = 0.5 to 10 and a force multiplier $\alpha = 0.60$. The value of D_f increases from $D_f = 1/(1+\alpha)$ at $r = R_*$ and reaches $D_f = 1$ at $r = \infty$ because there the point source approximation is exact. The small value of D_f

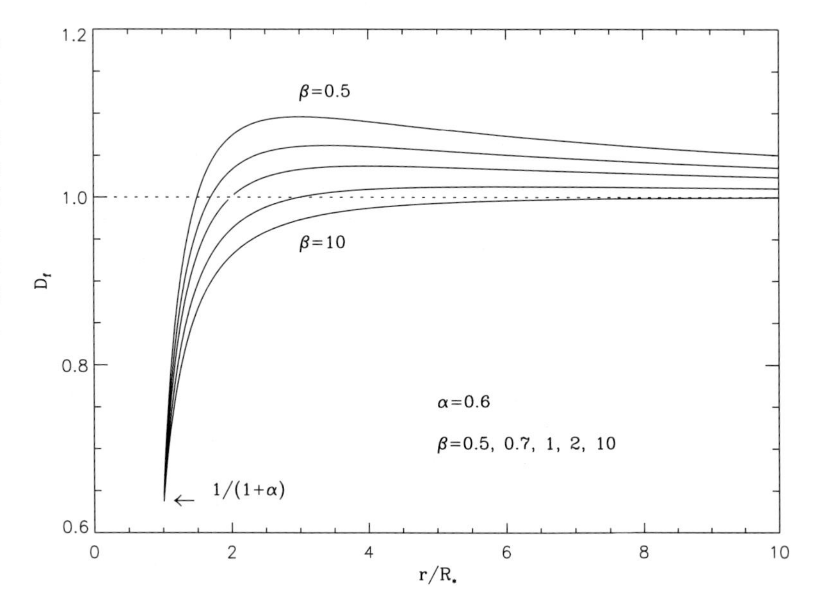

Figure 8.15 The finite disk correction factor, D_f, for a star as a function of the distance for a force multiplier $\alpha = 0.6$ and different β-type velocity laws with β = 0.5, 0.7, 1.0, 2.0 and 10.

near $r \simeq R_*$ is due to the fact that the radiation close to the star reaches r from a wide angle so the radiative acceleration is smaller than in the case of a radial photon flux. The increase to $D_f > 1$ is due to the fact that the Sobolev optical depth of the lines in the transverse direction is smaller than in the radial direction at large distances. As the radiative acceleration of the lines is proportional to $(1-e^{-\tau(\mu)})/\tau(\mu)$ a decrease in the optical depth results in an increase in the radiative acceleration. At $r > 2R_*$ the *increase* of g_L due to the decrease of τ as a function of μ has a larger effect than the *decrease* of g_L due to the reduction of the flux by a factor μ compared to the point source limit and so $D_f > 1$ for $\beta \leq 2$.

8.9.2 The effect of the finite disk on the mass loss rate and velocity

If the finite disk is taken into account, the momentum equation (8.96) for line driven winds becomes

$$\left(1-\frac{a^2}{v^2}\right) r^2 v \frac{dv}{dr} = -GM_*(1-\Gamma_e) + 2a^2 r + C\, D_f \left\{\frac{10^{-11} n_e}{W}\right\}^{\delta} \left(r^2 v \frac{dv}{dr}\right)^{\alpha} \quad (8.148)$$

This equation is more complicated, due to the factor D_f, than Eq. (8.96) for a point source. There is no analytic solution to it, which implies that the momentum equation has to be solved numerically. This is done in a way similar to that described in § 8.7.3. Kudritzki *et al.* (1989) have described a simple program for calculating the mass loss rate and terminal velocity of stellar winds with the finite disk correction taken into account. With this program we have calculated a set of wind models for stars of $T_{\rm eff}$ = 20 000 and 40 000 K, with force multiplier parameters $k = 0.40$, $\alpha = 0.52$ and $\delta = 0.10$ and for a β-type velocity law with $\beta = 0.8$. We adopted a mass-luminosity relation of $L_*/L_\odot = 10 \times (M_*/M_\odot)^{2.5}$ in the range of $10^3 < L_*/L_\odot < 10^6$, and singly ionized He at 20 000 K and twice ionized He at 40 000 K. The results are shown in Figs. (8.16) and (8.17).

The results of finite disk models differ from the models in the point source limit in three ways:

(i) The mass loss rate is smaller by about 0.2 dex at $10^6\ L_\odot$ and 0.6 dex at $10^3\ L_\odot$. This is because the radiative acceleration close to the star is reduced. We have seen in Chapter 3 that a reduction of the acceleration in the subcritical region always leads to a decrease in $\dot{M}$, because the density scale height in the subcritical region becomes smaller which results at a lower density at the critical point.

(ii) The terminal velocity of the wind is larger by a factor of 1.85. This arises from two effects. (*a*) Since the mass loss rate is smaller, the density in the wind is smaller and the Sobolev optical depth of the lines is smaller. A larger radiative acceleration then occurs because $g_L \sim t^{-\alpha}$, and this leads to a higher terminal velocity.

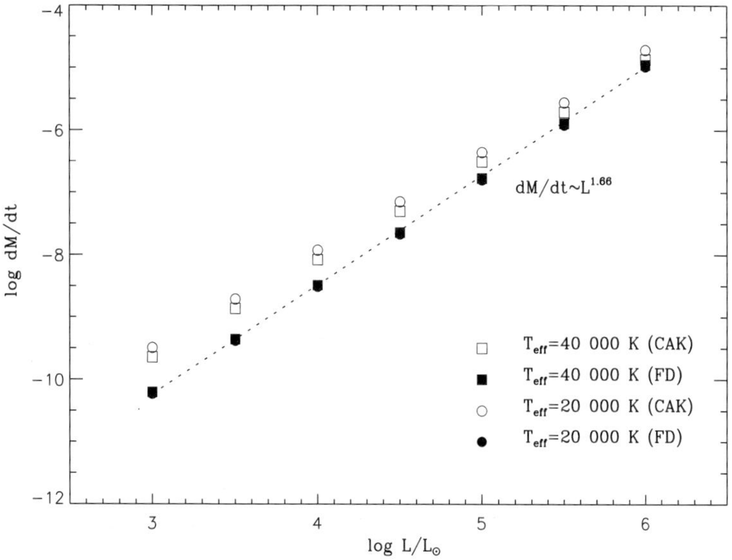

Figure 8.16 The effect of the finite disk correction on $\dot{M}$. Filled and open symbols refer to models with and without the finite disk correction respectively. The mass loss rates are calculated with force multipliers $k = 0.4$, $\alpha = 0.52$ and $\delta = 0.10$. The mass loss rate scales as $L_*^{1.66}$.

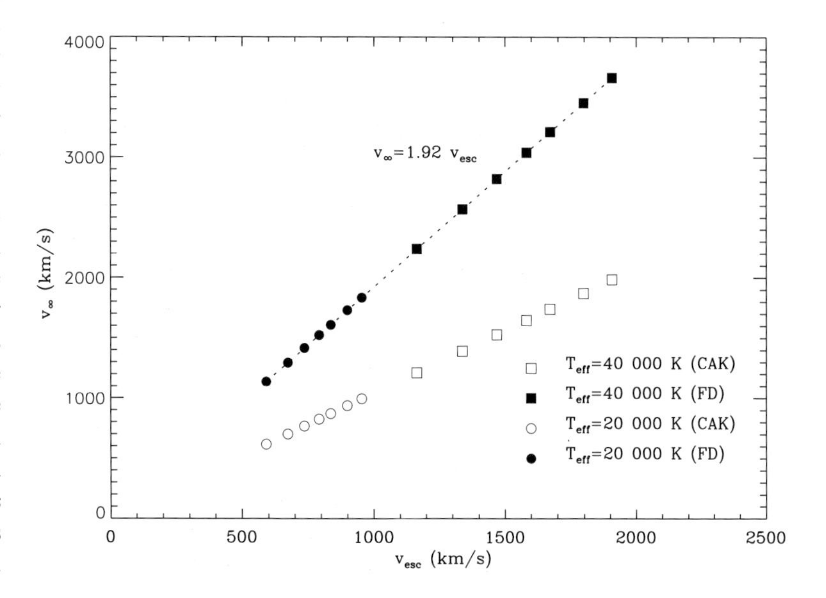

Figure 8.17 The effect of the finite disk correction on the terminal velocities of line driven winds. The calculations are for the same models as in the previous figure. The models with the finite disk correction (filled symbols) have 1.85 times larger terminal velocities than the point source models. The predicted terminal velocities scale with the photospheric escape velocity as $v_\infty \simeq 1.92 v_{esc}$.

This effect explains an increase in v_∞ by about a factor 1.5. (*b*) The correction factor D_f is larger than unity beyond the critical point because of the reduction of the Sobolev depth of the lines for $\sigma < 0$. This produces an additional increase in v_∞.

(iii) The velocity law is slightly 'softer' as it rises more gradually. It is approximately a β-law with $\beta \simeq 0.8$ rather than $\beta = 0.5$ in the point source limit. The upper part of Fig. (8.18) shows the calculated velocity law in a wind of an O4 star. Notice the drastic difference of the velocity laws calculated in the Sobolev approximation with or without the finite disk correction factor. The velocity with the finite disk is extremely similar to a simple β-law

$$\frac{v(r)}{v_\infty} = \left\{1 - \frac{0.9983R_*}{r}\right\}^{0.83} \tag{8.149}$$

The lower part of the figure compares the velocity law with the finite disk correction to the one calculated with the more accurate 'comoving frame method', developed by Mihalas *et al.* (1975). Notice the excellent agreement between the results of the comoving frame method and the Sobolev method with the finite disk correction.

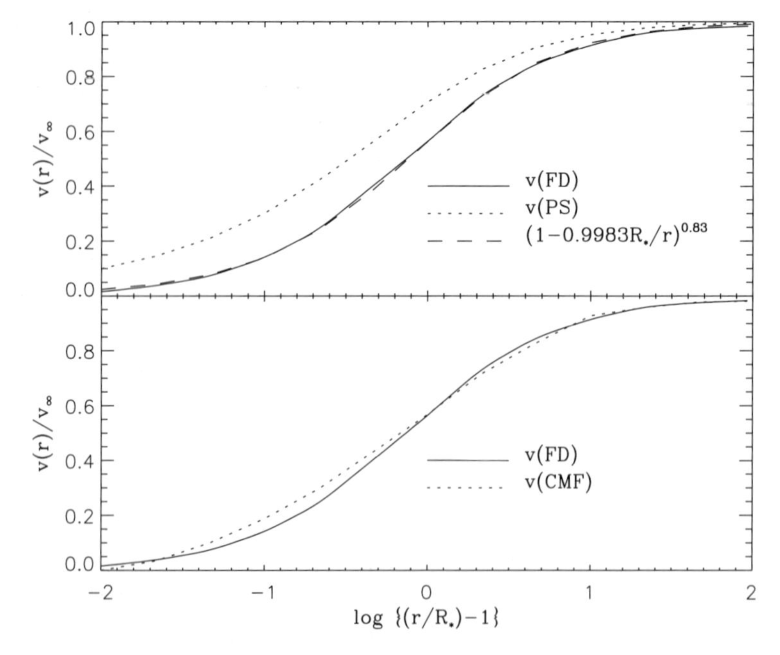

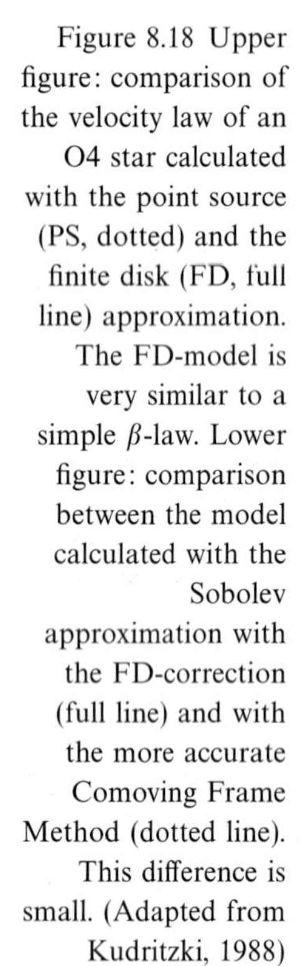
Figure 8.18 Upper figure: comparison of the velocity law of an O4 star calculated with the point source (PS, dotted) and the finite disk (FD, full line) approximation. The FD-model is very similar to a simple β-law. Lower figure: comparison between the model calculated with the Sobolev approximation with the FD-correction (full line) and with the more accurate Comoving Frame Method (dotted line). This difference is small. (Adapted from Kudritzki, 1988)

8.10 Multiple scattering

In the theory of line driven winds discussed so far we have assumed that photons can be scattered only once in the wind. This is a reasonable assumption if there are only a few atomic transitions which produce the radiative acceleration or if there are many transitions but their spectral lines are widely spaced. The reason is that photons which are absorbed and re-emitted (scattered) in a spectral line at some distance in the wind cannot be absorbed again in that same line anywhere else in the wind. Due to the spherical expansion of the wind there are no two points along any ray through the wind which have the same velocity component along that ray. Therefore a photon emitted or scattered in the wind in an atomic transition will escape from the wind or will be absorbed by the photosphere, unless it can be absorbed in *another* atomic transition. The absorption by photons in different spectral lines is called *multiple scattering.*

Multiple scattering is important if the mean spacing in frequency of the line transitions is less than $(2v_\infty/c)\nu$. If two lines are separated by $(2v_\infty/c)\nu$, a photon emitted at a large distance from the star, where $v(r) = v_\infty$, by the line with the higher frequency in the direction of the star (but just passing it) can only be absorbed again at the opposite side of the wind, where $v(r) = v_\infty$ in the other direction, by the line with the lower frequency. A photon emitted by the line with the lower frequency cannot be absorbed again by the line with the higher frequency. This implies that for photons experiencing multiple scattering each scattering occurs at a spectral line of lower frequency until there is a gap in the frequency distribution of the absorption lines. Then the photon will escape. Hence, *gaps in the frequency distribution of the absorption lines play a crucial role in limiting the effect of multiple scattering.*

The condition for multiple scattering of a photon is shown in Fig. (8.19). A photon emitted by a transition of frequency ν_1 at a distance r_1 in a certain direction can only be absorbed or scattered by another transition of frequency ν_2, with $\nu_2 < \nu_1$, at r_2 if the difference in the velocity components at r_1 and r_2 corresponds to $c(\nu_1 - \nu_2)/\nu_1$, so

$$v(r_1)\cos\alpha_1 - v(r_2)\cos\alpha_2 = c(\nu_1 - \nu_2)/\nu_1 \tag{8.150}$$

We have seen in § 8.6 that the radiative acceleration in the winds of hot stars is provided by about 10^5 lines in total and about 10^2 to 10^3 effective optically thick lines. This implies that the average spacing of the lines in velocity is of the order of 10^{-5} c which is only a few $\mathrm{km\,s^{-1}}$. Since this is much smaller than $2v_\infty$ and of the order of the intrinsic width of the absorption coefficients, multiple scattering must occur in the winds of hot stars.

The process of multiple scattering is shown schematically in Fig. (8.20). Photons from the photosphere travel outwards until they reach a resonance zone (i.e. a shell with a velocity such that the photon can be scattered by a line transition), where they may experience numerous nearly istropic scatterings. Each scattering occurs in a different spectral transition: every next scattering is in a line of lower rest frequency. The net effect of the momentum deposition is determined by the angle of incidence into the shell and the angle at which the photon leaves the shell. This is shown in Fig. (8.21).

Multiple scattering is the subsequent transfer from one resonance shell to another with each shell being associated with a different line transition. The escape from a Sobolev resonance shell is equally likely in the forward and backward directions. So the transfer between shells can be considered analogous to isotropic scattering. It can be further simplified by asuming that the opacity at each resonance shell is the same. In the limit of a very dense distribution of line transitions, the opacity can be considered as a continuum opacity and the multiple scattering is then similar to that described in § 7.2 for radiative transfer through a dust shell. There we found that the radiation pressure produces a wind momentum of

$$\dot{M}\, v_\infty = \tau \frac{L_*}{c} \tag{8.151}$$

with $\tau \leq 2c/v_\infty$. Compare that with the upper limit for *single* scattering of all the stellar photons: $\dot{M}\, v_\infty \leq L_*/c$, Eq. (8.19). We see that multiple scattering in a wind with a *very dense* distribution of lines over wavelengths can produce a considerably higher mass loss rate than the single scattering limit.

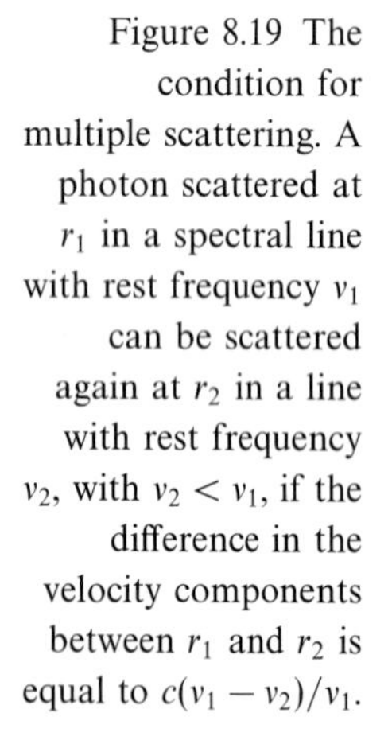

Figure 8.19 The condition for multiple scattering. A photon scattered at r_1 in a spectral line with rest frequency ν_1 can be scattered again at r_2 in a line with rest frequency ν_2, with $\nu_2 < \nu_1$, if the difference in the velocity components between r_1 and r_2 is equal to $c(\nu_1 - \nu_2)/\nu_1$.

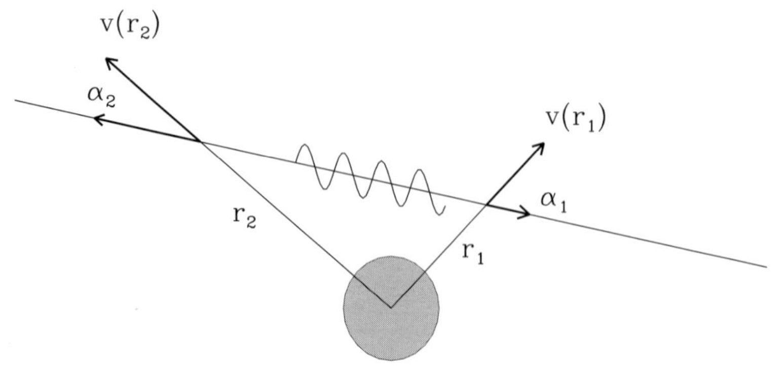

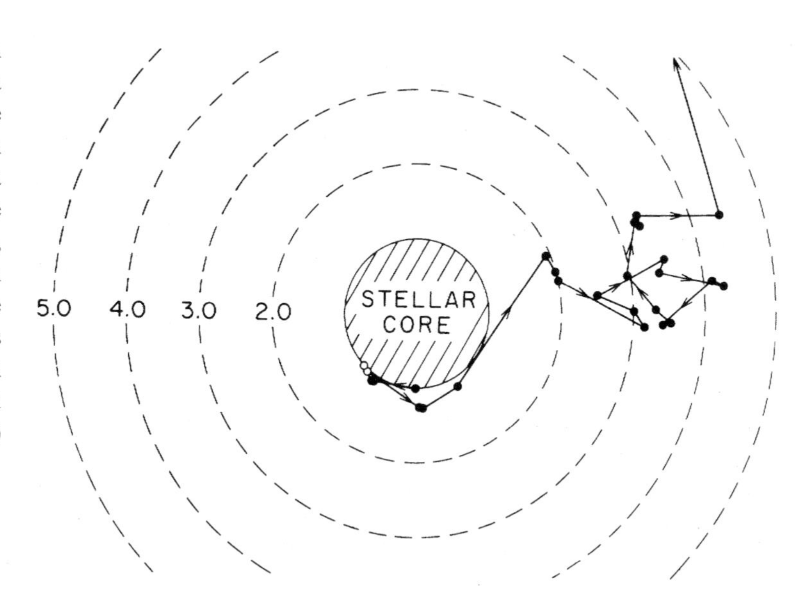

Figure 8.20 The path of a photon in a wind with multiple scattering. Each scattering occurs in a different spectral line at a lower frequency, because from each point in the wind the surrounding is receding. (From Abbott and Lucy, 1985)

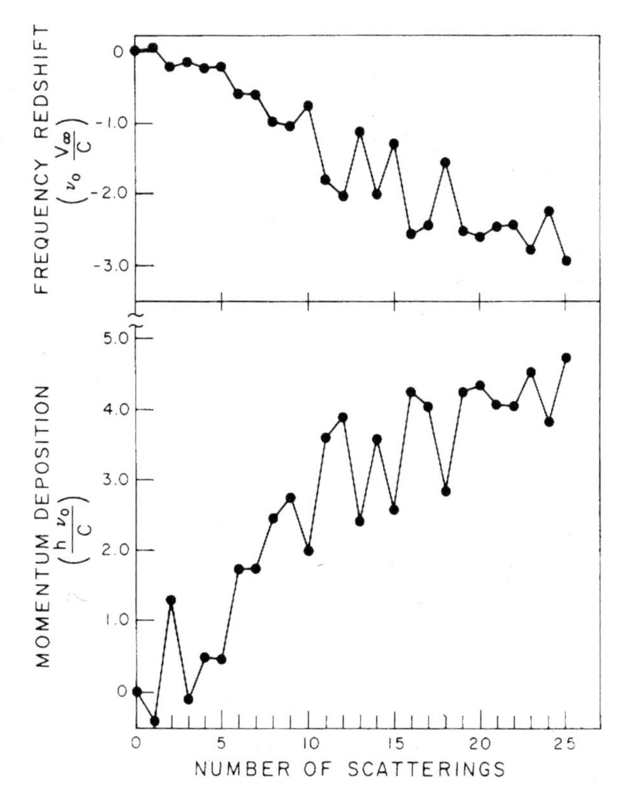

Figure 8.21 Upper figure: the cumulative redshift of the photon along the multiple scattering path shown in the previous figure. Lower figure: the momentum deposition into the wind. (From Abbott and Lucy, 1985)

Some insight can be gained by considering the gray isotropic scattering analogy. The photons can be considered as diffusing out of the atmosphere. The number of scatterings suffered by a photon before it escapes from the outer layers of the wind where the velocity is v_∞ is

$$N_{\rm scat} = \left\{\frac{v_\infty}{\Delta v}\right\}^2 \tag{8.152}$$

where Δv is the redshift experienced by the photon after each scattering. The total redshift experienced by a photon during its diffusion outwards is $N_{\rm scat} \times \Delta v$, so the energy of the photon transfered to the gas is

$$\frac{\Delta E}{E} = \left\{\frac{v_\infty}{\Delta v}\right\}^2 \frac{\Delta v}{c} = \frac{{v_\infty}^2}{c\Delta v} \tag{8.153}$$

The momentum deposited into the wind by one photon is

$$\Delta P = \frac{\Delta E}{c} = \frac{h\nu_0}{\Delta v}\left\{\frac{v_\infty}{c}\right\}^2 \tag{8.154}$$

where $h\nu_0$ is the original energy of the photon. So if all the photons from the star were diffusing outward through the wind, the total momentum deposited into the wind per second would be about

$$\dot{M}v_\infty \simeq \frac{L_*}{\Delta v}\left\{\frac{v_\infty}{c}\right\}^2 \tag{8.155}$$

This gives a maximum mass loss rate for multiple scattering of

$$\dot{M}_{\rm max} \simeq \frac{L_*}{c^2}\frac{v_\infty}{\Delta v} \tag{8.156}$$

So if the scattering lines are very close together with a typical separation of Δv, the mass loss rate can be very much higher than the single scattering limit of $\dot{M} < L_*/v_\infty c$ (Eq. 8.19) by a factor ${v_\infty}^2/c\Delta v$. For $v_\infty = 2000$ km s^{-1} and $\Delta v = 1$ km s^{-1} the maximum mass loss rate of multiple scattering is about a factor of 13 higher than for single scattering.

The effect of multiple scattering on the wind depends strongly on the location where most scatterings occur. We have seen in chapter 3 that adding momentum to the wind in the subcritical region, i.e. below the critical point, results in an increase in the mass loss rate, whereas the addition of momentum beyond the critical point results in an increase in the terminal velocity of the wind.

Accounting for the realistic properties of the line opacities, Friend and Castor (1983), Abbott and Lucy (1985), Puls (1986) and Gayley *et al.* (1995) have shown that multiple scattering can provide an enhancement of the wind momentum over that from single scattering by a factor of about six for the dense winds of Wolf-Rayet stars. For the O stars the effect is much smaller and only about a factor of two.

8.11 Wind blanketing

Multiple scattering not only increases the mass loss rate and the momentum of the wind but it can also alter the structure of the photosphere. Some of the photons from the photosphere will be scattered back into the photosphere. If more than about 10 percent of the radiative energy from the photosphere is scattered back to the star, the temperature structure of the upper photospheric layers may change drastically and hence the emitted energy distribution will be affected. This effect is called '*wind blanketing*' because the wind acts like a blanket and heats the photosphere by reflecting its radiation.

We can easily estimate the effect of wind blanketing in the single scattering limit. For each atomic line ℓ in the wind the fraction of radiation that is backscattered from a distance r into the photosphere is $W(r)(1-e^{-\tau_\ell})$, if we assume for simplicity that the photons emerging from the star are directed radially outwards. The factor $1 - e^{-\tau_\ell}$ is the fraction of the radiation that is scattered in the line. This fraction is unity if the Sobolev optical depth of the line at r is very large and the fraction is zero if $\tau_\ell = 0$. The factor $W(r)$ is the geometrical dilution factor, Eq. (8.65), and describes the solid angle subtended by the photosphere as seen from the point where the scattering occurs. The total fraction of the radiation that is reflected back to the star by scattering in the wind is in the single scattering limit

$$f_{\rm refl} \simeq \int_0^{v_\infty} \sum_\ell \frac{\nu_\ell \mathscr{F}_{\nu_\ell}}{\mathscr{F}} \left\{1 - e^{-\tau_\ell(r)}\right\} W(r) \frac{dv}{c} \tag{8.157}$$

The factor $(\nu_\ell/c)dv$ describes the frequency interval in which each line, ℓ, can be scattered. The parameter $f_{\rm refl}$ is called *the reflectance of the wind.* Remember that in the point source limit of radially directed photons the contribution by one line to the radiative acceleration is [Eq. (8.77)]

$$g_\ell = \frac{\nu_\ell \mathscr{F}_{\nu_\ell}}{c} \left\{1 - e^{-\tau_\ell(r)}\right\} \frac{dv}{dr} \cdot \frac{1}{c\rho} \tag{8.158}$$

and that the summation of the contribution of all the lines is

$$g_L = \sum_\ell g_\ell \simeq \frac{\sigma_e^{\rm ref} \mathscr{F}}{c} \cdot k t^{-\alpha} \tag{8.159}$$

where we have ignored the weak dependence on n_e/W in Eq. (8.87).

Substitution into Eq. (8.157) gives

$$
\begin{aligned}
f_{\rm refl} &= \sigma_e^{\rm ref} k \int_{R_*}^{\infty} \rho t^{-\alpha} W(r) dr \\
&= k \frac{v_\infty}{v_{\rm th}} \left\{ \frac{\sigma_e^{\rm ref} v_{\rm th} \dot{M}}{4\pi R_* {v_\infty}^2} \right\}^{1-\alpha} \int_0^1 \left\{ \frac{1}{x^2 w} \frac{dx}{dw} \right\}^{1-\alpha} W(x)\, dw \\
&= k \frac{v_\infty}{v_{\rm th}} \left\{ \frac{\sigma_e^{\rm ref} v_{\rm th} \dot{M}}{4\pi R_* {v_\infty}^2} \right\}^{1-\alpha} I(\beta) \qquad (8.160)
\end{aligned}
$$

where $w = v/v_\infty$ and $x = r/R_*$. For $\sigma_e = 0.30$ cm^{2}g^{-1}, $k = 0.40$, $\alpha = 0.52$, $\delta = 0.0$ and $\beta = 0.8$ we find $I(\beta) = 1.67$ and so for $v_{\rm th} = 20$ km s^{-1} we find

$$
f_{\rm refl} \simeq 0.50\, I(\beta) \left\{ \frac{\dot{M}/10^{-6}}{R_*/10R_\odot} \right\}^{0.48} = 0.81 \left\{ \frac{\dot{M}/10^{-6}}{R_*/10R_\odot} \right\}^{0.48} \qquad (8.161)
$$

The reflectance of the wind can affect the temperature structure of the photosphere and the energy distribution if $f_{\rm refl} \gtrsim 0.3$, which is the case for $\dot{M} \gtrsim 10^{-8} R_*/R_\odot$. So for luminous O stars and Wolf-Rayet stars the wind blanketing will affect the structure of the photosphere. It has been calculated by Abbott and Hummer (1985) for the winds of O stars.

8.12 Rotating line driven winds

If a star is rotating at a significant fraction of its 'critical rotational' (or 'break-up') velocity, the centrifugal force provides an additional acceleration to the wind. The critical rotational velocity is defined as the photospheric rotation velocity at which the centrifugal acceleration $\omega^2 R_* = v_{\rm rot}^2/R_*$ is equal to the effective acceleration of gravity, so

$$
v_{\rm crit} = \sqrt{GM_*(1-\Gamma_e)/R_*} \qquad (8.162)
$$

This may affect the mass loss rate and the velocity structure of the radiation driven winds. The calculation of rapidly rotating radiation driven winds is considerably more complicated than for nonrotating winds because the models are no longer spherically symmetric. The centrifugal force is high in the equatorial plane and vanishes along the polar axis. We will describe the effect of rotation on the mass loss and the terminal velocity in the equatorial plane only, following the results of Friend and Abbott (1986). The effects at other latitudes are smaller.

The centrifugal force in the equatorial plane is

$$
f_{\rm cent} \geq \rho(r) \frac{v_\phi^2(r)}{r} = \rho(r) v_{\rm rot}^2 \frac{R_*^2}{r^3} \qquad (8.163)
$$

where v_ϕ is the rotational velocity component in the equatorial plane. This velocity is determined by the rotational velocity at the equator of the photosphere, v_{rot}, and by the conservation of angular momentum

$$v_\phi(r) = v_{\mathrm{rot}} \frac{R_*}{r} \tag{8.164}$$

The momentum equation in the equatorial plane has an extra term due to the centrifugal force and it becomes [see Eq. (8.96)]

$$\left(1 - \frac{a^2}{v^2}\right) r^2 v \frac{dv}{dr} = -GM_*(1 - \Gamma_e) + 2a^2 r + C\ D_f \left(r^2 v \frac{dv}{dr}\right)^\alpha + \frac{v_{\mathrm{rot}}^2 R_*^2}{r^3} \tag{8.165}$$

The parameter C was defined by Eq. (8.97) and contains the mass loss rate, and D_f is the correction factor for the finite disk (Eq. 8.147). In the case of a nonspherical wind the expression for the mass loss rate has to be replaced by an expression containing the mass flux, F_m in g $\mathrm{cm}^{-2}\mathrm{s}^{-1}$, at the photosphere. Hence the constant C for the momentum equation in the equatorial plane becomes

$$C = \frac{\sigma_e^{\mathrm{ref}} L_* k}{4\pi c} \left\{\sigma_e^{\mathrm{ref}} v_{\mathrm{th}} F_m^e R_*^2\right\}^{-\alpha} \tag{8.166}$$

where F_m^e is the mass flux at the equator.

The solution of Eq. (8.165) is more difficult because the extra r^{-3} term complicates the conditions at the critical point. The numerical solutions of the equation by Friend and Abbott (1986) have shown that it differs from the case of no rotation in three ways.

Firstly, the critical point moves outward. Secondly, the terminal velocity of the wind in the equatorial plane decreases. This is due to the fact that v_∞ scales with v_{esc} [Eq. (8.104)] and the effective escape velocity goes down due to the centrifugal force. Thirdly, the mass flux in the equatorial plane increases. This is because the centrifugal force acts like a reduction of the density scale height in the subcritical region along the equatorial plane, and we have seen in chapter 3 that this leads to a higher density at the critical point and hence to a higher mass flux.

The results of the rotating radiation driven winds for models with $\alpha = 0.6$ (from Friend and Abbott, 1986) are shown in Figs. (8.22) and (8.23) for the effects on the terminal velocity in the equatorial plane and on the equatorial mass flux. Note that the effects of rotation become important if $v_{\mathrm{rot}} \gtrsim 0.5\ v_{\mathrm{crit}}$.

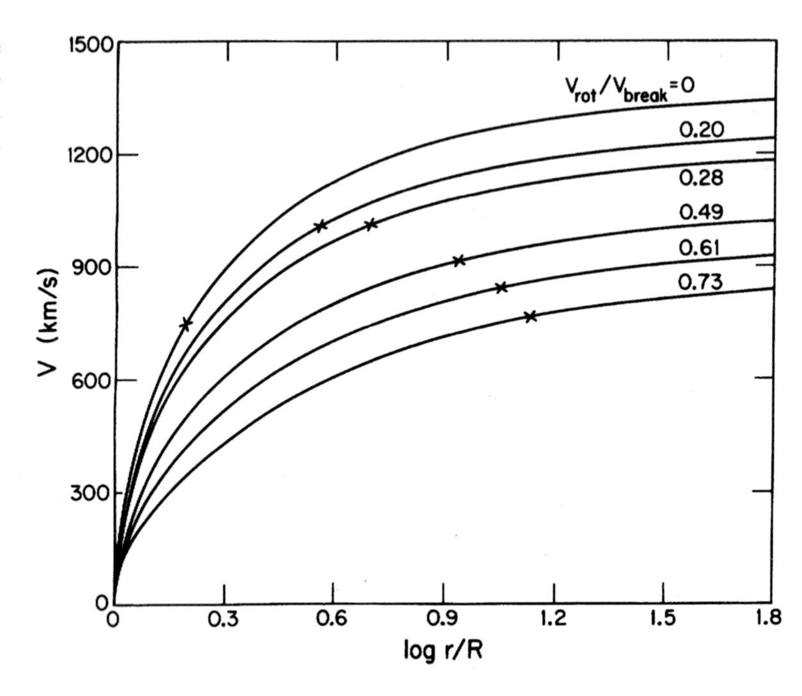

Figure 8.22 The velocity law in the equatorial plane of a rotating star as a function of the ratio between $v_{\rm rot}$ and the critical or break-up velocity. The crosses indicate the location of the critical point. (From Friend and Abbott, 1986)

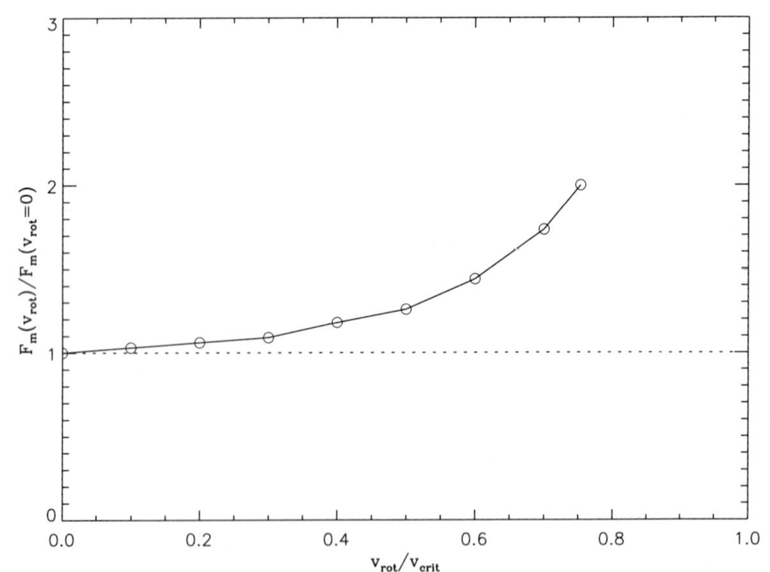

Figure 8.23 The effect of rotation on the equatorial mass flux of a rotating radiation driven stellar winds with $\alpha = 0.6$ as a function of the ratio $v_{\rm rot}/v_{\rm crit}$. The figure shows the ratio between the equatorial mass flux with rotation and without rotation. (After Friend and Abbott, 1986)

8.13 The instability of line driven winds

Up to now we have assumed that the line driven winds are homogeneous and stationary. The predictions of line driven wind models in terms of mass loss rates, terminal velocities and velocity laws agree very well with the observations. This implies that the stationary models are adequate for describing the global structure of the line driven winds. However, they fail to explain some of the important details of the observations of the winds of hot stars. In particular: the X-rays from the winds of hot stars; superionization of the winds to ions such as O VI which clearly require an extra source of high energy photons; and the presence and variability of the discrete absorption components in the P Cygni profiles formed in the winds of hot stars. These observations suggest that the radiation-driven winds of O stars are unstable. We will show by a simple argument why the line driven winds are unstable.

Suppose a hot star has a stationary line driven wind with a smooth distribution of velocity, density and temperature. Assume that at a certain moment the velocity distribution is slightly perturbed and that this perturbation has the shape of a sine-wave, between r_1 and r_4, as shown in Fig. (8.24).

Due to this sine-wave the velocity gradient in the layers where the perturbation occurs changes. Define r_2 and r_3 as the points where the perturbed velocity has the same gradient as the unperturbed velocity law. In the layers $r_1 < r < r_2$ and $r_3 < r < r_4$ the velocity gradient is larger than in the unperturbed wind, and in the layer $r_2 < r < r_3$ it is smaller.

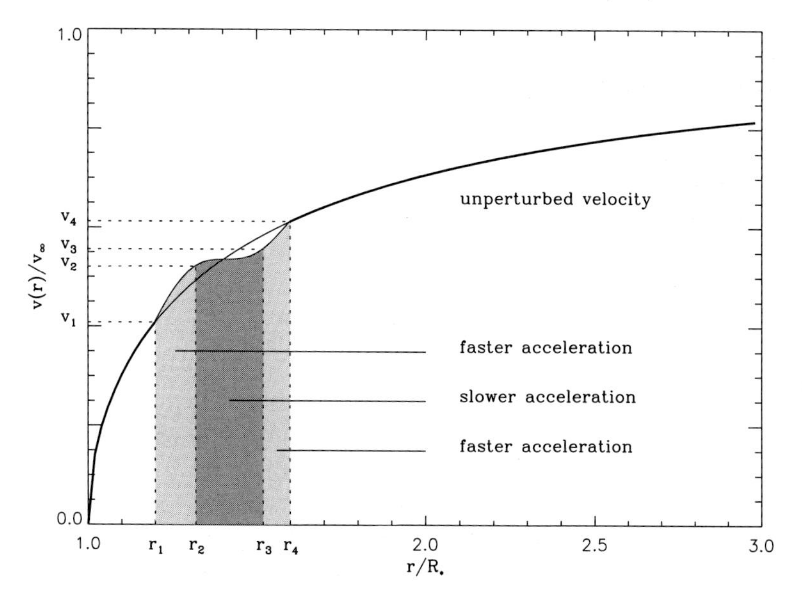

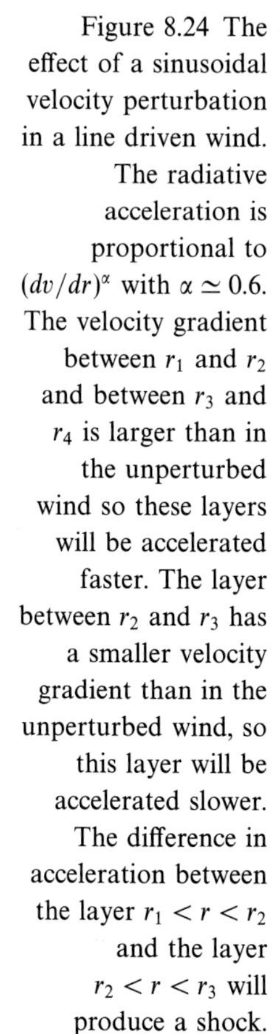
Figure 8.24 The effect of a sinusoidal velocity perturbation in a line driven wind. The radiative acceleration is proportional to $(dv/dr)^\alpha$ with $\alpha \simeq 0.6$. The velocity gradient between r_1 and r_2 and between r_3 and r_4 is larger than in the unperturbed wind so these layers will be accelerated faster. The layer between r_2 and r_3 has a smaller velocity gradient than in the unperturbed wind, so this layer will be accelerated slower. The difference in acceleration between the layer $r_1 < r < r_2$ and the layer $r_2 < r < r_3$ will produce a shock.

The radiative acceleration depends on the velocity gradient

$$g_L \sim kt^{-\alpha} \sim \rho^{-\alpha}\left(\frac{dr}{dv}\right)^{-\alpha} \sim \rho^{-\alpha}\left(\frac{dv}{dr}\right)^{\alpha} \tag{8.167}$$

Therefore the layers $r_1 < r < r_2$ and $r_3 < r < r_4$ will be accelerated faster than the unperturbed wind at that distance and in the layer $r_2 < r < r_3$ the acceleration will be smaller. A layer which has a steeper velocity gradient than normal will be accelerated even further, and the reverse is true for a layer with a lower velocity gradient. This implies that velocity perturbations will grow in the wind.

There is another effect that enhances perturbations. A local increase in velocity (rather than the velocity gradient) moves the line out of the shadowing by the lower layers. So a parcel of gas that is moving faster than the rest of the wind at that distance will be able to absorb or scatter more photons coming from the star, and hence it is accelerated even faster. Both effects will occur, i.e. the one due to the velocity gradient and the one due to the velocity. The relative importance of the effect depends on the optical depth over the length of the perturbations. If the length is large, the optical depth is large and the velocity gradient is more important (see § 8.5.5). If the length of the perturbations is small, the optical depth is small and the line force is independent of the velocity gradient, but it is sensitive to the decreased shadowing. Owocki *et al.* (1988) have shown that the growth of the instability is mainly due to perturbations of small length, i.e. to the decreased shadowing.

The amplitude of a velocity instability grows exponentally with time as $e^{\Omega . t}$, where Ω is the growth rate with

$$\Omega \sim \frac{L}{v} \sim \frac{v(dv/dr)}{v_{\rm th}} \tag{8.168}$$

where $L = v_{\rm th}/(dv/dr)$ is the Sobolev length. If we take $t \simeq R_*/v_\infty$ as the characteristic flow time through the wind and $dv/dr \simeq v_\infty/R_*$ as the characteristic velocity gradient, we find that

$$\Omega t \simeq \frac{v_\infty}{v_{\rm th}} \tag{8.169}$$

This shows that velocity perturbations can grow by a factor of e^{100} or so. So the wind will be unstable.

The effects of instabilities in radiation-driven winds have been studied by various groups and have been reviewed by Owocki (1994). Figure (8.25) shows the results of a time dependent model of the wind of an O star for which the lower boundary was slightly perturbed. Notice the presence of many shocks. The velocity law shows the general trend of an outward increase, but with spikes of up to 500 km s^{-1} above the average. The density structure also shows a spiky distribution with drops of about a factor of 10^4.

The effects of line driven instabilities are

(i) The mean, i.e. time-averaged, mass loss rate of unstable line driven winds is the same as that found in the stationary solutions. The time-averaged terminal velocity of the wind and the time-averaged velocity law are also very similar to the stationary case.

(ii) Velocity and density perturbations can grow on a timescale which is short compared to the characteristic flow time R_*/v_∞, of the wind. Thus a considerable fraction of the wind, i.e. more than about 10 percent of its mass, will have gone through shocks. The velocity perturbations can grow to about 500 km s^{-1}. This results in a broadening of the absorption and emission parts of the P Cygni profiles formed in the winds.

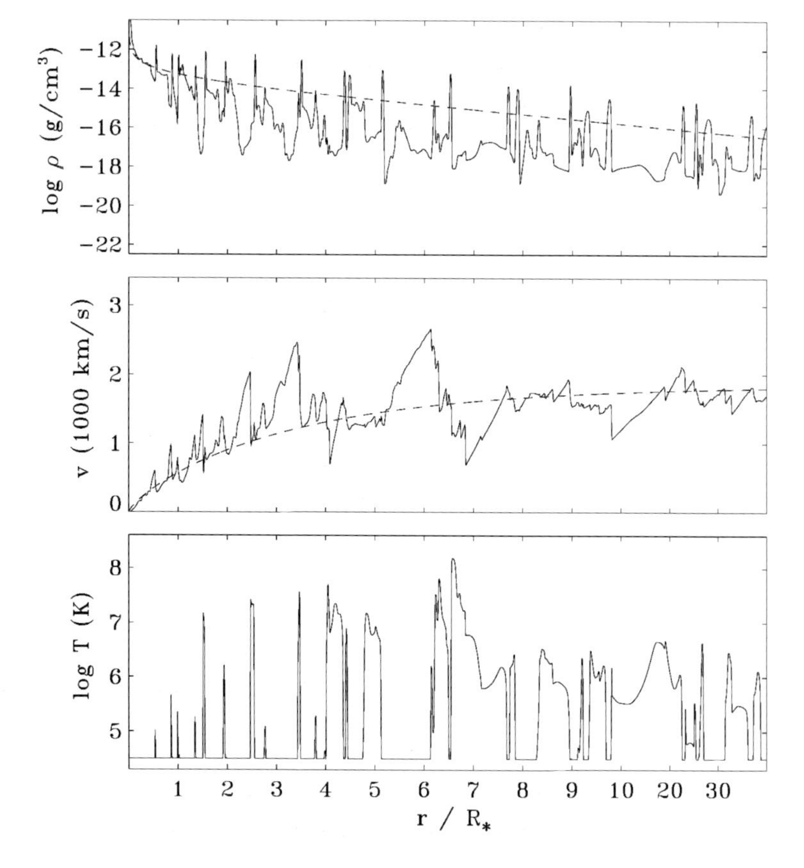

Figure 8.25 The effect of shocks on the density and velocity structure of a wind at a certain snapshot time, showing the presence of many shocks in the wind. Although the variations in density and velocity are large, the time-averaged velocity and density structure is very similar to that of the unperturbed wind (dashed line). (From Feldmeier, 1997 Priv. Comm.)

(iii) The shocks can reach temperatures up to 10^7 K. This explains the observed X-ray flux from the winds of hot stars.
(iv) The X-rays created in the shocks can superionize part of the wind to high stages of ionization. In particular, ions such as O VI and N V can be created by Auger ionization in which the absorption of a soft X-ray photon results in the ejection of two electrons, e.g. O IV + $h\nu \rightarrow$ O VI + $2e^-$, as first suggested by Cassinelli and Olson (1979).

8.14 Conclusions

We have shown in this chapter that the transfer of momentum from the photospheric radiation to the gas via the absorption or scattering in spectral lines can drive the stellar winds of hot stars very efficiently. This transfer can most easily be described and calculated in the Sobolev approximation if the intrinsic width of the absorption line is small compared to the width that results from the Doppler broadening due to the velocity gradient over the line forming region. In the Sobolev approximation the radiative acceleration due to a spectral line depends on the product of local quantities: $(\rho dr/dv)$. With this approximation it is possible to calculate the radiative acceleration due to an ensemble of a very large number, 10^5 to 10^6, of spectral lines. It turns out that the radiative acceleration can be approximated to an accuracy of better than about 10 percent by a simple power law dependence of $(\rho dr/dv)$.

The momentum equation of line driven winds is dominated by the acceleration of gravity and by the radiative acceleration due to the spectral lines. The equation has the characteristic form of an equation with a critical point. However, the equation has nonlinear terms due to the fact that the radiative acceleration is proportional to $(\rho dr/dv)^{-\alpha}$ with $\alpha \simeq 0.6$. The requirement that the solution of the momentum equation goes from subsonic at the photosphere to supersonic at large distances and that it passes smoothly through the critical point uniquely determines the mass loss rate and the velocity law.

A first order solution can be derived very easily in an analytic way if two assumptions are made: the forces due to the gas pressure are negligible and the radiation from the photosphere comes out radially (i.e. the point source limit). The result is very similar to the physically more correct one which takes into account the gas pressure but requires a numerical solution of the momentum equation. The mass loss rates predicted with this simple theory are lower than observed and the terminal velocities are higher than observed. This is due to the assumption of a stellar point source. The effect of the finite size of the star can be taken into account by a simple correction term in the expression for the radiation pressure. The momentum equation then no longer has an analytic solution, and the calculation of the mass loss rate and the velocity law requires a numerical solution.

The results are in very good agreement with the observations of the winds of O stars. They also agree very well with the more elaborate calculations where the radiative transfer is solved without the Sobolev approximation in the comoving frame. The mass loss rates of A-type supergiants also agree with the predictions from the radiation driven wind theory (Achmad *et al.*, 1997).

We discussed the effect of multiple scattering of photons in line driven winds. Multiple scattering may increase the mass loss rate above the single scattering limit. This effect may be important for the winds of very luminous O stars and for Wolf-Rayet stars which have a total momentum about 10 to 100 times as high as the momentum of the radiation. Stellar rotation may also enhance the mass loss from a star, especially from the equatorial regions. This will be discussed in Chapter 11.

A simple description shows that line driven winds are intrinsically unstable because velocity or density perturbations will grow rapidly into shocks. These shocks are responsible for the X-rays observed from O stars and for the superionization of the winds. Fortunately they do not affect the overall structure of the wind in terms of time-averaged mass loss rates and wind velocities.

8.15 Suggested reading

Cassinelli, J.P. 1979 'Stellar winds', *An. Rev. Astr. Ap.* **17**, 275
(A review of stellar wind theories)

Kudritzki, R.P. 1988 'The atmospheres of hot stars: modern theory and observations', in *Radiation in moving gaseous media*, eds. Y. Chmielewski & T. Lanz (Geneva Observatory) p. 1
(A review of the atmospheres of hot stars and the radiation driven wind theory)

Kudritzki, R.P., Pauldrach, A.W.A., Puls, J. & Abbott, D.C. 1989 'Radiation driven winds of hot stars: VI Analytic solutions for wind models including the finite cone angle effect', *A & A* **219**, 205
(This paper describes 'the cooking recipe': a simple method to predict $\dot{M}$ and v_∞ for hot stars.)

Lamers, H.J.G.L.M. 1997 'The line driven wind theory', in *Stellar atmospheres and winds* ed. J.P. de Greve, Lecture Notes in Physics, in press.
(A review of the observations and theory of line driven winds)

Owocki, S.P. 1994 'Theory review: line driven instability and other causes of structure and variability of hot star winds', *Ap. & Space Sc.* **221**, 3
(A review on instabilities of line driven winds)

9 Magnetic rotator theory

In this chapter the combined effects of rotation and magnetic fields are studied. Rotation alone, in the absence of a magnetic field, has relatively little effect on the mass loss rate unless the star rotates at nearly the maximum velocity, at which the equatorial centrifugal force equals gravity. However, even a moderate magnetic field that is frozen in with the wind plasma can lead to an increase of several orders of magnitude in the mass loss rate, wind momentum, and wind energy. One of the most important effects of such a wind is the enhanced loss of stellar angular momentum, or 'spin down' of the star. Weber and Davis (1967) developed the basic equations for magnetic rotator theory, and showed that the time scale for the sun to lose its current angular momentum is comparable to the sun's age. This is surprising because the total mass that is lost by way of the solar wind corresponds to only $10^{-4}M_{\odot}$ over the sun's lifetime. The efficient loss of angular momentum occurs because the magnetic field leads to an 'effective co-rotation' of the solar wind out to the 'Alfvén radius', r_A, at $>$ 10 solar radii. The surface rotation speed of the sun is small (2 $\mathrm{km\,s^{-1}}$) relative to the maximal equatorial speed of $v_{\mathrm{max}} = \sqrt{GM_{\odot}/R_{\odot}}$ =440 $\mathrm{km\,s^{-1}}$. In contrast, the early-type stars often have rotation speeds faster than 100 $\mathrm{km\,s^{-1}}$. In going from early type to late type stars, there is a steep drop in the rotation speeds at about F5V. The slow rotation of the late type stars provided some of the earliest evidence that stars other than the sun have stellar winds (Brandt, 1970). Stars later than F5 have convective outer envelopes, in which acoustic or magneto-acoustic energy fluxes can be produced. So it was argued that these energy fluxes produce coronae and coronal driven winds which could lead to a loss of stellar angular momentum, and hence the observed slow rotationg speeds. Kraft (1967) found that in young clusters, the late type stars have faster rotation speeds than do stars in older clusters, or the sun. This indicates that stars lose angular momentum on evolutionary time scales. Furthermore, Belcher and MacGregor (1976) showed from their theoretical models that the winds from young stars are likely to be enhanced by the transfer of rotational energy and angular momentum from the star by way of the magnetic fields.

In this chapter we study how stellar rotational energy can be converted to an equatorial expansion of the outer atmosphere in the form of a 'magnetic rotator wind'. We relate the terminal kinetic energy of the outflow to the Poynting vector flux at the base of the wind, and derive relations for the angular momentum loss rate. Our goal is to develop expressions for the constants of the motion and magnetic field structure and then apply these to the structure of *'fast magnetic rotator'* (FMR) winds. Extreme version of FMR winds are the *'centrifugal magnetic rotator'* (CMR) winds. For a CMR wind, the mass loss is fully determined by centrifugal acceleration at the base of the wind, whereas the terminal velocity is determined by the surface magnetic field. This separation of the centrifugal and magnetic effects also holds in the case of hybrid wind models such as the *'luminous magnetic rotator'* model, for which radiation forces are also operating in driving the wind.

We begin with a derivation of the radial and azimuthal momentum equations of Weber and Davis' equatorial wind theory. In § 9.3 we find that each equation leads to a constant of the motion. The conservation of angular momentum constant is used in § 9.4 to eliminate angular quantities in terms of the radial variables. The wind equation for $v(r)$ has properties that are quite different from those discussed so far in this book. For example, instead of the single transonic point of the Parker wind theory, there are three singular points through which the radial velocity solutions must pass. Introducing the Poynting vector in § 9.5 allows us to develop some insight into the energy deposition that is required for the outflow to occur. The solution of the purely radial equation is described in § 9.7. This is applied to the case of a centrifugal magnetic rotator in § 9.8. Finally, in § 9.9, we briefly discuss luminous magnetic rotator models which combine the line driving forces of Chapter 8 with those discussed in this chapter.

Although the theory discussed in this chapter is for a flow in the equatorial plane only, the general concepts developed are useful for more general rotationally driven outflows. Understanding the equatorial equations is essential for studying latitude dependent winds and the production of bipolar outflows that arise from the angular momentum transfer during the star formation process. Models for winds from rapidly rotating stars, but which do not have dynamically important magnetic fields, will be discussed in Chapter 11.

9.1 The equatorial wind theory

The rotation of the star causes the magnetic field and the velocity field to develop azimuthal components. Let (r, θ, ϕ) be the spherical polar coordinates, where θ is the angle from the polar axis, and ϕ is the azimuthal angle. For the equatorial flow, the velocity and magnetic field vectors have the form

$$\mathbf{v} = V_r(r)\hat{\mathbf{e}}_r + V_\phi(r)\hat{\mathbf{e}}_\phi \tag{9.1}$$

$$\mathbf{B} = B_r(r)\hat{\mathbf{e}}_r + B_\phi(r)\hat{\mathbf{e}}_\phi \tag{9.2}$$

where $\hat{\mathbf{e}}_r$ and $\hat{\mathbf{e}}_\phi$ are unit vectors in the radial and azimuthal directions.† Because of the assumed rotational symmetry, the components V_r, V_ϕ, B_r and B_ϕ are functions of radius only, and because of symmetry above and below the equatorial plane, we may assume that there are no θ components to $\mathbf{v}$ and $\mathbf{B}$, in the equatorial plane. We shall often find it convenient to use the Alfvén velocity $\mathbf{A}$ to express the components of the magnetic field in velocity units.

$$|\mathbf{A}|^2 = A_r{}^2 + A_\phi{}^2 = \frac{B_r{}^2}{4\pi\rho} + \frac{B_\phi{}^2}{4\pi\rho} \tag{9.3}$$

To visualize the differences between the velocity and magnetic structures, consider a purely radial outflow $\mathbf{v} = V_r(r)\hat{\mathbf{e}}_r$. Because a magnetic field line is rooted on the star, the star's rotation causes the magnetic field to twist into a spiral pattern. The very different radial velocity field configuration and the spiral magnetic field can nevertheless satisfy the 'frozen-in' field constraint that ionized matter does not cross the magnetic field lines. This situation can be pictured by considering the movement of the needle on a phonograph record. If we rotate the record counterclock-wise, the needle will move radially outward, and it is this motion that corresponds to the radially moving plasma, while the grooves correspond to the field line directions.

There are some important differences between the stellar case and the record groove analogy. For a rotating star the magnetic field lines leave the star radially rather than at a constant pitch angle of record grooves. This radial field causes the gas to co-rotate with the star

† In this chapter we use the symbol V instead of v for the wind velocity, because we wish to distinguish the V_ϕ and V_r components from velocity descriptors such as v_∞ and $v_{\rm esc}$.

out to some distance called the ‘co-rotation radius’. Unlike the record needle, the plasma velocity does not follow a radial path inside the co-rotation radius, but has a V_ϕ component that increases linearly with r. The tendency of the field to cause co-rotation leads to an outward transfer of angular momentum from the field to the gas. The reaction of that torque, which is transmitted along the magnetic field lines, removes angular momentum from the star.

The vector form of the momentum equation is

$$\rho \mathbf{v} \cdot \nabla \mathbf{v} + \nabla p + \rho \frac{GM_*}{r^2} \hat{\mathbf{e}}_r - \frac{1}{c} \mathbf{J} \times \mathbf{B} = 0 \tag{9.4}$$

The first three terms are the velocity gradient, pressure gradient and gravity force. The last term is the Lorentz force (in Gaussian units).

In the stationary frame of reference, the current density, $\mathbf{J}$, satisfies Ohm's law in the form,

$$\mathbf{J} = \sigma_c (\mathbf{E} + \frac{\mathbf{v} \times \mathbf{B}}{c}) \tag{9.5}$$

where σ_c is the coefficient of conductivity. Since the coefficient σ_c is extremely large ($\approx$ infinite) in a plasma, the $\mathbf{E} + \mathbf{v} \times \mathbf{B}/c$ term must be zero, so that $\mathbf{J}$ can remain finite. Therefore we get the well known ‘frozen-in field’ approximation that

$$\mathbf{E} = -\frac{\mathbf{v} \times \mathbf{B}}{c} \tag{9.6}$$

Now using this in Faraday's law

$$\nabla \times \mathbf{E} = -\frac{1}{c} \frac{\partial \mathbf{B}}{\partial t} \tag{9.7}$$

and, assuming steady conditions, $\partial \mathbf{B}/\partial t = 0$, we find that

$$\nabla \times (\mathbf{v} \times \mathbf{B}) = 0 \tag{9.8}$$

Both the radial and azimuthal components of this vector are zero. From the ϕ component, we get

$$\frac{1}{r} \frac{d}{dr} \{ r(V_r B_\phi - V_\phi B_r) \} = 0 \tag{9.9}$$

This is readily integrated, after multiplication by r. The constant of integration is evaluated by considering the conditions at the reference radius r_0 at the base of the wind (perhaps just slightly above R_*). Let Ω be the angular speed of the star at r_0. The tangential velocity $V_{\phi,0} = r_0 \Omega$, is large; $V_{r,0} \ll V_{\phi,0}$, because the radial velocity at the base of the wind is small, and $B_{\phi,0} \ll B_{r,0}$ because the field lines leave the stellar surface radially. So $V_{r,0}B_{\phi,0} \ll V_{\phi,0}B_{r,0}$ and the integration constant is $-r_0 V_{\phi,0}B_{r,0}$. Equation (9.9) becomes:

$$r(V_r B_\phi - V_\phi B_r) = \text{constant} = -r_0{}^2 \Omega B_{r,0} \tag{9.10}$$

This can be further simplified by using Maxwell's equation, $\nabla \cdot \mathbf{B} = 0$, which in spherical coordinates provides the conserved quantity $\mathscr{F}_B$,

$$\mathscr{F}_B = r^2 B_r = r_0{}^2 B_{r,0} \tag{9.11}$$

since the B components are functions of only r. Using this, in combination with (9.10), we get the ratio of B_ϕ to B_r at any r to be

$$\frac{B_\phi}{B_r} = \frac{V_\phi - r\Omega}{V_r} \tag{9.12}$$

This is a useful expression relating the angular and radial components of $\mathbf{B}$ to those of $\mathbf{v}$ and the solid body azimuthal velocity at r, which is $r\Omega$.

Equation (9.12) is used quite often in our derivations, and is referred to as the 'consequence of Faraday's law'. From Fig. (9.1) we see that B_ϕ/B_r is the tangent of the angle, ψ, that the magnetic field makes relative to the radial direction at r. Note that $\psi = 0$, if the rotation is solid body. This situation tends to occur close to the star where the field lines can be radial and cause co-rotation. Farther out in radius,

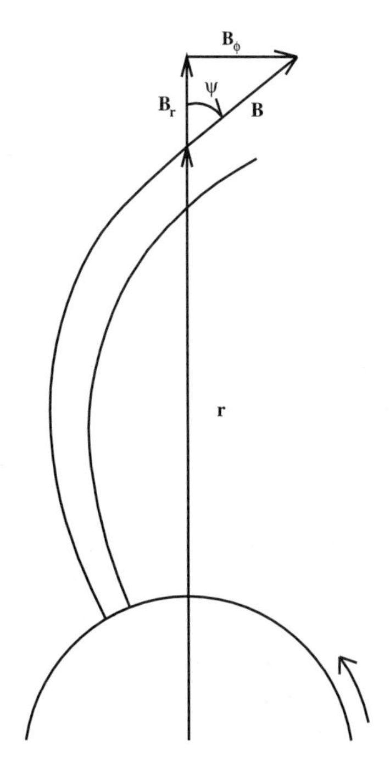

Figure 9.1 This figure shows the magnetic field **B** relative to the radial direction from the star.

$V_\phi \ll r\Omega$, and we see from Eq. (9.12) that B_ϕ has a large negative value, i.e. $\mathbf{B}$ is directed in the $-\hat{\mathbf{e}}_\phi$ direction, which gives the 'trailing' spiral structure of the field lines.

In the equation of motion (9.4), we now use Ampère's law

$$\nabla\times\mathbf{B} = \frac{1}{c}4\pi\mathbf{J} \tag{9.13}$$

to express the Lorentz force as $(\mathbf{J}\times\mathbf{B})/c = (\nabla\times\mathbf{B})\times\mathbf{B}/4\pi$. This vector expression can be expanded to provide terms in the three component directions, which can then be simplified by using the assumptions associated with the equatorial flow: the quantities in $\partial/\partial\phi$ and $\partial/\partial\theta$ are zero, and in the equatorial plane we also have $B_\theta = 0$ and $\sin\theta = 1$. We thus find the Lorentz force in the equatorial plane to be

$$\frac{1}{c}(\mathbf{J}\times\mathbf{B}) = -\frac{1}{4\pi}\left[\frac{B_\phi}{r}\frac{d}{dr}(rB_\phi)\right]\hat{\mathbf{e}}_r + \frac{1}{4\pi}\left[\frac{B_r}{r}\frac{d}{dr}(rB_\phi)\right]\hat{\mathbf{e}}_\phi \tag{9.14}$$

Note that the Lorentz force provides both a radial component for accelerating the matter outward, and an azimuthal term, which produces a torque on the outflow.

Using a vector transformation to polar coordinates, the first term of Eq. (9.4) can be written as

$$\begin{aligned}(\mathbf{v}\cdot\nabla)\mathbf{v} &= \tfrac{1}{2}\nabla v^2 - \mathbf{v}\times(\nabla\times\mathbf{v})\\ &= \left\{\tfrac{1}{2}\frac{d}{dr}(V_r^2 + V_\phi^2) - \frac{V_\phi}{r}\frac{d}{dr}(rV_\phi)\right\}\hat{\mathbf{e}}_r \\ &\quad + \left\{\frac{V_r}{r}\frac{d}{dr}(rV_\phi)\right\}\hat{\mathbf{e}}_\phi\end{aligned} \tag{9.15}$$

Combining Eqs. (9.15), (9.14) and (9.4), and separating the terms in $\hat{\mathbf{e}}_r$ and $\hat{\mathbf{e}}_\phi$, we can obtain equations of motion in the radial and azimuthal directions.

The equation of motion in the radial direction is

$$\frac{1}{2}\frac{d}{dr}(V_r^2+V_\phi^2)+\frac{1}{\rho}\frac{dp}{dr}+\frac{GM_*}{r^2}-\frac{V_\phi}{r}\frac{d}{dr}(rV_\phi)+\frac{B_\phi}{4\pi\rho r}\frac{d}{dr}(rB_\phi) = 0 \tag{9.16}$$

The derivative of the azimuthal velocity V_ϕ, appearing in this radial momentum equation, can be removed by using the identity

$$\frac{V_\phi}{r}\frac{d}{dr}(rV_\phi) = \frac{V_\phi^2}{r} + \frac{1}{2}\frac{d}{dr}V_\phi^2 \tag{9.17}$$

and the radial momentum equation (9.16) becomes

$$V_r\frac{dV_r}{dr} + \frac{1}{\rho}\frac{dp}{dr} + \frac{GM_*}{r^2} - \frac{V_\phi^2}{r} + \frac{B_\phi}{4\pi\rho r}\frac{d}{dr}(rB_\phi) = 0 \tag{9.18}$$

The first three terms are identical to the non-rotating wind momentum equation treated early in this book. The fourth term corresponds to centrifugal acceleration. The last term is the radial acceleration produced by the slinging effect of the rotating magnetic field. This produces an outward accleration if B_ϕ decreases with radius more gradually than r^{-1}.

In Weber and Davis' original treatment of the magnetic rotating solar wind, a polytropic energy equation was used to eliminate the gas pressure from the wind equation. Here, we choose to use the simpler isothermal approximation, because our interest is in strong winds for which the gas pressure gradient is not the dominant acceleration. The differences between isothermal and polytropic equations have already been explored in Chapter 4. Assuming that the flow is isothermal we can eliminate the gas pressure from the wind momentum equation using $p = \rho\, a^2$ where a^2 is constant.

The equation of motion in the azimuthal direction, found from Eqs. (9.4) and (9.15), is

$$\rho V_r \frac{d}{dr}(rV_\phi) = \frac{B_r}{4\pi}\frac{d}{dr}(rB_\phi) \tag{9.19}$$

This equation states that the rate of change of the angular momentum per unit volume is the result of the torque per unit volume produced by the magnetic field, $r(\mathbf{J}\times\mathbf{B})_\phi/c$.

In addition to the momentum equations we also need the conservation of mass equation, $\nabla \cdot (\rho \mathbf{v}) = 0$. After using the symmetries associated with equatorial outflow, the mass conservation equation becomes

$$\mathscr{F}_m = \rho V_r r^2 = \text{ constant} \tag{9.20}$$

where $\mathscr{F}_m$ is the mass loss per steradian leaving the star at the equator. (Since the mass flux depends on θ, the total mass loss rate from the star is not simply $4\pi\mathscr{F}_m$). Wherever possible, we will express the wind equations in terms of the constants of the motion $\mathscr{F}_B$ and $\mathscr{F}_m$, using Eqs. (9.11) and (9.20), respectively. These constants are especially useful for deriving equations of conservation of energy and angular momentum in the flow.

9.2 The angular momentum and energy constants

The azimuthal and radial equations of motion can each be expressed as perfect differentials, and thus they directly provide us with two constants for the motion. Let us first consider the azimuthal equation of motion (9.19). From Eqs. (9.11) and (9.20), we see that the ratio

$r^2B_r/(\rho V_r r^2)$ is constant. Therefore, after dividing by ρV_r, Eq. (9.19) can be expressed as

$$\frac{d\mathscr{L}}{dr} = 0 \tag{9.21}$$

where

$$\mathscr{L} = rV_\phi - \left(\frac{rB_rB_\phi}{4\pi\rho V_r}\right) = \text{constant} \tag{9.22}$$

This is 'the constant of angular momentum per unit mass'. The first term in this expression is the angular momentum per unit mass that is carried by the gas. The second term in Eq. (9.22) is the angular momentum carried by the magnetic stresses. To develop an interpretation of this, let us consider the product, $\mathscr{L}V_r$. Using the Alfvén velocity to eliminate B_r and B_ϕ from Eq. (9.22), we find

$$\mathscr{L}V_r = (rV_\phi)V_r + (-rA_\phi)A_r \tag{9.23}$$

Note that A_ϕ is negative so the second term is positive. In Eq. (9.23), the term $\mathscr{L}V_r$ is the total angular momentum per unit mass carried outwards with speed V_r. This equals the angular momentum per unit mass (rV_ϕ) carried radially outwards with speed V_r and the angular momentum per unit mass $(-rA_\phi)$ of the magnetic field carried radially outwards with the Alfvén speed A_r.

We can also derive a useful constant of the motion from the radial momentum Eq. (9.16). The last two terms of this equation can be combined by using Eq. (9.19) to eliminate $d(rV_\phi)/dr$, and the consequence of Faraday's law, Eq. (9.12), to eliminate the term V_rB_ϕ, thus forming the relation,

$$-\frac{V_\phi}{r}\frac{d}{dr}(rV_\phi) + \frac{B_\phi}{4\pi\rho r}\frac{d}{dr}(rB_\phi) = -\frac{d}{dr}\left\{\frac{r\Omega B_\phi\mathscr{F}_B}{4\pi\mathscr{F}_m)}\right\} \tag{9.24}$$

Hence the radial equation of motion (9.16) can be written as

$$\frac{d\varepsilon}{dr} = 0 \tag{9.25}$$

where

$$\varepsilon = \frac{1}{2}(V_r{}^2 + V_\phi{}^2) + a^2\ln\rho - \frac{GM_*}{r} - \frac{r\Omega B_rB_\phi}{4\pi\rho V_r} = \text{constant} \tag{9.26}$$

This is 'the constant energy per unit mass of the flow'. The first term is the kinetic energy, the second is the thermal energy, the third is the gravitational potential energy, and the last term is the magnetic energy per unit mass.

Combining the two constants of the motions, Eqs. (9.26) and (9.22), we find that

$$\varepsilon = \frac{1}{2}(V_r{}^2 + V_\phi{}^2) + a^2 \ln\rho - \frac{GM_*}{r} - r\Omega V_\phi + \mathscr{L}\Omega \tag{9.27}$$

in which the explicit dependence on **B** has been eliminated from ε. This latter form for ε will be especially useful for developing a wind momentum equation that involves only the radial velocity.

9.3 Angular momentum transfer

The effect of the magnetic term in the conservation of angular momentum per unit mass, Eq. (9.22), can be understood by considering a marble being flung by a flexible tube. Part of the angular momentum of that system would be in the stress in the tube, and so even if the rotation and the base of the tube were stopped, the unflexing of the tube would continue to transfer angular momentum to the marble during the straightening of the tube.

Using the consequence of Faraday's law (9.12) to eliminate B_ϕ from Eq. (9.22), we find from the conservation of angular momentum that

$$V_\phi = r\Omega \frac{\dfrac{V_r{}^2 \mathscr{L}}{r^2\Omega} - A_r{}^2}{V_r{}^2 - A_r{}^2} \tag{9.28}$$

Here we have eliminated B_r in terms of the square of the radial component of the Alfvén velocity, $A_r{}^2 = B_r{}^2/(4\pi\rho)$. Note that Eq. (9.28) has a singular point where the denominator goes to zero. This occurs at the 'Alfvén radius', r_A. From the vanishing of the denominator we find at r_A that

$$V_r(r_A) = A_r(r_A) \equiv V_A \tag{9.29}$$

The vanishing of the numerator of Eq. (9.28) determines the value of the Alfvén radius and leads to the requirement that

$$\mathscr{L} = r_A{}^2\Omega \tag{9.30}$$

Since the product, $r^2\Omega$, corresponds to the specific angular momentum of matter co-rotating with the star, Eq. (9.30) states that the angular momentum per unit mass of the stellar wind behaves *as if* the star rotates as a solid body out to the Alfvén radius. We say 'as if' because the near solid body co-rotation actually ceases at a radius well within r_A. Thus, there must be another contributor to the angular momentum of the rotating wind. This is the stress of the magnetic field, and it operates in the way we described in the tube analogy above.

Using Eq. (9.28) along with the ratio for B_ϕ/B_r from Eq. (9.12), we can write an equation for the azimuthal magnetic field B_ϕ in terms of purely radial quantities

$$\frac{B_\phi}{B_r} = \frac{r\Omega\left(4\pi\rho V_r \frac{\mathcal{L}}{r^2\Omega} - 4\pi\rho V_r\right)}{4\pi\rho V_r^2 - B_r^2} \tag{9.31}$$

Note that this equation has a singularity at the Alfvén radius where $V_r(r_A) = A_r(r_A)$, as was the case for Eq. (9.28).

9.4 The purely radial wind equations

Equations (9.28) and (9.31) express the azimuthal variables V_ϕ and B_ϕ in terms of radial variables only. Thus our equatorial wind problem can now be reduced to purely radial wind equations. We will find that the Alfvén radius, r_A, plays a crucial role in magnetic rotator winds. To begin, let us introduce the radial 'Alfvén Mach number'

$$M_A = \frac{V_r}{A_r} \tag{9.32}$$

At the Alfvén radius, $M_A(r_A) = 1$, while at other radii, the radial Alfvén Mach number is related to local properties by the useful relations

$$M_A{}^2 = \frac{4\pi\rho V_r{}^2}{B_r{}^2} = \frac{4\pi\mathcal{F}_m}{\mathcal{F}_B{}^2} r^2 V_r \tag{9.33}$$

Evaluating the constants of the flow $\mathcal{F}_m$ and $\mathcal{F}_B$ at r_A, Eq. (9.33) yields

$$M_A{}^2 = \frac{r^2 V_r}{r_A{}^2 V_A} \tag{9.34}$$

where V_A is the Alfvén speed at r_A. Since the radial Alfvén Mach number can be expressed in terms of the ratios r/r_A and V_r/V_A, it is convenient to introduce the dimensionless variables

$$x = \frac{r}{r_A} \quad \text{and} \quad u = \frac{V_r}{V_A} \tag{9.35}$$

Equation (9.34) now becomes

$$M_A{}^2 = x^2 u \tag{9.36}$$

Using this expression and Eq. (9.30), the azimuthal velocity (9.28) and the magnetic field ratio (9.31), expressed in terms of (x, u), are

$$V_\phi(x) = \frac{r_A \Omega \, x(1-u)}{1-x^2 u} \tag{9.37}$$

$$\frac{B_\phi}{B_r} = -\frac{r_A \Omega \, x}{V_A} \frac{1-x^2}{1-x^2 u} \tag{9.38}$$

The angular momentum, $\mathscr{L}$, of Eq. (9.22), can also be written in terms of (x, u) in such a way that the partitioning of the angular momentum between particles and the stress in the field can be seen more clearly.

$$\mathscr{L} = r_A{}^2 \Omega \left[\frac{x^2(1-u)}{1-x^2 u} + \frac{1-x^2}{1-x^2 u} \right] \tag{9.39}$$

$$= \mathscr{L}_{\text{gas}} + \mathscr{L}_{\text{mag}} \tag{9.40}$$

Note that at large radii where $x \to \infty$ and the flow velocity reaches its asymptotic value ($V_r = v_\infty$) the first term in Eq. (9.40) approaches $-(1-u)/u$ and the second approaches $1/u$ with $u \to v_\infty / V_A$. The gas and magnetic stress portions of $\mathscr{L}$ are thus seen to be given in this large r limit by

$$\mathscr{L}_{\text{gas}} = r_A{}^2 \Omega \left(1 - \frac{V_A}{v_\infty} \right) \tag{9.41}$$

$$\mathscr{L}_{\text{mag}} = r_A{}^2 \Omega \frac{V_A}{v_\infty} \tag{9.42}$$

Calculations show that V_A/v_∞ ratios are typically from about 2/3 to nearly unity. That being the case these equations give the interesting result that at large distances from the star, *most of the angular momentum loss by the star is carried away by the stresses in the magnetic field and not by the gas.* Examples of the radial distributions of $\mathscr{L}_{\text{gas}}$ and $\mathscr{L}_{\text{mag}}$, (as fractions of $\mathscr{L}_{\text{tot}} = r_A{}^2\Omega$) are shown in Fig. (9.2). Notice that most of the change in these distributions occurs near the Alfvén radius.

9.5 Magnetic energy transfer

For stellar winds in general, it is interesting to understand how the gas acquires the energy required to escape from the star. The gas is bound to the star at the base (i.e. has a negative energy per unit mass) and it becomes unbound (i.e. positive energy) farther out as a result of the wind driving process. Here we explore the energy transfer that occurs in magnetic rotator winds. For this it is useful to isolate the gas and magnetic components of the energy per unit mass constant, (Eq. 9.26).

$$\varepsilon = \varepsilon_{\text{gas}} + \varepsilon_{\text{mag}} = \text{ constant} \tag{9.43}$$

The gas energy term is

$$\varepsilon_{\text{gas}} = \frac{1}{2}(V_r{}^2 + V_\phi{}^2) + a^2 \ln\rho - \frac{GM_*}{r} \tag{9.44}$$

and this is the sum of the gas kinetic energy, enthalpy and potential energy per unit mass. This gas energy, ε_{gas}, is negative near the star because V_r is subsonic and very small, V_ϕ is less than the escape speed, and we are assuming that the enthalpy term is not a major contributor to the energy per gram. Hence, the gravitational potential (which is negative) is the dominant term in the energy equation, so the gas is gravitationally bound to the star at small r.

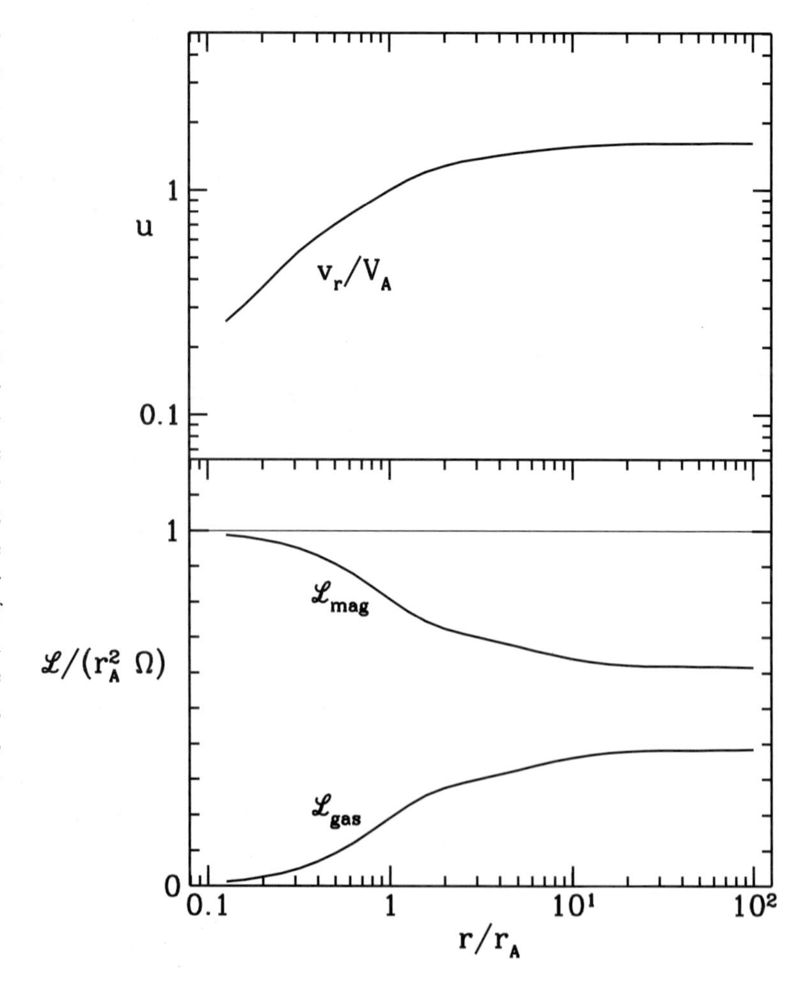

Figure 9.2 Upper panel shows the velocity distribution u versus x of a Weber and Davis (1967) wind model for a 1 $M_\odot$ star. (From Belcher & MacGregor, 1976, model A) It is used along with Eq. (9.40) to find the angular momenta carried by either magnetic stresses or by azimuthal gas motions that are shown in the lower panel as functions of radial distance in Alfvén radii. For the model shown, r_A=28.5 $R_\odot$, V_A=963 km s^{-1}.

The magnetic energy term is

$$\varepsilon_{\mathrm{mag}} = -\frac{r\Omega B_r B_\phi}{4\pi\rho V_r} \tag{9.45}$$

This is a positive energy because B_ϕ is negative, as discussed in § 9.1. By comparing Eqs. (9.22) and (9.45), we find $\varepsilon_{\mathrm{mag}} = \Omega\mathscr{L}_{\mathrm{mag}}$. Therefore, $\varepsilon_{\mathrm{mag}}(r)$ decreases with $\mathscr{L}_{\mathrm{mag}}(r)$, which is illustrated in Fig. (9.2).

The gas energy $\varepsilon_{\mathrm{gas}}$ becomes positive at large r because of the transfer of energy from $\varepsilon_{\mathrm{mag}}$ to $\varepsilon_{\mathrm{gas}}$. There is no need to invoke a seperate source of energy or momentum deposition to explain the energetics of magnetic rotator outflows.

9.5.1 The Poynting vector

The coupling of the magnetic and gas energy terms can be described by considering the electromagnetic flux density, or Poynting vector, which is defined as

$$\mathbf{S} = \frac{c}{4\pi}(\mathbf{E}\times\mathbf{B}) \tag{9.46}$$

The magnitude of this Poynting vector can be obtained by combining Eq. (9.46) with the frozen-in field condition $\mathbf{E} = -(\mathbf{v}\times\mathbf{B})/c$. We find by using vector properties that

$$\mathbf{S} = \mathbf{B}\times(\mathbf{v}\times\mathbf{B})/4\pi \tag{9.47}$$

$$= (\mathbf{v}B^2 - \mathbf{B}(\mathbf{v}\cdot\mathbf{B}))/4\pi \tag{9.48}$$

$$= \frac{B_\phi}{4\pi}[V_r B_\phi - V_\phi B_r]\hat{\mathbf{e}}_r + \frac{B_r}{4\pi}[V_\phi B_r - V_r B_\phi]\hat{\mathbf{e}}_\phi \tag{9.49}$$

$$= -\frac{r\Omega B_\phi B_r}{4\pi}\hat{\mathbf{e}}_r + \frac{r\Omega B_r{}^2}{4\pi}\hat{\mathbf{e}}_\phi \tag{9.50}$$

where we have used the ratio B_ϕ/B_r of Eq. (9.12) in the last step. Comparing (9.50) with the magnetic part of the energy equation (9.45), we see that

$$\varepsilon_{\mathrm{mag}} = \frac{S_r}{\rho V_r} \tag{9.51}$$

Hence $\varepsilon_{\mathrm{mag}}$, which is an energy per unit mass, is the ratio of the radial component of the Poynting energy flux to the mass flux.

By comparing the value of S_r at both small and large distances from the star, we can find the magnitude of the energy transfer from the field to the gas. Eliminating B_ϕ from Eq. (9.50) and using Eq. (9.12), we get

$$S_r = \frac{r\Omega B_r{}^2}{4\pi V_r}(r\Omega - V_\phi) \tag{9.52}$$

Now converting to the (x, u) variables of Eq. (9.35), and using $B_r^2 = 4\pi\rho V_r^2/M_A^2$ from Eqs. (9.33), (9.37) and (9.38), we find an expression for S_r as a function of distance from the star.

$$S_r = \rho V_r \, (r_A\Omega)^2 \frac{1 - x^2}{1 - x^2 u} \tag{9.53}$$

Near the star, where $u \ll 1$, we see from Eqs. (9.51) and (9.53) that

$$\varepsilon_{\rm mag}(r_0) = (r_A\Omega)^2 \, (1 - x_0^2) \tag{9.54}$$

Here the subscript 0 refers to the lower boundary, r_0, where $x = x_0$. For large x, where $u = u_\infty$, we similarly find that

$$\varepsilon_{\rm mag}(\infty) = (r_A\Omega)^2 \left(\frac{V_A}{v_\infty}\right) \tag{9.55}$$

Thus between $r = r_0$ and $r = \infty$ each gram of wind material has gained gas energy from the magnetic energy by the amount

$$\left(\frac{S_r}{\rho V_r}\right)_0 - \left(\frac{S_r}{\rho V_r}\right)_\infty = (r_A\Omega)^2 \left(1 - x_0^2 - \frac{V_A}{v_\infty}\right) \tag{9.56}$$

Note that the magnitude of the energy transfer depends upon the radius and velocity at the Alfvén point.

9.5.2 The slow and fast magnetic rotator approximations

We can use the transfer of energy to distinguish between two limiting classes of magnetic rotator winds: the *'slow magnetic rotator'*, or SMR, and the *'fast magnetic rotator'*, or FMR . These categories were introduced by Belcher and MacGregor (1976) to describe the loss of angular momentum during the evolution of the sun. For the sun, wind models show that $V_A \approx (2/3)v_\infty \approx 200$ km s^{-1}, $r_A \simeq 10R_\odot$ (so $x_0 \approx 0.1$) and $r_A\Omega = 0.1v_\infty$. Therefore, from Eq. (9.56), we see that the energy transferred to the solar wind from the solar magnetic field is less than 1 percent of the wind's total kinetic energy per gram. This small percentage is mostly because $r_A\Omega$ is small. The main source of the gas energy in coronal models of the solar wind is the enthalpy due to high temperature, and the wind is accelerated by the gradient of the gas pressure. The solar wind is an example of a slow magnetic rotator, which means that the radial component of the wind structure is not changed significantly by the presence of the magnetic field.

For a fast magnetic rotator, the dominant source of energy is the transfer of Poynting flux energy, and the radial acceleration by the field is much larger than can be accounted for by the gas pressure. Thus for our discussion of FMR winds we can ignore $a^2 \ln\rho$ in Eq. (9.44).

The FMR case is especially useful because it allows us focus on the effects of the magnetic field and to ignore gas pressure in the region of the wind where the acceleration to high speeds is occuring.

From Eq. (9.52) we can see that the magnitude of the Poynting energy flux is of the order $(r\Omega)^2 B_r^2/4\pi V_r$. By making use of the conserved quantities $\mathscr{F}_m$ and $\mathscr{F}_B$, and assuming $V_\phi \ll r\Omega$ (which is valid at large r), the Poynting energy flux can be expressed as

$$S_r \approx \rho \left(\frac{\Omega^2 \mathscr{F}_B{}^2}{4\pi \mathscr{F}_m} \right) = \rho V_M{}^3 \tag{9.57}$$

Here we have expressed the energy flux in terms of a quantity known as the *Michel velocity*, V_M, (Michel, 1969)

$$V_M = \left(\frac{\Omega^2 \mathscr{F}_B{}^2}{4\pi \mathscr{F}_m} \right)^{\frac{1}{3}} = \left(\frac{r^4 B_r{}^2 \Omega^2}{\dot{M}_E} \right)^{\frac{1}{3}} \tag{9.58}$$

where we have let $\dot{M}_E = 4\pi \mathscr{F}_m$.† Notice that the Michel velocity does not depend on radius, because $\mathscr{F}_B$, $\mathscr{F}_m$ and Ω are constants, that are defined by parameters given at the base of the wind. The Michel velocity will be useful for relating the terminal conditions in an FMR wind to those at the base. As we shall see later, for an FMR wind, $v_\infty \simeq V_M$.

Expressing the total energy per unit mass ε in terms of V_M gives

$$\varepsilon = \frac{1}{2}(V_r{}^2 + V_\phi{}^2) + a^2 \ln\rho - \frac{GM_*}{r} + \frac{V_M{}^3}{V_r}\left(1 - \frac{V_\phi}{r\Omega}\right) \tag{9.59}$$

We can also find a relation between the Michel velocity and the radius and velocity at the Alfvén point, by noting that the ratio $\mathscr{F}_B{}^2/\mathscr{F}_m$ appears in both Eq. (9.58) and in the Alfvén Mach number expression Eq. (9.33). Evaluating the latter ratio at r_A, and substituting Eq. (9.58), we find

$$r_A{}^2 \Omega^2 = \frac{V_M{}^3}{V_A} \tag{9.60}$$

Combining this with Eq. (9.53), we find that the magnetic energy per unit mass, Eq. (9.51), is

$$\varepsilon_{\text{mag}} = \frac{V_M{}^3}{V_A} \left[\frac{1 - x^2}{1 - x^2 u} \right] \tag{9.61}$$

† $\dot{M}_E$ is the mass loss rate associated with the flow from the equator. The mass loss rate $\dot{M}$ of the whole star is smaller than $\dot{M}_E$.

Therefore, close to the star $\varepsilon_{\rm mag} \simeq V_M{}^3/V_A$, while at very large distances

$$\varepsilon_{\rm mag}(\infty) = \frac{V_M{}^3}{v_\infty} \tag{9.62}$$

Finally, we can also relate the angular momentum flux, $\mathscr{L}$, to the Michel velocity, using Eqs. (9.40) and (9.60), and we get the result that at large radii in FMR winds

$$\mathscr{L}_{\rm gas} = \frac{V_M{}^3}{V_A\Omega}\left(1 - \frac{V_A}{v_\infty}\right) \tag{9.63}$$

The dependence on the Alfvén point conditions occurs because much of the conversion of angular momentum from magnetic stresses to gas azimuthal motion occurs in the Alfvén point region, as illustrated in Fig. (9.2).

9.6 Solution of the radial equation of motion

Let us now consider the radial momentum equation (9.18) and the process by which r_A, V_A and the entire radial velocity structure can be determined. We will first consider the general magnetic rotator equations and then again will shift attention to the FMR limiting case.

The derivative of B_ϕ appearing in the momentum equation (9.18) can be found by using Eq. (9.31) in the form

$$rB_\phi = (r^2B_r)\Omega(4\pi\rho V_r r^2)\frac{\mathscr{L}/\Omega - r^2}{(4\pi\rho V_r r^2)r^2V_r - (r^4B_r{}^2)} \tag{9.64}$$

where the parentheses in this equation contain the conserved quantities $r^2B_r = \mathscr{F}_B$ and $4\pi\rho V_r r^2 = 4\pi\mathscr{F}_m$. After inserting these two constants in Eq. (9.64), the only non-constants in the equation are r^2 and r^2V_r, so one can differentiate the equation to obtain

$$\frac{d}{dr}(rB_\phi) = \frac{-2V_r\,(V_rB_\phi + r\Omega B_r) - rB_\phi V_r\dfrac{dV_r}{dr}}{V_r{}^2 - A_r{}^2} \tag{9.65}$$

in which we have used the radial component of the Alfvén speed, Eq. (9.3), to replace B_r in the denominator. Equation (9.65) can be further simplified using $(V_rB_\phi + r\Omega B_r) = V_\phi B_r$, from Eq. (9.12).

The magnetic force is proportional to $d(rB_\phi)/dr$ given in Eq. (9.65). We should note that this introduces another term to the radial momentum equation that contains $V_r dV_r/dr$. This means that there will be another term in the 'denominator' of the wind equation. By substituting Eq. (9.65) and into Eq. (9.18), and gathering the coefficients of

dV_r/dr we obtain

$$\left(V_r^2 - a^2 - \frac{A_\phi^2 V_r^2}{V_r^2 - A_r^2}\right) \frac{r}{V_r} \frac{d}{dr} V_r = 2a^2 - \frac{GM_*}{r} + V_\phi^2 + 2\frac{A_r A_\phi V_r V_\phi}{V_r^2 - A_r^2} \tag{9.66}$$

This is a general expression for the velocity gradient in isothermal magnetic rotator theory (Hartmann & MacGregor, 1982). It contains the modifications introduced by the centrifugal and Lorentz forces. It will be useful in our study of the properties of the wind critical points. First, let us find a momentum equation that involves only radial quantities, such as V_r and A_r.

9.6.1 The dimensionless velocity equation

The radial momentum equation (9.66) does not yet have the azimuthal terms completely removed, because both A_ϕ and V_ϕ appear. The complete separation of the ϕ and r dependent quantities can be achieved by now eliminating V_ϕ by using Eq. (9.37) and using $A_\phi/A_r = B_\phi/B_r$, as given in Eq. (9.38). Making these substitutions in Eq. (9.66) using the Alfvén radius dependent expression for $M_A{}^2 = x^2 u$, yields

$$\begin{aligned} \left\{1 - \frac{\mathcal{V}}{u^2} + \mathcal{W}\frac{x^2(1-x^2)^2}{(1-x^2u)^3}\right\} x^2 u \frac{du}{dx} = \\ +2\mathcal{V}x - \mathcal{U} + \mathcal{W}\left\{\frac{x^3(1-u)^2}{(1-x^2u)^2}\right\} \\ +\mathcal{W}\left\{\frac{2x^3u(1-u)(1-x^2)}{(1-x^2u)^3}\right\} \end{aligned} \tag{9.67}$$

where the dimensionless constants are

$$\mathcal{U} = \frac{GM_*}{r_A V_A{}^2}, \quad \mathcal{V} = \frac{a^2}{V_A{}^2}, \quad \mathcal{W} = \frac{(r_A\Omega)^2}{V_A{}^2} \tag{9.68}$$

These constants have the following physical significance: $\sqrt{2\mathcal{U}}$ is the ratio of the escape speed at the Alfvén point to the flow speed at that point, $\sqrt{\mathcal{V}}$ is the ratio of the sound speed to the flow speed at the Alfvén point, and $\sqrt{\mathcal{W}}$ is the solid body rotation speed divided by the flow speed at the Alfvén point.

The second to the last term in Eq. (9.67) corresponds to the centrifugal term, while the other coefficients of $\mathcal{W}$ arise from the Lorentz force. The topology of the radial momentum equation (9.67) was first developed by Weber and Davis (1967) for the case of the solar wind. In contrast with the wind models treated in earlier chapters, the magnetic rotator momentum equation has *three* singular points. Weber and Davis showed that these three critical points correspond to the locations at which V_r is equal to (a) the radial phase speed of the slow magnetosonic wave, (b) the radial Alfvén velocity V_A, and (c) the speed of the fast magnetosonic wave. The radial distances from the center of the star to these three locations are r_s, r_A, and r_f, respectively. The slow (r_s) and fast (r_f) points are each X-type singularities quite similar to those discussed in Chapters 3 and 4. The Alfvén point, however, is not an X-type singularity, as we shall now see, and it does not impose any additional conditions on the solution for V_r, but it does set the constant $\mathscr{L}$ from Eq. (9.30).

9.6.2 The nature of the Alfvén point

Equation (9.67) is not a simple one, but we can develop some understanding of the nature of its solution by considering the critical points. Equation (9.67) can be expressed as

$$x^2uu' \; s(x,u) + i(x,u) = F(x,u,u') = 0 \tag{9.69}$$

where $u' = \frac{du}{dx}$. This equation corresponds to a straight line on an F versus x^2uu' graph, and $s(x,u)$ is the slope of the line and $i(x,u)$ is the intercept on the F axis. The definitions of the functions s and i can be determined from the comparison of Eq. (9.69) with Eq. (9.67).

Let us use Eq. (9.69) to explore the nature of the Alfvén point. The Alfvén point was the location of the singularity that appeared in the *azimuthal* momentum equation, (9.28). However, our problem now involves *radial* quantites only, so the significance of an azimuthally defined singularity is not obvious. Note that $(1 - x^2u)$ (which is the same as $1 - {M_A}^2$) occurs in the denominator of three terms (those involving $\mathscr{W}$) in Eq. (9.67). So, very near to the Alfvén point, these three terms are the dominant ones, and we can ignore the other terms in the equation. If we make a linear expansion about the Alfvén point of the coefficients of Eq. (9.69) using $x = 1 + \Delta x$ and $u = 1 + \Delta u$, we find that in the limit of small Δx and Δu that the centrifugal term can be neglected relative to the magnetic terms, and the coefficients $s(x,u)$ and $i(x,u)$ take the following values

$$s(\Delta x, \Delta u) = -\frac{4\mathscr{W}\Delta x^2}{(2\Delta x + \Delta u)^3} \tag{9.70}$$

$$i(\Delta x, \Delta u) = +\frac{4\mathscr{W}\Delta x \Delta u}{(2\Delta x + \Delta u)^3} \tag{9.71}$$

Approaching the Alfvén point from the sub-Alfvénic region, i.e. Δx being negative, we find from Eqs. (9.70) and (9.71) that $(s,i) \to (+\infty, -\infty)$, respectively. On the other hand, approaching the Alfvén point from the super-Alfvénic side, i.e. with Δx positive, we find $(s, i) \to (-\infty, +\infty)$. What this means is that as we follow a solution of the equation outward through the Alfvén point both the slope and intercept flip discontinuously and revert back to the values they had deep in the subsonic portion of the flow, with *slope* $\approx -\infty$ and *intercept* $= +\infty$. This reversion to values similar to the original ones, allows the solution to go through a second X-type singular point, but at r_f, much farther out in radial distance from the star than the initial critical point. The Alfvén radius itself does not correspond to an X-type singular point, but conditions at the Alfvén radius make it possible for the second critical point to exist in the wind solution.

Another contrast between the Alfvén point and the X-type singularities is that there is not a unique slope to a solution that passes through the Alfvén point. In general we have $x^2 u du/dx = -i/s$ from Eq. (9.69). But at r_A we also have $x = 1$, $u = 1$ and we can get i/s from Eqs. (9.70) and (9.71), so we find at the Alfvén point

$$\frac{d\Delta u}{d\Delta x} = \frac{\Delta u}{\Delta x} \tag{9.72}$$

This has the solution $\Delta u =$ constant $\times\ \Delta x$. Since the constant here is arbitrary, this solution means that the slope du/dx can have any positive value at the Alfvén point. In other words, in the graphical plane u versus x, for every solution, $u(x)$, that approaches the Alfvén point singularity from the left there is a solution with the same slope leaving towards the right. It is not necessary to impose the condition on the solution that it pass through the Alfvén point because any solution that passes through both the slow and fast points automatically passes through the Alfvén point. This situation is very different from the standard X-type singularities that we encountered in Chapter 3, which had a unique velocity and velocity gradient at the critical point.

From an algebraic point of view, our solution is determined by the three constants $\mathscr{U}$, $\mathscr{V}$, and $\mathscr{W}$. These do not involve either the stellar radius R_* nor the velocity or magnetic field at the base of the wind. So, we need to discuss how we can apply the magnetic rotator differential equation to a specific star. Suppose we seek to find the critical solution to Eq. (9.67) that passes smoothly from small velocities u at small x to values of $u > 1$ as $x \to \infty$. We will find that the solution has the following properties: at $x = 1$, $u = 1$, the slope $(du/dx)_A$ is fixed by the condition that the inward integrations pass through the inner X-type critical point. So one of the constants $\mathscr{U}$, $\mathscr{V}$, or $\mathscr{W}$ cannot be a free parameter, because the slope $(du/dx)_A$ determined from the inward solution will not, in general, lead to a solution in the outward direction. So the solution determines one of the three constants, say $\mathscr{W}$, given the two others $\mathscr{U}$ and $\mathscr{V}$.

Now let us apply this to a star. We find that we need to specify six constants to identify a wind model for a particular star. As an example, we could specify $\mathscr{U} = GM_*/r_A V_A^2$; $\mathscr{V} = a/V_A$; $r_0 = R_*$; $B_r(R_*)$; M_*, the stellar mass; and T, the wind temperature. The mathematical solution gives us $\mathscr{W} = r_A \Omega^2 / V_A^2$ and $u(x)$. Combining these we get equations for $M_* = M(\mathscr{U}, r_A, V_A)$, $T = T(\mathscr{V}, V_A)$ and $\Omega = \Omega(\mathscr{W}, r_A, V_A)$. The inner boundary occurs at $x_0 = r_0/r_A$, where the solution gives $u = u_0$, so the radial velocity at the base is $V_r(r_0) = V_A u_0$ and then, the mass flux is $\mathscr{F}_m = \rho V_r(r_0) r_0^2$, and magnetic flux $\mathscr{F}_B = r_0^2 B_r(r_0)$.

In conclusion, there are effectively six equations and six unknowns. The six equations arise from the disappearance of the numerators and denominators at two X-type critical points, plus the conservation of energy and angular momentum equations. For the next discussion it is convenient to consider the six unknowns to be the locations and velocities at the three critical points: r_s, $V_{r,s}$, r_A, V_A, r_f and $V_r(r_f)$. In the next section we describe the solution for each of these quantities, for the case of an FMR wind.

Models for the solar wind at a range of rotation rates and surface magnetic fields were calculated by Belcher and MacGregor (1976) in their studies of the evolutionary development of the wind. Figure (9.3) shows the radial and azimuthal velocities versus radius at a given rotation speed ($r_0\Omega = 0.01 v_{\rm esc}$) and for three different magnetic fields. The vertical tickmarks on the velocity solutions correspond to the locations of the Alfvén point, $Z_{\rm A}$, and fast point, $Z_{\rm f}$. Note that as the surface field strength increases, the co-rotation is maintained to a larger radius, and the wind speed increases. Beyond the Alfvén radius, V_ϕ decreases $\propto 1/r$, as expected from conservation of angular momentum.

Figure (9.4) shows the dependence of V_0 and V_r on the rotation speed of the star. Since ρ_0 is specified in these models the value of V_0 determines the mass loss rate. Note that at low rotation rates both the mass loss rate and terminal velocity are not sensitive to Ω. As we will emphasize later, in the slow rotation (weak field) regime the mass loss rate and terminal velocity are determined by the 'primary mechanism' (in this case the coronal mechanism). However, when we enter the fast magnetic rotator regime, where V_M exceeds the terminal velocity of the primary mechanism, both the terminal velocity and the mass loss rate increase with increasing values of Ω. Figure (9.5) shows the locations of the critical radii as a function of Ω and B. Note that for the slow magnetic rotators the fast point and the Alfvén point are nearly coincident. These two radii separate by an amount that increases with both magnetic field strength and rotation rate.

9.7 The fast magnetic rotator equations

Fast magnetic rotator winds are those for which the magnetic and centrifugal accelerations play a dominant role in determining the mass loss and radial velocity distribution in the wind. A star is considered to be in the FMR regime if the Michel velocity is larger than the terminal velocity that would be derived in absence of the magnetic rotator forces. For example, in Chapters 3 and 4 we found that the terminal velocity for thermally driven winds is of order a, while for

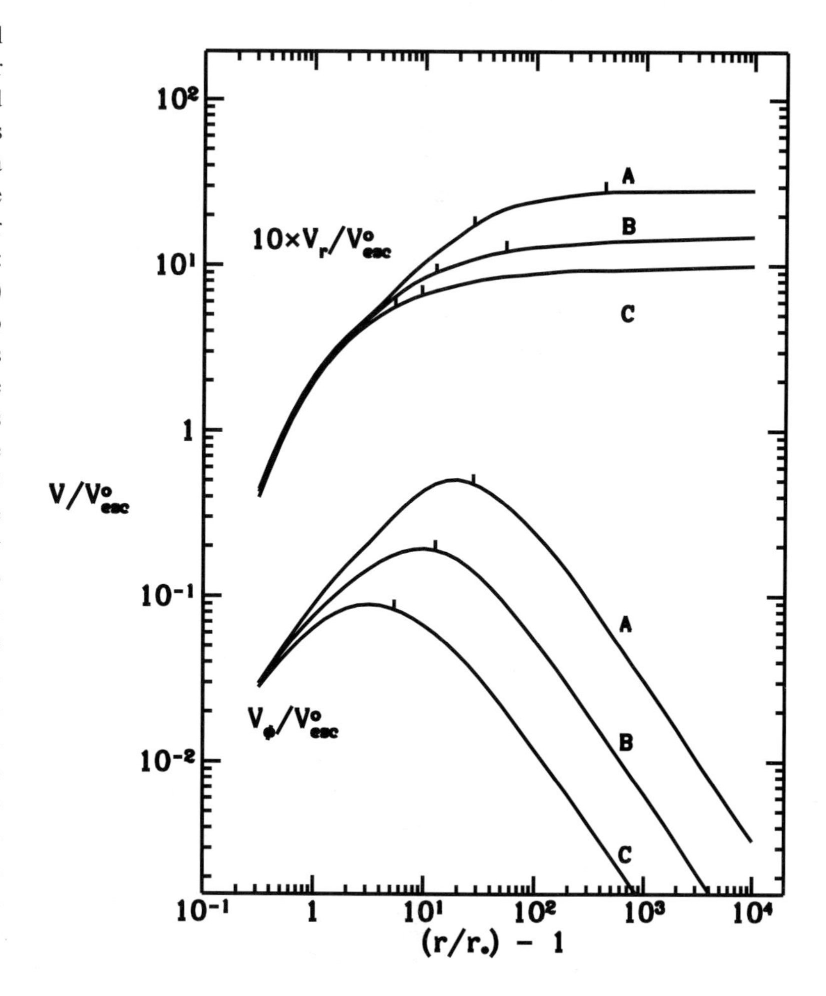

Figure 9.3 Radial velocities (upper profiles) and azimuthal velocities (lower profiles) for a fixed rotation rate $\Omega r_0/v_{esc} = .01$ and for three field strengths: with (A,B,C) corresponding to (4.6,1.5,0.46) Gauss respectively. The Weber & Davis (1967) model for the sun assumes B=1.5 Gauss. The azimuthal velocity increases nearly up to the Alfvén radius, indicated by the vertical tickmarks. The tickmarks on each of the radial velocity curves indicate the locations of the Alfvén radius and the fast critical point radius. (From Belcher & MacGregor, 1976)

the line driven winds the terminal velocity is given approximately by $\sqrt{\alpha/(1-\alpha)}\ v_{\rm esc}$ with α being determined by the line strength distribution. In this section we assemble the equations that have been derived in the earlier sections of this chapter. We then find the solutions for the velocities and the locations of the three critical points, using the approach developed by Hartmann and MacGregor (1982).

The radial momentum equation (9.66) is conveniently written as

$$\frac{r}{V_r}\frac{dV_r}{dr} = \frac{(V_r^2 - A_r^2)(2a^2 + V_\phi^2 - \frac{GM_*}{r}) + 2V_rV_\phi A_rA_\phi}{(V_r^2 - A_r^2)(V_r^2 - a^2) - V_r^2A_\phi^2} \tag{9.73}$$

The desired solution of this equation passes continuously through both of the points where the numerator and denominator simultaneously vanish, the slow and fast critical points, located at r_s and r_f, called

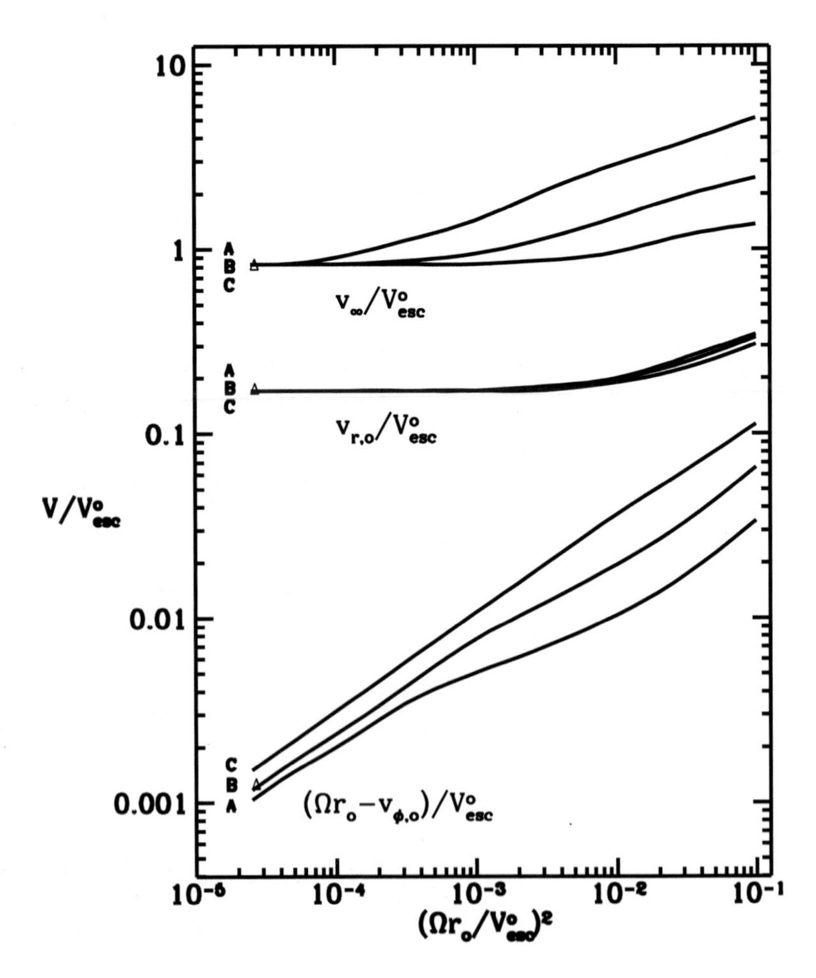

Figure 9.4 The terminal velocity, base velocity and azimuthal speeds, (all normalized by $v_{\rm esc}$ at r_0) as a function of rotation rate, for models using the three B fields in Fig. (9.3). The mass loss rate is proportional to $V_{r,0}$. This figure illustrates the transition from SMR to FMR wind solutions. The Weber & Davis (1967) SMR results for the present-day sun are indicated by the △ symbols. (From Belcher & MacGregor, 1976)

the 'slow point' and 'fast point' for short (these points are discussed in § 9.6.1).

Let us assume that we know the the field strength $B_{r,0}$, and the density ρ_0 at the reference radius r_0 at the base of the wind, as well as the stellar parameters M_*, R_*, T and Ω. Since we will now be using the base radius r_0 as a reference point, it is convenient to introduce the dimensionless variable $Z \equiv r/r_0$. So our goal is to get Z_s, Z_A, Z_f and the velocities at those three points.

At the slow point, we assume in FMR theory that the Alfvén Mach number is small, and that the rotational velocity is nearly in solid body co-rotation out to at least r_s, so $V_r \ll A_r$ and $V_\phi \approx r\Omega$, and

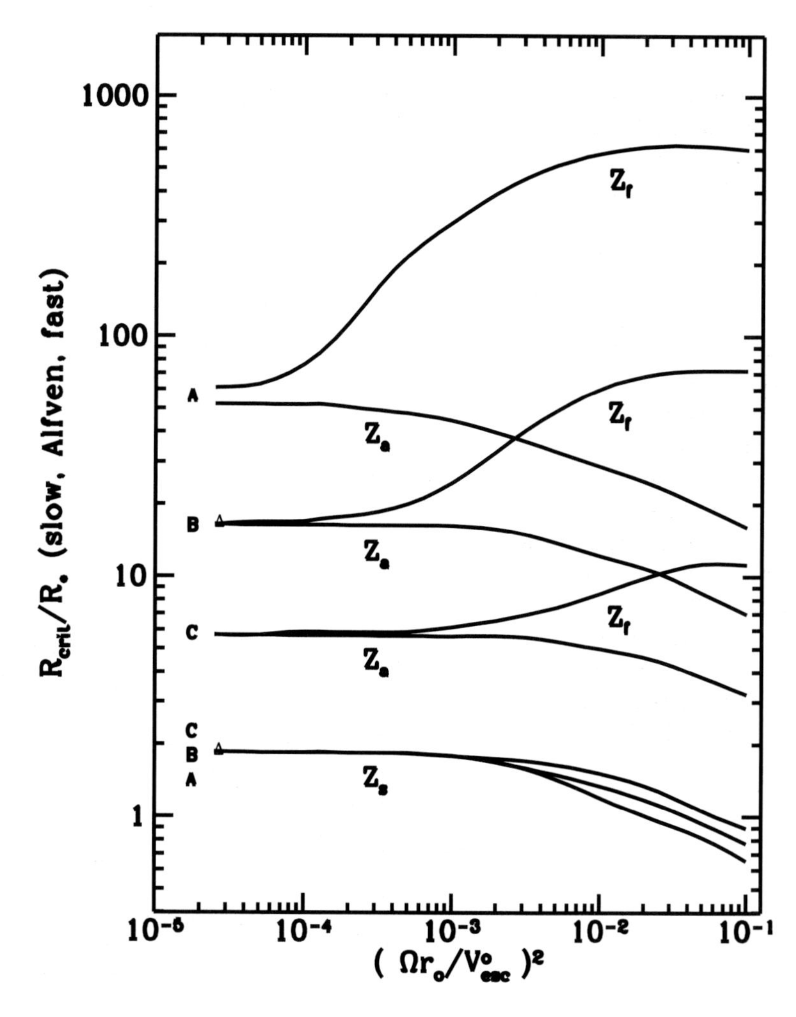

Figure 9.5 The locations of the slow (Z_s), Alfvén (Z_A), and fast (Z_f) critial points as a function of rotation rates, for three values of the magnetic field (A, B, C) The radii are given as the ratio $Z = r/r_0$. (Adapted from Belcher & MacGregor, 1976)

using Eq. (9.12) this leads to

$$\frac{V_r{}^2 A_\phi{}^2}{A_r{}^2} = (V_\phi - r\Omega)^2 \approx 0 \tag{9.74}$$

Therefore, the vanishing of the denominator in Eq. (9.73) requires

$$V_{r\,s}^{\,2} \equiv V_{r,s} = a^2 \tag{9.75}$$

The radial velocity at the slow point is the sound speed, so the slow point is also the sonic point radius. Now from the vanishing of the numerator of Eq. (9.73) at the sonic point we get

$$-A_r{}^2 \left(2a^2 + V_\phi{}^2 - \frac{GM_*}{r}\right) + 2V_r V_\phi A_r A_\phi = 0 \tag{9.76}$$

Dividing this equation by $A_r{}^2$ and eliminating the last term using Eq. (9.74), we obtain an expression valid at the inner critical point, at r_s

$$2a^2 - \frac{GM_*}{r_s} + r_s{}^2\Omega^2 = 0 \tag{9.77}$$

The solution of this provides the radius of the inner critical point. For problems involving stellar rotation, it is convenient to express angular speeds relative to the maximal rotation speed at the reference point,

$$\Omega^2_{\max} = GM_*/r_0{}^3 \tag{9.78}$$

so we define ω to be the ratio

$$\omega = \frac{\Omega}{\Omega_{\max}} \tag{9.79}$$

If we also let r_p be the 'Parker radius', $r_p = GM_*/2a^2$, which is the radius at the sonic point for an isothermal wind, we have, after expressing radii in r_0 units,

$$Z_p = \frac{r_p}{r_0} = \frac{GM_*}{2a^2 r_0} \tag{9.80}$$

and the vanishing of (9.77), the numerator in the transsonic region now becomes

$$Z_s{}^3 + \frac{1}{\omega^2 Z_p} Z_s - \frac{1}{\omega^2} = 0 \tag{9.81}$$

which is a cubic equation for the radus of the inner critical point.

9.7.1 The centrifugal magnetic rotator limit

Analysis of Eq. (9.81) shows that it has one real root. For the case in which winds are dominated by centrifugal and magnetic forces a is small, so Z_p is large $\gg 1$, so we get

$$Z_s{}^3\omega^2 \approx 1 \tag{9.82}$$

This result for Z_s has a simple interpretation. Consider a radial distance, r, at which the co-rotation speed, $r\Omega$, is equal to the speed of a particle in a circular Keplerian orbit (where centrifugal and gravitational forces are equal) $V_\phi = \sqrt{(GM_*/r)}$. There we find that $(r/r_0)^3 = (1/\omega)^2$, which is clearly equivalent to Eq. (9.82). Thus we can conclude that Eq. (9.82) means that *the inner critical point, which is the sonic point, occurs at the radius where the co-rotation speed has reached the circular orbital speed.* We refer to this limiting case as a centrifugally dominated subsonic region, or refer to the wind as a 'centrifugal magnetic rotator', or CMR.

In the centrifugal magnetic rotator, the magnetic field lines act as rigid radial flow tubes out to radii larger than the inner critical point. It is particularly interesting that for a CMR we can derive the radius of the critical point if we are given only the rotation rate, Ω, and the mass and radius of the star! For no other case in wind theory is critical point radius so easily determined.

9.7.2 The mass loss rate of fast magnetic rotators

In a rapidly rotating star, the mass flux is likely to depend strongly on the polar angle, θ, reaching a maximum in the equator at $\theta = \pi/2$. The magnetic rotator theory as discussed in this chapter provides an equatorial mass flux per steradian, $\mathcal{F}_m$, and it is often convenient to convert this to a 'spherical mass loss rate equivalent', $\dot{M}_E$, by multiplying by 4π

$$\dot{M}_E = 4\pi\mathcal{F}_m \tag{9.83}$$

While this is useful for comparing the magnitudes of mass loss rates caused by different driving forces, we should realize that a correction factor of typically between 0.1 to 0.5 should be applied to the equatorial $\dot{M}_E$ result to get the actual mass loss occuring from the entire surface of the star. This is because the mass loss in the equatorial region is much higher than that from other parts of the stellar surface.

As in the basic wind theory of Chapters 3 and 4, the mass loss rate is determined by conditions in the subsonic region. In the case in which we can specify the density, ρ_0, at the reference level r_0, the mass flux per solid angle is determined by the radial velocity at the reference point $V_{r,0}$; so $\mathscr{F}_m = \rho V_{r,0} r_0^2$. The base velocity can be derived from the energy Eq. (9.27) by equating the energy per unit mass at r_0 to that at r_s. Assuming the solid body expression $V_\phi = r\Omega$ in the subsonic region, we obtain

$$\varepsilon = \frac{1}{2}V_r^2 - \frac{1}{2}(r\Omega)^2 + a^2 \ln\rho - \frac{GM_*}{r} + \mathscr{L}\Omega \tag{9.84}$$

Equating $\varepsilon(r_0)$ to $\varepsilon(r_s)$, and using $V_{r,s} = a$, we find an expression for the velocity at the base of the wind.

$$\left[\frac{1}{2}\left(\frac{V_{r,0}}{a}\right)^2 - \ln\left(\frac{V_{r,0}}{aZ_s^2}\right)\right] = +\frac{1}{2} + 2Z_p\left[\left(1 - \frac{1}{Z_s}\right) - \frac{1}{2}\omega^2(Z_s^2 - 1)\right] \tag{9.85}$$

Given Z_s from Eqs. (9.80) to (9.82), we now have an implicit expression for $V_{r,0}$, and the mass loss rate can be determined.

It is useful to compare the density of the subsonic region with the rotating hydrostatic equilibrium structure $\rho_{HS}(r)$. This hydrostatic structure is obtained by ignoring the V_r terms in the momentum equation, so that we find

$$\frac{1}{\rho}\frac{dp}{dr} + \frac{GM_*}{r^2} - \frac{V_\phi^2}{r} = 0 \tag{9.86}$$

Assuming again that the magnetic fields enforce co-rotation throughout the subsonic region, we have $V_\phi = r\Omega$, so

$$\frac{V_\phi^2}{r} = \omega^2 \frac{GM_*}{r_0^3} r = \frac{GM_*}{r^2}(\omega^2 Z^3) \tag{9.87}$$

and letting $p = \rho a^2$, we find

$$\frac{1}{\rho}\frac{d\rho}{dr} = -\frac{GM_*}{a^2 r^2}(1 - \omega^2 Z^3) \tag{9.88}$$

which has the solution

$$\rho_{HS} = \rho_0 \exp\left\{-\frac{GM_*}{a^2 r_0}\left(1 - \frac{1}{Z}\right)\right\} \times C_{cent} \tag{9.89}$$

Thus the hydrostatic structure is the product of the density distribution of a non-rotating atmosphere and the 'centrifugal factor', C_{cent}, as given by

$$C_{cent} = \exp\left\{+\omega^2 \frac{GM_*}{2a^2 r_0}(Z^2 - 1)\right\} \tag{9.90}$$

The first two factors of Eq. (9.89), i.e. ρ_{HS}/C_{cent}, give the isothermal hydrostatic equilibrium structure that was discussed in Chapter 3. The important property of the rotational correction factor is that it increases exponentially with ω^2. For values of ω less than but near unity, the density structure of the rotating hydrostatic atmosphere is dominated by the centrifugal force effect.

Now let us consider the density structure of an expanding, rotating atmosphere. If we include $V_r dV_r/dr$ on the left hand side of Eq. (9.86), eliminate the pressure gradient using $p = \rho a^2$ and $\rho V_r r^2 = \mathscr{F}_m$, we can find the subsonic density structure for the expanding wind case

$$\left\{1 - \frac{a^2}{V_r{}^2}\right\} V_r \frac{dV_r}{dr} = \frac{2a^2}{r} - \frac{GM_*}{r^2}(1 - \omega^2 Z^3) \tag{9.91}$$

The left hand side of this can be integrated as described in Chapter 3. After exponentiating each side we get

$$\frac{1}{V_{r,0} r_0{}^2} \exp\left\{\frac{1}{2}\frac{V_{r,0}{}^2}{a^2}\right\} = \frac{1}{V_r r^2} \exp\left\{+\frac{GM_*}{a^2 r_0}\left(1 - \frac{1}{Z}\right)\right\} \times C_{\text{cent}} \tag{9.92}$$

and on using Eq. (9.89) and the constant $\mathscr{F}_m$ we get

$$\rho(r) = \rho_{\text{HS}}(r) \exp\left\{\frac{V_{r,0}{}^2 - V_r{}^2}{2a^2}\right\} \tag{9.93}$$

Evaluating this at $r = r_s$, where $V_{r,s} = a$ gives, in the limit $V_{r,0} \ll a$

$$\rho(r_s) = e^{-1/2} \rho_{\text{HS}}(r_s) = 0.607 \rho_{\text{HS}}(r_s) \tag{9.94}$$

This result was also derived in Chapter 3 in our comparison of the subsonic structure of a Parker wind with a hydrostatic non-rotating star. The density in the subsonic region thus remains very similar to the hydrostatic structure even when centrifugal forces are accounted for.

We see from Eqs. (9.88) and (9.93) that the subsonic density structure also contains the centrifugal factor, $C_{\text{cent}}(\omega)$. Therefore the density throughout the subsonic region increases exponentially with ω^2. There is a maximal value of ω, that can be derived from Eq. (9.91) setting the right hand side equal to zero, and setting $Z = Z_s = 1$. This gives

$$\omega_{\text{max}}^2 = \frac{Z_p - 1}{Z_p} \tag{9.95}$$

with Z_p defined by Eq. (9.80). For large Z_p the value of ω_{max}^2 is approximately unity.

For a star to have a large mass loss rate produced by the centrifugal magnetic rotator model, we also expect the sonic point radius to be close to the stellar radius. From Eq. (9.82) we find $\omega_s^2 = 1/Z_s{}^3$, so if Z_s is approximately 1.1, we get $\omega_s = 0.87$. This rapid rotation is typical of the value required in fast magnetic rotator wind models that yield large mass loss rate increases.

The mass loss rate of a centrifugal magnetic rotator is determined only by the speed of sound and by the rotation rate parameter ω. The 'equivalent mass loss rate' (see Eq. 9.83) for the centrifugally dominated case is

$$\dot{M}_E \equiv 4\pi V_r r^2 \rho \tag{9.96}$$
$$= 4\pi a\, r_0{}^2 Z_s{}^2 0.607 \rho_0 \exp\left\{-\frac{GM_*}{a^2 r_0}\left(1-\frac{1}{Z_s}\right)\right\} \times C_{\mathrm{cent}}(Z_s)$$

This shows the dependence of the equatorial mass flux on ρ_0, Z_s, a, and on the centrifugal factor $C_{\mathrm{cent}}(Z_s)$.

It is perhaps surprising that the mass loss rate has no explicit dependence on the magnetic field of the star. This is because in the strong field limit the field causes the subsonic material to rotate as a solid body, and any further increase in the field can have no further effect on the subsonic structure. We will find that the actual value of the magnetic field does determine the structure of the supersonic portion of the wind, as can be illustrated by deriving the velocity at the outer critical point.

9.7.3 The wind speed of fast magnetic rotators

Let us now consider conditions far out in the wind, where the radial velocity approaches the terminal speed. Let us assume that $V_r \gg a$, $r_f \gg r_A$. We can also assume $V_r{}^2 \gg A_r{}^2$ or $M_A^2 \gg 1$. This is because $M_A^2 = x^2 u$, so it increases with x at least as rapidly as x^2. Under these conditions the radial velocity equation (9.73) reduces to

$$\frac{r}{V_r}\frac{dV_r}{dr} = \frac{2a^2 + V_\phi{}^2 - \frac{GM_*}{r} + 2A_r\, A_\phi\, V_\phi/V_r}{V_r{}^2 - A_\phi{}^2} \tag{9.97}$$

The denominator of Eq. (9.97) gives the condition at the fast point that the radial velocity at the fast point is equal to the azimuthal component for the Alfvén speed at r_f,

$$V_r(r_f) = A_\phi(r_f) \tag{9.98}$$

To evaluate $A_\phi(r_f)$, we can use Eq. (9.38) in the limit $x^2 u \gg 1$, thus

$$\frac{A_\phi(r_f)}{A_r(r_f)} = -\frac{r_f \Omega}{V_r(r_f)}\left(1 - \frac{1}{x_f^2}\right) \tag{9.99}$$

Now on using the definition of the radial Alfvén speed, i.e. $A_r = B_r/(\sqrt{4\pi\rho})$ evaluated at r_f we find

$$A_\phi(r_f)^2 = \frac{r_f{}^2\Omega^2}{V_r(r_f)^2}\left(\frac{B_{r,f}^2}{4\pi\rho_f}\right)\left(1 - \frac{1}{x_f^2}\right)^2 \tag{9.100}$$

$$= \frac{\mathcal{F}_B{}^2\Omega^2}{4\pi\mathcal{F}_m V_r(r_f)}\left(1 - \frac{1}{x_f^2}\right)^2 \tag{9.101}$$

Using Eq. (9.98), we get at large r_f where $x_f^2 \gg 1$,

$$V_r^3(r_f) = \frac{\mathcal{F}_B{}^2\Omega^2}{4\pi\mathcal{F}_m} = \frac{\mathcal{F}_B{}^2\Omega^2}{\dot{M}_E} = V_M^3 \tag{9.102}$$

This states that the velocity of the flow at the fast point is the Michel velocity that we introduced in Eq. (9.58) with regard to the wind energy deposition. Now, evaluating the energy constant, ε, (Eq. 9.59) at infinity where $V_\phi \propto r^{-1} \to 0$ and $GM_*/r \to 0$, $a^2\ln\rho \to 0$ and $V_r \to v_\infty$, we find that the total energy per gram in the wind as $r \to \infty$ is

$$\varepsilon_{\text{total}}(\infty) = \frac{v_\infty{}^2}{2} + \frac{V_M{}^3}{v_\infty} = \frac{3}{2}v_\infty{}^2 \tag{9.103}$$

where we have used $V_r(r_f) \approx v_\infty$. Equation (9.103) shows that in the FMR regime of a magnetic rotator, there is twice as much energy per unit mass in the magnetic energy as there is in the terminal kinetic energy of the flow.

Equation (9.102) is a very useful result of fast magnetic rotator theory. Since the fast point occurs where the flow velocity has nearly reached the terminal wind speed, v_∞, the equation says that v_∞ is determined by the basic stellar parameters of $\mathcal{F}_B$, $\mathcal{F}_m$ and Ω. In the previous section we argued that the mass loss rate constant $\mathcal{F}_m$ depends on only the rotationally modified subsonic density structure. Now we see that with $\mathcal{F}_m$ known, the terminal speed is determined by the product $\mathcal{F}_B\Omega$. So the fast magnetic rotator wind nicely separates into a rotational part that determines $\dot{M}$ and a magnetic part that determines the speed that is reached by the flow. These results allow us to understand the rapid rotation segments of the velocity curves shown in Fig. (9.4). The faster the rotation, the larger is the mass loss rate (nearly independently of the field strength). Also, for a given Ω, the stronger the field, the larger is the Alfvén radius, and the faster is the wind.

One interesting application of Eq. (9.102) is to derive the magnetic field strength, $B_{r,0}$, that would allow a known stellar mass loss rate

Table 9.1 *Fast magnetic rotator wind properties*

			Star and stellar wind parameters				
Stars:	Sun	WR star	ζ Pup	B[e]	Be	AGB	K1 III
$M/M_\odot$	1	10	52	40	8	5	3
$\log L/L_\odot$	0.0	5.00	5.92	6.00	3.70	4.15	1.70
$\log T_{\rm eff}$ (K)	3.76	4.78	4.62	4.30	4.28	3.45	3.66
$R_*/R_\odot$	1.0	2.9	19.0	86.	6.6	510	11
$\log (\dot{M})$	-13.7	-4.50	-5.22	-4.70	-9.00	-5.00	-8.00
v_∞(km s^{-1})	300	1700	2300	500	1000	20	100.
$\log \rho_0$	-17.0	-9.0	-11.1	-11.4	-14.0	-12.7	-11.4
$\log \Omega_{\rm max}$	-3.20	-3.45	-4.32	-5.53	-3.98	-6.93	-4.54
$B_{\rm min}$(G)	.0609	6370.	985.	125.	11.5	.484	1.43
			FMR results[a] *with* ω*=0.8,* B*=10* $\times$ $B_{\rm min}$				
Z_s	1.09	1.09	1.09	1.09	1.09	1.09	1.09
$V_{r,s}$(km s^{-1})	126.	213.	165.	49.9	138.	12.0	65.1
Z_A	2.78	8.71	15.2	10.9	7.93	2.08	1.97
$V_{r,A}$	429.	2780.	3774.	818.	1630.	24.8	118.
Z_f	11.9	90.0	265.	140.	74.	8.22	7.71
V_M	739.	4180.	5660.	1230.	2460.	49.2	246.
$\dot{M}_E$[b]	-12.07	-3.89	-4.58	-4.06	-8.37	-4.36	-7.37
$\log \tau_J$(yr)	11.9	2.8	3.7	3.4	7.2	4.2	8.0

[a] For these sample model results, the coronal temperature is chosen such that Z_p=6, for each of the stars.
[b] The base density used is related to the stellar wind parameters as: $\rho_0 = 30\ \dot{M}/(4\pi v_{\rm esc} R_*^2)$

and terminal velocity to be explained by FMR forces. Substituting the given mass loss rate and the maximal rotation rate for a star into Eq. (9.102) and using the given v_∞ for the Michel velocity, we find

$$B_{\rm min} = \frac{(\dot{M}_E v_\infty{}^3)^{1/2}}{R_*{}^2 \Omega_{\rm max}} \tag{9.104}$$

where $\Omega_{\rm max}$ is the critical rotation rate, Eq. (9.78). Values for the minimal field necessary for a star to have a FMR wind are given in Table (9.1) for various types of stars across the HR diagram (provided no other force is acting such as the radiative force in OB stars).

If we also have an estimate of the rotation rate, Ω_*, of a given star in addition to its measured stellar mass loss rate and terminal velocity, we can derive a 'maximal surface field' for a star (Nerney, 1980), and obtain

$$B_{\max} = \frac{(\dot{M}_E v_\infty{}^3)^{1/2}}{R_*{}^2 \Omega_*} \tag{9.105}$$

This and other estimates of minimal and maximal magnetic fields in rotating stars are discussed by Maheswaran and Cassinelli (1992).

The location of the fast critical point can be determined by using the vanishing of the numerator of Eq. (9.97). This expression involves $V_\phi(r_f)$ which we can derive from Eq. (9.37), in the limit $x_f^2 u_f \gg 1$. This gives

$$V_\phi(r_f) = \frac{r_A{}^2 \Omega}{r_\mathrm{f}} \left(1 - \frac{V_A}{V_r}\right) \tag{9.106}$$

or, using the definition of $\mathscr{L}$ in Eq. (9.22) and the relation (9.99) between A_r and A_ϕ for large x

$$V_\phi(r_f) = r_\mathrm{f}\Omega \left[\frac{\mathscr{L}}{r_\mathrm{f}{}^2\Omega} - \frac{A_r(r_f)^2}{V_r(r_f)^2}\right] \tag{9.107}$$

At large x, Eqs. (9.98) and (9.99) give

$$\frac{A_r(r_f)}{V_r(r_f)} = -\frac{V_r(r_f)}{r_\mathrm{f}\Omega} \tag{9.108}$$

therefore

$$V_\phi(r_f) = \frac{\mathscr{L}}{r_\mathrm{f}} - \frac{V_M{}^2}{r_\mathrm{f}\Omega} \tag{9.109}$$

So the vanishing of the numerator of the momentum equation gives an expression for determining the location of the fast point

$$Z_\mathrm{f}{}^2 + \left(\frac{GM_*}{2a^2 r_0}\right) Z_\mathrm{f} + \frac{1}{2a^2 r_0{}^2}\left(\mathscr{L} - \frac{V_M{}^2}{\Omega}\right)^2 - \frac{V_M{}^2}{a^2 r_0{}^2 \Omega}\left(\mathscr{L} - \frac{V_M{}^2}{\Omega}\right) = 0 \tag{9.110}$$

The solution for the structure of the wind is not yet complete, however, because Eq. (9.110) involves $\mathscr{L}$, which is determined by the location of the Alfvén radius r_A (i.e. $\mathscr{L} = r_A{}^2\Omega$). So another equation is needed to determine r_A. This is provided by setting the energy at the fast point equal to the value at r_0. We get from this equality,

$$Z_\mathrm{A}{}^2 = \frac{1}{2\omega^2 Z_\mathrm{p}} \times \left\{\frac{3}{2}\left(\frac{V_M}{a}\right)^2 + \ln\left(\frac{V_{r,0}}{Z_\mathrm{f}{}^2 V_M}\right) - \frac{V_{r,0}{}^2}{2a^2}\right\} + \left\{\frac{1}{\omega^2}\left(1 - \frac{1}{Z_\mathrm{f}}\right) + \frac{1}{2}\frac{Z_\mathrm{f}}{Z_\mathrm{p}}\right\} \tag{9.111}$$

These last two equations can be solved simultaneously to find both

Z_f and Z_A. The solution also determines $\mathscr{L}$. Finally, the speed at the Alfvén point is given by setting $V_{r,A} = A_{r,A}$, yielding

$$V_{r,A} = \frac{\mathscr{F}_B{}^2}{r_A{}^2 \dot{M}_E} \tag{9.112}$$

This completes the solution for the six values that determine the structure of a fast magnetic rotator wind.

This solution has brought to the forefront the limiting case of the centrifugal magnetic rotator in which the determination of the mass loss rate and terminal speed are particularly simple. The equatorial mass flux is determined by conditions in the subsonic region, and for a CMR is given by Eq. (9.97). The terminal velocity in fast magnetic rotator theory is given to a good approximation by Eq. (9.102).

Table (9.1) gives some solutions for centrifugal magnetic rotators, using a variety of stellar parameters. We can notice there that the mass loss rate and terminal velocity can both be increased significantly by the magnetic rotator forces operating in the equatorial plane.

9.8 The spin down times

In discussing these analytic results, we have chosen to focus on the fast and centrifugal magnetic rotator cases instead of the slow magnetic rotator because our main interest throughout the book is on stars which have strong winds, i.e. with large mass loss rates. In the case of slow magnetic rotators like the sun the mass loss rate and terminal velocity are not affected much by the magnetic field, but the field plays a crucial role in the 'spin down' or loss of angular momentum by the sun.

The loss of angular momentum is certainly one of the more important effects of stellar winds. The late-type main sequence stars tend to rotate slowly because of the angular momentum lost through their long lifetime. Early type stars rotate much more rapidly on average, and this occurs at least in part because the stars are more massive, have a larger initial angular momentum, and the lifetimes are relatively short compared to those of the F and G main sequence stars.

For the late type stars and perhaps the Wolf-Rayet stars, the rate of loss of angular momentum is interesting, because it determines how long the outward transfer of angular momentum can drive fast magnetic rotator winds. We also must know the angular momentum available in the star. Observationally one can estimate the stellar rotation velocity from a careful study of the photospheric line profiles. This analysis can provide the value of $V_\phi \sin i$. Here V_ϕ is the tangential speed at the equator of the rotating star and i is the inclination angle between the spin axis and the line of sight. The stellar age can be derived from the 'turn-off point' of the cluster in which the star lies.

The time scale for the loss of angular mometum is determined by the ratio of the total angular momentum $\mathcal{J}$, to the rate of angular momentum loss, $d\mathcal{J}/dt$. The e-folding time, or 'spin down time', is then given by

$$\tau_J = \frac{\mathcal{J}}{d\mathcal{J}/dt} \tag{9.113}$$

For a star with a mass loss rate $\dot{M}$ occurring from the equatorial zone only, the rate at which the star loses angular momentum is $d\mathcal{J}/dt = \dot{M}\mathcal{L} = \dot{M}r_A^2\Omega$, where we have used Eq. (9.30) for $\mathcal{L}$. However, for the perhaps more common case of mass loss occuring from the entire spherical surface of the star, the angular momentum loss rate is given by Weber and Davis (1967) as

$$\frac{d\mathcal{J}}{dt} \approx \frac{2}{3}\dot{M}r_A{}^2\Omega \tag{9.114}$$

If we use here the angular momentum of a uniform density spherical star, $\mathcal{J} = (2/5)M_*R_*{}^2\Omega$, then we find an estimate of the spin down time of

$$\tau_J \approx \frac{3}{5}\frac{M_*R_*{}^2}{\dot{M}r_A{}^2} \tag{9.115}$$

This expression is commonly used for the spin down time. The mass loss rate and the Alfvén radius are properties of the stellar wind. Some values are given for the models in Table (9.1).

Here, let us consider the Alfvén radius, and assume the mass loss rate is known, and find approximate expressions for r_A and V_A in the FMR wind regime.

Equate the energy per unit mass at large radius (Eq. 9.103) to the value at r_0. Let us assume that for the FMR regime the energy at the base is dominated by the magnetic term. So on equating ε_0 to ε_∞, we find

$$\frac{3}{2}V_M{}^2 = -\left(\frac{\Omega r A_r A_\phi}{V_r}\right)_0 \tag{9.116}$$

and thus

$$A_{\phi 0} = -\frac{3}{2}V_M{}^2\frac{V_{r0}}{r_0\Omega A_{r0}} \tag{9.117}$$

Now equate the angular momentum constant $\mathcal{L}$, from Eq. (9.22), evaluated at r_0, to $r_A{}^2\Omega$ (Eq. 9.30), and if we further assume that

${r_A}^2 \gg {r_0}^2$, then

$$\mathscr{L} \approx -r_0 A_{r0} A_{\phi 0}/V_{r0} = {r_A}^2 \Omega \tag{9.118}$$

which upon using Eq. (9.117), gives an approximate expression for the Alfvén radius

$$r_A \Omega = \sqrt{\frac{3}{2}} V_M \approx \sqrt{\frac{3}{2}} v_\infty \tag{9.119}$$

Remember that the Michel velocity V_M is defined in terms of the basic wind parameters, ${V_M}^3 = ({\mathscr{F}_B}^2 \Omega^2/\dot{M}_E)$. Note that if all but the mass loss rate is held constant, as $\dot{M}_E$ increases, r_A moves closer to the star. Given this approximate expression for r_A, we can get the velocity at the Alfvén point from Eq. (9.60), and find that

$$V_r(r_A) = \frac{2}{3} V_M = \frac{2}{3} v_\infty \tag{9.120}$$

The spin down time can be found from Eqs. (9.115) and (9.119).

$$\tau_J \approx \frac{2}{5} \frac{M_*}{\dot{M}} \frac{R_*^2 \Omega^2}{v_\infty^2} \approx \frac{2}{5} M_* \left\{ \frac{\Omega^2}{R_*^2 B_{r,0}^4 \dot{M}} \right\}^{1/3} \tag{9.121}$$

where we have used $v_\infty \approx v_M$. Thus we see that for the FMR regime there are rather simple expressions for the terminal velocity, the Alfvén radius and velocity, and the spin down time.

9.9 Magnetic rotators in combination with other wind forces

9.9.1 Hybrid wind models

It is generally useful to consider stars with equatorially enhanced winds as having hybrid winds, involving other accleration mechanisms. The magnetic rotator combination of the **B** field and the centrifugal forces always provides a way for a star to lose angular momentum. However, the magnetic rotator forces can also lead to a enhancement of the radial acceleration of the wind and to an increased mass loss rate. Very luminous stars such as the Wolf-Rayet stars have extremely large mass loss rates, and their momentum fluxes have been difficult to explain by radiative driving alone. So the combination of magnetic rotator and radiation forces has been considered for these stars, and they have been called *'luminous magnetic rotators'*.

In our discussion of Eq. (9.104) we found it convenient to consider each star to have a 'primary mechanism' that can drive a wind in the absence of magnetic and centrifugal forces. In the case of the basic Weber and Davis model, this primary mechanism is the gas pressure gradient in the solar corona. If the star also has a magnetic field larger than $B_{\min}$, then the magnetic rotator forces can lead to an FMR wind structure.

Figure (9.6) shows a a plot of $\dot{M}$ versus v_∞ that is useful for discussing hybrid winds. The 'primary' wind mechanism for a given star would provide a specific $\dot{M}$ and v_∞, or the point P, on the plot. The Michel velocity relation (Eq. 9.102) provides lines with *slope* $= -3$ on this plot, for specific fields B (assuming $\omega = 1$ as defined in Eq. 9.79). Hence, there is a Michel velocity line that passes through the point, P, that is paramaterized by the field strength, $B_{\min}$. For any point on the diagram there is, of course another field and iso-Michel velocity line that crosses through it. For magnetic fields larger than $B_{\min}$, the Michel velocity lines lie further to the right on this plot.

Given this background, let us consider the effect on the wind of our model star if we spin up the star, assuming that the star has a field $B_r(r_0) > B_{\min}$. At small ω the rotation will have little effect on either the mass loss rate or v_∞. This situation corresponds to slow magnetic rotators (SMRs). As ω increases even further, the wind

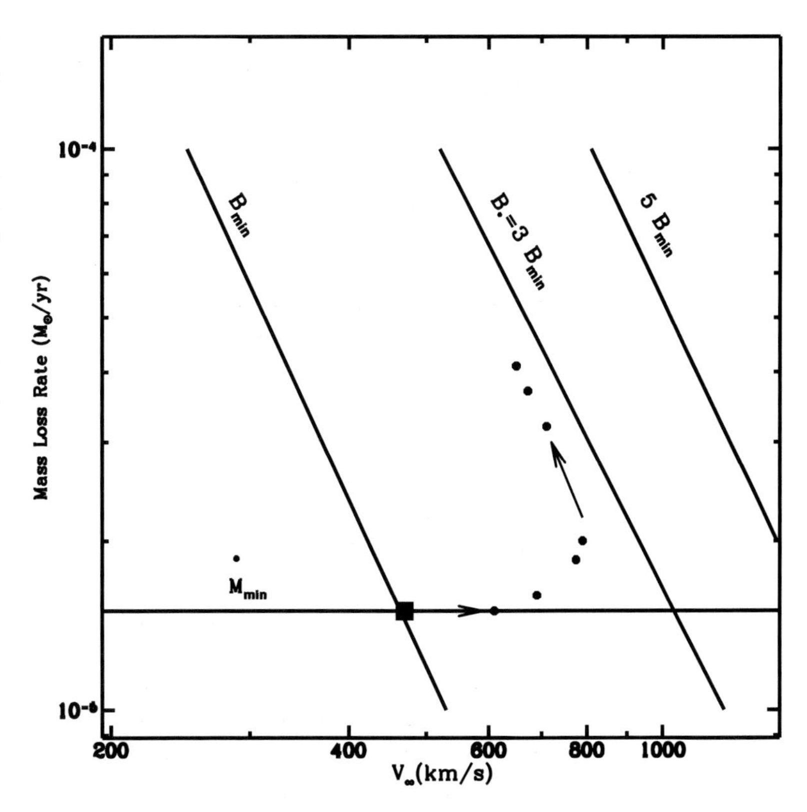

Figure 9.6 A plot of the mass loss rate versus v_∞, for a hybrid wind involving magnetic rotator forces. The black square indicates point P (see text) with $\dot{M}$ and v_∞ from the primary mechanism. The arrows indicate the effect of an increasing rotation velocity.

enters the FMR regime in which the speed of the wind increases but with little change in the mass loss rate, as the latter is still determined by the primary mechanism. This is indicated by the horizontal arrow in Fig. (9.6). At even larger values of ω the star enters the centrifugal magnetic rotator (CMR) regime. Now, as ω increases, the mass loss rate increases to values well above that provided by the primary mechanism. The point on the plot that represents the wind moves up and along the Michel velocity line, associated with the magnetic field of the star. In the CMR regime the mass loss rate is detemined by ω alone, as we saw in § (9.7.2), and the terminal velocity is equal to the Michel velocity.

Thus we find the $\dot{M}$ versus v_∞ plot useful for distinguishing between the types of magnetic rotators that have been discussed in this chapter: the SMR, FMR and CMR winds. However, the plot is also useful for deriving magnetic and rotational properties of the stars for which the winds seem to have excessively large mass loss rates or velocities, relative to what would be expected from the primary mechanism.

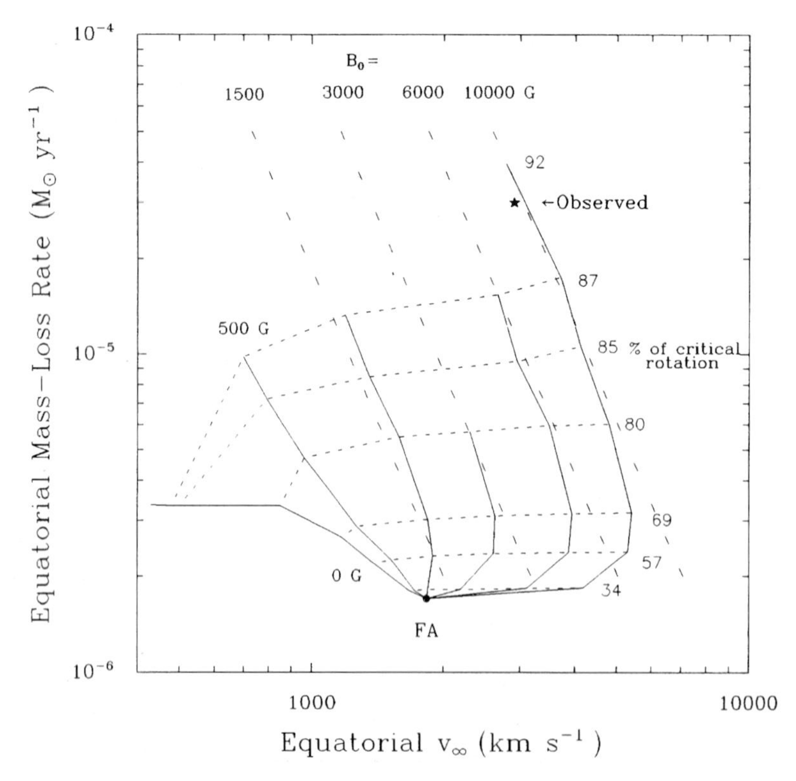

Figure 9.7 Plot of the mass loss rate versus v_∞, for a hybrid wind involving magnetic rotator forces, for a Wolf-Rayet star. The star indicates the $\dot{M}$ and v_∞ values derived from radio continuum and UV line profiles, respectively. (From Poe, Friend & Cassinelli, 1989)

9.9.2 Luminous magnetic rotators

A possible application of magnetic rotator theory is to luminous stars near the top of the HR diagram. We have already seen that line driving forces are important for luminous early type stars. However, the combination of magnetic field and rotation can provide a mechanism for advecting angular momentum from the star, that leads to a further acceleration of the outflow or to an increased mass loss rate. The combination of radiation forces and magnetic rotator forces were initially explored by Friend and MacGregor (1984). Models of luminous magnetic rotators have been described and calculated by Poe and Friend (1986) for Be stars, and by Poe, Friend, and Cassinelli (1989) for Wolf-Rayet stars. The Be stars are rapidly rotating near-main sequence B stars that show strong emission features at H α and show intrinsic polarization, evidence for a equatorially enhanced outflow. The Wolf-Rayet stars are noted for having wind momentum factors ($\eta = \dot{M}v_{\infty}/(L/c)$) that are larger than 10 and hence pose special problems for line driven wind models (§ 2.7.2).

Figure (9.7) illustrates results of models for a Wolf-Rayet star, which assume the primary mechanism is the line driving wind mechanism. One sees that a wide range of mass loss rates and terminal velocities can be achieved through various combinations of stellar magnetic fields and rotation rates. The mass loss rate can be increased by an order of magnitude if the rotation rate is about 90 percent of the maximal value. Also shown is the result for models in which the surface field is zero. In this case the main effect of increasing rotation is to decrease the terminal wind speed. This is because the wind speed is proportional to the surface escape speed for line driven winds, and as the star rotates faster the effective escape speed in the equatorial region is decreased by centrifugal effects.

9.10 Conclusions

Magnetic fields are present in plasmas throughout the universe. In this chapter we have explored the effects that the fields can have on the winds of rotating stars. There are several situations in which this coupling of magnetic fields and rotation is important. In the case of protostars the collapse can procede only if initial angular momentum can be shed from the collapsing cloud. The coupling of the rotation and the magnetic field is thought to lead to the angular momentum loss through the production of bipolar outflows from the protostars.

There can be a further loss of angular momentum during the pre-main sequence and early main sequence evolution of stars. This is expecially noticable in studies of the angular speeds of stars of spectral type later than about F5V, for which there is a distinct correlation of the rotation of the stars with their age. The loss of angular momentum that occurs is thought to be due to the effects of their magnetic rotator winds. The basic mechanism that we have discussed was derived by Weber and Davis (1967) for the case of the sun. The mechanism can explain the fact that the sun now has an equatorial rotation speed of only about 2 km s^{-1}, while it should have a speed of several hundred km s^{-1}, assuming that it co-formed with the planets, and suffered no angular momentum loss.

In this chapter we derived the basic magnetic rotator equations, and gave special emphasis to the angular momentum transfer and the role of the Poynting flux in making the outflow energetically possible. We found that much of the angular momentum that is lost is not in the form of mass but in the stresses in the magnetic field lines. For the purposes of deriving the dependence of the mass loss rate and terminal velocity we chose to focus on a greatly simplified version of the theory that we call the centrifugal magnetic rotator. It was found that the mass loss rate is determined primarily by the rotation rate of the star, while the terminal velocity is determined by the strength of the magnetic field. This result even holds in hybrid models in which other forces such as line driving forces are included. By acounting for magnetic rotator effects the mass loss rate can in principle be increased by an order of magnitude and the flow speed can either be increased or decreased. This is determined by the stellar quantities that occur in the definition of the Michel velocity.

9.11 Suggested reading

Belcher, J., & MacGregor, K.B. 1976, 'Magnetic Acceleration of Winds from Solar-Type Stars', *Ap. J.* **210**, 498

(The basic equations for fast magnetic rotator theory and its applicaton to the evolutionary change of the solar angular momentum.)

Brandt, J.C. 1970, *Introduction to the Solar Wind* (San Francisco: Freeman)

(This book contains a chapter on the general influence of the magnetic field on the solar wind.)

Hartmann, L., & MacGregor, K.B. 1982, 'Protostellar Mass and Angular Momentum Loss', *Ap. J.* **259**, 180

(A concise derivation of the fundamental equations and a discussion of application of the theory to star formation.)

Hundhausen, A.J. 1972 *Solar Wind and Coronal Expansion* (Heidelberg: Springer-Verlag)
(A discussion of the observations of the sun regarding the magnetic rotator structure.)

Weber, E.J. & Davis, L. 1967, 'The Angular Momentum of the Solar Wind', *Ap. J.* **148**, 217
(The basic magnetic rotator theory was first developed in this paper.)

10 Alfvén wave driven winds

In the last chapter we have seen that if a star has an open magnetic field in the equatorial region and is also rapidly rotating, a very strong stellar wind can be produced. In this chapter we consider the effects of the magnetic field in absence of rotation. If oscillations are induced in the field at the base of the wind, transverse 'Alfvén' waves will be generated. The dissipation of energy and momentum associated with the wave propagation can lead to the acceleration of the outer atmosphere in the form of an 'Alfvén wave driven wind'. Open field regions can arise in a variety of configurations, depending on the circulation currents or dynamo properties of the interior of the star. Furthermore, the strength and geometry of the magnetic field can vary significantly from one location on the star to another, and the wind flow tubes will vary accordingly.

In the absence of a magnetic field, a star that has a spherically symmetric hot corona will produce a steady, radial, structureless wind, driven by the thermal gas pressure gradients in the corona (Parker, 1958), as discussed in Chapter 5. Within a few years after the solar wind was predicted by Parker, interplanetary space probes proved that indeed there is a wind from the sun that occurs at all times. However, the wind was found to be far from steady and structureless. To understand the spatial and temporal variability of the wind, Parker (1965) considered outflow in open magnetic field structures. He showed that Alfvén waves could perform work on the wind, and the nature of the acceleration and mass flux would depend strongly on the conditions of the field at the base of the wind.

To understand the basic mechanism, we will consider a very idealized situation in which there is a radial magnetic field frozen in a plasma flow. We then consider the flow equations along one narrow flow tube. By assuming that all tubes are alike, and that the flow emerges from the entire surface of the star, we will estimate the mass loss rate and terminal velocity that can be driven from a star with a given surface magnetic field and wave energy flux.

One of the important properties of Alfvén wave driven winds is that the flow speed can be fast. In the case of the solar wind as it flows past the earth, high speed streams are often detected with $v \approx 700\ \mathrm{km\,s^{-1}}$, which is about a factor of two faster than the time averaged solar wind speed at the earth. Furthermore the fast solar flow originates in coronal hole regions that are cooler than average coronal temperatures. This combination of minimal thermal acceleration, and high speed outflow is a characteristic of Alfvén wave driven winds.

The Alfvén wave driving wind mechanism that is discussed in this chapter, and the radiation driving mechanisms discussed in Chapters 7 and 8, are the most commonly occuring classes of wind mechanisms across the HR diagram. Alfvén waves are a valuable addition to the list of mass loss mechanisms, because the waves can produce a significant mass loss even in stars which have neither hot coronal zones nor strong radiative fluxes. Consequently, Alfvén waves have been considered as a possible cause for the mass loss in every portion of the HR diagram where the radiative driving mechanisms are ineffective. This includes stars along the main sequence later than about spectral class B, low mass pre-main sequence stars, and post-main sequence giant and supergiant stars with effective temperatures between 3000 K and 15 000 K. (Cool luminous stars with $T_{\mathrm{eff}} < 3000$ K and $L > 10^4 L_\odot$ may have dust driven winds as discussed in Chapter 7.)

One of the most difficult problems for stellar wind theory has been to explain the mass loss from red giant stars. These stars have slow wind speeds of less than about 30 $\mathrm{km\,s^{-1}}$, which are well below the escape speed from the surface of the stars. Hartmann and MacGregor (1980) proposed that the winds are driven by Alfvén waves, and to reduce the terminal speed to the observational values they introduced wave damping. However, to achieve the low speeds, the model required a unacceptably delicate adjustment of the wave damping parameter (Holzer, Flå & Leer, 1983). Modifications to the theory have been made, such as considering radial changes in the flow tube area, or wave reflection effects, and it now appears that some combination of Alfvén waves plus these auxiliary effects will be required to solve the wind problem of cool stars that have no dust. We will not deal with these auxiliary effects, but focus on the basic equations and concepts regarding the Alfvén wave mechanism.

The Alfvén waves discussed here differ from the purely acoustic waves discussed in Chapter 6 in several ways. First, the Alfvén wave mechanism requires that the star have a magnetic field, extending out from the star through the wind acceleration region. Second, the Alfvén waves are transverse, i.e. the oscillation is perpendicular to the direction of the propagation of the wave, in contrast with the compressional sound waves. Third, transverse Alfvén waves are incompressive, whereas the propagation of the acoustic waves occurs because of the compressive changes in density of the medium.

As we did in the discussion of magnetic rotator winds, we assume that the field is frozen-in with the wind plasma. We also assume, for simplicity, that the wind is fully ionized, although this may not be the case throughout the winds of K and early M giants and supergiants.

10.1 The wind equations with a varying transverse B field

The are many similarities between Alfvén wave driven winds and the magnetic rotator winds of the previous chapter, and the two mechanisms are not mutually exclusive. The curved field tubes that we discussed for magnetic rotators can also transmit Alfvén wave disturbances. For the magnetic rotators these flow tubes are defined by the magnetic field and co-rotate with the star. In this chapter we ignore stellar rotation, so the flow tubes are basically radial, but with an oscillatory transverse motion imposed at the base of the tube.

10.1.1 The three types of waves

In the presence of a magnetic field, there are three modes of magneto-acoustic wave propagation: the slow, intermediate, and fast modes. The slow mode corresponds to a compressional wave propagating down a flow tube, and if the transverse Alfvén waves are absent, the thermal gas pressure is the restoring force and the speed of the slow mode is essentially that of sound. The fast mode is also a compressional and longitudinal wave, that has the magnetic pressure gradient as the restoring force. The intermediate velocity wave mode is the one that is usually referred to as the 'Alfvén wave' mode, and it differs from the other two in that the displacements are transverse. For the Alfvén wave mode, the restoring force is perpendicular to the direction of propagation and arises from the magnetic tension. The familiar example of a transverse wave is a wave travelling along a string that is shaken at one end. Such a wave is driven by the tension in the string. In our case the analogue of the string is the magnetic flux tube. In additional to being transverse, Alfvén waves are non-compressional, which means that the cross section of the tube retains the r^2 radial dependence that holds for the undisturbed tube. However, the wavelength, speed and amplitude of the wave depend on the distance from the star. The geometry is illustrated in Fig. (10.1).

10.1.2 The momentum equation

As a result of our assumptions regarding the magnetic field and flow geometry, the terms in the momentum equation arising from the gas pressure gradients and gravity are identical to corresponding terms in the spherical winds treated in earlier chapters of this book. So as to benefit from the magnetic field description of the magnetic rotator theory, we assume that the imposed oscillation is in the ϕ direction, i.e. parallel to the equator. Consequently, our equations will again involve only the r and ϕ components of the magnetic field.

The velocity, $\mathbf{v}$, and the magnetic field, $\mathbf{B}$, each have a radial component and perturbations, δv and δB, owing to the transverse Alfvén

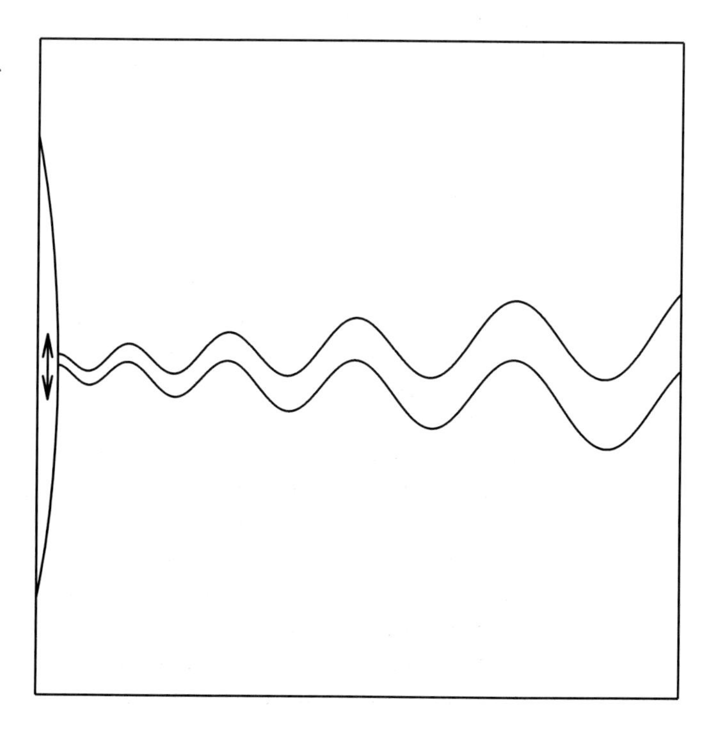

Figure 10.1 Schematic drawing of the propagation of a transverse Alfvén wave. The amplitude and speed of the wave depend on the distance from the star.

waves, thus

$$\mathbf{v}(r,t) = v(r)\,\hat{\mathbf{e}}_{\rm r} + \delta v(r,t)\,\hat{\mathbf{e}}_{\phi} \tag{10.1}$$

$$\mathbf{B}(r,t) = B(r)\hat{\mathbf{e}}_{\rm r} + \delta B(r,t)\hat{\mathbf{e}}_{\phi} \tag{10.2}$$

where $\hat{\mathbf{e}}_{\rm r}$ and $\hat{\mathbf{e}}_{\phi}$ are unit vectors in the radial and transverse directions, respectively. These equations are appropriate for one flux tube, but later we assume the star is covered with the foot points of identical tubes to derive global properties of the wind.

The vector form of the mass and momentum conservation equations are:

$$\frac{\partial \rho}{\partial t} + \nabla \cdot (\rho \mathbf{v}) = 0 \tag{10.3}$$

$$\rho \frac{\partial \mathbf{v}}{\partial t} + \rho(\mathbf{v} \cdot \nabla \mathbf{v}) + \nabla p + \rho \nabla \Phi - \frac{1}{c}(\mathbf{J} \times \mathbf{B}) = 0 \tag{10.4}$$

where Φ is the gravitational potential, and the last term in (10.4), $\ell = -(\mathbf{J} \times \mathbf{B})/c$, is the Lorentz force (in Gaussian units). Upon using Ampère's law,

$$\nabla \times \mathbf{B} = \frac{4\pi}{c}\mathbf{J} \tag{10.5}$$

the Lorentz force becomes

$$\ell = \frac{1}{4\pi}(\mathbf{B} \times \nabla \times \mathbf{B}) = +\frac{1}{8\pi}\nabla(\mathbf{B} \cdot \mathbf{B}) - \frac{1}{4\pi}(\mathbf{B} \cdot \nabla)\mathbf{B} \tag{10.6}$$

where the last equality follows from a vector identity.

The radial component of the momentum equation can be found by substituting the components of $\mathbf{v}$ and $\mathbf{B}$, from Eqs. (10.1) and (10.2), into Eq. (10.4). This gives the radial momentum equation

$$\frac{\partial v}{\partial t} + v\frac{\partial v}{\partial r} - \frac{\delta v^2}{r} + \frac{1}{\rho}\frac{\partial p}{\partial r} + \frac{GM_*}{r^2} + \frac{1}{8\pi\rho r^2}\frac{\partial}{\partial r}(r^2(\delta B)^2) = 0 \tag{10.7}$$

Here we have formally included the time derivative $\partial v/\partial t$ because of the presence of the time dependent terms δv^2 and δB^2. However, we are interested only in the time averaged motion so we replace this equation with its time average

$$v\frac{dv}{dr} - \frac{\langle(\delta v)^2\rangle}{r} + \frac{1}{\rho}\frac{dp}{dr} + \frac{GM_*}{r^2} + \frac{1}{8\pi\rho r^2}\frac{d}{dr}(r^2\langle(\delta B)^2\rangle) = 0 \tag{10.8}$$

where the brackets $\langle\ \rangle$ indicate a time average over one wave period. Squared quantities such as $\langle(\delta v)^2\rangle$ are positive throughout the period and thus remain positive after the time averaging is carried out. Note in Eq. (10.8) that the transverse wave motion leads to both the centrifugal and the magnetic terms that appear in this equation.

The ϕ and θ components of the momentum equation (10.4) are respectively,

$$\rho \frac{\partial \delta v}{\partial t} + \rho \frac{v}{r} \frac{\partial}{\partial r}(r \delta v) - \frac{1}{4\pi} \frac{B}{r} \frac{\partial}{\partial r}(r \delta B) = 0 \tag{10.9}$$

$$\left((\delta v)^2 - \frac{(\delta B)^2}{4\pi\rho} \right) \frac{\cot\theta}{r} = 0 \tag{10.10}$$

Equation (10.9) provides an expression for the time variation of the velocity displacement. To analyze the wave propagation we need a similar expression for the time dependence of δB. This can be obtained from Faraday's law.

$$\frac{1}{c} \frac{\partial \mathbf{B}}{\partial t} = -\nabla \times \mathbf{E} \tag{10.11}$$

Eliminating **E** by using the frozen-in field approximation (as discussed in Chapter 9),

$$\mathbf{E} = -\frac{1}{c} (\mathbf{v} \times \mathbf{B}) \tag{10.12}$$

Eq. (10.11) can be written as

$$\frac{\partial \mathbf{B}}{\partial t} = \nabla \times (\mathbf{v} \times \mathbf{B}) \tag{10.13}$$

After substituting the components of **v** and **B** into this vector expression and setting $\partial/\partial\phi = \partial/\partial\theta = 0$ because of symmetry conditions, Faraday's law (Eq. 10.13) becomes

$$\frac{\partial \delta B}{\partial t} - \frac{1}{r}(B\delta v - v\delta B) - B\frac{\partial \delta v}{\partial r} + v\frac{\partial \delta B}{\partial r} - \delta v\frac{dB}{dr} + \delta B\frac{dv}{dr} = 0 \tag{10.14}$$

This equation can be further simplified by using conservation of mass and magnetic flux constraints. The conservation of mass equation (10.3) can be written as

$$\frac{1}{r^2} \frac{d}{dr}(\rho v r^2) = 0 \tag{10.15}$$

Note that, unlike (10.3), this does not include $\partial\rho/\partial t$, because transverse Alfvén wave disturbances are non-compressive and the density is a function of r alone. From Eq. (10.15) we obtain the mass conservation constant

$$\mathscr{F}_m = \rho v r^2 = \text{ constant} \tag{10.16}$$

Using Eq. (10.2), Maxwell's equation $\nabla \cdot B = 0$ becomes

$$\frac{1}{r^2}\frac{d}{dr}(r^2 B) = 0 \tag{10.17}$$

so that

$$\mathscr{F}_B = r^2 B = \text{ constant} \tag{10.18}$$

The constants $\mathscr{F}_m$ and $\mathscr{F}_B$ can now be used to eliminate the derivatives dv/dr and dB/dr from Eq. (10.14). Faraday's law for our problem becomes

$$\frac{\partial}{\partial t}(\delta B) - B\frac{\partial}{\partial r}(\delta v) + v\frac{\partial}{\partial r}(\delta B) - \frac{1}{r}(v\delta B - B\delta v) - \frac{v\delta B}{\rho}\frac{d\rho}{dr} = 0 \tag{10.19}$$

Equations (10.9) and (10.19) provide us with expressions for the variation with time of the transverse wave quantities, δv and δB. These equations will be used to find the wave propagation speed. They will also provide expressions for the time averages, $\langle(\delta B)^2\rangle$ and $\langle(\delta v)^2\rangle$, that are needed in the wind radial momentum equation (10.8).

10.2 The propagation of Alfvén waves

In general a system that undergoes forced oscillations will respond sinusoidally, but with a phase delay. The wave amplitudes δv and δB will have amplitudes and phase delays that are dependent on the radial position along the tube. Consequently we expect an approximate solution of the form $a(r) \exp i(s(r) - \omega t)$, where $a(r) = \delta B(r)$ or $\delta v(r)$ for the magnetic and velocity perturbations respectively, and $s(r)$ is the position dependent phase of the wave relative to the phase at the base of the tube. This assumption for the form of pertubations allows us to eliminate the time dependent terms that we derived in the previous sections in terms of the frequency of the wave, ω. The wave number ($k = 2\pi/\lambda$) is given by

$$k(r) = \frac{ds(r)}{dr} \tag{10.20}$$

and the frequency ω is a constant. We linearize Eqs. (10.9) and (10.19), so as to obtain an approximate solution. The wave perturbations are assumed to have a frequency such that the wavelength, λ, is much smaller than the scale length, h, over which the physical quantities v, ρ, and B vary in the wind (or $k \gg h$). (This is often referred to as the WKB approximation). Thus the physical quantities are assumed to be constant over the scale of a wavelength. To implement the linearization procedure we expand our equations in a power series over the quantity

$$\mu = \frac{\lambda}{h} = \frac{2\pi}{kh} \qquad (10.21)$$

where μ is assumed to be small and arbitrary, and we obtain

$$\delta B(r,t) = \{b(r) + b'(r)\mu + \ldots\} \exp i\,(s(r) - \omega t) \qquad (10.22)$$

$$\delta v(r,t) = \{u(r) + u'(r)\mu + \ldots\} \exp i\,(s(r) - \omega t) \qquad (10.23)$$

In these expressions, u and b are the velocity and magnetic amplitudes of the transverse Alfvén wave, and u' and b' are their derivatives with respect to r.

Expressions (10.22) and (10.23) are now substituted in the time dependent Eqs. (10.9) and (10.19), and we keep only terms up to first order in μ. The expansion yields real and imaginary coefficients of both μ^0 and μ^1. Since the value of μ is arbitrary, each of the coefficients in the expansions must be equal to zero. Given that there are two equations, two powers of μ, and both real and imaginary coefficients, one finds that the expansion leads to eight equations. In the zeroth order approximation, we now assume that μ is negligibly small, and find the following four relations for the wave amplitudes u and b:

$$i\,\{(kv - \omega)u - (kB/4\pi\rho)b\} = 0 \qquad (10.24)$$

$$i\,\{(kv - \omega)b - (kB)u\} = 0 \qquad (10.25)$$

$$\frac{v}{r}u - \frac{B}{4\pi\rho r}b + v\frac{du}{dr} - \frac{B}{4\pi\rho}\frac{db}{dr} = 0 \qquad (10.26)$$

$$-\frac{v\,b}{r} + \frac{B\,u}{r} + v\frac{db}{dr} - B\frac{du}{dr} - \frac{v\,b}{\rho}\frac{d\rho}{dr} = 0 \qquad (10.27)$$

Multiplying Eq. (10.24) by b and Eq. (10.25) by u and subtracting one equation from the other we obtain a relation between u^2 and b^2

$$u^2 = b^2/4\pi\rho \qquad (10.28)$$

These squared quantities are of interest because time dependent averages of them are needed in the radial momentum equation. (Time dependent averages of the wave amplitudes u and b themselves are zero because they are sinusoidal in time, so the negative and positive phases cancel over each period of the wave.)

Substituting $u = -\sqrt{b^2/4\pi\rho}$ and then $u = +\sqrt{b^2/4\pi\rho}$ into

Eq. (10.24), we obtain the 'Alfvén wave dispersion relations':

$$\omega = k\left(v + \frac{B}{\sqrt{4\pi\rho}}\right) \quad \text{for } u = -\sqrt{\frac{b^2}{4\pi\rho}} \tag{10.29}$$

$$\omega = k\left(v - \frac{B}{\sqrt{4\pi\rho}}\right) \quad \text{for } u = +\sqrt{\frac{b^2}{4\pi\rho}} \tag{10.30}$$

Once solutions for v and B are known, these equations give $k(r) = \omega/(v \pm A)$. The solution for $s(r)$ then follows from $s(r) = \int k(r)dr$. However, our main interest here is in the amplitudes u and b of the waves.

The speed of propagation of a wave is given by the product of the frequency times the wavelength. Equations (10.29) and (10.30), relating the angular frequency and wavenumber of the Alfvén waves, give the group speed of disturbances that are propagating in the outward and inward directions respectively, via the relation $v_g = d\omega/dk$, (where v_g is called the group velocity)

$$v_g = v + \frac{B}{\sqrt{4\pi\rho}} \quad \text{for an outward wave} \tag{10.31}$$

$$v_g = v - \frac{B}{\sqrt{4\pi\rho}} \quad \text{for an inward wave} \tag{10.32}$$

We conclude from this that the speed at which the disturbance propagates is a superposition of the flow speed $v(r)$ and the wave speed $A(r)$ where

$$A = \frac{B}{\sqrt{4\pi\rho}} \tag{10.33}$$

This is the 'Alfvén speed' for transverse waves in a plasma. Recall that this speed was also an important quantity in the magnetic rotator theory of Chapter 9. Using Eq. (10.33) to eliminate the density from the expression for u in Eq. (10.29) for an outward wave, we find a relation between the wave velocity displacement and the wave magnetic field displacement,

$$\frac{u}{A} = -\frac{b}{B} \tag{10.34}$$

The minus sign indicates that, for an outwardly propagating wave, the velocity and magnetic field wave displacements are out of phase by 180 degrees. This is illustrated in Fig. (10.2) which shows a segment of an outward propagating Alfvén wave. As the wave travels outward with an amplitude δB and a speed $v + A$ it pushes on the slower moving wind material and imparts the transverse velocity component

δv, as is shown in the figure. The phase difference between δB and δv can also be explained by considering the base of the wind where the wave originates. If at some instant the base of the magnetic flux tube is given a velocity displacement δv in the positive ϕ direction, the magnetic field recieves an increment δB in the negative ϕ direction.

The other two equations from the linearization procedure, Eqs. (10.26) and (10.27), provide useful relations for the change in the wave amplitudes as a function of radius. The terms that are linear in u and b can be eliminated by multiplying Eq. (10.26) by $\sqrt{4\pi\rho}$, and subtracting it from Eq. (10.27). After using $u = -b/\sqrt{4\pi\rho}$, these steps lead to an expression involving derivatives of b

$$\sqrt{4\pi\rho}\frac{d}{dr}\left(\frac{b}{\sqrt{4\pi\rho}}\right)(v + A) + \frac{db}{dr}(v + A) - bv\frac{d\ln\rho}{dr} = 0 \qquad (10.35)$$

which simplifies to

$$\frac{d\ln b}{dr} = \frac{1}{4}\left(\frac{3v + A}{v + A}\right)\frac{d\ln\,\rho}{dr} \qquad (10.36)$$

After introducing the Alfvén Mach number $M = v/A$ and solving for the quantity b^2, which is needed for the radial momentum equation,

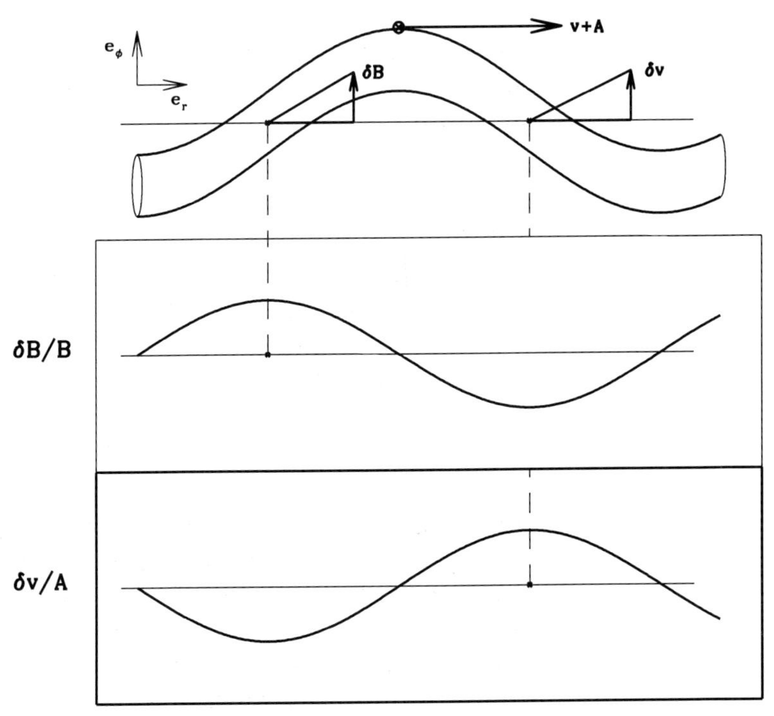

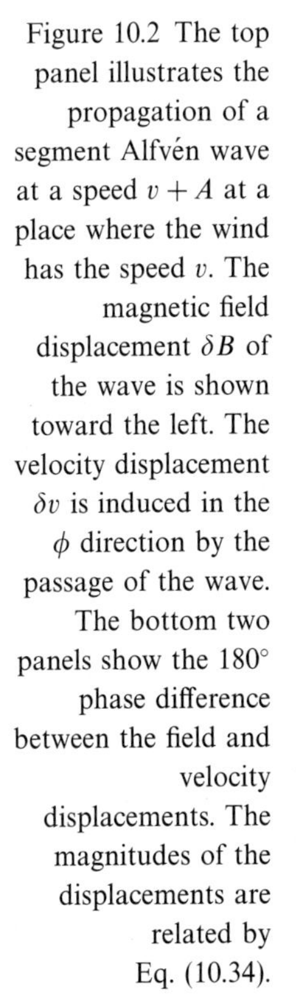
Figure 10.2 The top panel illustrates the propagation of a segment Alfvén wave at a speed $v + A$ at a place where the wind has the speed v. The magnetic field displacement δB of the wave is shown toward the left. The velocity displacement δv is induced in the ϕ direction by the passage of the wave. The bottom two panels show the 180° phase difference between the field and velocity displacements. The magnitudes of the displacements are related by Eq. (10.34).

we obtain

$$\frac{d\ln b^2}{dr} = \frac{1}{2}\left(\frac{3M+1}{M+1}\right)\frac{d\ln \rho}{dr} \tag{10.37}$$

The right hand side of this equation can be converted to a function of M alone by using the following relation between M and ρ,

$$M = \frac{vr^2}{Ar^2} = \frac{vr^2}{Br^2}\sqrt{4\pi\rho} = \frac{\sqrt{4\pi}\mathscr{F}_m}{\mathscr{F}_B\sqrt{\rho}} = \frac{\text{constant}}{\sqrt{\rho}} \tag{10.38}$$

This equation allows us to relate the Alfvén Mach number to local conditions in the wind. It also leads to the following identities,

$$d\ \ln(M+1) = \frac{M}{M+1}d\ \ln\ M = -\frac{1}{2}\frac{M}{M+1}d\ \ln\rho \tag{10.39}$$

that allow us to express Eq. (10.37) as

$$\frac{d\ln b^2}{dr} = -\frac{d\ln\,(M+1)^2}{dr} - \frac{d\ln M}{dr} \tag{10.40}$$

As all three terms in Eq. (10.40) are perfect differentials, this equation integrates to provide us with an extremely useful conservation relation for the amplitude of the Alfvén waves.

$$b^2(M+1)^2M = \text{ constant } \equiv W \tag{10.41}$$

We call W the *'wave constant'*. According to this conservation relation, if the magnitude b^2 of the wave is specified at the base of the flow tube, and if we can find the Alfvén Mach number as a function of radius, then we will know the amplitude of the wave at all radii. Thus, we now want to find a solution for $M(r) = v(r)/A(r)$.

10.3 The radial momentum equation

The three components of the vector wind momentum equation (10.4) are given in Eqs. (10.8) to (10.10). We have already used the ϕ momentum equation (10.9) in deriving the linearization procedure above.

The radial momentum equation (10.8) will now allow us to find a solution for $v(r)$, but first we need to know the time averages of the time dependent squares of the wave displacements δv and δB. The time average over one period of the waves yields

$$\langle(\delta v)^2\rangle = \frac{u^2}{2} \quad \text{and} \quad \langle(\delta B)^2\rangle = \frac{b^2}{2} \tag{10.42}$$

Thus Eq. (10.28) provides us with the result

$$\langle(\delta v)^2\rangle = \frac{\langle(\delta B)^2\rangle}{4\pi\rho} = \frac{b^2}{8\pi\rho} \tag{10.43}$$

Substituting Eq. (10.43) into Eq. (10.10), we find that there is no additional information in the θ momentum equation.

In the radial momentum equation (10.8), the Alfvén wave disturbance introduces the terms involving time averages of δv^2 and δB^2. We identify the sum of these as the wave force per unit mass (or wave acceleration), given by

$$\langle g_w\rangle = \frac{\langle(\delta v)^2\rangle}{r} - \frac{\langle(\delta B)^2\rangle}{4\pi\rho r} - \frac{1}{\rho}\frac{d}{dr}\left(\frac{\langle(\delta B)^2\rangle}{8\pi}\right) \tag{10.44}$$

The three wave acceleration terms correspond to a centrifugal term, a magnetic stress term and a gradient of $\langle(\delta B)^2\rangle$. There is a cancellation of the first two terms in $\langle g_w\rangle$ because of Eq. (10.43). Physically this is because in the small amplitude approximation, the magnetic tension provides the restoring force for the wave.

The radial momentum equation (10.8) now becomes

$$v\frac{dv}{dr} + \frac{1}{\rho}\frac{dp}{dr} + \frac{1}{\rho}\frac{d}{dr}\left(\frac{\langle\delta B^2\rangle}{8\pi}\right) + \frac{GM_*}{r^2} = 0 \tag{10.45}$$

This equation is very similar to the momentum equation used in the early chapters of this book, with the important exception that there is now a 'magnetic pressure gradient' term, from which we identify the magnetic pressure as

$$p_B = \frac{\langle(\delta B)^2\rangle}{8\pi} = \frac{b^2}{16\pi} \tag{10.46}$$

The ratio, $\beta \equiv p/p_B$ is often used in plasma physics, and the case for which the gas pressure is small relative to the magnetic pressure is referred to as a 'low β' plasma.

A gradient in the magnetic pressure will accelerate the wind in much the same way as does a gas pressure gradient. Since we know a relation between b^2 and the Mach number from Eq. (10.41), we can eliminate the derivative of $\langle(\delta B)^2\rangle$ from the radial momentum equation using Eq. (10.42). The value of the constant W is evaluated by boundary conditions giving B_0, v_0, ρ_0 and b_0^2 at the base of the wave driven wind, r_0. Hence Eq. (10.41) provides b as a function of r

$$b^2(r) = b^2(r_0)\frac{M_0}{M}\frac{(M_0+1)^2}{(M+1)^2} \tag{10.47}$$

Now if we use $M = M_0\sqrt{\rho_0/\rho}$ from Eq. (10.38), b^2 becomes a function of the local density $\rho(r)$

$$b^2(r) = b^2(r_0)\left(\frac{\rho}{\rho_0}\right)^{3/2} \frac{\left(1+\frac{A_0}{v_0}\right)^2}{\left\{1+\frac{A_0}{v_0}\left(\frac{\rho}{\rho_0}\right)^{1/2}\right\}^2} \tag{10.48}$$

This equation provides the description of the change in the wave amplitude, b, as a function of the local density and the conditions at the base of the wind.

An expression for the magnetic pressure gradient can be derived using Eq. (10.37), and we find

$$\frac{1}{\rho}\frac{dp_B}{dr} = \frac{b^2}{16\pi\rho}\frac{d\ln b^2}{dr} \tag{10.49}$$

$$= \frac{b^2}{32\pi\rho}\left(\frac{3M+1}{M+1}\right)\frac{d\ln\rho}{dr} \tag{10.50}$$

The density gradient in this expression can be eliminated in favor of the velocity, using the mass conservation constant, $\mathscr{F}_m$, in which case the radial wind momentum equation (10.45) takes on the numerator over denominator form so common in wind theory (see Chapter 3)

$$\frac{d\ln v}{dr} = \frac{\frac{2}{r}\left(a^2+w^2\right) - \frac{GM_*}{r^2}}{\left(v^2-a^2-w^2\right)} \tag{10.51}$$

in which

$$w^2 = \left(\frac{b^2}{32\pi}\right)\frac{1}{\rho}\left(\frac{3M+1}{M+1}\right) \tag{10.52}$$

The net effect of the Alfvén waves is to add the term w^2 to the square of the thermal speed in both the numerator and denominator of the wind equation. The quantity (a^2+w^2) is the square of the sound speed in the presence of Alfvén waves. The two terms a^2 and w^2 behave differently; a^2 tends to be constant or decrease with radius along with the temperature, whereas w^2 increases with radius. Thus the effect of the Alfvén wave corresponds roughly to an increase in the temperature of the atmosphere from which the wind is driven.

Detailed solutions of Eq. (10.51) were first carried out by Belcher (1971) and Alazraki and Couturier (1971). They found that the Alfvén waves allow the star to drive a wind that has both a larger mass loss rate and higher velocity than is the case in the absence of the extra driving terms. Equation (10.51) has the general form of the wind equations discussed in § 3.4. Note that both the numerator and denominator are changed relative to the isothermal wind equation. The change in the numerator relative to the isothermal case implies that the critical point radius is changed. The change in the denominator means that the speed at the critical point is also affected. Both modifications can be traced to the fact that the Alfvén wave force is dependent on the local density.

10.4 Cold Alfvén wave driven winds

Let us consider the simplified case of a 'cold atmosphere' (i.e. a low β plasma) in which the thermal gas pressure gradient dp/dr can be ignored, in the radial momentum equation. This assumption allows us to focus our attention on the effects of the Alfvén waves and will lead to expressions for the mass loss rate and terminal velocity in wave dominated winds.

10.4.1 The cold hydrostatic structure

In Chapter 3 we found that, to a fairly good approximation, the subsonic portion of a thermally driven wind has a density structure that is quite close to the hydrostatic density distribution. If that is the case for an Alfvén wave dominated wind, we will already have a good understanding of the source of mass loss of a wave driven wind.

The hydrostatic equilibrium equation is obtained by eliminating the velocity derivative from the radial Alfvén wave momentum equation (10.45), and eliminating the gas pressure gradient because of the cold atmosphere approximation, we obtain

$$\frac{1}{\rho}\frac{dp_{\rm B}}{dr} = -\frac{GM_*}{r^2} \tag{10.53}$$

As in Eq. (10.50), we can obtain the magnetic pressure gradient by using the wave constant W. Since we are here dealing with the region where the expansion velocity is negligible, we use $M \ll 1$, and then it follows from Eqs. (10.41) and (10.38) that

$$\frac{b^2}{\sqrt{4\pi\rho}} = \text{constant} \qquad \text{for } M \ll 1 \tag{10.54}$$

Now using Eqs. (10.42) and (10.43), we also find

$$\langle(\delta B)^2\rangle = \langle(\delta B_0)^2\rangle\sqrt{\frac{\rho}{\rho_0}} \qquad \text{for } M \ll 1 \tag{10.55}$$

and

$$\langle(\delta v)^2\rangle = \langle(\delta v_0)^2\rangle\sqrt{\frac{\rho_0}{\rho}} \qquad \text{for } M \ll 1 \tag{10.56}$$

Given these proportionalities, we can find several useful relations for the magnetic pressure gradient from Eq. (10.49) that are valid where $M \ll 1$.

$$\frac{1}{\rho}\frac{dp_B}{dr} = \frac{1}{16\pi\rho}\frac{b^2}{2}\frac{d\ln\rho}{dr} \tag{10.57}$$

$$= \frac{\langle(\delta v)^2\rangle}{4}\frac{d\ln\rho}{dr} \tag{10.58}$$

$$= -\frac{1}{2}\frac{d\langle(\delta v)^2\rangle}{dr} \tag{10.59}$$

On using Eq. (10.59), the hydrostatic equation (10.53) becomes the perfect differential

$$\frac{1}{2}\frac{d}{dr}\{\langle(\delta v)^2\rangle + v_{\rm esc}^2(r)\} = 0 \tag{10.60}$$

where $v_{\rm esc}(r) = \sqrt{2GM_*/r}$ is the escape speed at r. The quantity in the brackets in this equation is constant throughout the hydrostatic atmosphere, so $\{\langle(\delta v)^2\rangle + v_{\rm esc}^2(r)\} = \{\langle(\delta v_0)^2\rangle + v_{\rm esc}^2(r_0)\}$. Using Eq. (10.56) to eliminate $\langle(\delta v)^2\rangle$ in terms of $\langle(\delta v_0)^2\rangle$, we obtain the hydrostatic density distribution for an atmosphere supported by the pressure gradient,

$$\rho(\text{hydrostatic}) = \rho_0 \left\{1 + \frac{v_{\rm esc}^2(r_0)}{\langle(\delta v_0)^2\rangle}\left(1 - \frac{r_0}{r}\right)\right\}^{-2} \tag{10.61}$$

Note that this is not an exponential density distribution, as occurs in a hydrostatic atmosphere supported by gas pressure. So, depending on the initial magnitude of the wave amplitude $\langle(\delta v_0)^2\rangle$, this wave supported density distribution can be much more extended than one that is thermally supported.

Now let us compare this hydrostatic density distribution with that in a wind that is Alfvén wave dominated. The enhanced density scale height produced by the waves in the hydrostatic case already provides us with an indication that the waves will lead to a mass loss rate that is much larger than a coronal wind.

10.4.2 The cold wind structure

For wave dominated winds the critical point and all of the subcritical region occur where the flow is very sub-Alfvénic. Consequently in this section we again assume that $M \ll 1$.

If we use the radial momentum equation for a cold wind

$$v\frac{dv}{dr} + \frac{1}{\rho}\frac{dp_B}{dr} + \frac{GM_*}{r^2} = 0 \tag{10.62}$$

along with Eq. (10.59), we get the wind velocity equation

$$\frac{d\ \ln v}{d\ \ln r} = \frac{\frac{1}{2}\left\{\langle(\delta v)^2\rangle - v_{\rm esc}^2(r)\right\}}{\left\{v^2 - \langle(\delta v)^2\rangle/4\right\}} \tag{10.63}$$

This equation has the familiar form of a wind momentum equation with a critical point. At the critical point, $r_{\rm c}$, the numerator and denominator of this equation both vanish, leading to the critical point conditions

$$\langle(\delta v_c)^2\rangle = v_{\rm esc}^2(r_{\rm c}) = v_{\rm esc}^2(r_0)\frac{r_0}{r_{\rm c}} \tag{10.64}$$

$$v_c^2 = \frac{1}{4}\langle(\delta v_c)^2\rangle \tag{10.65}$$

In addition to these two critical point equations, the momentum equation (10.62) leads to a perfect differential, analogous to Eq. (10.60). This provides the energy per unit mass constant, ϵ, for $M \ll 1$

$$\epsilon = \frac{1}{2}\left\{v^2 - \langle(\delta v)^2\rangle - v_{\rm esc}^2(r)\right\} = \text{constant} \tag{10.66}$$

$$= \frac{1}{2}\left\{{v_0}^2 - \langle(\delta v_0)^2\rangle - v_{\rm esc}^2(r_0)\right\} \tag{10.67}$$

These expressions for ϵ do not contain the thermal energy because we assumed a low β plasma, i.e. the wind is cold, so $kT \ll \langle(\delta v)^2\rangle$. Evaluating the energy constant at $r_{\rm c}$, and eliminating v_c^2 and $\langle(\delta v)^2\rangle_c$ from it using Eqs. (10.64) and (10.65), we find that the location of the critical point is

$$\frac{r_{\rm c}}{r_0} = \frac{7}{4}\left\{1 + \frac{\langle(\delta v_0)^2\rangle - {v_0}^2}{v_{\rm esc}^2(r_0)}\right\}^{-1} \equiv \frac{7}{4}\xi \tag{10.68}$$

which provides the simple result for the radius of the critical point

$$\frac{r_{\rm c}}{r_0} = \frac{7}{4} \quad \text{if} \quad \frac{\left|\langle(\delta v_0)^2\rangle - {v_0}^2\right|}{v_{\rm esc}^2(r_0)} \ll 1 \tag{10.69}$$

where the condition arises from setting $\xi = 1$ in Eq. (10.68).

To obtain the mass loss rate for the cold Alfvén wave driven wind we need the velocity and density and radius at the critical point. The velocity is given by Eq. (10.65) and can be expressed in terms of the escape speed at the base of the wind as

$$v_c = \frac{1}{2}v_{\rm esc}(r_0)\sqrt{\frac{r_0}{r_{\rm c}}} \tag{10.70}$$

Combining Eqs. (10.64) and (10.56) to evaluate the critical point value for $\langle(\delta v)^2\rangle$ we find the density at the critical point to be

$$\frac{\rho_c}{\rho_0} = \left\{\frac{7}{4}\xi \frac{\langle(\delta v_0)^2\rangle}{v_{\text{esc}}(r_0)^2}\right\}^2 \tag{10.71}$$

Equations (10.61) and (10.71) provide equations for wave dominated density structures in the hydrostatic and cold wind limits. Evaluating ρ(hydrostatic) in Eq. (10.61) at $r = (7/4)r_0$, and comparing it with Eq. (10.71), we find $\rho_c(\text{wind}) = 0.56\rho_c(\text{hydrostatic})$ for the density at r_c for the case in which $\langle(\delta v_0)^2\rangle \ll v_{\text{esc}}^2(r_0)$. Thus, as was the case for the thermally driven winds in Chapter 3, and for the centrifugal magnetic rotator of Chapter 9, we see that the density follows the hydrostatic distribution very closely in the region below the critical point (and most of the departure occurs very near to the critical point).

Because of mass conservation the value of $\dot{M}$ evaluated at r_c and at r_0 must be the same. Using Eqs. (10.69), (10.70) and (10.71), to provide r_c, v_c and ρ_c, respectively, we find

$$\dot{M} = 4\pi\rho_0 {r_0}^2 v_0 = 4\pi\rho_0 {r_0}^2 \frac{1}{2}\frac{\langle(\delta v_0)^2\rangle^2}{v_{\text{esc}}^3(r_0)}\left(\frac{r_c}{r_0}\right)^{3.5} \tag{10.72}$$

Since both ρ_0 and r_0 cancel from Eq. (10.72), we have an expression for the velocity at the base of the wind

$$v_0 = \frac{1}{2}\frac{\langle(\delta v_0)^2\rangle^2}{v_{\text{esc}}^3(r_0)}\left(\frac{r_c}{r_0}\right)^{3.5} \tag{10.73}$$

Physically, the mass loss is driven by the disturbance of the stellar magnetic field at, or below, the photosphere. However, in deriving the above expressions for the mass loss rate and the initial expansion velocity we have used the condition $M \ll 1$, or the strong field limit according to Eq. (10.38). As a result we find that an explicit dependence on the field does not appear. The mass loss rate is formally determined only by the amplitude of the velocity disturbance at the base of the flow. This is somewhat analogous to the centrifugal magnetic rotator result that the mass loss rate does not depend explicitly on the field but rather on the rotational speed of the foot points. The star could rotate with or without the field being present. However, the transverse waves exist only because of the field. So now let us consider the mass loss as determined by the rate at which the Alfvén wave energy is supplied at the base.

The wave energy flux $f_w(r_0) \equiv f_{w,0}$ at the base of the wind is given by the product of the magnetic energy density, $\varepsilon = \langle(\delta B)^2\rangle/(4\pi) =$

$\rho_0 \langle (\delta v_0)^2 \rangle$, and the Alfvén speed, A_0, at r_0

$$f_{w,0} = \rho_0 \langle (\delta v_0)^2 \rangle A_0 = \frac{\langle (\delta B_0)^2 \rangle}{4\pi} A_0 \tag{10.74}$$

Using this flux in place of $\langle (\delta v_0)^2 \rangle$ in Eq. (10.72) and expressing A_0 in terms of the magnetic field and density at the base of the wind, we get the mass flux at the base of the wind to be

$$\rho_0 \, v_0 = 2\pi \left(\frac{r_c}{r_0} \right)^{3.5} \frac{{f_{w,0}}^2}{{B_0}^2} \frac{1}{v_{esc}^3} \tag{10.75}$$

Thus, after using Eq. (10.68) for r_c, the mass loss rate, $4\pi \rho_0 v_0 {r_0}^2$, for an Alfvén wave driven wind is given by

$$\dot{M} = 8\pi^2 \left(\frac{r_0 f_{w,0}}{B_0} \right)^2 \left(\frac{r_0}{2GM_*} \right)^{3/2} \left(\frac{7}{4} \xi \right)^{7/2} \tag{10.76}$$

Note that for a given stellar magnetic field and Alfvén wave energy flux, the mass loss rate varies inversely with the cube of the escape velocity and is proportional to the surface area of the star. These proportionalities explain why very large mass loss rates can be driven from supergiants and cool giant stars by Alfvén waves. A numerical form for the mass loss rate in Eq. (10.76) is given by

$$\dot{M} = 1.8 \times 10^{-15} \left(\frac{f_{w,5}}{B_0} \right)^2 \left(\frac{{R_*}^{3.5}}{{M_*}^{1.5}} \right) \xi^{7/2} \;\; M_\odot \, \mathrm{yr}^{-1} \tag{10.77}$$

where R_* and M_* are the stellar radius and mass in solar units, B_0 is the surface field in Gauss, and $f_{w,5} = f_{w,0}/10^5$ is the wave energy flux at the base of wind normalized by the value comparable to those that occur at the base of solar coronal hole regions (10^5 erg cm^{-2} s^{-1}).

It is also useful to express the incident wave energy flux in terms of the ratio

$$\alpha = \frac{\langle (\delta B)^2 \rangle}{{B_0}^2} \tag{10.78}$$

as this ratio must certainly be less than unity for our linearization analysis to be valid (a maximal value for α is about 0.1). Thus it is useful to have the mass loss rate expressed in terms of α. The Alfvén wave flux and α are related by

$$f_{w,0} = \frac{\alpha \, {B_0}^3}{\sqrt{(4\pi)^3 \rho_0}} = 7.1 \times 10^4 \frac{\alpha \, {B_0}^3}{\sqrt{10^{13} \rho_0}} \tag{10.79}$$

where ρ_0 is in units of g cm^{-3}. The typical value for the density at the base of a wind is $\rho_0 = 10^{-13}$ g cm^{-3} which corresponds to a number density of $n_0 = 0.9 \times 10^{11}$ atoms cm^{-3}. The mass loss rate is

$$\dot{M} = 9.1 \times 10^{-16} \frac{\alpha^2 {B_0}^4 R_*^{3.5}}{M_*^{1.5}(10^{13}\rho_0)} M_\odot \, \mathrm{yr}^{-1} \tag{10.80}$$

When we express the wave flux as a maximal amount possible for a star of a given magnetic field, we find the high sensitivity to the stellar surface magnetic field displayed in this equation. This is in contrast with what we concluded from the lack of a field dependence in Eq. (10.72), where only the velocity amplitude appeared. Results for the mass loss rate for several sets of stellar parameters are given in Table (10.1).

In this section we derived the mass loss rate by considering just the conditions close to the star. To derive the terminal speed it is necessary to consider the accumulative effect of energy deposition through the region beyond the critical point.

10.4.3 The wind energy and terminal velocity

For the region far beyond the critical point, we can no longer use the limit that the Alfvén Mach number is small because the flow will become super-Alfvénic. However, for any value of the Mach number, M, it is possible to find an integral of the radial momentum equation. Using $p_B = b^2/16\pi$, and eliminating b^2 in terms of M and the wave constant W of Eq. (10.41), we obtain

$$\frac{1}{\rho}\frac{dp_B}{dr} = \frac{W}{16\pi\rho}\frac{d}{dr}\left(\frac{1}{(M+1)^2\,M}\right) \tag{10.81}$$

Now we express M in terms of the density using Eq. (10.38) which we write as $M = 1/(C\sqrt{\rho})$, where C is constant. If we let $x = \sqrt{\rho}$, Eq. (10.81) can be expressed as a perfect differential by using the identity

$$\frac{1}{x^2}\frac{d}{dr}\left(\frac{x^3}{(1+Cx)^2}\right) = \frac{d}{dr}\left(\frac{x(3+2Cx)}{(1+Cx)^2}\right) \tag{10.82}$$

Thus we find the perfect differential form for the pressure gradient,

$$\frac{1}{\rho}\frac{dp_\mathrm{B}}{dr} = \frac{1}{16\pi}\frac{d}{dr}\left(\frac{3M+2}{\rho M}b^2\right) \tag{10.83}$$

$$= \frac{d}{dr}\left(\frac{3M+2}{M}\frac{\langle(\delta v)^2\rangle}{2}\right) \tag{10.84}$$

Substituting this result into Eq. (10.45) we derive the energy per unit mass constant

$$\epsilon = \frac{1}{2}v^2 - \frac{1}{2}v_{\rm esc}^2(r) + \left(\frac{1}{M} + \frac{3}{2}\right)\langle(\delta v)^2\rangle = \text{constant} \tag{10.85}$$

Let us consider the asymptotic value of the energy per unit mass, ϵ, as r becomes large. The density decreases towards zero, so as seen from Eq. (10.38) M, which is proportional to $\rho^{-1/2}$, becomes large. The wave constant, from Eq. (10.41), then varies as $W = b^2M^3$. Using this we find from $\langle(\delta B)^2\rangle = b^2/2$ that $\langle(\delta B)^2\rangle = W/(2M^3) \propto \rho^{3/2}$ and $\langle(\delta v)^2\rangle \propto \rho^{1/2}$ so that $\langle(\delta v)^2\rangle \to 0$ as $r \to \infty$. At infinity $v_{\rm esc}(r)$ also vanishes, thus

$$\epsilon = v_\infty^2/2 \tag{10.86}$$

where v_∞ is the terminal speed of the wind.

Since ϵ is the terminal kinetic energy per unit mass of the wind at infinity, the terminal velocity can be derived by evaluating ϵ at a point closer to the star. At r_0 where $M = M_0 \ll 1$, the magnetic term in ϵ can be written in terms of the flux $f_{w,0}$ as follows. The quantity $\langle(\delta v)^2\rangle$ can be eliminated from Eq. (10.85) in terms of $f_{w,0}$, ρ_0 and A_0 using Eq. (10.74). Then eliminating M_0 in terms of the mass and magnetic flux constants using Eq. (10.38), we find that the magnetic term in Eq. (10.85) becomes simply $f_{w,0}/(\rho_0 v_0)$. So, after equating the expressions for ϵ, evaluated at large and small r, we find

$$v_\infty^2 = 2\frac{f_{w,0}}{\rho_0 v_0} + {v_0}^2 - v_{\rm esc}^2(r_0) \tag{10.87}$$

Using $(\rho_0 v_0)$, as given in Eq. (10.75), leads to the numerical result

$$v_\infty^2 = v_{\rm esc}^2(r_0)\left\{28.\left(\frac{{B_0}^2}{f_{w5}\xi^{3.5}}\right)\left(\frac{M_*}{R_*}\right)^{\frac{1}{2}} - 1\right\} \tag{10.88}$$

with ξ defined in Eq. (10.68). Values for v_∞ and other Alfvén wave driven wind properties as derived by Hartmann and MacGregor (1980) are given in Table (10.1). Cold wave dominated winds tend to have terminal velocities larger than the escape speed. This agrees with what is observed in the high speed streams from the sun.

10.4.4 The wind luminosity and the Poynting flux

Some additional insight into the Alfvén wave driving process can be obtained by considering the energy flow through the wind. Note that in our derivations of $\dot{M}$ and v_∞ we have not appealed to a dissipation of the wave energy. In this regard the Alfvén wave driven wind appears to be similar to the radiation driving of winds by line

Table 10.1 *Alfvén wave driven wind models*

M_* $(M_\odot)$	R_* $(R_\odot)$	T (K)	N_0 (cm^{-3})	B_0 (G)	$f_{w,0}$[a]	$\dot{M}$[b] $(M_\odot\,\mathrm{yr}^{-1})$	v_∞[b] $(\mathrm{km\,s}^{-1})$
16	400	10^4	10^{11}	1	6.7×10^3	1.1×10^{-10}	823.
16	400	10^4	10^{11}	5	8.4×10^5	4.3×10^{-8}	508.
16	400	10^4	10^{11}	10	6.7×10^6	5.5×10^{-7}	408.
1.33	27	10^4	10^{11}	10	6.7×10^6	1.8×10^{-9}	484.

[a] $f_{w,0} = \frac{\langle(\delta B_0)^2\rangle}{4\pi} A_0$ = Alfvén wave flux at the base of the wind in erg $\mathrm{cm}^{-2}\,\mathrm{s}^{-1}$, the values for $f_{w,0}$ correspond to $\alpha = 0.1$.
[b] $\dot{M}$ and v_∞ values are from Hartmann & MacGregor (1980) for cases with no wave damping.

opacity, where no loss of radiative luminosity is assumed. However, in the Alfvén wave driving case there is in fact a change in the wave luminosity that has been veiled by our use of the wave constant W in Eq. (10.41).

Let us consider the total luminosity of the wind, including the energy flux of the Alfvén waves. This is the energy constant per unit mass, ϵ, times the mass loss rate $4\pi r^2 \rho v$. The energy constant is given by Eq. (10.85) which consists of wind terms and a magnetic term. The wind term is

$$\epsilon_w = \frac{v^2}{2} - \frac{GM_*}{r} + \frac{\gamma}{\gamma - 1}\frac{p}{\rho} \tag{10.89}$$

which contains the kinetic energy, the potential energy and the enthalpy. For more generality we have included the gas pressure term for a polytropic relation $p \propto \rho^\gamma$ (see Chapter 4). The Alfvén wave term of the energy constant in (10.85) is

$$\begin{aligned} \epsilon_A &= \left(\frac{1}{M} + \frac{3}{2}\right)\langle(\delta v)^2\rangle \\ &= \frac{A+v}{v}\langle(\delta v)^2\rangle + \frac{\langle(\delta v)^2\rangle}{2} \\ &= \frac{A+v}{v}\frac{\langle(\delta B)^2\rangle}{4\pi\rho} + \frac{\langle(\delta v)^2\rangle}{2} \end{aligned} \tag{10.90}$$

where we have substituted $\langle(\delta v)^2\rangle = \delta B/(4\pi\rho)$ (Eq. 10.43). So the total wind luminosity is

$$L_{\mathrm{tot}} = L_w + L_A = \text{ constant} \tag{10.91}$$

gradients are of no importance. We have found that for a cold wave-dominated wind the density structure of the sub-critical point region differs very little from that of a hydrostatic magnetically supported atmosphere. Since the vanishing of the numerator and denominator of the wind equation provides information about the velocity and density at the top of the near hydrostatic region, we have been able to derive simple expressions for the mass loss rate and terminal velocities of Alfvén wave driven winds. Consideration of the energy equation has shown that matter that is gravitationally bound near the star can escape farther out because a transfer of energy occurs by way of the Poynting energy flux. Recall that this was also the case for the magnetic rotator winds discussed in Chapter 9. There we presented the hybrid luminous magnetic rotator model. A hybrid radiation plus Alfvén wave model has been developed for Wolf-Rayet Stars by dos Santos, Jatenco-Pereira, and Opher (1993).

Although this chapter provides a necessary background for understanding the driving of winds from red giant stars, the results do not adequately match the observations. The winds that are predicted by the theory presented here tend to be very fast. This may be appropriate for the open field regions in the sun. However, for the red giant stars with winds strong enough to be diagnosed from spectral observations, the wind speeds tend to be slow, of order 10 to 50 $\mathrm{km\,s^{-1}}$, and these speeds are well below the escape speed at the base of the wind. The slow winds pose major problems, as were described by Hartmann and MacGregor (1980) and Holzer, Flå and Leer (1983). To achieve the lower speeds the damping or reflection of Alfvén waves must be accounted for (MacGregor and Charbonneau, 1997).

In regard to its application across the HR diagram, the Alfvén wave mechanism has been mainly considered for stars which have winds that cannot be explained by radiative driving. The winds produced by either line opacity or dust grain continuum opacity tend to occur only for stars with luminosities above about $10^4\ L_\odot$. The Alfvén mechanism could be the dominant wind driving process for stars in the remaining portions of the HR diagram.

10.6 Suggested reading

Alazraki, G. & Couturier P. 1971, 'Solar Wind Acceleration Caused by the Gradient of the Alfvén Wave Pressure' *A & A* **13**, 380

Belcher, J.W. 1971, 'Alfvénic wave Pressures and the Solar Wind', *Ap. J.* **168**, 509

(The basic theory for Alfvén wave driven winds)

Holzer, T.E., Flå, T. & Leer, E. 1983, 'Alfvén Waves in Stellar Winds', *Ap. J.* **275**, 808

(The basic theory expanded)

Holweg, J.V. 1973, 'Alfven Waves in a Two-Fluid Model of the Solar Wind' *Ap. J.* **181**, 547

Hartmann, L. & MacGregor, K.B. 1980, 'Momentum and Energy Deposition in Late-Type Stellar Atmospheres and Winds', *Ap. J.* **242**, 260

MacGregor, K.B. & Charbonneau, P. 1994, 'Stellar Winds with Non-WKB Alfvén Waves I. Wind Models for Solar Coronal Conditions', *Ap. J.* **430**, 387

11 Outflowing disks from rotating stars

Early-type stars often show rotationally broadened photospheric lines that indicate that they are rotating with equatorial speeds in the range 100 to 400 $\mathrm{km\,s^{-1}}$. These stars have radiatively driven winds owing to the strong line opacities in their outer atmospheres, as described in Chapter 8. The rotation of the stars leads to interesting effects, the most prominent of which is the tendency to concentrate the outflowing material toward regions near the equatorial plane. The equatorial material is moving outwards from a star whose surface is rotating at a speed below the critical speed. Therefore these disks are called *outflowing disks* or *de-cretion disks*, in contrast to the 'accretion disks' around pre-main sequence stars or around the gaining stars in binary systems with mass transfer.

In this chapter we consider only the formation of outflowing disks. For a star that has a stellar wind and also an outflowing disk, the contrast in density from equator to pole is typically about a factor of ten or so. We discuss two basic pictures for producing such a contrast. The first is a piece-wise spherical outflow in which the equatorial density is enhanced because the mass flux from the near-equatorial latitudes is larger or the wind velocity is lower than those in the polar regions. Such a wind could be the result of the *'rotation induced bi-stability'* (RIB) model of Lamers and Pauldrach (1991). The second is the wind compression picture in which the streamlines of the gas from both hemispheres of a rotating line driven wind are bent towards the equatorial plane. This leads to a compression of the wind in the equatorial plane. The result is a decrease in the density of the wind at higher lattitudes and a corresponding increase in density near the equatorial plane. This is the effect underlying the *'wind compressed disk'* (WCD) model of Bjorkman and Cassinelli (1993).

We first discuss the evidence for the existence of outflowing disks around B stars in § 11.1. In § 11.2 we explain the bi-stability mechanism and the formation of outflowing disks around rapidly rotating B stars by the RIB mechanism. In the subsequent sections we discuss the wind compression effects in rotating stars. In § 11.3 we use the supersonic approximation of the WCD model and introduce the kinematic equations that allow us to derive the structure of the wind and the disk. We will make a distinction between the very flat wind compressed disks, which are formed by means of shocks, and the wind compressed zones, where shocks do not occur. Properties of the wind compressed disks and zones are discussed in § 11.4. In § 11.5 results of hydrodynamical calculations that lead to a circulation of disk material back onto the star are discussed. Conclusions and the consequences of outflowing disks are given in § 11.6.

11.1 Stars with ouflowing disks

There are two classes of early-type emission line stars that most clearly show the presence of equatorially enhanced winds. These are the near-main sequence Be stars and the much more luminous B[e] (called 'B - bracket e') supergiants. (Reviews of the observational data regarding Be stars and B[e] supergiants can be found in Balona *et al.* (1994), and Zickgraf (1992), respectively.)

Both types of stars show spectroscopic evidence for a fast low density wind with a velocity of typically 500 to 1500 $\mathrm{km\,s^{-1}}$ by the shapes and width of the UV resonance lines. These values are quite similar to those found for normal B stars and B supergiants. However, the Be stars and the B[e] supergiants have much stronger Balmer emission lines and much stronger infrared excess due to free-free emission than normal early type stars with line driven winds. Since the Balmer lines are mainly formed by recombination, the emissivity is proportional to the square of the density. The same applies to the emissivity of the free-free radiation (see § 2.3 and § 2.4). So the presence of strong Balmer emission and strong infrared excess in a wind with normal UV lines indicates that there must be a region in the wind that has a much higher density and a much higher emission measure than normal winds. The profiles of the Balmer lines clearly indicate rotation (e.g. by double peaked profiles) and rather small outflow velocities of a few hundred $\mathrm{km\,s^{-1}}$. This means that the high density region is concentrated along the equatorial plane and the outflow velocity in this plane is much smaller than that of the wind at higher lattitudes.

This picture is confirmed by polarization observations. The intrinsic polarization of stars is a diagnostic that is especially useful for detecting rotationally distorted stellar atmospheres and winds. The polarization is produced in the case of hot stars by electron scattering of the photospheric light. If the wind is circularly symmetric as seen from the direction of the observer, the starlight will not be polarized because the polarized intensity produced by scattering at one sector of the envelope will be cancelled by the polarized intensity from a sector 90 degrees away. However, if a star has a rotationally distorted wind, and is seen more nearly equator on, most of the scattering of the star light occurs in the sectors containing the equatorial zone. This enhanced scattering from the equatorial region can lead to a net polarization of the stellar radiation by up to about 2 percent, which is large enough to be measured to high accuracy. Some Wolf-Rayet stars also show intrinsic polarization. This implies that their winds are also rotationally distorted, but the equatorial compression seems to be smaller than for Be stars and B[e] supergiants. (For a reviews of the observed properties of WR stars, see van der Hucht and Williams, 1995).

Empirical modelling of Be stars and B[e] supergiants has led to a two component picture of their wind structures. There is a slow outward moving dense equatorial or 'disk' component, and a fast tenuous wind over the remainder of the star. The fast component is consistent with a line driven wind models for the stars. The possible causes of the slow and dense equatorial outflows is the topic of this chapter.

11.2 Outflowing disks formed by rotation induced bi-stability

In this section we discuss the formation of outflowing disks formed by the bi-stability mechanism, applied to rotating stars. We first explain the concept of bi-stable *spherically symmetric* radiation driven winds. Then we apply this concept to winds from rotating stars to formulate the rotation induced bi-stability mechanism.

11.2.1 The bi-stability of radiation driven winds

From detailed calculations of line driven winds for the luminous blue variable star P Cygni (B1 Ia$^+$, $T_{\rm eff}$=19300 K), Pauldrach and Puls (1990) made an interesting discovery. Relatively slight changes in the basic parameters of the star, such as the effective gravity ($g_{\rm eff} = GM_*(1-\Gamma_e)/R_*^2$), or the effective temperature or the luminosity can lead to very different wind structures. The star has either a tenuous wind with a high terminal velocity or a denser wind with a much slower velocity.

Pauldrach and Puls (1990) found that when the Lyman continuum optical depth of the wind at a wavelength of 912 Å is less than about unity, the wind is fast, whereas the wind is slow if the optical depth is larger than about unity. So for a star like P Cygni, whose wind has an optical depth very close to unity, a small change in the stellar parameters can result in a drastic change in the wind structure. They showed that in the case of P Cygni the wind might even switch between the two solutions. That is why they called it a 'bi-stable wind'. Since then, the mechanism is called the 'bi-stability' mechanism, although other stars do not really switch from one solution to the other. For other stars the bi-stability manifests itself as a sudden change in the structure of the wind at certain spectral types or luminosities. This is called the *'bi-stability jump'*.

The very sensitive dependence of the wind characteristics on the stellar parameters of P Cygni is shown in Fig. (11.1). This figure shows the changes in the mass flux per unit surface area from the star, $F_m = \dot{M}/4\pi R_*^2 = \rho(R_*)v(R_*)$, and in the terminal velocity of the wind as a function of the effective gravity g_{eff}. The radiation driven wind theory predicts that as the effective gravity decreases, the mass loss rate and hence the mass flux increases proportional to $\dot{M} \sim g_{\text{eff}}^{(\alpha-1)/\alpha}$ and the terminal velocity decreases proportional to $v_\infty \sim v_{\text{esc}} = \sqrt{2R_* g_{\text{eff}}}$. In these expressions α is one of the force multiplier parameters that describe the radiative acceleration due to lines, with α typically about 0.6 (see § 8.6.1). This explains the gradual trends in Fig. (11.1) of increasing F_m and decreasing v_∞ as g_{eff} decreases. However, the wind models show a clear discontinuity near $g_{\text{eff}} = 1.5$. We have marked the models with different symbols for different optical depths in the Lyman continuum at 912 Å. The models with $\tau_{912} < 1$ are on one branch of the solution, and those with $\tau_{912} > 3$ on another branch. The transition occurs at $1 < \tau_{912} < 3$.

The difference between the two sets of solutions is the following: in the models with optically thin winds, $\tau_{912} < 1$, the ionization is 'high' and the line driving of the wind is mainly done by high ionization stages of ions with lines in the Lyman continuum and in the Balmer continuum shortward of about 1700 Å. Since the line driving is due to a mixture of optically thick and optically thin lines, the value of α is relatively high, $\alpha \simeq 0.6$. This results in a wind velocity of about 600 km s^{-1} for P Cygni.

In the models with the optically thick wind, the lines that drive the wind are of lower ionization stages, mainly in the Balmer continuum at $\lambda > 1700$ Å. These lines are generally less optically thick and hence the value of α is smaller. This results in a smaller terminal velocity. The drop in α is accompanied by a jump in k, the force multiplier that determines the mass loss rate as $\dot{M} \sim k^{1/\alpha}$ (see § 8.7.1). This is because there are many more weaker lines in the optically thick model, which can absorb or scatter a larger fraction of the photons from the star.

The jump between the two sets of models is rather sudden, because once the wind starts to recombine, the wind velocity decreases and the mass loss rate increases. Together this gives a higher density in the

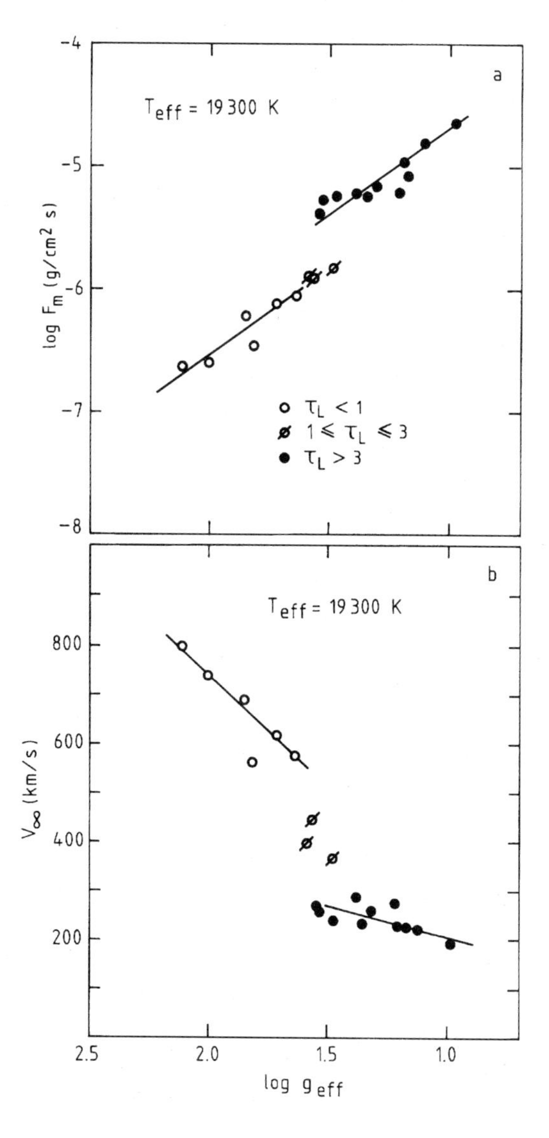

Figure 11.1 The bi-stability of the radiation driven wind models calculated for the star P Cygni by Pauldrach and Puls (1990). Different symbols refer to different optical depths of the wind in the Lyman continuum at 912 Å. The jump from one set of solutions to another occurs where the wind becomes optically thick in the Lyman continuum. (From Lamers and Pauldrach, 1991)

wind which results in even more recombination. This quickly forces the wind into a new line driven equilibrium. Figure (11.1) shows that the bi-stability jumps in v_∞ and in the mass loss rate are both about a factor of three. Since the wind density is proportional to $\dot{M}/v$, it jumps by almost a factor of ten!

The existence of the bi-stability jump was also found for normal supergiants by Lamers, Snow and Lindholm (1995) using their extensive survey of wind line profiles obtained with the IUE satellite. They tested the bi-stability prediction that, for stars with spectral type near B1, i.e. $T_{\text{eff}} \simeq 21\ 000$ K, slight differences in effective temperatures lead to very different wind properties. Figure (11.2) shows the observational result that there is an abrupt change in the ratio of v_∞/v_{esc} near $T_{\text{eff}} \simeq 21\ 000$ K. The jump is in the direction predicted by the bi-stability mechanism. For stars with log $(T_{\text{eff}}) > 4.35$, the ratio has the value $v_\infty/v_{\text{esc}} \approx 2.6$, while for stars with log $(T_{\text{eff}}) < 4.30$ the ratio is $v_\infty/v_{\text{esc}} \approx 1.3$. Thus, a rather small change in the effective temperature leads to a wind speed decrease by about a factor of two (see also § 2.7). Also in agreement with the bi-stability predictions is the finding that stars on the hot side of the jump have more highly ionized winds, while the stars on the cool side have winds of lower ionization state. The hotter stars with the faster winds have strong C IV wind lines and no wind produced C II lines, whereas the slightly cooler stars with the slower winds have strong C II P Cygni profiles

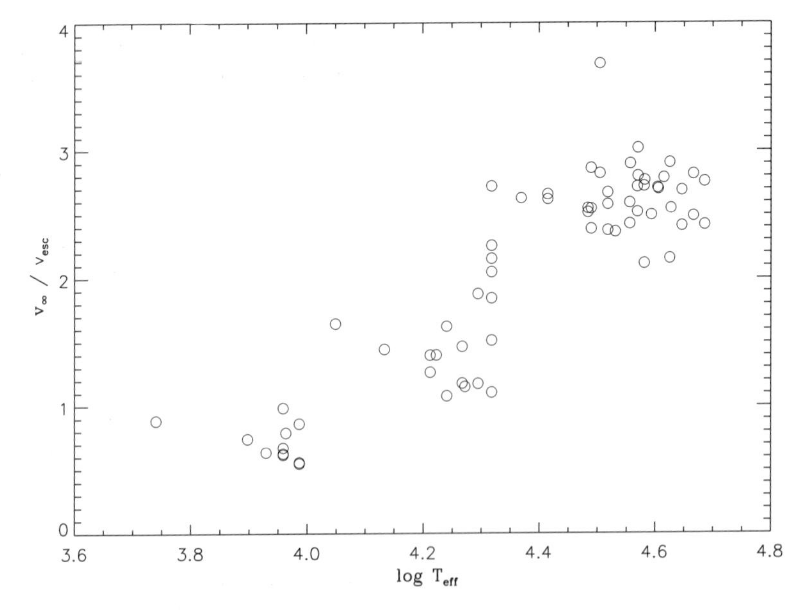

Figure 11.2 The ratio v_∞/v_{esc} as a function of T_{eff}. The data show a bi-stability jump for stars with temperatures near 21 000 K. Stars on the hot side have a higher value for v_∞/v_{esc} than do stars on the cool side of the jump. There is evidence for a second bi-stability jump near 10 000 K. (From Lamers, Snow and Lindholm, 1995)

and weak C IV lines. A third prediction of the bi-stability mechanism is that the stars on the cool side of the jump should have a larger mass loss rate. This has not yet been confirmed as it is more difficult to derive mass loss rates from line profiles than wind speeds and ion identifications (see § 2.2).

The change in the driving lines that cause the bi-stability jump can be seen in Fig. (8.10), where the model of $T_{\rm eff}$=30 000 K, as an example of a star with $T_{\rm eff} > 21\ 000$ K, is compared with the model of $T_{\rm eff}$=20 000 K. The figure shows that the driving lines in the hotter model are in the Lyman continuum and in the Balmer continuum at $1300 < \lambda < 1700$ Å, whereas the driving lines of the cooler model are at $912 < \lambda < 1500$ Å and at $1700 < \lambda < 2300$ Å. The figure also shows that a larger fraction of the luminosity is used for driving the wind in the cooler model than in the hotter model. This explains the increase in $\dot{M}$ at the bi-stability jump.

Lamers and Pauldrach (1991) investigated whether the bi-stability may produce *outflowing disks* around rapidly rotating early B stars, but understanding this prediction first requires an understanding of the general effects of rotation on the underlying star.

11.2.2 The effects of rotation on stars

Rotation has interesting effects on the structure of the outer atmosphere and stellar wind. First, let us consider the near hydrostatic photospheric region of a star rotating as a solid body with an angular velocity, Ω. In the simplest models, called Roche models, the gravitational force is everywhere assumed to be radial and the hydrostatic equilibrium equation is

$$\frac{1}{\rho}\nabla P = -\nabla(\Phi_G + \Phi_\Omega) \equiv -\nabla\Phi \tag{11.1}$$

where Φ is the sum of the gravitational potential and the centrifugal potential. Assuming solid body rotation for Φ_Ω (Clayton, 1968), we get

$$\Phi = -\frac{GM_*}{r} - \tfrac{1}{2}(r\Omega\sin\theta)^2 \tag{11.2}$$

where θ is the polar angle on the star ($\theta = 0$ along the polar axis).

The rotation leads to a distortion of the equipotential surfaces as shown in Fig. (11.3). These surfaces are found from solutions to the cubic equation

$$R(\theta)^3 - \frac{2GM_*}{R_0\,\Omega^2\sin^2\theta}R(\theta) + \frac{2GM_*}{\Omega^2\sin^2\theta} = 0 \tag{11.3}$$

where R_0 is the polar radius of the isopotential surface corresponding to $\Phi = GM_*/R_0$ = constant. Equation (11.3) can be derived from expression (11.2) for the potential by setting $\Phi = \Phi(R_0)$ (see Problem 11.1).

From Eq. (11.1) we see that the pressure gradient is perpendicular to surfaces of constant potential and hence the surfaces are equi-pressure surfaces. Equi-potential implies that there are no pressure gradients along the surface that would cause motions of the gas along the surface. By taking the curl of Eq. (11.1) we find that $(\nabla\rho \times \nabla P)$=0, so the surfaces are also equi-density surfaces. On using the perfect gas law equation of state, we find that the equi-potential surfaces are also surfaces of equi-temperature.

The temperature stratification has implications regarding the radiation field as a function of latitude. For the case in which the Planck function B is the radiation source function and is constant on surfaces of equi-potential, the flux at the base of a zone in the atmosphere is given by the diffusion approximation

$$\mathscr{F} = -\frac{4\pi}{3}\frac{1}{\kappa\rho}\frac{dB}{d\Phi}\nabla\Phi \tag{11.4}$$

where κ is the Rosseland mean opacity and B is the integrated Planck function ($\propto T^4$). Taking the local 'net gravity' to be $g_{\rm net} = -\nabla\Phi$, we see that *the radiative flux is proportional to the local net gravity.* This is the von Zeipel theorem (1924). The plausibility of this can be seen by inspection of Fig. (11.3). Note that the surfaces of equi-

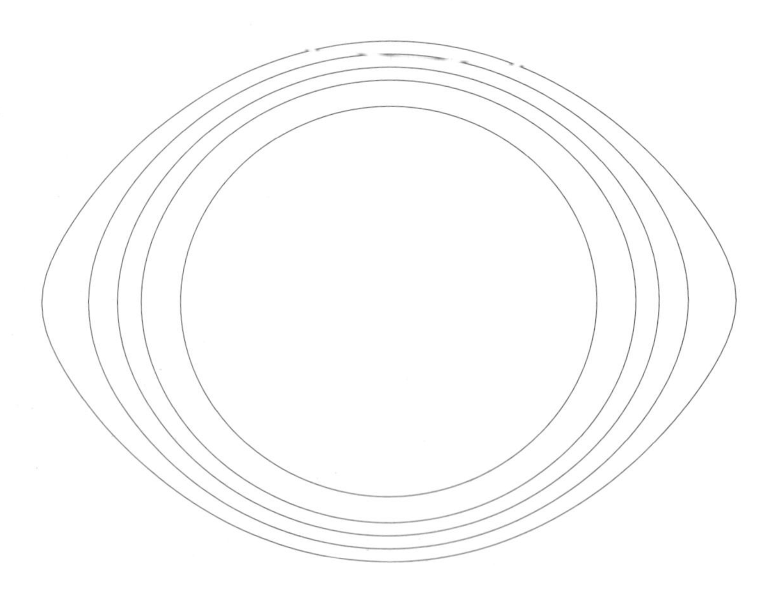

Figure 11.3 Projected equipotential surfaces of a rigidly rotating star with $\Omega R_{\rm polar} = 0.54\sqrt{GM_*/R_{\rm polar}}$.

temperature are closer together in the polar region, thus the gradient of the temperature is steepest near the pole and the radiative flux is thus largest at the pole. The von Zeipel dependence of the radiative flux on latitude is also called 'polar brightening' (or 'gravity darkening' when the interest is on the equator). A more detailed discussion, given by Clayton (1968), shows that a star with the flux determined by the von Zeipel theorem, Eq. (11.4), can not have a static atmosphere, and Eddington-Sweet circulation currents are set up in the envelope of the star, which modifies the temperature structure and flux distribution derived in the von Zeipel model. Nevertheless the von Zeipel theorem is a convenient starting point for investigating rotational effects.

Another effect of rotation that is obvious from the Fig. (11.3) is that the surface of the star has a larger radius at the equator than at the pole. The maximal equatorial rotation speed is called the 'breakup' speed, or more correctly the 'critical' speed, $v_{\rm crit}$.† If we account for the reduction of gravity by the radiative acceleration on electrons; $g_e = \sigma_e L_*/(4\pi c) = \Gamma_e G M_*/r^2$ with $\Gamma_e = \sigma_e L_*/(4\pi c G M_*)$ (see § 8.6.1), we obtain an expression for $v_{\rm crit}$:

$$\begin{aligned} v_{\rm crit} &= \sqrt{\frac{GM_*(1-\Gamma_e)}{R_{\rm eq}}} \\ &= v_{\rm esc}/\sqrt{2} \end{aligned} \tag{11.5}$$

where $R_{\rm eq}$ is the stellar equatorial radius and $v_{\rm esc}$ is the escape speed at the equator. Since $v_{\rm crit}$ is an upper limit to the rotational speed for a star, it is convenient to express the star's equatorial rotation speed in terms of the ratio ω as

$$\omega \equiv \frac{v_{\rm rot}}{v_{\rm crit}} = v_{\rm rot}\sqrt{\frac{R_{\rm eq}}{GM_*(1-\Gamma_e)}} \tag{11.6}$$

The quantity ω has a maximal value of unity.

11.2.3 The rotation induced bi-stability model

In § 11.2 we have seen that line driven winds from early B stars can change from low density and high velocity winds to high density and low velocity winds if the optical depth of the wind in the Lyman continuum becomes larger than unity. The basic idea of the RIB model is that the same transition can occur in the wind of a rotating early B

† The expression 'breakup speed' can be confusing, because the matter at the surface is *not* moving at the speed needed for escape, as the word *breakup* implies.

star between the pole and the equator. This is because the reduction of the effective gravity between the pole and the equator due to rotation enhances the optical depth in the wind from the pole to the equator. If the optical depth changes from a value smaller than unity at the pole, to a value larger than unity at the equator, the bi-stability jump will occur at some intermediate latitude. If this happens, the equatorial wind will be slow and dense, whereas the polar wind will be fast and tenuous. This is very similar to the observed characteristics of B stars with outflowing disks. An attractive feature of the RIB mechanism is that it is expected to work only for early B stars, because the models and the observations show that the bi-stability can occur around $T_{\rm eff}$ about 20 000 K. (The exact value of $T_{\rm eff}$, where the bi-stability jump occurs, depends on the gravity and the rotational velocity.)

Since the bi-stability depends on the optical depth of the wind in the Lyman continuum, we consider the various effects of stellar rotation on the wind parameters, following Lamers and Pauldrach (1991).

The net surface gravity, $g_{\rm net}$, is given as a function of θ by

$$g_{\rm net}(\theta) = \frac{GM_*(1-\Gamma_e)}{R(\theta)^2}(1-\omega^2\sin^2\theta) \tag{11.7}$$

Following Lamers and Pauldrach, we ignore the rotational distortion of the star for simplicity, so that we can adopt $R(\theta) = R_*$. The von Zeipel theorem, discussed above, shows that the radiative flux at the stellar surface is proportional to the local gravity, so

$$T_{\rm eff}{}^4(\theta) \sim (1-\omega^2\sin^2\theta) \tag{11.8}$$

The changes in gravity and radiation temperature as a function of θ affect the wind in several ways.

(a) The wind speed in the radiation driven wind theory is proportional to the escape speed. If we make the piecewise spherical approximation, the wind speed at stellar lattitude θ is

$$v_\infty(\theta) \sim v_{\rm esc}(\theta) \sim \sqrt{R_* g_{\rm net}(\theta)} \sim (1-\omega^2\sin^2\theta)^{0.5} \tag{11.9}$$

So the wind speed decreases from the pole to the equator.

(b) The mass flux of a radiation driven wind, F_m, depends on the local radiative flux and on the local net surface gravity

$$F_m \sim \left\{T_{\rm eff}{}^4(\theta)\right\}^{1/\alpha} g_{\rm net}^{(\alpha-1)/\alpha} \sim (1-\omega^2\sin^2\theta)^{1.0} \tag{11.10}$$

(Eq. 8.105). This means that the mass flux decreases from the pole to the equator. Detailed models give a slower decrease (Lamers *et al.*, 1999).

(c) The Lyman continuum flux is very sensitive to the flux of the radiation field. The brightness temperature in the Lyman continuum, T_L, varies approximately as ${T_{\rm eff}}^{1.6}$ for B stars (Kurucz, 1979), and since ${T_{\rm eff}}^4(\theta) \sim g_{\rm net}$, we find that T_L varies approximately as $T_L \sim (1 - \omega^2 \sin^2\theta)^{0.4}$. The flux in the Lyman continuum is a very steep function of the brightness temperature T_L. For example, near $T \simeq 20\,000$ K the flux at 800 Å varies as ${T_L}^{9.1}$, and so the flux at, say 800 Å, is expected to vary with latitude as

$$F(800\mathring{A}) \sim (1 - \omega^2 \sin^2\theta)^{3.6} \tag{11.11}$$

So the ionizing UV flux will decease very strongly from the pole to the equator.

(d) The optical depth in the Lyman continuum through the wind, τ_L, is proportional to the column density of neutral H. As this is determined by recombination, τ_L is proportional to $\rho^2 \sim (F_m(\theta)/v_\infty(\theta))^2$. The optical depth is also inversely proportional to the flux in the Lyman continuum, because that determines the ionization rate. Accounting for the dependence of the optical depth on both the ionizing flux and the column density, we find that the optical depth of the wind in the Lyman continuum varies with polar angle approximately as

$$\tau_L \sim \left\{\frac{F_m(\theta)}{v_\infty(\theta)}\right\}^2 \frac{1}{F(800\mathring{A})} \sim (1 - \omega^2 \sin^2\theta)^{-2.6} \tag{11.12}$$

for $\alpha \simeq 0.5$. This relation implies that an early B star that is rotating will have a much larger wind optical depth in the Lyman continuum in the equatorial region than near the polar region. The difference will be a factor of 14 for a star that is rotating with $\omega = 0.8$ and a factor of 3 if $\omega = 0.6$.

In §11.2.1 we have shown that stellar winds of early B stars can make a bi-stability jump to a much higher density and lower velocity, if the optical depth of the wind in the Lyman continuum changes from $\tau_L < 1$ to $\tau_L > 3$. In a rapidly rotating early B star this same jump in wind density can occur between the polar region and the equatorial region, because τ_L is a very steep function of the polar angle θ. This will produce a disk-like structure, with a high density and low velocity wind from the equator and a low density and high velocity wind from the polar regions. Whether this will happen depends on two characteristics: (1) the optical depth τ_L of the wind from the pole must be smaller than unity (but not too small), and (2) the increase in τ_L from the pole to the equator must be large enough to make $\tau_L > 1$ at the equator.

Lamers and Pauldrach calculated the optical depth for a star with the stellar parameters of a typical B[e] supergiant: $T_{\rm eff} = 20\,000$,

$L_* = 10^{5.7} L_\odot$, $R = 59 R_\odot$ and $M = 35 M_\odot$. They found that $\tau_L = 0.05$ at the poles and $\tau_L > 1$ at the equator if $\omega > 0.6$. So a star with these parameters will have a rotation induced bi-stability disk. They also estimated this effect quantitatively for the classical Be stars, i.e. the rapidly rotating near-main sequence B stars. They found that for these stars, which are much less luminous than the B[e] supergiants, the RIB mechanism by itself is not sufficient to explain the presence of the disks. This is because the Be stars have mass loss rates that are far too small to make the wind optically thick in the Lyman continuum.

The rotation induced bi-stability model illustrates that a rapidly rotating B[e] supergiant with a temperature near 20 000 K can have a wind which is optically thick in the Lyman continuum near the equator (due to the smaller ionizing flux, higher mass loss rate, and lower wind velocity) and a wind which is optically thin near the poles (due to the higher ionizing flux, a lower mass loss rate, and a higher v_∞). So, the bi-stability will produce a drastic change in the wind property at some stellar latitude between the pole and the equator. The mass loss flux at the equator will be about a factor of three higher than at the poles and the terminal wind velocity will be about a factor of three smaller. This produces a density contrast ($\rho \propto \dot{M} v_\infty$) of about an order of magnitude between the equator and the poles, with a rather steep jump at the latitude where the Lyman continuum in the wind is of order unity. The result is an equatorial outflowing disk. Figure (11.4) illustrates the nature of the wind predicted by the RIB model for B[e] supergiants. This picture is quite similar to the empirical model of these stars derived by Zickgraf *et al.* (1986).

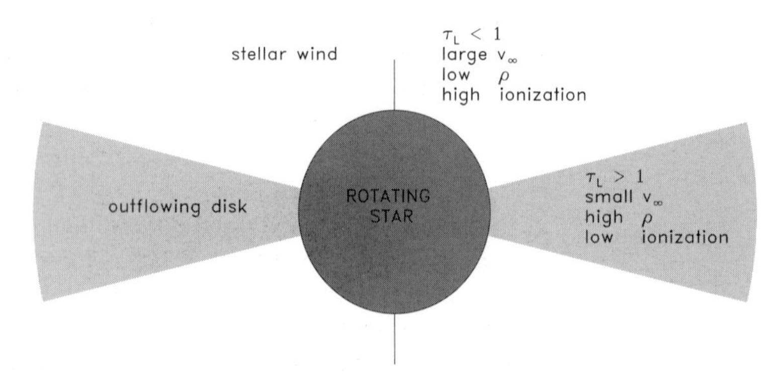

Figure 11.4 A schematic picture of a rapidly rotating B[e] supergiant formed by the bi-stability mechanism. The wind is optically thin, ($\tau_L < 1$) in the polar regions, and so the wind there has a high velocity, high ionization state and low density. The wind is thick ($\tau_L > 1$) in the equatorial region so the wind there has a low velocity, lower ionization, and a higher density. (From Lamers and Pauldrach, 1991)

11.3 The wind compression model

In the remainder of this chapter we consider the two-dimensional flow pattern originating from just above the photosphere of a rotating star, and explore the confluence of the flow from higher latitudes as it converges toward the equatorial region. The wind compressed disk model was developed by Bjorkman and Cassinelli (1993). The basic conclusions for the case of radial external forces were confirmed by the numerical hydrodynamical models by Owocki, Cranmer and Blondin (1994), who also derived additional information concerning the properties of disks. Here we will describe the fundamental ideas underlying wind compression models.

A major simplification is provided by the 'supersonic approximation' in which we ignore the gas pressure gradients at and beyond the sonic point. This approximation has been used in several previous chapters (Chapters 8 to 10). (It is justifiable if the velocity reaches values much larger than the sound speed, a.) This approximation of ignoring gas pressure gradients in the supersonic region is made because we have seen in Chapter 2 that there can be a significant reduction in the effects of gas pressure gradients beyond the sonic point. Here we are assuming that the transition is abrupt from gas pressure being dominant to being negligible.

The supersonic approximation allows us to discuss the outflow as if the particles are on gas-pressure-free trajectories. Here, we are especially concerned with the rotation of the star and the tangential velocity which is imparted to the outflowing particles at the sonic point.

The reason there is a flow toward the equatorial region can be pictured by considering the analogy of the wind particles with satellites lauched from the earth. Consider two identical satellites launched simultaneously from the northern and southern hemispheres at the same longitude and latitude. The trajectory of each satellite orbit lies in an orbital plane that includes the launch point and the center of the earth. Thus the two orbital planes have an inclination relative to the equatorial plane of the earth. All three planes cross each other at a line of nodes. The trajectories of the two satellites will take them to the equatorial plane at the same place and same time and so a collision will occur. The collision corresponds to the shock compression of equatorial gas that we will be discussing below. In the case of a wind, the 'launches' of the gas parcels occur continuously from all locations on the star and at all times, so the equator crossing trajectories inevitably lead either to disks or to some 'equatorial wind compression' of the gas near the equator. We discuss two forms of the equatorial density enhancement: one involves shocks, and is called a *'wind compressed disk'* (or WCD) model, the other is a milder situation, with no shocks, called a *'wind compressed zone'* (WCZ) model.

The ram pressures associated with the shock interface between gas flows from the two hemispheres can produce large equatorial density enhancements, by as much as a factor of 10^2 to 10^3, relative to the densities in the wind from the poles. Such large density increases can produce even larger enhancements in the continuum and line radiation from the wind because the emissivities depend on ρ^2. Thus, rapidly rotating stars with only moderate mass loss rates, such as the Be stars, can have strong Hα emission and large infrared continuum excesses arising from their equatorial disks.

The distortion and polar brightening effects on a star discussed in § 11.1 require rather rapid rotation, with $\omega > 0.7$. This is more rapid than is typically estimated for Be stars. However, wind compression effects can occur even at rotation rates of 0.5 or less, so we will simplify our discussion in this section by assuming that the star is spherical and has a radial external force field, such as a radial radiation field. So we now assume $R_{eq} = R_{pole} \equiv R_*$, and that the mass flux from the sonic point is only weakly dependent on polar angle.†.

11.3.1 The effects of a centrifugal force on the equatorial flow

Before we can discuss the wind compression arising from a flow toward the equatorial plane, we need to understand the more basic centrifugal effects for a wind in the equatorial plane alone. These results will then allow us to treat the orbital effects that lead to a flow towards the equator as occurs in the WCD model. As we have seen in our discussion of magnetic rotator winds in Chapter 9, there are major simplifications associated with equatorial flows. Because of symmetry about the equatorial plane there is no dependence of the momentum equation on the polar coordinate, θ, and because of azimuthal symmetry there is no dependence on the azimuthal angle ϕ. Here, we are ignoring magnetic effects so there are no torques on the flow as it moves away from the star. Therefore, the azimuthal component of velocity, $v_\phi(r)$, is determined by conservation of angular momentum per unit mass,

$$r v_\phi(r) = R_* v_{rot} = \text{constant} \tag{11.13}$$

† The effects of rotational distortion are described by Owocki, Cranmer and Blondin (1994)

The radial motion is determined by the radial external forces and gravity. In the case of early-type stars the dominant wind driving mechanism is the line driving mechanism discussed in Chapter 8. Following Bjorkman and Cassinelli (1993) we will assume that this is radial, and we consider a kinematical model in which the radial velocity of the wind is specified beforehand. A useful parameter in the description of the radial velocity field is the terminal velocity v_∞. Friend and Abbott (1986) and Pauldrach *et al.* (1986), studied the equatorial wind as modified by stellar rotation, as we discussed in § 8.7. Friend and Abbott (1986) gave convenient fits to the terminal velocity derived from their numerical models as a function of $v_{\rm rot}$

$$v_\infty = \zeta\ v_{\rm esc} \left(1 - \frac{v_{\rm rot}}{v_{\rm crit}}\right)^\gamma \tag{11.14}$$

with $\gamma = 0.35$ and with ζ being a constant that depends on stellar parameters. Empirically, for early-type stars, $\zeta = 2.6$ for $T_{\rm eff} > 21\ 000$ K, and $\zeta = 1.3$ for $T_{\rm eff} < 21\ 000$ K as shown in Fig. 11.2. In the equatorial models there are two components of the velocity, the radial component, $v_r(r)$, and the azimuthal component, $v_\phi(r)$. A convenient form for the radial component is the β velocity law

$$v_r(r) = v_\infty \left(1 - \frac{bR_*}{r}\right)^\beta \tag{11.15}$$

(see Eq. 2.3). Setting $v_r(R_*) = a$, the speed of sound, in Eq. (11.15) gives

$$b = 1 - (a/v_\infty)^{1/\beta} \tag{11.16}$$

The solution to the line driving wind equations provides a mass loss rate, which in the equatorial wind case is to be considered the product of the equatorial mass flux, F_m, and the total surface area, $4\pi R_*^2$. Thus we have

$$\dot{M} \equiv 4\pi R_*^2\ F_m = \tfrac{1}{2}\dot{M}_{\rm CAK} \left(1 - \frac{v_{\rm rot}}{v_{\rm crit}}\right)^\xi \tag{11.17}$$

where $\dot{M}_{\rm CAK}$ is given by Eq. (8.105). The factor $\frac{1}{2}$ approximates the finite disk correction to CAK theory as discussed in § 8.9. A fit to the mass loss results of Friend and Abbott for rotating stars gives $\xi = -0.43$.

Let us assume that the solution for a line driven wind that is affected by centrifugal forces is adequately described by the set of equations (11.14) to (11.17). The sonic point is an important reference location for the base of the stellar wind, and for simplicity we assume that the sonic point lies just above the photospheric radius, so $r_s \approx R_*$. The boundary conditions of the equatorial wind at $r = R_*$ and $\theta = \pi/2$ are

$$\begin{aligned} v_r(R_*) &= a \\ v_\theta(R_*) &= 0 \\ v_\phi(R_*) &= v_{\rm rot} \end{aligned} \tag{11.18}$$

Now let us consider trajectories at other latitudes of the star.

11.3.2 Trajectories in the orbital plane

To understand the wind compressed disk model we have to study the trajectories of the gas emitted by the star at any latitude, $\pi/2 - \theta_0$. Therefore we consider the forces and the motion of the gas so as to obtain the velocity, $\mathbf{v}(r, \theta)$, as a function of both the radial distance, r, and the polar angle, θ, measured from the spin axis. The star-centered spherical coordinate system (r,θ,ϕ) is shown in Fig. (11.5). We assume that there is azimuthal symmetry so the wind quantities do not depend on the azimuthal angle ϕ.

Let us consider a fluid element that leaves the star at some latitude other than the equator, with polar coordinates, R_*, θ_0. If we take the base of the outflow to be the sonic radius, we can ignore the effects of gas pressure in the wind, so the only external forces acting on the parcel of matter are gravity and the radiation pressure gradients due to line and continuum opacity. Since we are assuming that each of these forces operates in the radial direction, there is no torque on the material leaving the star, so conservation of angular momentum holds. Therefore, just as is the case for the Keplerian motion of particles, the fluid element will remain in a single plane. This plane passes through the center of the star and contains the initial velocity vector $\mathbf{v}(R_*,\theta_0)$ in the direction of the rotation. We call the plane on which the motion occurs the 'orbital plane',† because of the analogy with Keplerian motion, and we refer to the point of origination of the flow stream as the 'launch point'. Figure (11.6) shows the orbital plane.

Let $\mathbf{L}$ be the constant vector angular momentum per unit mass, as given by

$$\mathbf{L} = \mathbf{r} \times \mathbf{v} = \text{ constant} \tag{11.19}$$

and directed perpendicular to the orbital plane. The component of

† The name *orbital plane* does not imply that the trajectories actually orbit around the star anymore than does a hyperbolic 'orbit' in Keplerian motion imply circumstellar motion.

the angular momentum along any fixed axis is also constant. In particular, the component along the stellar polar axis is the constant L_z, given by

$$L_z = rv_\phi(r)\sin\theta_0(r) = \text{ constant} \tag{11.20}$$

The initial conditions thus determine the orientation of the orbital plane, the value for **L**, and provide the boundary conditions for the velocity structure as a function of radial distance, r. The inital conditions at $r = r_s \approx R_*$ and $\theta(R_*) = \theta_0$, are $v_r(R_*) = a$, $v_\theta \approx 0$, and $v_\phi(R_*) = v_{\rm rot}\sin\theta_0$. The last of these is the azimuthal velocity at the launch point due to the rotation of the star. In regards to v_θ, we assume that below the sonic point the gas pressure gradients are sufficient to keep the matter co-rotating with the star while moving radially outward. So there is no θ-component to the velocity at and

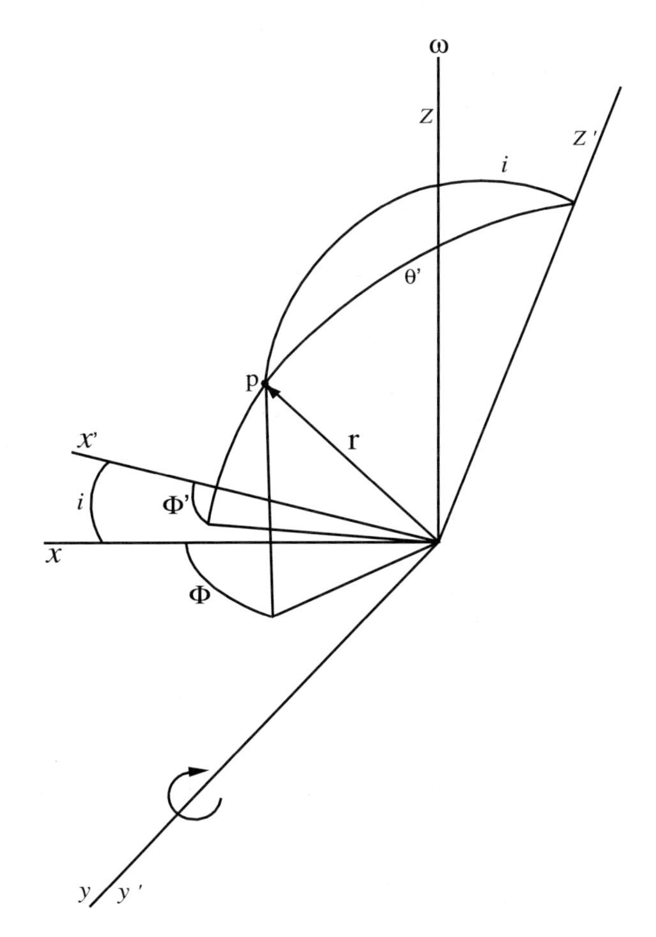

Figure 11.5 Coordinate systems used to describe flow from a rotating star: (1) the (x, y, z) frame is a star-centered system with z along the spin axis. The corresponding spherical polar coordinates are (r, θ, ϕ). (2) The motion of the gas is described in the (x', y', z') orbital frame, formed by a rotation about the $y = y'$ axis by the angle $i = \pi/2 - \theta_0$.

below the sonic point. Then beginning at the sonic point, the gas pressure gradients are considered to be ignorable, so a $v_\theta(r)$ component can begin to develop.

The orbital plane passes through the center of the star and through the launch point. Thus, relative to the equatorial plane, the orbital plane is inclined by an angle

$$i = \frac{\pi}{2} - \theta_0 \tag{11.21}$$

as is shown in Fig. (11.6). Figure (11.5) shows two coordinate systems, the stellar frame (x, y, z) with z in the polar direction, and the orbital plane coordinate frame (x', y', z'). The latter is formed by a rotation about the $y = y'$ axis by the angle i as shown. The polar coordinates in the orbital coordinate system are (r',θ',ϕ'), and are related to quantites in the stellar frame as shown. Clearly, $r' = r$, and $\theta' = \pi/2$ in the orbital plane. The azimuthal angle, ϕ', at any point along the orbit is given by the 'trajectory' in the orbital plane, which we now discuss.

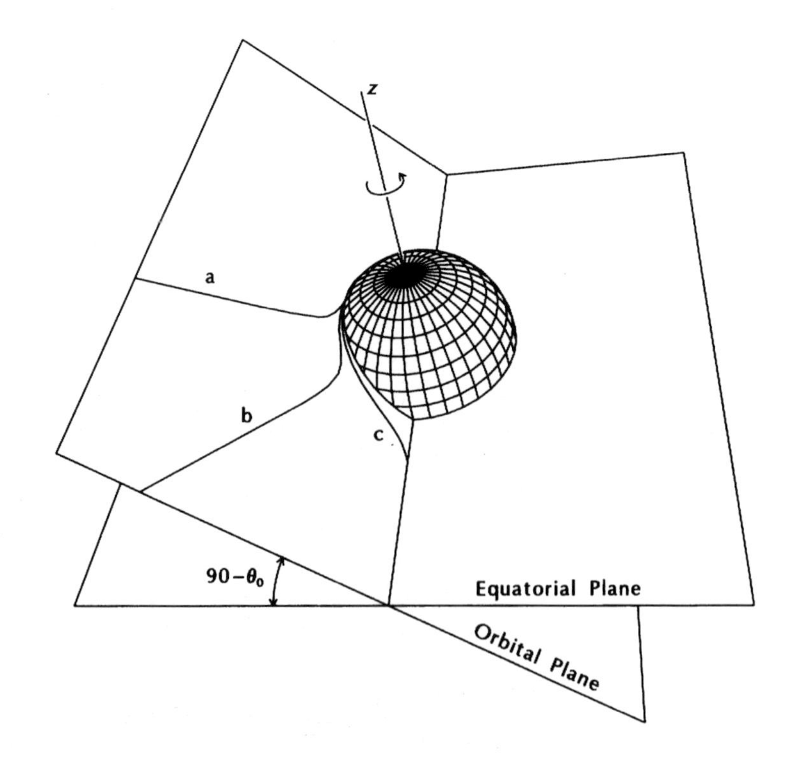

Figure 11.6 The orbital plane for streamlines originating at θ_0. The orbital plane is inclined to the equatorial plane by the angle, $i = \pi/2 - \theta_0$, and ϕ' is the azimuthal angle in the orbital plane coordinate system. Shown are three streamlines: a is the streamline for very small rotation, b corresponds to moderate rotation, c is the trajectory for a streamline crossing the equatorial plane, at $\phi' = \pi/2$ and $\theta = \pi/2$.

The initial velocity components in the orbital plane are

$$v_r'(R_*) = a$$
$$v_\theta'(R_*) = 0 \tag{11.22}$$
$$v_\phi'(R_*) = v_{\rm rot}\sin\theta_0$$

These conditions are for an arbitrary polar angle θ_0. Note that these are nearly identical to those for the motion in the equatorial plane, Eqs. (11.18), except for a modification of v_ϕ by the factor $\sin\theta_0$. Thus the initial conditions in the orbital plane correspond to those in the equatorial plane solution, but for a model with a reduced value for the parameter $v_{\rm rot}$. Furthermore, the external forces of gravity and radiation pressure gradients should be very similar to those used in the equatorial plane solution.† We thus obtain the solution in the inclined plane directly from the equatorial solutions by simply replacing $v_{\rm rot}$ with $v_{\rm rot}\sin\theta_0$, and v_∞ with $v_\infty(\theta_0)$. This gives

$$v_\infty(\theta_0) = \zeta\, v_{\rm esc}\left(1 - \sin\theta_0\,\frac{v_{\rm rot}}{v_{\rm crit}}\right)^\gamma$$
$$v_r'(r,\theta_0) = v_\infty(\theta_0)\left(1 - \frac{bR_*}{r}\right)^\beta \tag{11.23}$$
$$b = 1 - \left(\frac{a}{v_\infty(\theta_0)}\right)^{1/\beta}$$
$$v_\phi'(r,\theta_0) = v_{\rm rot}\sin\theta_0\left(\frac{R_*}{r}\right)$$
$$4\pi R_*{}^2 F_m{}'(R_*,\theta_0) = \tfrac{1}{2}\dot{M}_{\rm CAK}\left(1 - \sin\theta_0\,\frac{v_{\rm rot}}{v_{\rm crit}}\right)^\xi$$

where $F_m{}'(R_*,\theta_0)$ is the mass flux at the initial point of the streamline.

Because there is a v_ϕ' component to the velocity, the flow is not purely radial. We need to find the ϕ'(r) coordinate of the stream trajectory to find whether the flow moves towards the equatorial plane (see Fig. 11.7). The azimuthal angle, ϕ', is measured from the x' axis toward the y' axis, and the y' axis is in the equatorial plane. So the larger the angle ϕ'(r), the closer the material gets to the equator, increasing the wind compression. We call ϕ' the 'displacement angle' of the streamline, and the variation of ϕ'(r) with radial distance gives the trajectory of the parcel of material launched from θ_0.

† The forces along a particular trajectory are not strictly the same as in an equatorial plane trajectory, because the radiation force depends on density, which tend to be higher at low latitudes. We ignore this effect in the trajectory derivation.

To evaluate the effect that the tilted orbital planes have on the overall flow structure, we need to find a solution for the trajectory as a function of r for the mass released at θ_0. A differential equation for $\phi'(\mathrm{r})$ is found by dividing v_r' $(= dr/dt)$ by $v_\phi'(= r d\phi'/dt)$, which gives

$$\frac{dr}{d\phi'} = \frac{r^2}{R_*}\left[\frac{v_\infty(\theta_0)}{v_{\mathrm{rot}}\sin\theta_0}\right]\left(1-\frac{bR_*}{r}\right)^{\beta} \tag{11.24}$$

Integrating this, we find

$$\phi'(r) = \left(\frac{v_{\mathrm{rot}}\sin\theta_0}{v_\infty(\theta_0)}\right)\frac{1}{b(1-\beta)}\left[\left(1-\frac{bR_*}{r}\right)^{1-\beta} - (1-b)^{1-\beta}\right] \tag{11.25}$$

Letting $r \to \infty$, we find that ϕ' of Eq. (11.25) has an asymptotic value, ϕ'_{max} given by

$$\phi'_{\mathrm{max}} = \left(\frac{v_{\mathrm{rot}}\sin\theta_0}{v_\infty(\theta_0)}\right)\frac{1}{b\,(1-\beta)}\left[1-(1-b)^{1-\beta}\right] \tag{11.26}$$

The fact that ϕ' has an asymptotic value means that at large radius the trajectory becomes radial in the orbital plane due to the radiation forces.

Figure (11.7) illustrates trajectories calculated with increasingly large rotation rates. The trajectories consist of two components: a tangential component that tends to make a trajectory wrap around the star, and a radial component due to line radiation forces. For a small rotational velocity or a large radiation force, the trajectories quickly become radial (e.g., trajectory a). For a large rotational velocity or a small radiation force, the trajectories wrap farther around the star before they become radial (e.g., trajectories b and c).

As a consequence of the asympototic behaviour of the trajectory, Eq. (11.26), we can identify solutions which are distinguished by small versus large values for ϕ'_{max}. The trajectories develop from being radial, with $\phi'_{\mathrm{max}} = 0$ for $v_{\mathrm{rot}} = 0$, to being more strongly curved as seen in trajectories (a), (b) and (c), which correspond to rotation rates of 20 %, 40 %, and 60 % critical, respectively. Trajectory (c) crosses the equator when the parcel has travelled in the ϕ' direction to $\phi' = \pi/2$. This is because the launch point is on the x' axis and the equator is along the y' axis, which is of course an angle $\pi/2$ away from the x' axis. All of the trajectories eventually become radial along the ϕ'_{max} direction. This because the angular velocity decreases ($\sim r^{-1}$) as the gas moves away from the star with constant angular momentum per unit mass. At the distance where $v_r = v_\phi$, the radial flow begins to dominate over the tendency for the trajectory to wrap towards the equator. Trajectories that are immediately deflected in the radial direction by the outward radiation forces will never reach the equator, while trajectories that start with long segments with $v_r < v_\phi$ wrap farther around the star and may reach $\phi' > \pi/2$ before being deflected in the purely radial direction.

We can identify four ways to obtain large versus small values of $\phi'_{\rm max}$, by considering the terms in Eq. (11.26):

(1) Increase the rotation speed, $v_{\rm rot}$. The larger the rotation rate, the larger the deflection ϕ', as we see in going from curves (a) to (c) in Fig. (11.6) and in the trajectories in Fig. (11.7).

(2) Decrease the launch latitude. Trajectories originating nearer the equator have larger values of $\sin\theta_0$, so these are more likely to cross the equator than are trajectories starting nearer the pole where $\theta_0 = 0$. A launch point nearer the equator has a larger initial transverse velocity.

(3) Decrease the terminal velocity. A trajectory with a smaller terminal velocity, has a more extended region over which its speed is less than the escape speed, so it is more affected by gravity, which causes the material to flow toward the equatorial region. O stars which have large terminal velocities (typically $\sim 3\ v_{\rm esc}$) are likely to have trajectories such as (a) or (b), and significant wind compression occurs only if their rotation rates are extremely close to critical (breakup) speed. The main sequence B stars typically have smaller terminal speeds ($\leq v_{\rm esc}$) and are more likely to have equator crossing trajectories such as (c).

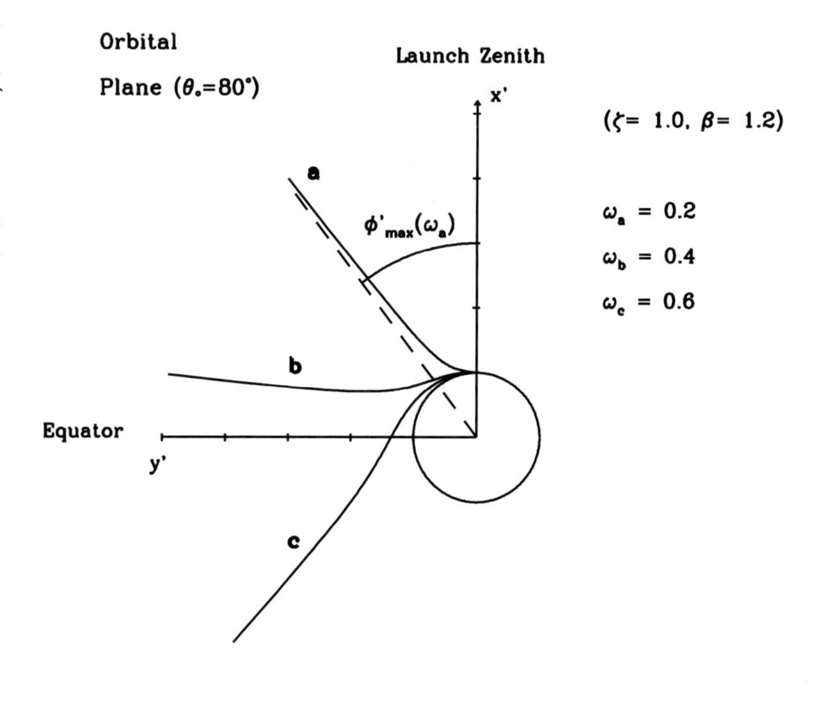

Figure 11.7 Trajectories, $\phi'(\rm r)$, in the orbital plane, of material that was launched at the base of the wind at R_* and at polar angle θ_0=80 deg from a B2V star with a $\zeta = 1, \beta = 1.2$ velocity law. The solid curved lines show the trajectories, and the dashed radial line shows the asymptotic direction of the a trajectory. The three trajectories have the $\omega = v_{\rm rot}/v_{\rm crit}$ values indicated on the right side of the figure.

(4) Increase the value of the velocity law exponent β. Large β values correspond to slow accelerations near the star. The slower the outward acceleration, the more the parcel is affected by gravity, and the more likely the trajectory will be like (c), while fast acclerations are more likely to produce trajectories like (a).

11.3.3 Transformation to stellar coordinates

With the orbital plane equations (11.23) and (11.25) we can describe the position and velocity components as a function of r and ϕ' in the orbital plane. We have found that the rotation leads to a flow toward the equator, so wind compression must be occuring. To derive the density distribution, $\rho(r, \theta)$, we need to determine the trajectory in the (unprimed) stellar coordinate system.

This transformation can be carried out using the spherical triangle that is shown in Fig. (11.8a). Using the laws of cosines and sines for spherical triangles, (and noting that $\theta' = \pi/2$) leads to the following relations.

$$\cos\theta = \cos\theta_0 \cos\phi' \tag{11.27}$$

$$\sin\phi = \frac{\sin\phi'}{\sin\theta} \tag{11.28}$$

$$\cos\psi = \frac{\sin\theta_0}{\sin\theta} \tag{11.29}$$

The transformation of the orbital velocities to the stellar coordinate system is illustrated in Fig. (11.8b). Using the angles on this figure we

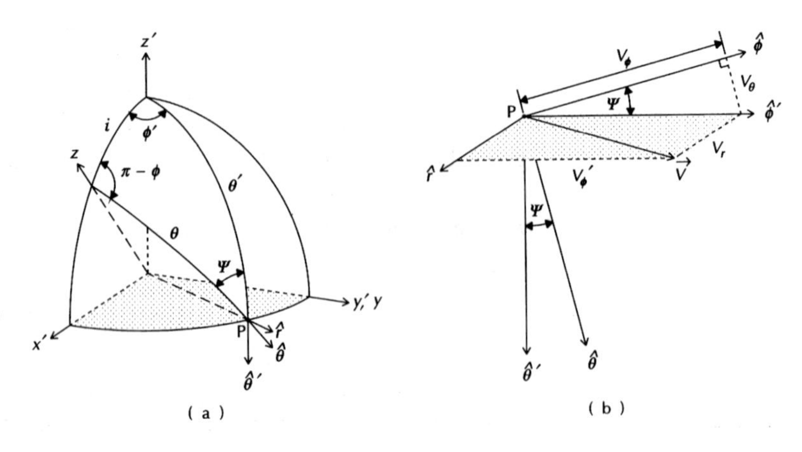

Figure 11.8 (a) The spherical triangle used to make the transformation from the orbital plane (primed) coordinates to the stellar (unprimed) coordinates. (b) The components of velocity in the orbital plane (dotted) and stellar coordinates.

obtain the following velocity components of the trajectories

$$
\begin{aligned}
v_r(r) &= v_r'(r) \\
v_\phi(r) &= v_\phi'(r)\cos\psi = \frac{v_{\rm rot}R_*}{r}\frac{\sin^2\theta_0}{\sin\theta} \\
v_\theta &= v_\phi'\sin\psi = \frac{v_{\rm rot}R_*}{r}\frac{\sin\theta_0\ \cos\theta_0}{\sin\theta}\sin\phi'
\end{aligned}
\tag{11.30}
$$

Here we have used the conservation of angular momentum to eliminate v_ϕ' in favor of $v_{\rm rot}$. The presence of a positive value for v_θ, in the stellar coordinate frame, means that there is a flow toward the equatorial plane, and we can see from Eq. (11.30) that the velocity of the flow depends on the factor $v_{\rm rot}\ R_*/r$. Thus the flow toward the equator is large for rapid rotators and at radii near the star.

With the equations above one can choose a location (r, θ, ϕ), in the flow in the star frame, and iteratively solve Eq. (11.25) and Eq. (11.27) for the coordinates $(R_*, \theta_0, 0)$, from which that parcel of the flow originated. This is of interest because the origin, or launch point, is where the mass flux along any streamline is specified.

Let us now find the density at locations (r, θ) in the stellar frame of reference. Since the fluid parcels that pass through the equator shock are no longer on free streamlines, we only treat the densities in the zones *before* the equatorial region is reached. We call the trajectories that never reach the equator the 'non-crossing trajectories', (e.g., trajectories a and b in Fig. 11.7).

11.3.4 Wind density in the non-crossing trajectories

The density, $\rho(r, \theta)$, is determined from the conservation of mass, given the velocity and the cross sectional area of the streamline at (r, θ). Because of the tendency for the material to flow toward lower latitudes (i.e. larger θ), the flow streams are not radial, nor does the density decrease as $r^{-2}v^{-1}$, as in the spherically symmetric wind models. At any given radius, r, the cross sectional area of a flow tube is given by $r^2 \sin\theta d\theta d\phi$. Multiplying this by the local value of the mass flux, ρv_r, and equating it to the rate at which matter enters the flow tube at the base leads to the mass conservation equation,

$$\rho v_r r^2 \sin\theta d\theta d\phi = \rho_0 v_{r0} R_*^2 \sin\theta_0 d\theta_0 d\phi_0 \tag{11.31}$$

$$= R_*^2 F_m'(\theta_0) \sin\theta_0 d\theta_0 d\phi_0 \tag{11.32}$$

The flow is azimuthally symmetric at all radii, so a flow tube that subtends an azimuthal angle $d\phi_0$ at the launch point will subtend the same azimuthal angle farther along in the flow, hence $d\phi = d\phi_0$. If we let $\mu = \cos\theta$, and solve Eq. (11.31) for the density, we find

$$\rho(r,\theta) = \frac{R_*^2 F_m'(\theta_0)}{r^2 v_r(r)} (d\mu/d\mu_0)^{-1} \tag{11.33}$$

The factor $(d\mu/d\mu_0)^{-1}$ corresponds to the decrease in the flow tube solid angle relative to the spherically symmetric flow case.

If we assume for simplicity that the wind straight out of the pole has the density distribution of a non-rotating star, and that the mass flux F_m' is independent of θ_0, the density contrast from pole to equator is given by

$$\rho_{\rm eq}(r) \approx (v_p(r)/v_{\rm eq}(r))(d\mu/d\mu_0)^{-1} \rho_p(r) \tag{11.34}$$

where the subscripts, 'eq′ and 'p′, refer to equator and pole respectively, and $v_{\rm eq}(r) \equiv v_r(r, \pi/2)$.

The density, Eq. (11.33), and the equator to pole density contrast, Eq. (11.34), that occur in the two-dimensional flow from a rotating star, both depend on $d\mu/d\mu_0$. This is because the flow tubes out from the equator increase in area more slowly with distance than $\sim r^2$, while flow tubes out from the polar region increase more quickly than that. Hence the density in a tube from the equator becomes larger than in tubes from higher latitude flows. An expression for the non-radial divergence of the tube solid angles can be derived from Eq. (11.27), giving

$$\frac{d\mu}{d\mu_0} = \cos\phi' - \mu_0 \sin\phi' \frac{d\phi'}{d\mu_0} \tag{11.35}$$

$$= \cos\phi' + \frac{\cos^2\theta_0}{\sin\theta_0} \frac{d\phi'}{d(\sin\theta_0)} \sin\phi' \tag{11.36}$$

The derivative in the last term in this equation can be found from a straightforward (but tedious) differentiation of the expression for ϕ' in Eq. (11.25). It is a lengthy expression† because θ_0 appears in the v_∞ and b factors in ϕ', Eq. (11.25). For the case with $b = 1$ we get

$$\frac{d\mu}{d\mu_0} = \cos\phi' + \cot^2\theta_0 \left(1 + \gamma \frac{\sin\theta_0 \, v_{\rm rot}/v_{\rm crit}}{1 - \sin\theta_0 v_{\rm rot}/v_{\rm crit}}\right) \phi' \sin\phi' \tag{11.37}$$

Using this with Eq. (11.33) gives the density as a function of r and θ.

At the equator, where $\theta_0 = \pi/2$, we find $(d\mu/d\mu_0)^{-1} = 1/\cos\phi'$. This is finite if the deflection angle, $\phi'(r, \pi/2)$, is less than $\pi/2$, in which case there will be a significant increase in the density contrast; Eq. (11.34). If $\phi'(r, \pi/2)$ approaches the value $\pi/2$, the equatorial density will tend

† given in Petrenz and Puls (1996)

to become infinite because $d\mu/d\mu_0 \to 0$. This situation corresponds to streamline trajectories crossing one another, so shock formation occurs. We will now consider models both with and without the shock formation.

11.3.5 Model results

The behaviour of the streamlines depends sensitively on the ratio $\zeta \equiv v_\infty/v_{\rm esc}$. If this ratio is large the compression of the flow toward the equatorial region is small, but if the ratio is small, there can be significant compression and streamline trajectories that cross the equator. Figure (11.9) shows streamlines that originate at 5 deg intervals on the stellar surface for six values of $\omega \equiv v_{\rm rot}/v_{\rm crit}$, with $\zeta = v_\infty/v_{\rm esc} = 1.4$. There is no compression in the azimuthal direction, so only the r and θ values along the streamline need to be shown to illustrate the compression as the motion toward the equator occurs. Note, the lines are not actually streamlines since the wrapping in the ϕ direction is not shown, but they denote the surfaces of revolution of the streamlines or 'flow sheets' from the star. For zero rotation the lines are radial. However, even for moderate values of ω there is significant compression toward the equatorial plane, even if equator crossing does not occur. Note that for ω larger than about 0.5, there is significant depletion of material in the polar regions and there is a zone of stellar latitudes from which the lines cross the equatorial plane.

The wind compression can be explained by a consideration of the trajectories as we have seen in the kinematic description of the previous section. However, it is also useful to consider the cause of the wind compression from the dynamical point of view, accounting for the forces that are operating on the gas as it flows from the star. Figure (11.10) shows four forces; the pressure gradient, radiation, gravity, and the centrifugal force, as a function of r for a model of a Be star. The upper panel shows these forces for a non-rotating star. There we see that the radiation forces already exceeds gravity at 1.01 R_*, where there is also a rapid decrease in the gas pressure because the flow is going transonic. To maintain the smoothly accelerating flow, the radiation force must support the flow once the gas pressure gradient has become negligible. In the case of the rotating star, shown in the lower panel, gravity exceeds radiation out to about 3 R_*, far beyond the sonic point. As a result there is a net component of the radial force toward the equator over a sufficiently large portion of the trajectory that the material can reach the equatorial plane. Gravity has a negative z-component, and the dominance of gravity is the physical explanation for why the gas parcels move toward the equator instead of being pushed outward.

To maintain the smoothly accelerating outflow in the rotating case, radiation does not have to quickly become the dominant force because the centrifugal force can also support the flow. Note that it is only after

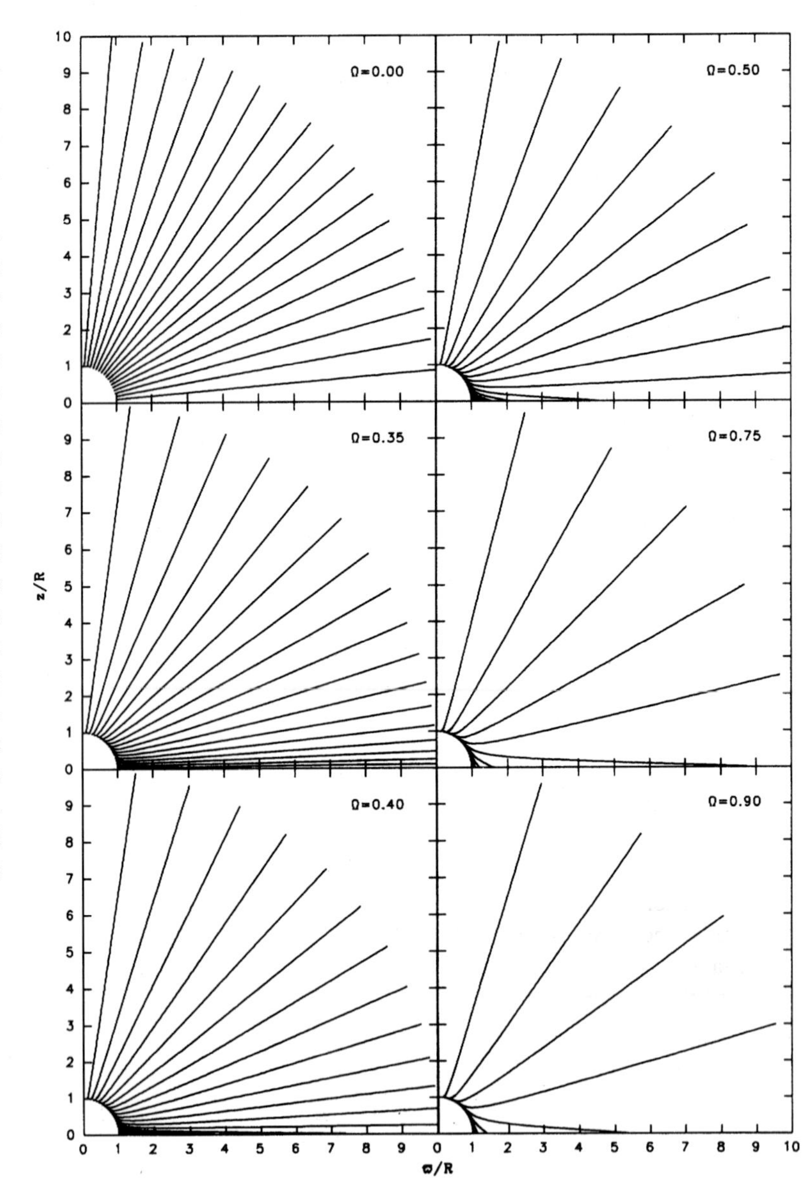

Figure 11.9 Streamlines for various values of the rotation rate $\omega \equiv v_{\rm rot}/v_{\rm crit}$ for a B2 star with $\zeta = 1.4$. The streamlines originate in 5 deg increments in latitude on the surface of the star. The distance between the streamlines thus indicates the compression produced by the rotation. Note that for $\omega > 0.4$ the streamlines cross the equator, forming a disk. For larger rotation rates, streamlines from a larger fraction of the star enter the disk. (From Bjorkman and Cassinelli, 1993)

the centrifugal force is lost that the radiation pressure provides the dominant force. By that time the trajectory has wrapped sufficiently around the star that the equator crossing and shock compression are inevitable. The centrifugal force is largest for material launched in the equatorial plane, and it decreases with increasing latitude. Thus by considering the dynamics of the outflow we again see that there is a range of latitudes from which we should expect equator crossing orbits.

Before discussing other results and applications of the equatorial wind compression, let us discuss the effects of the equator crossing trajectories that occur in the rapid rotation cases.

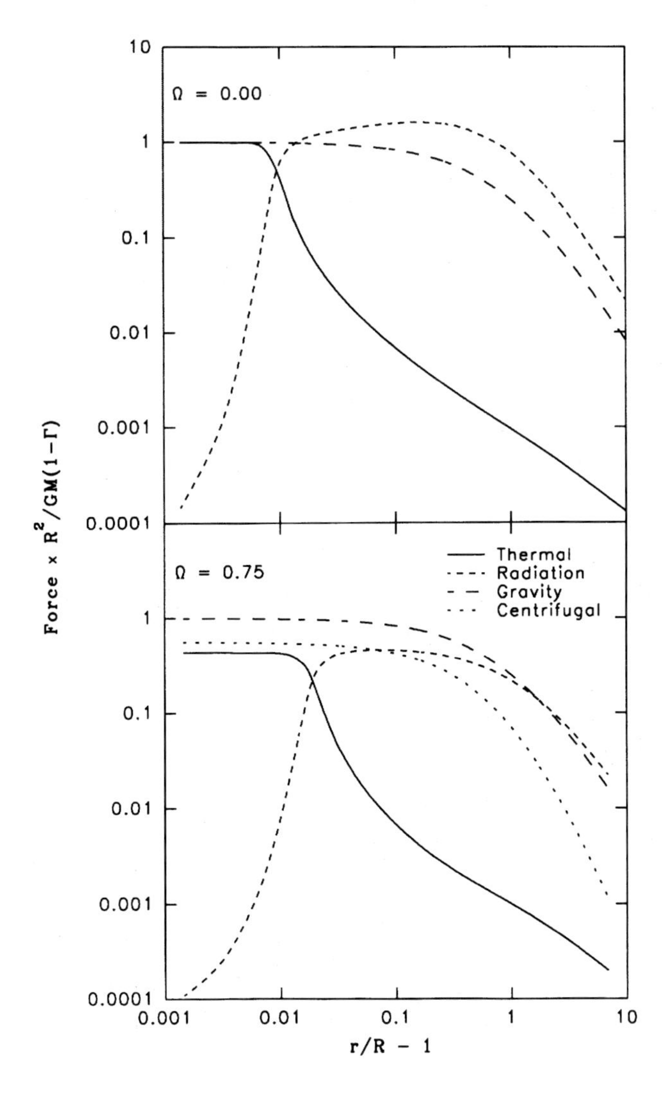

Figure 11.10 The forces versus radius in the equatorial plane. (a) Upper part shows forces for a non rotating star, the thermal (gas pressure), gravitational and line radiation forces. (b) Lower part shows the forces for a rotating model of the same star, but also shows the centrifugal force. Of particular interest is the radius out to which the gravitational force exceeds the outward force of radiation, because that radius determines the distance for which there is a net force taking material toward the equator. (From Bjorkman and Cassinelli, 1993)

11.4 Disk formation

In our consideration of isolated streamlines, we have found solutions in which the flow crossed the equatorial plane. This occurs if $\phi'_{max} > \pi/2$ (Eq. 11.26). Near the equatorial region, flow from the upper hemisphere would encounter flow from the lower hemisphere. Since streamlines cannot cross each other, the large density that is produced where the streamlines meet creates a pressure that deflects each of the streams. Where this occurs, our basic approximation of freely flowing fluid elements breaks down. Nevertheless, some properties of the interaction region can be derived based on considerations of the momentum of the incident winds. If the flow from the hemispheres has a supersonic component of velocity perpendiclar to the equatorial plane, a shock zone or 'disk' will develop, bounded on both the upper and lower hemispheres by oblique shocks. This is illustrated in Fig. (11.11). The pressure in the disk must balance the z-component of the momentum flux entering the disk. This will lead to a significant compression, in which the density in the disk exceeds the density in unshocked material at the same radial distance from the star by factors of 10^2 to 10^3. One can also expect that the shocks at the upper and lower boundary of the disk will produce temperatures well above the effective temperature of the star. This is of interest because it could explain the superionization seen in the UV spectra of Be stars.

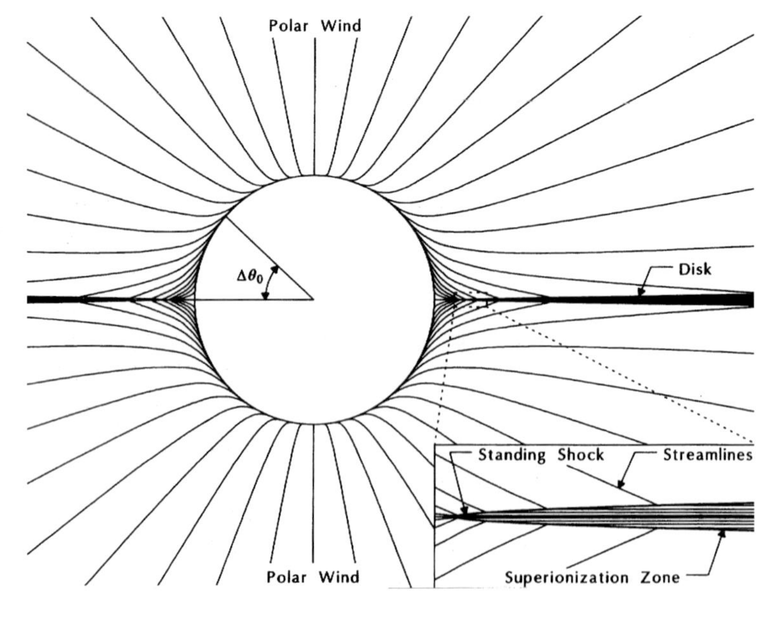

Figure 11.11 This figure illustrates the production of a dense disk by streams from opposite hemispheres that terminate at oblique shocks. These results are for the wind of a B2 star with $\omega = 0.5$. The shock would lead to temperatures of about 10^5 K and produce superionization near the equatorial plane. (From Bjorkman and Cassinelli, 1993)

An important property of WCD models is that the enhanced emission lines and continuum are produced *without* an enhancement in the mass loss rate from the star. It is the focusing of the wind material that originates at higher latitudes that leads to the disk. In the case of the Be stars the mass loss is produced by the line driven wind mechanism, so we have a good estimate of the total mass loss for a Be star of a given spectral type. If the star is rotating sufficiently rapidly to form a disk, most of the wind material will join the disk. Then, conservation of mass provides a constraint on the product of the disk density, ρ_d, velocity, v_d, and the solid angle, Ω_d, of the disk at some large radius, r.

$$\rho_d v_d \Omega_d \approx \rho_w v_w 4\pi \tag{11.38}$$

where the quantities on the right are the density, velocity and solid angle of the wind from the non-rotating model of the star. As an example, let us assume that the strong emission features of the Be star require that the density in the disk be enhanced by a factor of 10^3, relative to a non-rotating version of the star, and since the flow speeds in disks have been estimated to be about a factor of 10 lower than in the wind, we can then deduce from Eq. (11.38) that the opening angle of the disk corresponds to angles of order one degree! Detailed hydrodynamic models also give small opening angles. This prediction of small opening angles by WCD models is confirmed by polarization and interferometric observations of Be stars (Wood, Bjorkman and Bjorkman, 1997; Quirrenbach *et al.*, 1997).

11.4.1 Wind compressed zone (WCZ) models

Interesting wind compression effects can arise in stars that do not have radiation driven winds and in cases in which shock compressed disks do not form. Models in which rotation does not lead to equator crossing streamlines are called wind compressed zone (WCZ) models instead of wind compressed disk models. In the WCZ models there is no disk formation, but the density in a broad equatorial zone can still be enhanced due to the orbiting of the streamlines toward the equatorial latitudes. As a result, WCZ models can have observational consequences because of the ρ^2 dependences of emission in the Balmer lines and IR continuum. Also, by considering the more slowly rotating WCZ model, we find applicatons to stars that may be rotating at speeds below that required to form a disk.

In the case of the Be stars, the shocked disks seem to be a required feature because of the very large IR excesses and line emission fluxes they emit in spite of their having relatively weak winds intrinsically. However, there are other stars for which WCZ density enhancements can occur and may be important: (1) Wolf-Rayet stars that sometimes show intrinsic polarization which suggests rotational distortion; (2) the B[e] supergiants; (3) central stars of planetary nebulae that show bipolar structures (Frank & Mellema, 1994); (4) other stars such as novae and even asymptotic giant branch stars, that might be able to produce wind compressed zones. These are discussed by Ignace, Cassinelli and Bjorkman (1996).

Figure (11.12) shows the dependence of the wind compression effects on the velocity parameter β, in Eq. (11.23). The larger the value of β the more slowly the material accelerates away from the star 6 (Fig. (2.1)), so the more it can be affected by gravity. In the case of the WN5 star shown in the figure, note that for $\beta = 2$, the rotation rate needed to form a density enhancement by a factor of 3 is $\omega < 0.2$, and 20 % critical is not an implausible rotation rate for an evolved early type star.

Figure (11.13) shows the density enhancment relative to a non-rotating star as a function of polar angle θ_0. Note that there is a reduction of the density in the polar regions and an enhancement in the equatorial regions. Significant density increases occur even at relatively low rotation speeds.

The WCZ calculations have led to several general conclusions. If the wind accelerates rapidly away from the star, as is the case for most

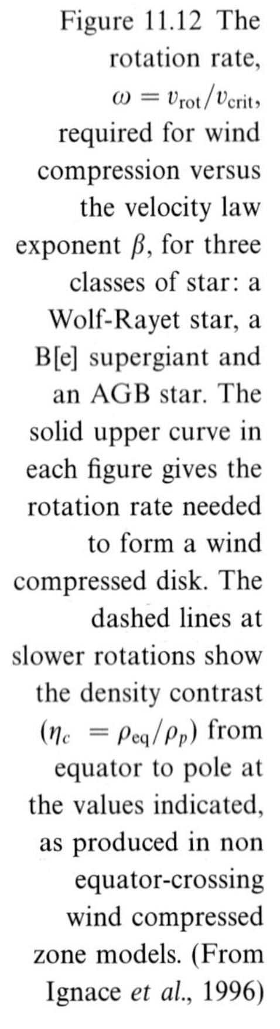
Figure 11.12 The rotation rate, $\omega = v_{\rm rot}/v_{\rm crit}$, required for wind compression versus the velocity law exponent β, for three classes of star: a Wolf-Rayet star, a B[e] supergiant and an AGB star. The solid upper curve in each figure gives the rotation rate needed to form a wind compressed disk. The dashed lines at slower rotations show the density contrast ($\eta_c = \rho_{\rm eq}/\rho_p$) from equator to pole at the values indicated, as produced in non equator-crossing wind compressed zone models. (From Ignace *et al.*, 1996)

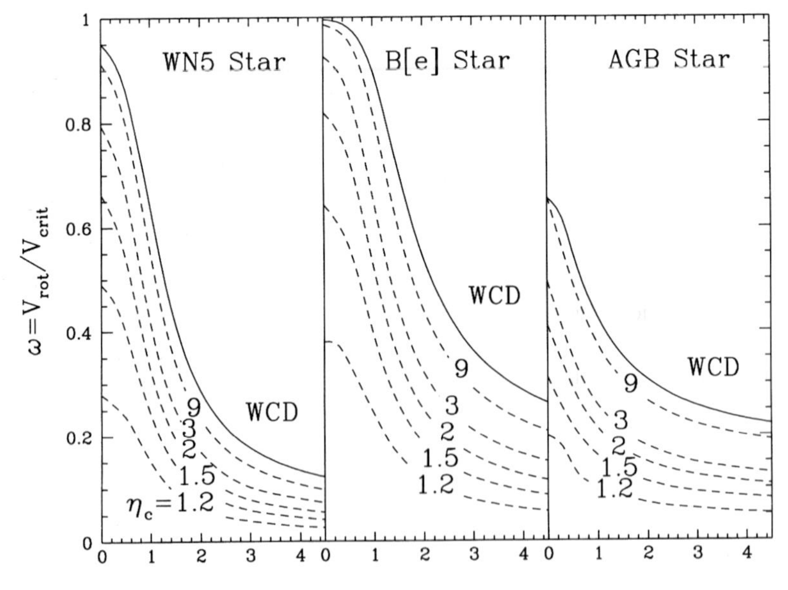

O-type stars, significant wind distortions will arise only if the star is rotating extremely rapidly, typically > 80 % critical, which for an O9V star corresponds to about 440 km s^{-1}. However, significant wind distortions can occur for slow to moderate stellar rotation rates if the wind accelerates slowly from the star. This can occur if the star has a small terminal velocity and has an extended region over which the wind speed is less than the velocity of escape. Such conditions hold for the Be stars and the B[e] supergiants. Even if v_∞ is large, as is the case for Wolf-Rayet stars, there can still be significant distortion of the wind if the velocity gradually increases with radius, as in a velocity law with a large β.

11.5 2-D hydrodynamical results

11.5.1 Results of hydrodynamical models assuming forces are radial

Once the material in a WCD model passes through the compression shock and enters the equatorial region, the supersonic approximation is no longer valid. After passing through a shock, the material is subsonic and gas pressure gradients determine the flow streamlines. A proper treatment of the disk properties requires a hydrodynamical study, as carried out by Owocki, Cranmer and Blondin (1994). Figure (11.14) illustrates the structure of the disk and wind region

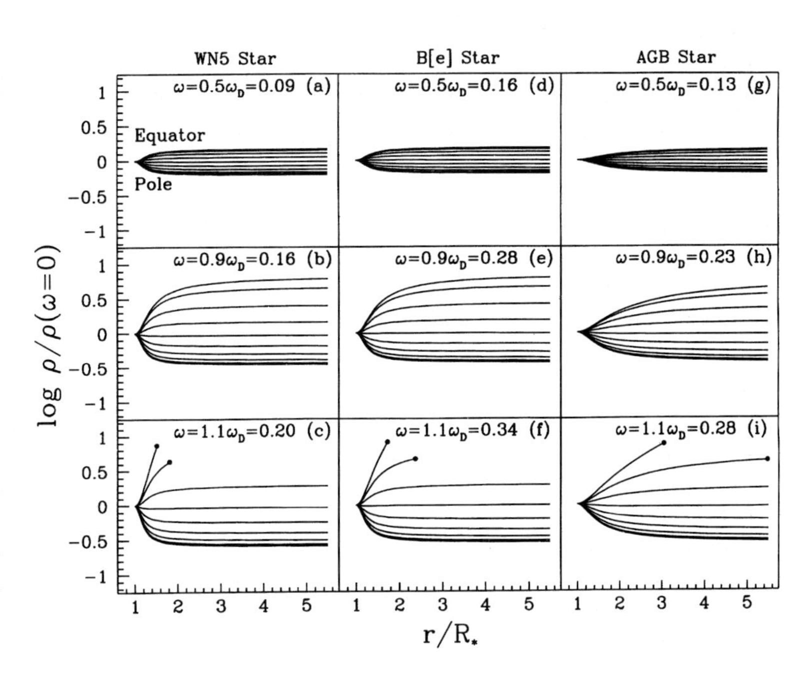

Figure 11.13 The ratio of the density in wind-compressed models to those in non-rotating models, as a function of radius, for $\beta = 3$, for the same stars as in Fig. (11.12). The density is computed at three rotation rates, ω, given both in terms of the rotation rate needed to form a disk, and as the (v_{rot}/v_{crit}) ratio. The ten lines in each panel correspond to streamlines originating at 10 deg increments in latitude, with the lower line corresponding to flow from the pole (θ_0=0 deg), and the upper to the equator. Note that even small rotation rates ($\omega \sim 0.2 - 0.3$) produce large equator to pole density contrasts. The dots on the lowest panels mark where streamlines enter a WCD disk. (From Ignace *et al.*, 1996)

of a star with the parameters of a B2.5e star derived from their 2-D hydrodynamical simulation. The gray scale indicates the density in the wind; superimposed are the flow vectors. The vectors above and below the high density equatorial region illustrate the wind focusing induced by the rotation.

A interesting property of the disk region that could not be derived from the kinematical/analytical approach, described in the previous sections, is the *inflow plus outflow* velocity structure in the disk. There is a stagnation point in the disk marked by the $\times$ where the radial flow is zero. Beyond the stagnation radius the material in the disk flows outward. However, in the inner region of the disk, material has a negative velocity, and therefore re-accretes back onto the star. A

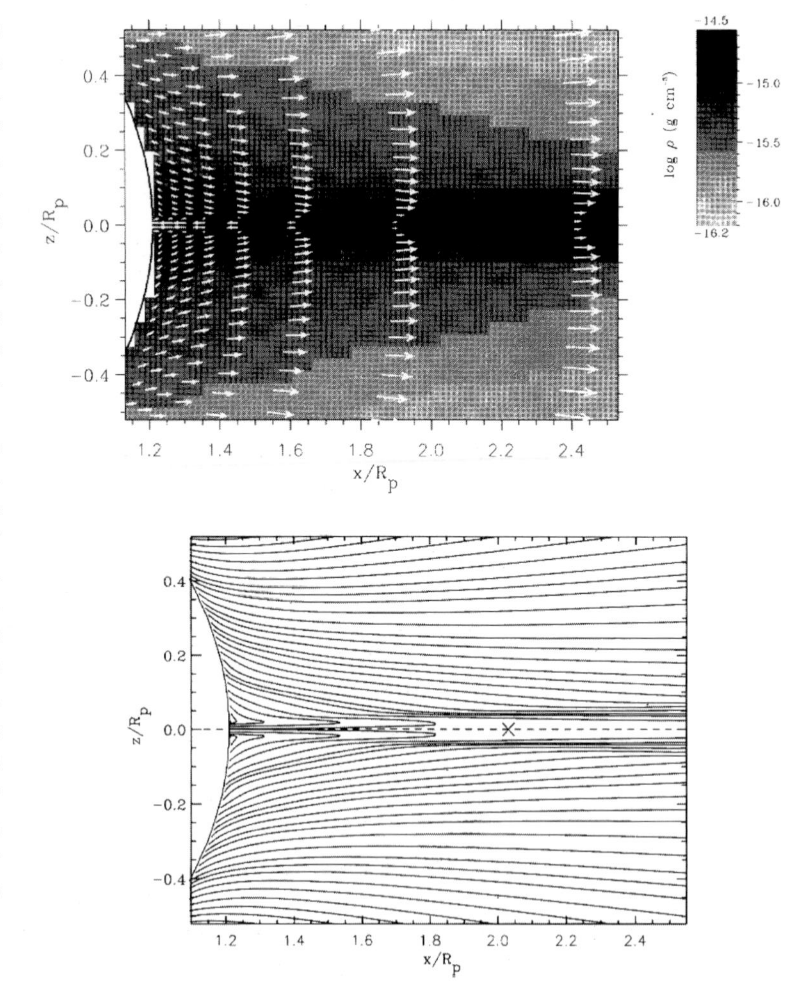

Figure 11.14 (a) The structure of the equatorial region of a model for a B2.5e star derived from a numerical hydrodynamic calculation. The gray scale shows the density, and the vectors show the magnitude and direction of the flow. (b) The streamlines for the same model showing the 'circulation flow' back onto the star in the equatorial disk. The streamlines originate at intervals of 1 deg in latitude. The $\times$ marks the stagnation radius where the radial flow speed is zero. (From Owocki, Cranmer and Blondin, 1994)

circulation flow exists in which all of the material from the star up to a latitude of 5 deg will flow back onto the star. The reason for the inflow and circulation is explained by considering the line acceleration g_L.

As discussed in Chapter 8, (Eq. 8.77), the line acceleration is inversely proportional to density, so in the higher density region near the equator the acceleration will be small. Thus the radiative acceleration is incapable of driving the wind out against the acceleration of gravity. Moreover, the radiative acceleration depends on the velocity gradient. For the disk region the velocity gradient becomes small or negative and that further reduces the line force. Figure (11.14b) shows the flow streamlines from the star. This clearly illustrates the circulation flow that takes material back onto the star.

A basic assumption of the wind compression theory is that the forces are radial. For line acceleration the forces depend sensitively on velocity gradients and these are especially complicated in rotating winds. Owocki *et al.* (1996) have shown that line accelerations have a component in the poleward direction that can inhibit the flow towards the equatorial plane. However, if the density in the equatorial plane is sufficiently increased by the rotation induced bi-stability of the wind, the line acceleration cannot prevent the formation of a wind compressed disk (Owocki *et al.*, 1998).

11.6 Conclusions

In this chapter we have described two models for explaining an enhancement of the density in the equatorial region. The first is the rotationally induced bi-stability (RIB) model of Lamers and Pauldrach (1991). This model uses the predicted and observed properties of line driven winds, which show that for early-B stars the wind can exist in two forms: (1) a fast low velocity wind with a high degree of ionization if the Lyman continuum is optically thin, (2) a slower high density wind with a low degree of ionization if the Lyman continuum is optically thick. The RIB model uses the properties of rotating stars such as the von Zeipel theorem to show that the bi-stability jump of a rotating star may occur between the pole and the equator. This will lead to a high density and low velocity wind from the equator and a low density and high velocity wind from the pole. The density contrast is about a factor of ten. This contrast is similar to what is observed for the B[e] supergiants. The RIB model cannot explain the disks of Be stars, which have a much higher density contrast.

The second model is the wind compressed disk model of Bjorkman and Cassinelli (1993), and the related wind compressed zone model of Ignace, Cassinelli and Bjorkman (1996). The wind compression models assume that the mass loss is the same from all latitudes of a rotating star, but use the wrapping of the gas trajectories toward the equator to produce a density contrast between the pole and equatorial region.

We derived equations for determining the two-dimensional structure of the wind from a rotating star. We have shown that it is possible to use the velocity and mass loss rates that have been obtained from purely equatorial wind models to describe the outflow as a function of latitude on a rotating star. The essence of the wind compressed disk model is that the fluid beyond the sonic point is treated as being in parcels that are not affected by gas pressure. The forces operating on a parcel are the inward force of gravity and the outward forces of line driven wind theory. To the extent that all of these forces are radial, the fluid follows trajectories that satisfy conservation of angular momentum. The parcels would then remain in their initial orbital plane, that was set by the launch conditions at the sonic point. The conclusion that the orbital planes can cross the equatorial plane of the star is what leads to wind compression.

Even for stars that have only trajectories of the non-equator crossing type, there can arise significant density enhancements relative to the polar regions of the star. Somewhat surprisingly, such density enhancements can occur in stars that are rotating at speeds well below the critical or break-up speed, if the radial acceleration of the wind is slow ($\beta > 1$).

The mass loss rate from a star with a wind compressed disk model is the *same* as that without a disk. This means that for early type stars the mass loss rate is essentially that of the radiative driven wind models. In the WCD and WCZ models the outflowing gas is only *redistributed* around the star, with a density concentration near the equator and a decrease in density near the poles.

The wind compression effects appear to go far toward explaining a long standing puzzle regarding the Be stars. Purely equatorial models, in which material was spun-out from the equator, showed that even moderate enhancements of the density near the equatorial plane would require the stars to be rotating close to breakup speed, $\omega > 0.8$ or so. However, observations persistently indicated actual rotation rates of $0.4 < \omega < 0.8$. Accounting for the two dimensional nature of the flow helps to solve the Be star problem, because the material in the equator does not simply come from the equatorial region of the photosphere but from wind emitted over a broad range of latitude.

The differences between the RIB and the WCD model are:

(a) In the RIB model the disk is due to the higher equatorial mass flux, whereas in the WCD model the disk is due to the compression of the wind towards the equator.

(b) The RIB model can give at most a factor of ten density contrast from equator to pole and works for B supergiants with high mass loss rates. The WCD model can produce a higher density contrast, by an orbital effect that is independent of mass loss rate.

The combination of the bi-stability mechanism and the WCD mechanism will result in even higher density contrasts: the bi-stability producing a higher mass flux from the equator, and the WCD effect concentrating the flow towards the equator.

11.7 Suggested reading

Collins G.W. 1989, *The Fundamentals of Stellar Astrophysics.* (San Francisco: Freeman)
(This book has sections on rotating stellar interiors and atmospheres, and discusses the von Zeipel theorem, Eddington-Sweet circulation currents and the intrinsic polarization of rotating stars.)

Clayton, D.D. 1968, *Principles of Stellar Evolution and Nucleosynthesis* (New York: McGraw Hill)
(Chapter 6 has a discussion the Properties of Rotating Stellar Envelopes.)

The rotation induced bi-stability model was developed in two papers:

Pauldrach, A.W.A. & Puls, J. 1990, 'Radiation-driven Winds of Hot Luminous Stars VIII, The Bistable Wind of the Luminous Blue Variable P Cygni (B1 Ia$^+$)' *A & A* **237**, 409
(The discovery of the bi-stability of stellar winds)

Lamers, H.J.G.L.M. & Pauldrach, A.W.A. 1991, 'The Formation of Outflowing Disks around Early-type Stars by Bi-stable Radiation-driven Winds,' *A & A* **244**, L5
(This paper describes the rotation induced bi-stability model, and applies it to B[e] supergiants.)

Four papers regarding the wind compressed disk theory:

Bjorkman, J.E. & Cassinelli, J.P. 1993, 'Equatorial Disk Formation Around Rotating Stars due to Ram Pressure Confinement by the Stellar Wind.' *Ap. J.* **409**, 429
(This paper presents the analytic WCD theory.)

Ignace, R., Cassinelli, J.P. & Bjorkman, J.E. 1996, 'Equatorial Wind-Compression Effects Across the H-R diagram.' *Ap. J.* **459**, 671
(The wind compressed zone model)

Owocki, S.P., Cranmer, S.R. & Blondin, J.M. 1994, 'Two- dimensional Hydrodynamical Simulations of Wind Compressed Disks around Rapidly Rotating B stars.' *Ap. J.* **424**, 887

Petrenz, P. & Puls, J. 1996, 'Hα Line Formation in Hot Star Winds; The Effects of Rotation' *A & A* **312**, 195

12 Winds colliding with the interstellar medium

Stars interact with the surrounding interstellar medium (ISM), both through their ionizing radiation and through the mass, momentum, and energy that is transferred by way of their winds. The extreme ultraviolet radiation from hot stars leads to ionized nebulae or H II regions around young stars. In the case of low mass stars about to become white dwarfs, the radiation leads to the ionization of planetary nebulae.

The mass loss in stellar winds leads to a recycling of matter back to the interstellar medium, and because of the nuclear processing that occurs in the interiors of stars, the matter which is returned is often chemically enriched. In the cases of late type giants and carbon rich Wolf-Rayet stars, dust grains are produced in the winds, so the outflows may carry grain enriched material into the interstellar medium. These grains could play a role in the next generation of star formation. There are also dynamical effects associated with wind-interstellar medium interactions. The collisions of the winds with their surroundings produce 'wind bubbles', and the momentum transfer helps to maintain the random velocities of interstellar clouds that otherwise would be damped out by the dissipative effects of cloud collisions.

The winds of 'massive stars' tend to have the greatest effect on the ISM, because their mass loss rates are large, and the massive stars that are hot also have winds that are very fast and carry large momentum fluxes. Massive stars are defined as stars that, even in the absence of binary effects, would end their lives as supernovae. This means that their masses are in the range from about 8 $M_\odot$ to about 100 $M_\odot$. Such stars will be able to ignite carbon burning in their cores, and soon thereafter terminate their evolution in supernovae explosions. Chapter 13 has a discussion of mass loss at different evolutionary phases.

The kinetic energy of a supernova from a massive star is about 10^{51} erg; roughly independent of the initial mass of the star. Abbott (1982) has shown that the wind of a very massive star ejects about the same amount of energy into the interstellar medium over its lifetime as a supernova. Moreover, massive stars tend to cluster, so there is a combined effect of the winds of many stars. The winds and supernovae explosions occuring from stars in the cluster, give rise to the '*super-bubbles*' observed in the Milky Way and other spiral galaxies. These have a major impact on the host galaxy. In particular, a 'star burst galaxy'; which has undergone an era of rapid star formation within the past few million years, will show evidence for an outflow due to superbubbles which will change the spectrum and appearance of the galaxy.

In describing the effects of stellar winds on the ISM, we consider only highly idealized cases in which the undisturbed ISM has a simple structure, with no clumps, no magnetic fields, no ionization phases, and other complications which occur in the actual ISM. We also assume that the winds are spherically symmetric. With these simplifications, we can hope to develop some basic understanding of wind-ISM interactions that occur in wind bubbles and the wind-wind interactions that occur in planetary nebulae.

We begin this chapter with a discussion of the overall structure of a spherical wind bubble and then trace the phases of its development as it expands and cools. We show that an important phase of the evolution of a bubble is the *snowplow phase*. In § 12.3 we present the equations that describe two snowplow phases which are called 'energy conserving' and 'momentum conserving'. In § 12.4 we discuss ring nebulae around massive stars. In § 12.5, we use momentum conserving snowplow theory to describe the formation of planetary nebulae from the interaction of the fast wind from the hot central star with the pre-existing slower wind from the same star when it was on the asymptotic giant branch. We end this chapter by discussing the deposition of mass, momentum and energy into the ISM, by stellar winds of both high mass and low mass stars.

12.1 The structure of wind-ISM interaction regions

The wind from a massive hot star carries not only mass but also wind kinetic energy. The rate of the energy loss by the wind is less than 1 percent of the stellar radiative luminosity (see Table 8.1). This can be shown by combining the single scattering upper limit on the wind momentum, $\dot{M}v_\infty = L_*/c$, with the definition of wind luminosity,

$L_w = \frac{1}{2}\dot{M}v_\infty{}^2$, which yields $L_w/L_* = \frac{1}{2}v_\infty/c \lesssim 0.005$. Although the wind luminosity is relatively small, the wind produces effects on the ISM that are quite different from those of the radiation field. This is because shocks are produced by the wind-ISM interaction, and they lead to a significant heating of both the shocked stellar wind and the swept up interstellar material. When the wind material runs into the ISM it produces a hot layer of both shocked wind and shocked ISM gas in between the fast moving wind and the undisturbed ISM. This hot layer wants to expand into two directions: forward into the ISM and backwards into the wind. It is this expansion of the hot layer that produces *two shocks*, one on the inside and one on the outside of the hot layer.

In the broad interaction region the kinetic energy of the wind is converted in heating to temperatures of order 10^7 K of a region called the 'wind bubble'. This temperature is to be compared with values of about 10^4 K that are produced in circumstellar nebulae by the radiation field of the star. Furthermore, since the wind bubble also contains zones in which the shocked gases have cooled, a wide range of temperatures, from about 10^4 to 10^7 K, are produced. Thus the wind-ISM interaction regions have a much wider range of ionization and excitation conditions than are present in classical ionized nebulae.

The basic theory for the wind-ISM interaction was developed by Castor, McCray and Weaver (1975), and we use that as a basis for much of this chapter. That theory has also been used to interpret observations of starbursts and other active regions in galaxies.

Figure (12.1) shows the basic onion-skin structure of a wind interacting with the surrounding ISM. The innermost of the concentric spheres is the wind emitting star, or in the case of a collective wind from many stars, it is the region from which gas and wind energy flow. Far outside is the undisturbed ISM. The regions in between is called the 'interaction region'. It is separated into four zones, as shown in the figure. From inside to outside w these are:

(1) Zone 1 is the freely flowing supersonic stellar wind. At the outer boundary of the supersonic wind is a 'driven wave', illustrated in Fig. (12.1), which itself forms the next three concentric zones.
(2) The innermost part of the driven wave, labelled Zone 2, is a region containing the wind gas that has been heated as it passed through an 'inward facing shock'.†

† The inward and outward facing shocks are identifiable by the temperature structure, i.e. material enters a shock front and is heated. The 'direction', inward or outward facing, is from the hot side toward the cold side.

(3) Zone 3 is a thin, dense shell or 'contact surface', separating material that had been in the wind from material that was originally in the ambient, interstellar medium. Usually, most of the material in the contact surface is interstellar material which was swept up at the outer shock, compressed, and then cooled radiatively.

(4) The outermost part of the driven wave, Zone 4, is swept up IS material, and this zone extends from the contact surface to the 'outward facing shock', through which ISM material passes and is compressed and heated.

(5) Zone 5 is the undisturbed interstellar medium that is not yet affected by the outwardly expanding driven wave.

12.1.1 The shock jumps

Here we consider the compression and heating of two parcels of matter as they enter the driven wave zone from opposite ends; one parcel starts as stellar wind material and one as undisturbed ISM.

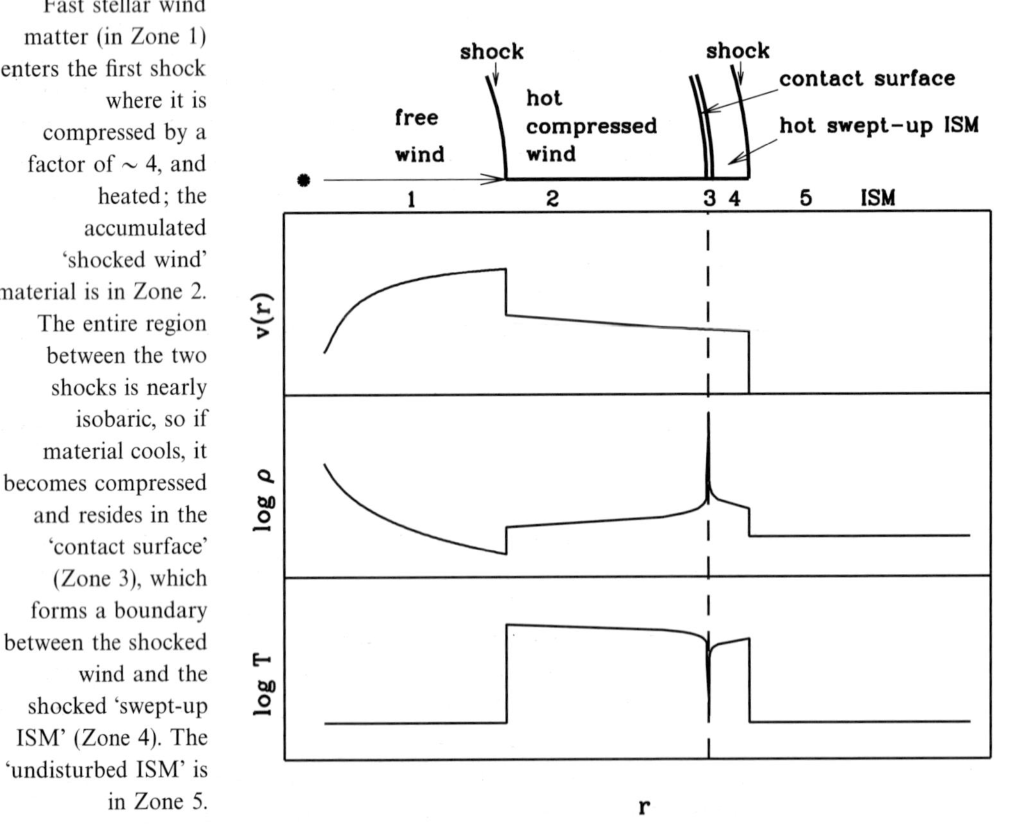

Figure 12.1 The formation of a pair of shocks in a 'driven wave'. (a) The upper part shows the wind interacting with the surrounding ISM. The interaction leads to a driven wave composed of shock compressed wind, a contact surface, and the swept-up ISM. (b) The lower part shows the structure of a driven wave in terms of v, ρ and T. Fast stellar wind matter (in Zone 1) enters the first shock where it is compressed by a factor of ~ 4, and heated; the accumulated 'shocked wind' material is in Zone 2. The entire region between the two shocks is nearly isobaric, so if material cools, it becomes compressed and resides in the 'contact surface' (Zone 3), which forms a boundary between the shocked wind and the shocked 'swept-up ISM' (Zone 4). The 'undisturbed ISM' is in Zone 5.

The behaviour of either parcel of gas passing through a shock can be derived from equations of conservation of mass, momentum and energy. The 'jump conditions' that are derived are collectively known as the Rankine-Hugoniot relations, and they are conveniently expressed in terms of velocities in the frame of the shock, for example at the inner shock, the flow in Zone 1 has the wind speed v_∞, so the velocity relative to the shock is $u_1 = v_\infty - v_s$, where v_s is the velocity of the shock in the rest frame. In Zone 2, on the other side of the infinitesmal shock, $u_2 = v_2 - v_s$, with v_2 being the flow speed of the gas that has just been shocked. With this notation the jump conditions associated with mass, momentum and energy conservation across a shock are respectively,

$$\begin{aligned} \rho_1 u_1 &= \rho_2 u_2 &&= C_m \\ \rho_1 u_1^2 + p_1 &= \rho_2 u_2^2 + p_2 &&= C_p \\ \tfrac{1}{2}u_1^2 + \tfrac{5}{2}\,p_1/\rho_1 &= \tfrac{1}{2}u_2^2 + \tfrac{5}{2}\,p_2/\rho_2 &&= C_e \end{aligned} \tag{12.1}$$

where C_m, C_p, and C_e are constants. The first of these, the C_m equation, states that relative to the shock the mass flux entering and leaving the shock front is constant, i.e. there is no 'pile up' of matter at the front. The C_p equation states that the sum of the dynamic pressure (ρu^2) and gas pressure, p, is the same on both sides of the shock. The C_e equation states that the sum of kinetic energy per gram plus the enthalpy per gram is constant across the shock. Thus there are no radiative energy losses occuring in the infinitesmal shock front. In the energy equation, we are also assuming that the shock involves a perfect gas, with $\gamma = 5/3$.

The following are well known adiabatic jump conditions for the density, velocity and pressure across a strong shock (see Problem 12.2).

$$\begin{aligned} \rho_2 &= 4\rho_1 \\ u_2 &= \tfrac{1}{4}u_1 \\ p_2 &= \tfrac{3}{4}\rho_1 u_1^2 \\ p_2/\rho_2 &= \tfrac{3}{16}u_1^2 \end{aligned} \tag{12.2}$$

We see from the first of these that the incident parcel of gas is compressed in passing through the shock, having its density increase by a factor of four. The second of these equations indicates that the parcel, after passing through the shock, continues to move onward but with a slower speed relative to the shock front. The parcel is also now moving subsonically relative to the ambient shocked medium. The third of these says that most (i.e. 75 %), but not all of the ram pressure, (ρu^2), of the incident material is converted to the gas pressure behind the shock. The last of these equations in combination with the perfect gas law, $p = \rho kT/(\mu m_H)$, provides an equation for the temperature of the shocked material.

So that we can apply the result to shocks in general we let $\Delta v \equiv u_1 = v - v_s$ be the relative speed of the material approaching the shock which itself is moving outward at the speed v_s. Then we obtain the useful expression

$$\begin{aligned} T_{\rm shock} &= \tfrac{3}{16}\frac{\mu m_H}{k}(\Delta v)^2 \\ &= 1.4 \times 10^5 \left(\frac{\Delta v}{100\ \mathrm{km\,s^{-1}}}\right)^2\ \mathrm{K} \end{aligned} \tag{12.3}$$

Thus, a jump in velocity by 270 km s^{-1} produces gas at a temperature of 10^6 K, and a jump of 840 km s^{-1}, which is still well below typical terminal wind speeds, produces a 10^7 K gas. The temperature immediately behind a shock clearly increases by a much larger factor than does the density. This is because of the major increase in the gas pressure produced by the incident ram pressure of the wind, which tends to be very large.

The adiabatic shock jump conditions given in Eqs. (12.2) allow us to understand the nearly discontinuous change in density, pressure and temperature across the inward facing and outward facing shocks shown in Fig. (12.1). However, there are further changes in the temperature and density once the material has entered the driven wave zone. For example, the temperature tends to decrease by radiation losses. Some of the effects produced by this cooling can be understood by considering 'isothermal shock jump conditions'. To be specific, let us consider the changes that occur in the density and pressure in the driven wave shock of Fig. (12.1) over the broad region from Zone 5 (the outer pre-shock region) to Zone 3 (the place at which the temperature has cooled back to its pre-shock value. The 'jump' in physical conditions between these two locations is called isothermal because it concerns two places with the same temperature. We can use the Rankine-Hugoniot mass and momentum constants, C_m and C_p of Eq. (12.1), to find the change in ρ and p (see Problem 12.3).

Referring to Fig. (12.1), let u_5 be the incident velocity relative to the outer shock front, and ρ_5 be the density of the ISM material about to enter the shock, and let u_3, ρ_3 and p_3 be the velocity, density and pressure at Zone 3. Also let $a = \sqrt{p_5/\rho_5}$ be the isothermal sound speed, and define $m = u_5/a$ to be the isothermal Mach number, which we assume to be large (i.e. $m \gg 1$). Then the the Rankine-Hugoniot mass and momentum constants, C_m and C_p, of Eqs. (12.1), lead to

$$\begin{aligned} u_3 u_5 &= a^2 \\ \rho_3 &= m^2 \rho_5 \\ p_3 &= \rho_5 u_5^2 (1 - 1/m) \approx \rho_5 u_5^2 \end{aligned} \tag{12.4}$$

The first of these equations gives $u_3 = u_5/m^2 \ll 1$, which means that the cooled gas moves at the same speed as the shock front. The second equation states that the density increases by a large factor. The last equation indicates that the pressure has increased from 3/4 of the incident ram pressure derived for the adiabatic shock jump to nearly 100 percent of the ram pressure. The pressure gradient implied by this increase causes the gas (as viewed in the rest frame) to accelerate from $3/4 v_s$ to v_s.

The reason the density can change so strongly across the driven wave zone (say in going from the outer part of Zone 4 to Zone 3 in Fig. (12.1) is that the pressure stays roughly constant (to within 25 %), as we have just seen, so $\rho \times T \sim$ constant. Radiative cooling leads to a decrease in T (say from 10^7 K to 10^4 K), therefore it leads to an increase in the density relative to the pre-shock value by a factor of 4×10^3; i.e. a very large compression of matter. Therefore, *if* cooling is important, then most of the material that has entered the driven wave resides in the cool and dense contact surface. We will find that this result is especially important for the swept up material, because a very large volume of swept up ISM material can be compressed into a thin shell.

The combination of the adiabatic jump conditions, Eqs. (12.2), and the isothermal jump conditions, Eqs. (12.4), allow us to understand the variations of T and ρ shown in Fig. (12.1). This combination of jump conditions will also be useful for describing the the several forms of the driven wave structure that occur during the evolution of a wind bubble.

In the following section we discuss three phases in the sequential development of the five zones shown in Fig. (12.1):

(I) the 'free expansion phase of the wind',

(II) the 'adiabatic phase', for the expansion of the interaction region,

(III) the 'snowplow phase'.

The last of these is of greatest importance from an observational point of view because it has a much longer duration than the other two, and so we will discuss it more thoroughly.

12.2 The evolution of a wind bubble

To understand the basic evolution of a wind-ISM interaction region, let us assume a simple picture in which a star is turned-on instantaneously when hydrogen burning begins. Let us also assume for numerical estimates that the wind has a mass loss rate $\dot{M} = 10^{-6}\ M_\odot\,\mathrm{yr}^{-1}$, a terminal velocity of v_∞=2000 km s^{-1}, and a kinetic energy of 10^{36} erg s^{-1}, and that the ISM around the star is homogeneous with a density $\rho_0 \simeq 2\ 10^{-24}$ (g cm^{-3}) corresponding to a typical ISM number density of $n_0 \simeq 1$ atom cm^{-3}.

(I) *First phase: free expansion of the wind*

The wind initially expands freely into the ISM, almost unopposed by the presence of the ISM, as is illustrated in Fig. (12.2). For this free expansion phase, note that for the two shocks illustrated in Fig. (12.1), the inner of the two shocks does not correspond to a significant decrease in the wind velocity. However, there is a strong outer shock which is originally moving outward at nearly the wind velocity, and so this is the speed at which the low density interstellar material is swept up. Because the outer shock moves into the ISM with almost the same speed as the wind, this phase is called 'the free expansion

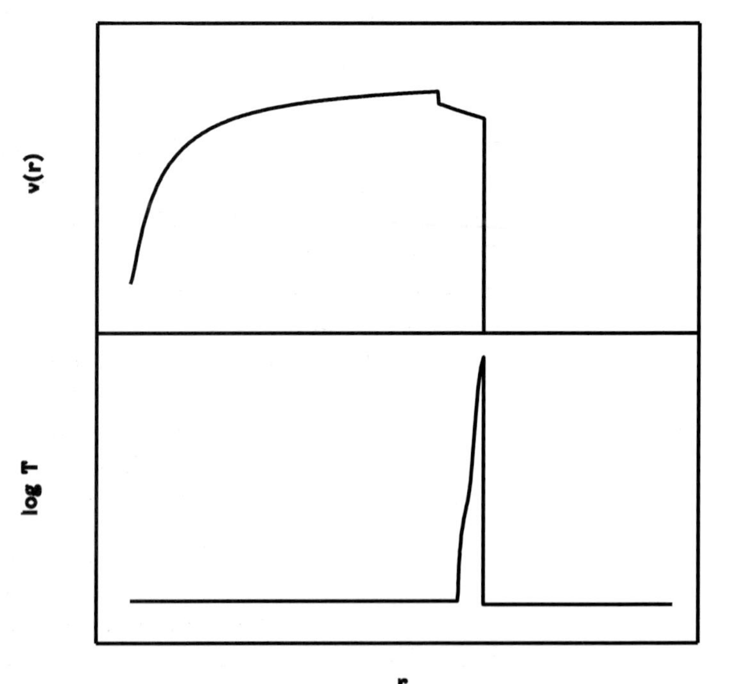

Figure 12.2
A schematic drawing of the free expansion phase of the wind.

phase'. As time progresses the amount of material swept up increases and the momentum from the wind can not drive the shell to such high speed. Thus the shell slows down. This implies that the inner shock jump will become increasingly strong because of a growing difference between the incident wind speed and the speed that material has after it joins the outwardly moving driven wave. This inner shock is the interface between Zones 1 and 2 in Fig. (12.1). The free expansion phase in considered to be terminated at a time τ_1, when the swept-up mass in Zone 4 is comparable to the mass in the driven wave from the wind in Zone 2. The mass lost by the star is $\dot{M}\tau_1$ and the mass of the swept up ISM is $\frac{4\pi}{3}(v_\infty \tau_1)^3 \rho_0$. These two masses are equal when

$$\tau_1 = \sqrt{\frac{3}{4\pi} \frac{\dot{M}}{\rho_0 {v_\infty}^3}} \tag{12.5}$$

which is about 100 years for the typical hot star wind bubble. During this time the wind bubble has reached a radius of $r = v_\infty \tau_1 \approx 0.2$ pc. At the end of this phase both Zone 2 and Zone 4 are assumed to be hot.

(II) *Second phase: adiabatic expansion of the bubble*

When the free expansion phase has ended the bubble complex will have the following structure, from inside to outside (see Fig. 12.1).

(1) The supersonic wind in free expansion out to radius R_1, where the wind encounters an inward facing shock.

(2) The region consisting of shocked *wind* material, with outer radius R_2. Kinetic energy of the wind is deposited in this region in the form of heat, because this gas has gone through a strong shock. This creates a temperature of about 10^7 K. All of the shocked stellar wind mass is in this region so $M_2 \approx \dot{M}\tau_1$, where $\dot{M}$ is the mass loss rate from the star. Because this region is hot, it causes the contact surface to expand outward with a speed $v_2 = dR_2/dt$.

(3) and (4) Surrounding Zone 2 is a shell of hot swept-up ISM gas that extends out to the outer boundary, R_4. Since it contains all of the swept-up gas its mass is $M_4 = \frac{4\pi}{3} {R_4}^3 \rho_0$. Note that the outer boundary is the outward facing shock.

Thus the structure is basically the same as that shown in Fig. (12.1) except that the temperature is hot throughout and there is no significant density enhancement at the contact surface (Zone 3).

During phase (II), material that has passed through either of the shocks is so hot that there is negligible energy loss by radiative cooling, hence this is called *'the adiabatic phase'*. The temperature is very high behind both the forward and reverse shocks and the emissivity of very hot gas is low.

Although radiative cooling is not yet important, the temperature of the bubble nevertheless will decrease with time because work is done in increasing the bubble volume. When Zone 4, the swept-up gas, has cooled adiabatically to a temperature of 10^6 K, the emission of line radiation becomes the dominant cooling process and the temperature of the shell rapidly decreases to 10^4 K. Therefore, we can say that the end of the adiabatic phase is marked by the production of post shocked gas with a temperature of just 10^6 K. The duration of the adiabatic expansion phase can thus be estimated by finding the time that it takes the expanding gas to cool from 10^7 K to 10^6 K. The temperature behind a shock it determined by the change in velocity jump, $\Delta v = v_s$, that occurs as the previously undisturbed interstellar matter enters the shock zone. Here v_s is the shock front speed, and we are assuming the undisturbed ISM is at rest. Using Eq. (12.3), we find that a change in temperature from 10^7 K to 10^6 K corresponds to a change in jump velocity by a factor of $\sqrt{10}$ (i.e. from about 840 to 270 km s^{-1}). In the next section we will study the change in the shock velocity versus time, and will find that during an adiabatic phase $v_s \propto t^{-3/5}$. Using that result tells us that the change in jump velocity corresponds to a ratio of ages of phase II to phase I of about 6. Thus, the age at the end of the adiabatic phase τ_2 is about $6 \times \tau_1$. Using τ_1 from Eq. (12.5), we conclude that the adiabatic phase ends in less than about 1000 years.

Although important in the dynamical development of the wind bubble, neither the freely expanding wind phase (I), nor the adiabatic phase (II) is important observationally because the duration of each phase is short (10^2 and 10^3 years respectively). Thus, it is improbable that we could observe a wind bubble in either phase.

(III) *Third phase: snowplow phase of a hot bubble and a cold shell*

We are most likely to observe a bubble when it is in the 'snowplow phase'. The ISM material that entered the outward facing shock was heated to a temperature below 10^6 K, so it cooled quickly to a temperature of about 10^4 K that can be maintained by the radiation field of the star. The mass of the swept up material is much larger than the amount of material in the bubble, and since it is cool it lies in a compressed region. This phase is called the 'snowplow phase' because there is a 'cool', compressed shell of material that is being pushed from behind, similar to what occurs for a snowplow.

Now the boundary between the outer shell and the ISM is essentially an isothermal shock. Since isothermal shocks have a high compression rate the density in the cold shell will be high ($n \approx 10^1$ to 10^3 cm^{-3}), and as a consequence the shell will be thin. The snowplow phase persists for as long as the star is able to sustain a powerful wind; which corresponds to (3 to 10) $\times 10^6$ years for massive stars.

12.3 The snowplow expansion of a wind bubble

The snowplow phase can be divided into two categories depending on the nature of the shocked wind material: the 'energy conserving' phase, and 'momentum conserving' phase.

In the energy conserving snowplow phase, the shocked wind material in Zone 2 remains hot and does not lose significant energy by radiative loses. So Zone 2 is extended because it contains a low density, high temperature gas. In the momentum conserving snowplow phase the material in Zone 2 has cooled radiatively, and hence Zone 2 is geometrically thin. Again, the basic difference between the two cases is whether the inward facing shock that thermalizes the wind produces material at temperatures above or below 10^6 K. Here we derive the physical conditions of the gas in both of the snowplow phases.

Let us consider cases in which the mass swept up is much greater than the mass of the wind material; $M_s(t) \gg \dot{M}t$, where $M_s(t)$ is the mass of the shell of swept up material, given by

$$M_s = (4/3)\pi R^3(t)\rho_0, \tag{12.6}$$

where $R(t)$ is the outer radius of the bubble at time t, and $v(t) = dR(t)/dt$ is the rate of expansion of the bubble.

The bubble expands because its gas pressure is higher than that of the surrounding ISM. Therefore, the expansion is described by the momentum equation

$$\frac{d}{dt}\{M_s(t)v(t)\} = 4\pi R^2 p_B \tag{12.7}$$

where p_B is the internal pressure of the bubble, which is caused by the wind. There are two assumptions regarding p_B, that lead to two classes of models in the snowplow phase:

(A) For the energy conserving snowplow phase p_B is determined by the gas pressure of the high temperature gas in Zone 2. This gas is heated by the kinetic energy of the incident wind material and it remains hot as long as the radiative cooling rate is low, which means that the temperature is above 10^6 K (see Fig. 12.3).

(B) For the momentum conserving snowplow phase, it is assumed that the thermal energy in Zone 2 has been lost due to radiative cooling, and that the wind impacts directly on the inner face of the shell. It is the direct transfer of the momentum of the wind that drives the further expansion of the bubble (see Fig. 12.4).

12.3.1 *Model A:* the energy conserving snowplow model

The structure of the energy conserving snowplow phase, case A, is shown in Fig. (12.3). The hot bubble (Zone 2) shown in the figure is still too hot to suffer significant radiative cooling and so it will continue to expand and cool adiabatically.

The total energy injected into the bubble by the wind over the lifetime, t, of the interaction is $L_w t$, where L_w is the wind kinetic luminosity which we assume to have been constant and equal to $\frac{1}{2}\dot{M}v_\infty{}^2$. The internal energy in the bubble is the product of the energy per gram of the material, or $\frac{3}{2}nkT/\rho_B = \frac{3}{2}p_B/\rho_B$, and the total mass of the bubble, $\frac{4}{3}\pi R^3 \rho_B$. Since the total internal energy from the bubble

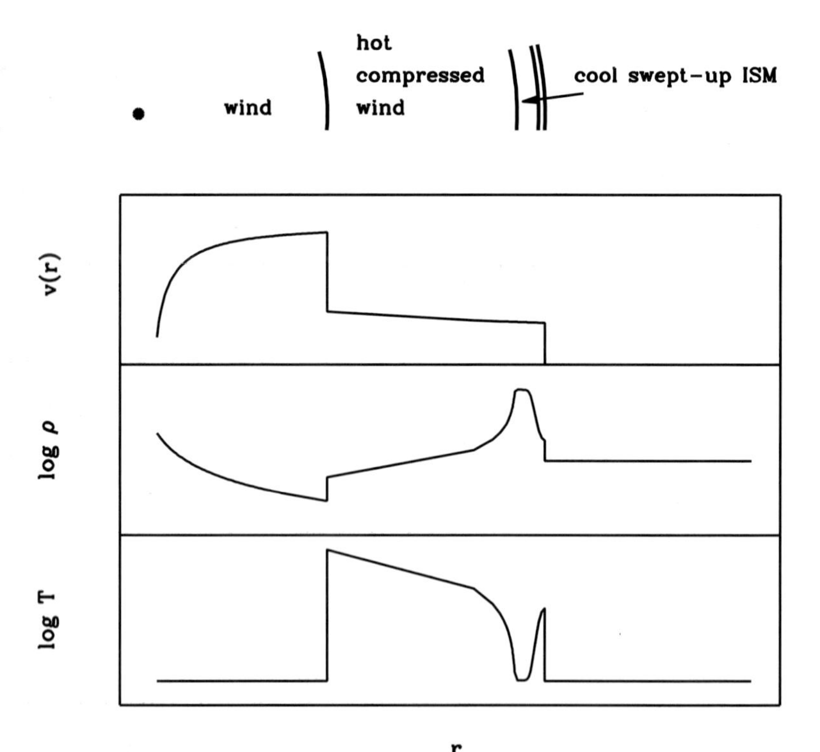

Figure 12.3 The structure of an *energy conserving* snowplow. In this type of snowplow, the wind material that entered the backward facing shock remains hot, but the material that entered the forward facing shock is cool. The cooled swept-up material is driven outward by the *high gas pressure* of the hot bubble.

comes from the energy of the wind over its lifetime, we find

$$p_B = \frac{L_w t}{2\pi R^3} \tag{12.8}$$

Using this expression in Eq. (12.7) gives the following result for the expansion of the bubble during the adiabatic phase.

$$\frac{R}{t}\frac{d}{dt}\left(R^3 \frac{d}{dt}R\right) = \frac{3L_w}{2\pi\rho_0} \tag{12.9}$$

where we have used $v(t) = dR/dt$, and $M_s(t) = (4/3)\pi R^3 \rho_0$. If, on the left hand side of Eq. (12.9), we let $R(t) = kt^n$, where k is a constant, we can obtain values for k and n resulting in

$$R(t) = \left(\frac{25L_w}{14\pi\rho_0}\right)^{1/5} t^{3/5} \tag{12.10}$$

This shows that the shell expands more slowly than would a freely expanding wind, for which $R(t) \propto t$. Expressing Eq. (12.10) in terms of $L_{36} = L_w/10^{36}$ erg s^{-1} and $t_6 = t/10^6$ yrs, we find the radii of

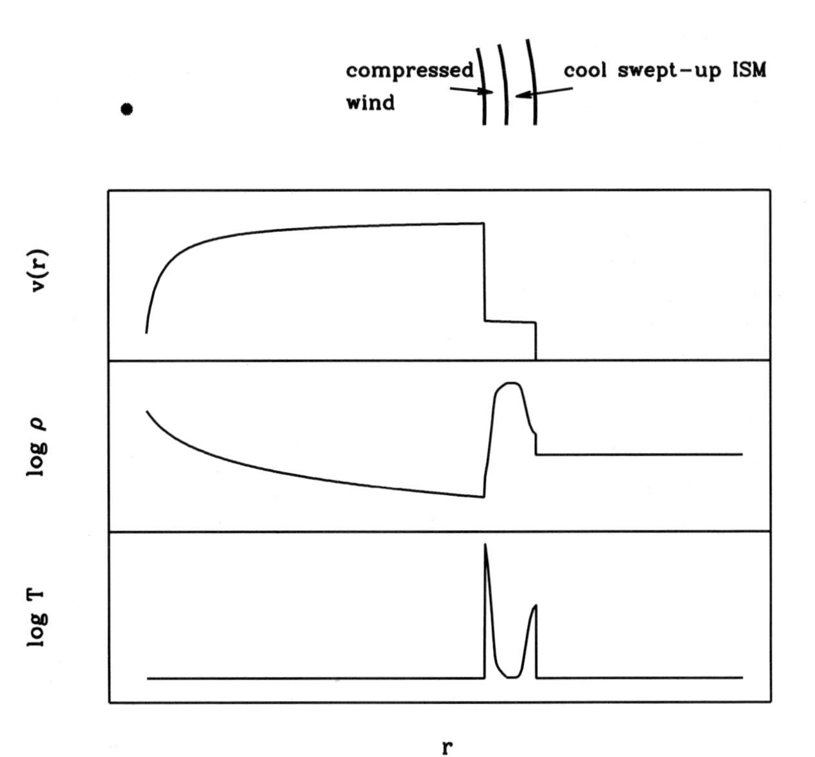

Figure 12.4 The structure of a *momentum conserving* snowplow. In this case the material that entered the backward facing shock cools soon after entering the shock. So the thermal energy from the shocked wind is fully lost to radiation. The shell is driven outward by the *dynamic pressure* of the wind.

Zones 2 and 3

$$R_2 \simeq R_3 = 28 \left(\frac{L_{36}}{\mu n_0}\right)^{1/5} t_6^{3/5} \quad \text{pc} \tag{12.11}$$

where n_0 is the particle denstiy in the undisturbed ISM. The time derivative of Eq. (12.11) provides the expansion speed of the bubble, $v(t)$.

$$v_2 \simeq v_3 = 17 \left(\frac{L_{36}}{\mu n_0}\right)^{1/5} t_6^{-2/5} \quad \text{km s}^{-1} \tag{12.12}$$

Castor *et al.* (1975) have shown that the temperature and density in the hot bubble region, Zone 2, are

$$T_2 = 1.6 \times 10^6 \left(L_{36}^8 n_0{}^2 t_6{}^{-6}\right)^{1/35} \quad \text{K} \tag{12.13}$$

$$n_2 = 0.01 \left(L_{36}{}^6 n_0{}^{19} t_6{}^{-22}\right)^{1/35} \quad \text{cm}^{-3} \tag{12.14}$$

So we see that a typical bubble of a star of 50 $M_\odot$ near the end of the energy conserving snowplow phase, which is at the end of the star's life after 4×10^6 years, has a radius of about 60 pc and an expansion velocity of 10 km s^{-1}.

At the boundary between the hot bubble and the cold shell the temperature gradient is so steep that thermal conduction by electrons will transport heat from Zone 2 to Zone 3. As a result of this, the heated inner layers of the cold shell will start to evaporate and gas will flow from the cold shell into the shocked wind region. This evaporation has no noticeable effect on the mass of the shell, but it can contribute significantly to the mass and composition in the mixed region.

As the hot bubble of wind material in Zone 2 expands, it also cools, and as the temperature decreases to about 10^6 K radiative losses become large and the energy conserving phase ends. The transition from an energy conserving snowplow to a momentum conserving snowplow occurs at a time $t_{\rm rad}$. We can find that transition time by setting $T_2 = 10^6$ K in Eq. (12.13), and find

$$t_{\rm rad} = 15.5 \times 10^6 \left(L_{36}^8 n_0{}^2\right)^{1/6} \quad \text{yrs} \tag{12.15}$$

12.3.2 *Model B*: the momentum conserving snowplow

If the cooling rate of Zone 2 by radiative losses is comparable with or larger than the heating rate by the wind luminosity input, then Zone 2 will become cooler and denser, and the thickness of Zone 2 will begin to compress. The wind bubble then enters the momentum conserving snowplow phase, model B, in which $R_1 \approx R_2$. The structure of the momentum conserving snowplow phase is illustrated in Fig. (12.4).

In the momentum conserving phase, the pressure on the inner side of the shell is the dynamic pressure of the wind or $\rho_w v_\infty^2$. After eliminating ρ_w using the wind mass conservation equation, we get from Eq. (12.7) the result

$$\frac{R}{t}\frac{d}{dt}\left(R^3 \frac{d}{dt}R\right) = \frac{\dot{M}v_\infty}{\frac{4}{3}\pi\rho_0} \tag{12.16}$$

which has the solution

$$R(t) = \left(\frac{3}{2}\frac{\dot{M}v_\infty}{\pi\rho_0}\right)^{1/4} t^{1/2} \tag{12.17}$$

Using the dimensionless variables introduced above, we find

$$R(t) = 16\left(\frac{L_{36}}{v_3 n_0}\right)^{1/4} t_6^{1/2} \quad \text{pc} \tag{12.18}$$

where $v_3 \equiv (v_\infty/1000 \text{ km s}^{-1})$. The time derivative of this provides the velocity of expansion of the shell

$$v(t) = 7.9\left(\frac{L_{36}}{v_3 n_0}\right)^{1/4} t_6^{-1/2} \quad \text{km s}^{-1} \tag{12.19}$$

In deriving Eqs. (12.18) and (12.19) we have neglected the initial radius of Zone 2 at the start of the momentum conservation snowplow phase. This is justified for wind bubbles for which the temperature $T_2 = 10^6$ K is reached after a short time (see Eq. 12.15), which depends on the value of $(L_{36}^8 n_0^2)$. For bubbles that spend most of their time in the energy conserving snowplow phase and only later enter the momentum conservation phase, the radius will be larger than that given by Eq. (12.18).

If we know the mass and the velocity of ring nebulae and the $\dot{M}$ and v_∞ history and duration of the central star, it is possible to determine whether a particular nebula is in the energy conserving or the momentum conserving snowplow phase. This can be done by deriving the two ratios:
ϵ = (kinetic energy in the shell)/(total wind energy provided)
π = (momentum in the shell)/(momentum imparted by the wind).
If $\epsilon \simeq 1$ the bubble is in the energy conserving phase, whereas it is in the momentum conserving phase if $\pi \simeq 1$. Observations of WR ring nebulae indicate they are in the momentum conserving snowplow phase. The momentum conserving snowplow phase will be discussed again in regard to the formation of planetary nebulae.

12.3.3 The terminal phase in the life of a wind bubble

After a time of about 4×10^6 years a massive, hot star will evolve off the main sequence and soon thereafter the wind luminosity will drop significantly, because the wind speeds associated with blue and red supergiants (500 km s^{-1} and 20 km s^{-1}) are much slower than the wind speed from an O star. The bubble could continue to expand until stalled by the pressure of the ISM. However, before that happens the star will have produced a supernova explosion, which will re-pressurize the bubble.

The extension of the wind bubble theory to 'superbubbles' associated with an OB association of hot stars and their supernovae shells has been developed by MacLow and McCray (1988). Observations and analyses of bubbles from 'galactic superwinds' that occur in starburst galaxies because of the collective affects of winds and supernovae are described by Heckman, Lehnert and Armus (1993).

12.4 Ring nebulae around Wolf-Rayet and newly formed O stars

The theory of wind blown bubbles makes many simplifying assumptions: spherical symmetry, a constant mass loss rate, a constant ρ_0, and that there are no neighboring O stars. Nevertheless, there are several very good candidates for wind blown bubbles in our galaxy. The three most common types are:

(1) Planetary nebulae around low mass stars in a very late evolutionary stage. These will be discussed in the next section.
(2) Ring nebulae around WR stars, which represents a late phase in the evolution of massive stars. Figure (12.5) shows the ring nebula RCW 58 that surrounds the WR star HD 96548 (WN8 spectral type). Ring nebulae around WR stars typically have radii of 3 to 10 pc, expansion velocities of 20 to 80 km s^{-1}, and shell masses of 5 to 20 solar masses (Chu, 1991). The WR ring nebulae are often enriched in nitrogen and helium, which indicates that the nebulae consists mainly of material from the mass losing star, rather than swept up interstellar matter.
(3) Ring nebulae are also seen around very young O stars that are still embedded in their natal molecular clouds. These nebulae are called ultracompact H II (UCHII) regions. Figure (12.6) shows radio contours of two examples of UCHII regions (Wood and Churchwell, 1989). The typical sizes of these are a few $\times 10^{16}$ cm (as opposed to size scales of order 10^{19} cm for H II regions), hence

the name ultracompact. From the large radio flux from these objects one can estimate the spectral type of the young O star inside. In the case of G5.89, (the object on the left of the figure) it is an O6.5 to O8 star. The intensity contours show evidence for a

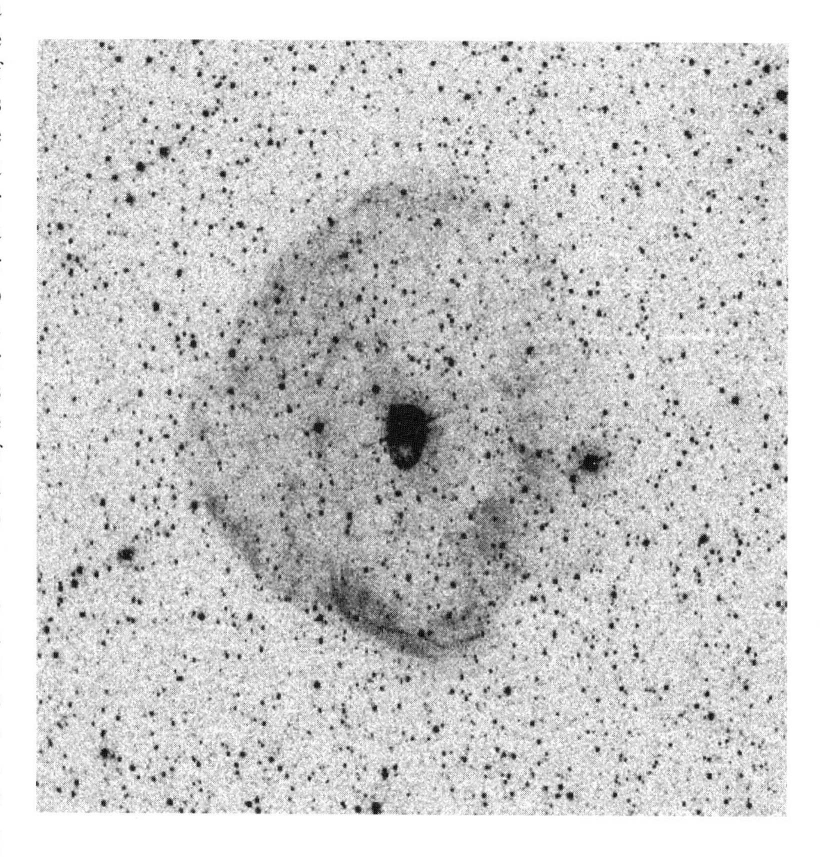

Figure 12.5 A high contrast print of the WR ring nebula RCW 58, at H α. The high N abundance of the nebula shows that most of the matter in this nebula was lost by the star in the form of a stellar wind, rather than being swept-up interstellar matter. The Wolf-Rayet star at the center is responsible for the photoionization of the nebula. (From Marston, 1995)

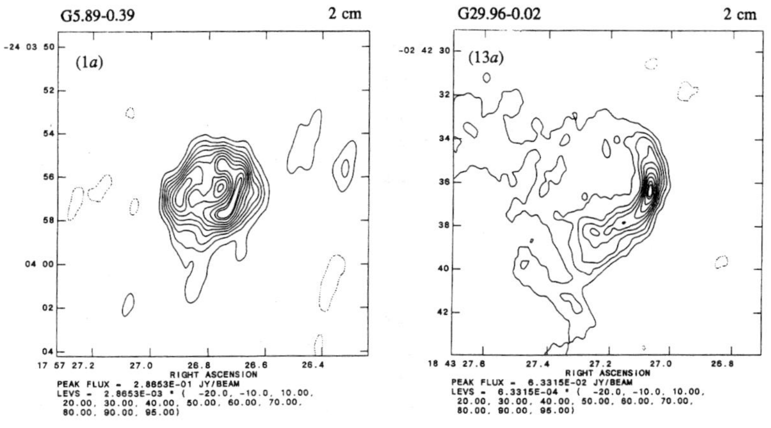

Figure 12.6 Two examples of ultra compact H II regions. The figures show the 2 cm radio emission contours. The object on the left, G5.89-0.39, shows the typical structure of a wind produced shell with the a bright outer ring. The object on the right, G29.96-0.02, is an example of a cometary H II region, due to the velocity of the star with respect to the ISM. (From Churchwell, 1990)

shell structure, thought to be caused by the the wind from the star. The object on the right, G29.96, shows an even more common type of UCHII, (around an O5.5 central star), with a cometary shape. This is thought to be caused by the action of the wind on the cloud material through which the star is drifting with a speed of order 10 $\mathrm{km\,s^{-1}}$. The motion causes the snowplow shock to assume the form of a bow shock, with the bow pointing in the direction of the star's motion. O stars are estimated to spend about 10 percent of their short lifetimes embedded in the molecular clouds in which they formed.

Ring nebulae are not prominent around older OB stars, even though these stars also have strong stellar winds. Probably the reason is that unlike the WR stars and very young O stars, which have thick gas zones around them, the older OB stars do not. In the case of the WR stars the winds are colliding with slower gas that the star ejected in an earlier evolutionary stage when it was either a red supergiant or a luminous blue variable. The interaction is then similar to that occuring in planetary nebulae that we will discuss in the next section. In the case of the UCHII regions the wind of the very young star collides with the remaining dense material of the natal cloud from which the star was formed.

12.5 Planetary nebulae

Planetary nebulae (PN) are ring nebulae around low mass stars that are in a very late stage of their evolution, making the transition to the white dwarf state. The central stars of planetary nebulae (here called CSPN) have masses of about 0.6 $M_{\odot}$, and effective temperatures in the range 3×10^4 to 2×10^5 K, and luminosities in the range 10^3 to 10^4 $L_{\odot}$. Figure (12.7) shows a picture of 'the Ring Nebula', NGC 6720.

A widely accepted model for the formation and dynamics of planetary nebulae is the 'interacting stellar winds' model, proposed by Kwok, Purton and FitzGerald (1978). The central stars of planetary nebulae show evidence in their P Cygni profiles for line driven winds with mass loss rates of order 10^{-7} $M_{\odot}\,\mathrm{yr^{-1}}$ and with wind speeds of about 1000 to 4000 $\mathrm{km\,s^{-1}}$(see § 2.7.3). However, prior to becoming a hot planetary nucleus star, it was a luminous red giant, most likely in the asymptotic giant branch (AGB) phase of evolution. Stars in this phase are known to eject a high density wind, sometimes referred to as a 'superwind' (Renzini, 1983) with an expansion speed of only about 10 $\mathrm{km\,s^{-1}}$ and a mass loss rates as high as 10^{-4} $M_{\odot}\,\mathrm{yr^{-1}}$(see § 2.8.2).

The fact that the slow wind precedes the faster wind of the CSPN means of course that the two concentric winds must interact where the fast wind overtakes the slower one. It is the interaction of the two winds that forms the site of the planetary nebula. This interacting stellar wind model supplanted an earlier idea for planetary nebulae, that they were formed by the impulsive ejection of the outer envelopes of red giant stars.

12.5.1 The formation of planetary nebulae

We consider the wind-wind interaction as an inelastic collision of two flows. The outer flow has a high density and a low velocity, whereas the inner flow from the wind of the CSPN has a lower density and higher velocity. Both wind flows have density distributions which vary as r^{-2} i.e. at constant velocity. This decrease of the density with radius is a major difference from the basic wind bubble model discussed in § 12.3, where we assumed that the outermost (ISM) region had a constant density. In that case, no matter how small the ISM density, eventually the swept-up ISM region would become more massive and denser than the incident fast wind. In the case of the planetary nebula

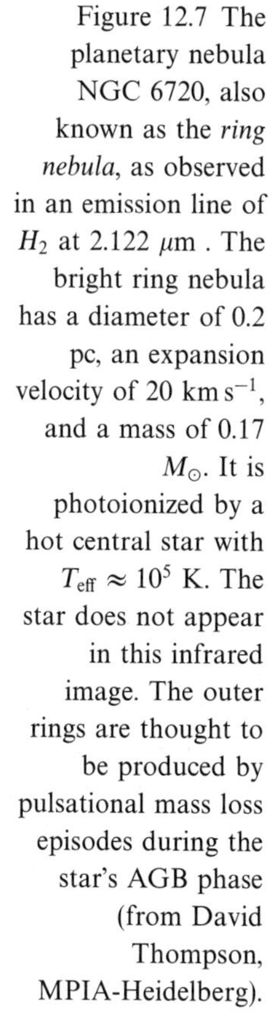

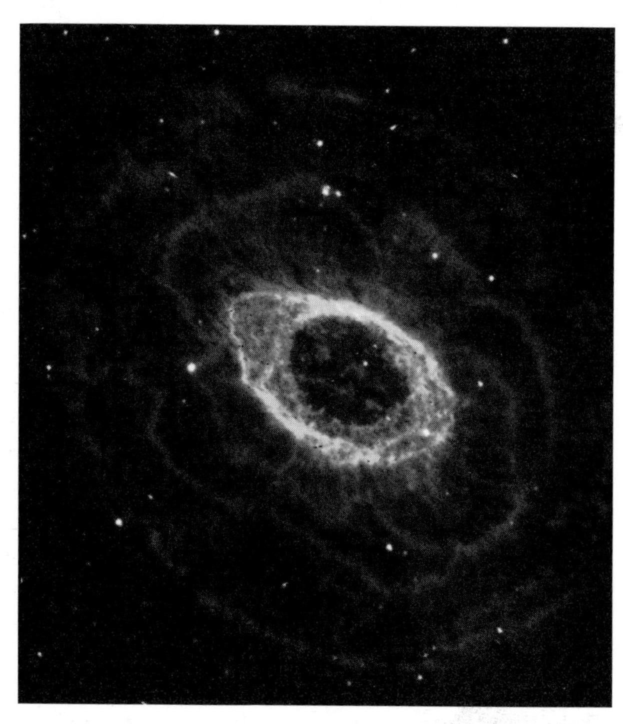

Figure 12.7 The planetary nebula NGC 6720, also known as the *ring nebula*, as observed in an emission line of H_2 at 2.122 μm . The bright ring nebula has a diameter of 0.2 pc, an expansion velocity of 20 km s^{-1}, and a mass of 0.17 $M_\odot$. It is photoionized by a hot central star with $T_{\rm eff} \approx 10^5$ K. The star does not appear in this infrared image. The outer rings are thought to be produced by pulsational mass loss episodes during the star's AGB phase (from David Thompson, MPIA-Heidelberg).

model, however, the ratio of the density on one side of the shell to that on the other remains a constant value, determined by the mass loss rates and velocities of the two winds.

The interaction of the fast wind with the pre-existing slower wind produces a shell containing both AGB ejecta and shocked fast wind material. At the same time that this shell accumulation is occuring, the ultraviolet radiation from the increasingly hot central star will photoionize the shell. The balance between the photoionizaton and recombination rates in the material leads to the prominent Hα emission of planetary nebulae. The densities also tend to be low enough that forbidden line radiation such as the [O III] 5007 line is prominent.

Let us assume that the wind from the AGB star and the wind from the central star of the planetary nebula are both spherically symmetric. For simplicity, let us ignore the details concerning the evolution of the star from the AGB to the CSPN phase, and just assume it is abrupt, with no intervening wind phases being important. Thus the wind is assumed to change discontinuously from one with low velocity and high density to one with high velocity and low density. We consider the conditions of the swept-up material at a time, t, after the transition occurs. We refer to the interaction region which consist of a ring of swept-up material as *the shell*. We describe the shell when it is in the momentum conserving snowplow phase in which the radiative cooling occurs quickly after the material enters the interaction region from both the outer and inner winds. This means that the shell is thin compared to its radius. So we can assume one radius $R_s(t)$ to represent the shell. The model is sketched in Fig. (12.8).

For the two winds, let $v_{\rm rg}$ and $\dot{M}_{\rm rg}$ be the velocity and mass loss rate of the star in the AGB phase (where the subscript stands for 'red giant'), and let $v_{\rm cs}$ and $\dot{M}_{\rm cs}$ be the respective values for the star in the CSPN phase. Let $M_s(t)$, $R_s(t)$ and $V_s(t)$, be respectively, the mass of the swept-up material that is in the shell, the radius, and velocity of the shell, at time t since the star switched on the fast wind. The radius of the star in its two phases is negligible compared to $R_s(t)$.

The mass in the shell $M_s(t)$ is composed of material from both winds and we must sum the contributions from each. In the case of the fast wind, the first particles emitted in the newly formed fast wind would have gone a distance $t \cdot v_{\rm cs}$, if there had been no interaction, so CSPN wind material that is in the shell would have been distributed from $R_s(t)$ to $t \cdot v_{\rm cs}$. Similarly, the last particles to leave the star in the form of the slow red giant wind would have gone a distance $t \times v_{\rm rg}$, in absence of the wind-wind interaction. Hence the material from $t \times v_{\rm rg}$ to $R_s(t)$ has been swept into the shell.

Hence the mass of the shell at time t is given by integrals over $4\pi\,\rho r^2 dr$ for the two winds. Using the mass conservation equations of the two winds to eliminate the densities $\rho_{rg}(r)$ and $\rho_{cs}(r)$, the mass of the shell becomes

$$M_s(t) = \int_{t \cdot v_{\rm rg}}^{R_s(t)} \frac{\dot{M}_{\rm rg}}{v_{\rm rg}} dr + \int_{R_s(t)}^{t \cdot v_{\rm cs}} \frac{\dot{M}_{\rm cs}}{v_{\rm cs}} dr \tag{12.20}$$

which yields

$$M_s(t) = \left(\frac{\dot{M}_{\rm rg}}{v_{\rm rg}} - \frac{\dot{M}_{\rm cs}}{v_{\rm cs}}\right) R_s(t) \; - (\dot{M}_{\rm rg} - \dot{M}_{\rm cs}) \cdot t \tag{12.21}$$

The momentum equation for the shell follows from the assumption that the collision of the two flows is completely inelastic. So there is a change in the momentum of the shell in the interval from t to $t + dt$, that is given by the sum of the momenta of the two wind contributions to the shell

$$\frac{d}{dt}\{M_s(t)\; V_s(t)\} = \dot{M}_{\rm cs}\{v_{\rm cs} - V_s(t)\} + \dot{M}_{\rm rg}\{v_{\rm rg} - V_s(t)\} \tag{12.22}$$

The change in the mass in the shell over that time is found by differentiating Eq. (12.21)

$$\frac{dM_s(t)}{dt} = \frac{\dot{M}_{\rm cs}}{v_{\rm cs}}\{v_{\rm cs} - V_s(t)\} + \frac{\dot{M}_{\rm rg}}{v_{\rm rg}}\{V_s(t) - v_{\rm rg}\} \tag{12.23}$$

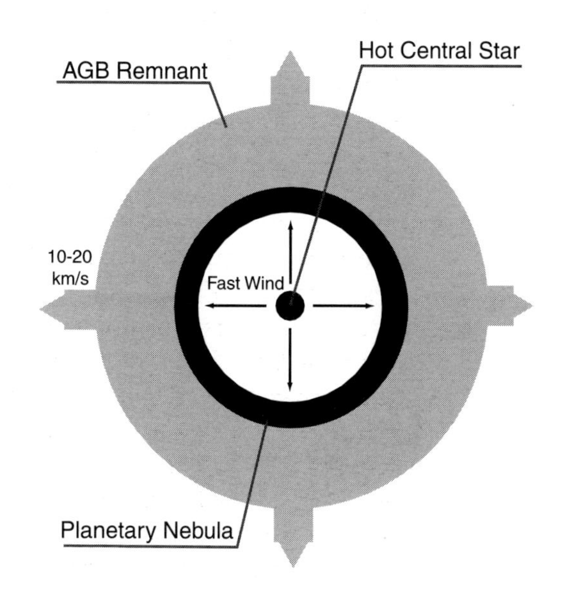

Figure 12.8 A schematic model of a planetary nebula formed by the interaction between a slow wind from the AGB phase and a fast wind from the CSPN phase. The structure consists of a sequence of concentric rings which show from inside out: the star, its fast wind, the interaction region or shell (black), the slow wind (gray). The names of these regions are noted on the figure.

Combining Eqs. (12.22) and (12.23) we find an expression for the acceleration of the shell

$$\begin{aligned} M_s(t)\frac{dV_s(t)}{dt} &= \frac{d\ M_s(t)V_s(t)}{dt} - V_s(t)\frac{dM_s(t)}{dt} \\ &= \frac{\dot{M}_{\rm cs}}{v_{\rm cs}}\{v_{\rm cs} - V_s(t)\}^2 + \frac{\dot{M}_{\rm rg}}{v_{\rm rg}}\{v_{\rm rg} - V_s(t)\}^2 \end{aligned} \tag{12.24}$$

Equations (12.21) and (12.24) may be combined to find $M_s(t)$ and $V_s(t)$ by means of numerical solutions.

The shell velocity rather quickly reaches a terminal velocity, because both the fast wind density and the slow wind density decrease with radius and the addition of mass and momentum become small as time increases. The asymptotic velocity is derived by Kwok (1983) to be

$$V_s(t) = \frac{(\dot{M}_{\rm rg} - \dot{M}_{\rm cs}) + (v_{\rm cs} - v_{\rm rg})\sqrt{\dot{M}_{\rm rg}\dot{M}_{\rm cs}/v_{\rm cs}v_{\rm rg}}}{(\dot{M}_{\rm rg}/v_{\rm rg}) - (\dot{M}_{\rm cs}/v_{\rm cs})} \tag{12.25}$$

With the expansion at this terminal speed the radius will be given approximately by $R_s(t) = t \cdot V_s(t)$. Thus, the mass of the shell can be derived from Eq. (12.24), giving

$$M_s(t) = t \cdot (v_{\rm cs} - v_{\rm rg})\sqrt{\frac{\dot{M}_{\rm rg}\dot{M}_{\rm cs}}{v_{\rm rg}v_{\rm cs}}} \tag{12.26}$$

The thickness of the shell can be found by assuming that it has the temperature typical of a photoionized nebula ($\approx 10^4$ K), and by balancing the gas pressure with the compressional pressure from the wind material incident on the shell. This pressure is $\rho_{cs}(v_{cs} - V_s)^2 + \rho_{rg}(V_s - v_{rg})^2$, because both the outside slow material and the inside fast material run into the expanding shell.

12.5.2 Models for planetary nebulae

Some model results are shown in Table (12.1) (from Kwok *et al.* 1978). Given plausible values for the mass loss rates and velocities of the winds, the models show that the speed of the shell is about 40 km s^{-1}, and that after a few thousand years the shell has a mass of about 0.1 $M_\odot$, a radius of about 0.1 pc, a thickness of 0.01 pc, and a density of 10^3 cm^{-3}. These are typical values for planetary nebulae. So the assumption that the shell is much thinner than its radius of curvature is justified.

Table 12.1 *Planetary nebulae gas shell parameters from the interacting wind model*

$\dot{M}_{cs}$	$\dot{M}_{rg}$	v_{cs}	v_{rg}	$V_s(t)$	$\frac{M_s(t)}{t}$	$\frac{R_s(t)}{t}$	$\frac{\Delta R_s(t)}{R_s(t)}$
$\frac{M_\odot}{yr}$	$\frac{M_\odot}{yr}$	$\frac{km}{s}$	$\frac{km}{s}$	$\frac{km}{s}$	$\frac{M_\odot}{10^3 yr}$	$\frac{pc}{10^3 yr}$	
3×10^{-7}	1×10^{-5}	1000	10	27	0.017	0.027	0.31
3×10^{-7}	3×10^{-6}	1000	10	40	0.009	0.041	0.12
1×10^{-6}	3×10^{-6}	1000	5	44	0.024	0.045	0.08

12.6 The deposition of mass and energy into the ISM

Stellar winds deposit mass and energy into the interstellar medium. The deposition of energy by stellar winds is comparable to that by supernovae. The mass that is deposited into the ISM by winds is chemically enriched by the nuclear processes in the stars. So stellar winds are major contributors to the energy and chemical evolution of the ISM.

The energy deposition and the mass deposition can be estimated from either observational data or from theoretical models of stellar winds. The various evolutionary channels by which wind matter re-enters the interstellar medium are known from calculations of stellar evolution with mass loss (see Chapter 13). For each channel the mass loss rate per star and the energy per unit mass are known. The rate at which mass is returned to the ISM per unit volume is

$$\left(\frac{dM(t)}{dt}\right)_{\text{ISM}} = \sum_j N_j(t)\dot{M}_j(t) \tag{12.27}$$

where the summation is over all the j types of stars. The rates of energy return to the ISM by massive stars can be estimated from the mass return rates by multiplying them by $v_\infty{}^2/2$.

$$\left(\frac{dE(t)}{dt}\right)_{\text{ISM}} = \sum_j N_j(t)\dot{M}_j(t)\frac{v_\infty{}^2}{2} \tag{12.28}$$

A useful estimate of the effects of many stars in a galaxy on its ISM is to assume an initial mass function, sum over the mass loss rate and the wind luminosity of each class of mass losing star, and to use the lifetime of each phase as derived from stellar evolution models.

12.6.1 Mass and energy deposition from massive stars

Almost all of the mass that forms a star of mass greater than 8 $M_\odot$ is returned to the interstellar medium, in one form or another, by the end of the star's evolution. Perhaps only 1 $M_\odot$ of the original tens of solar masses is left as a compact remnant such as a black hole or neutron star. One question of interest is how much is returned via the main sequence mass loss in the stellar wind of O and B stars, mass loss in a red supergiant wind, very rapid mass loss during the transitory phase as a luminous blue variable (LBV), and mass loss in the powerful winds of WR stars.

The amounts of the matter lost in wind phases depend on the evolutionary time scale and on the associated mass loss rates. The existence of nitrogen-rich ON stars, where the products of nuclear fusion appear at the surface, and of Wolf-Rayet stars, which are stripped of their outer hydrogen-rich layers, proves that wind mass loss is large for massive stars.

Stars with $M_i < 30$ $M_\odot$ lose most of their mass in the red supergiant phase, when the mass loss rate is high but the wind velocity is low, and in the WR phase, when the mass loss rate is high and the wind velocity is high. Stars more massive than about 40 $M_\odot$ lose a substantial fraction of their mass in the main sequence phase. They do not evolve to the red supergiant phase, so the post main sequence mass loss for these stars is seen to occur in the luminous blue variable phase and the WR phase.

Figure (12.9) shows the mass portions of the original mass that return to the interstellar medium in various forms (Castor, 1993). Mass loss during the main sequence phase is seen to be negligible for stars with initial mass, M_i, less than about 20 $M_\odot$, but this return route can grow to about 30 percent for stars with $M_i = 100$ $M_\odot$.

Figure (12.10) shows the mass return routes when weighted by the initial mass function, IMF, of Garmany, Conti and Chiosi (1982). Now we see that there is a much larger return from the lower mass stars simply because they are more numerous than the higher mass stars. If the supernova contribution is not counted, then we see again that more massive stars are important with the peak mass return occuring for stars with masses of about 40 $M_\odot$.

The Wolf-Rayet stars return a mass that is a factor 1.5 times that from stars on the main sequence. The total mass deposited by stars more massive than 15 $M_\odot$ is comparable to that by supernovae of type II.

12.6.2 Mass and energy deposition by low mass stars

For stars with masses between about 1 and 8 $M_{\odot}$, most of the mass loss occurs while the stars are on the asymptotic giant branch (AGB). These stars have completed the core helium-burning phase, and shell He-burning occurs during the early-asymptotic giant branch phase, or E-AGB. As the star continues to evolve a hydrogen burning shell can ignite and this double shell phase is the thermal-pulsating AGB, or TP-AGB, phase (Schaller *et al.* 1992, Habing, 1996). During the thermal pulse phase, stars can have large excursions in luminosity, and material from the helium fusion shell becomes convectively mixed to the surface of the star. This increases the carbon abundance in the atmosphere and in the wind. While on the AGB, the stellar radii reach large values, up to about 250 $R_{\odot}$, and low effective temperatures, $T_{\mathrm{eff}} < 2500$ K. The combination of several processes such as pulsations, Alfvén waves and acoustic waves, produce very extended density distributions in which dust grains can form. The stars with optically thin dust shells are the Mira variables and those with optically thick dust shells are

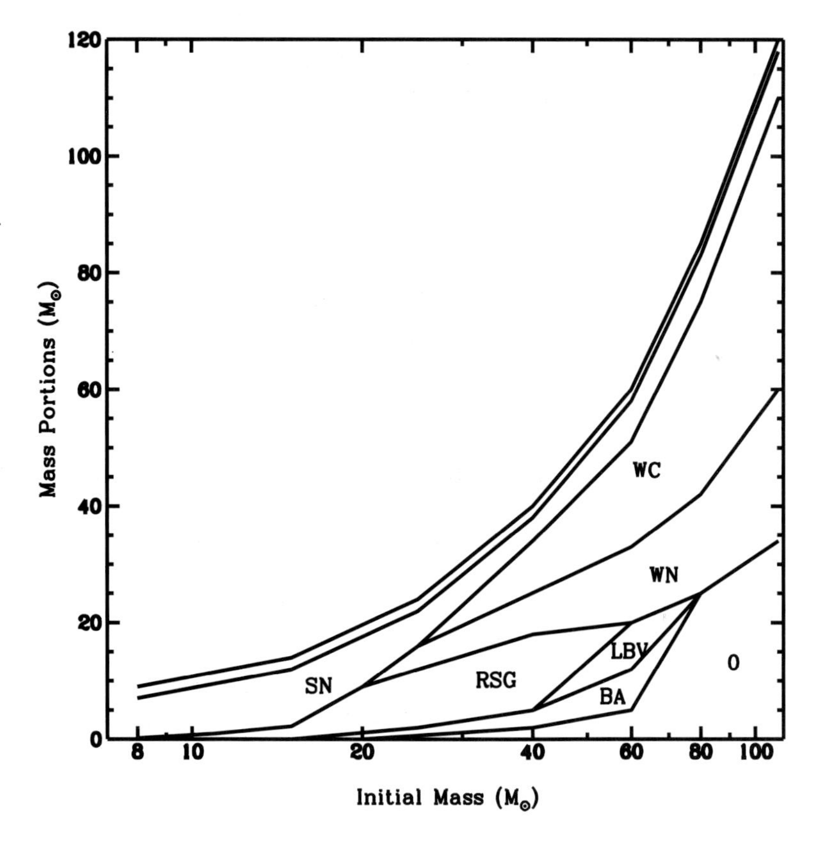

Figure 12.9 The fraction of the matter that returns to the interstellar medium during stellar evolution versus the initial mass of massive stars. The letters indicate the evolutionary phases when the mass loss occurs. The narrow strip at the top is the mass that remains in the compact stars. (From Castor, 1993)

the OH/IR stars or carbon-rich IR sources, depending on the relative abundance of oxygen and carbon. The winds of these luminous cool stars have mass loss rates in the range of 10^{-7} to 10^{-4} $M_\odot$ yr^{-1}, and low wind speeds of about 30 km s^{-1} or less (see § 2.8.2). Because of the low wind speeds the energy deposition into the ISM is small. So it is the recycling of chemically enriched matter to the ISM that is of most interest here.

One way by which the mass contribution of the low mass stars has been estimated is from the consideration of the initial mass, M_i, and final mass, M_f, of stars which have become white dwarfs. Weidemann (1987, 1993) developed semi-empirical estimates of the M_f versus M_i relation from studies of white dwarfs in open clusters. The initial masses are derived by subtracting the white dwarf cooling ages from the current cluster age so as to derive the main sequence turnoff mass of the stars that are now white dwarfs. The results for the remnant white dwarf mass are shown in Fig. 12.11. This figure shows the remnant mass and the AGB mass loss as a function of the initial mass for stars with $M < 8$ $M_\odot$. Thus we see that the low mass stars return

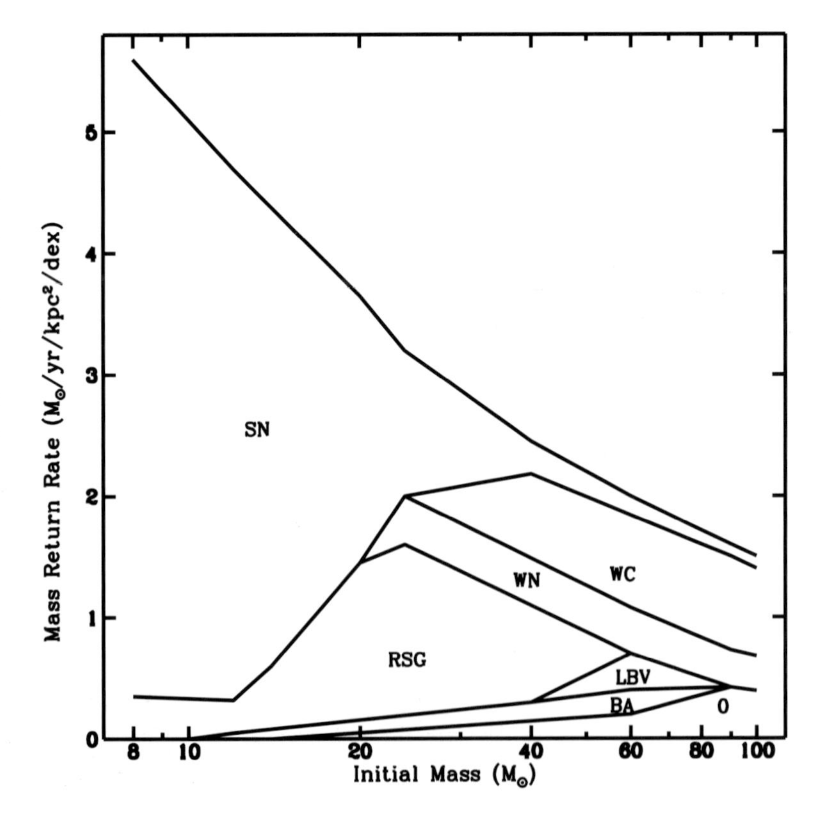

Figure 12.10 Shows, as a function of the initial stellar mass, the mass that returns to the interstellar medium weighted by the fraction of stars at each mass as given by the initial mass function. The units are $M_\odot$yr^{-1} kpc^{-2} for stars with masses in a $\Delta \log(M_*) = 1$ range.

a major fraction of their initial masses to the ISM. For a solar mass star 45 % is returned, leaving behind a 0.55 $M_\odot$ white dwarf. For stars in the range from 4 to 8 $M_\odot$, the mass return to the ISM is between 70 and nearly 90 %.

The nature of the material returned and the outflow process during the AGB phase have been studied by Jura and Kleinman (1989, 1990). The stars show a large range in mass loss rates, so the total mass returned to the ISM from the AGB stars is dominated by the relatively few objects with very high mass loss rates of near 10^{-4} $M_\odot$ yr^{-1} (see § 2.8). The optical light from these high mass loss stars is strongly attenuated by the grain opacity in the stellar outflow, so they appear as infrared sources and can be identified from IRAS satellite data. Using mass loss rates derived from 60 μm flux data (see Chapter 2), Jura and Kleinman (1989) concluded that: (a) the 63 AGB stars within 1 kpc of the sun return mass to the ISM at the flux rate of between 3 and 6 times 10^{-4} $M_\odot$ kpc^{-2} yr^{-1}, (b) the main sequence progenitors are typically about 1.5 $M_\odot$, (c) approximately half of the mass losing AGB stars are carbon rich, and they account for about

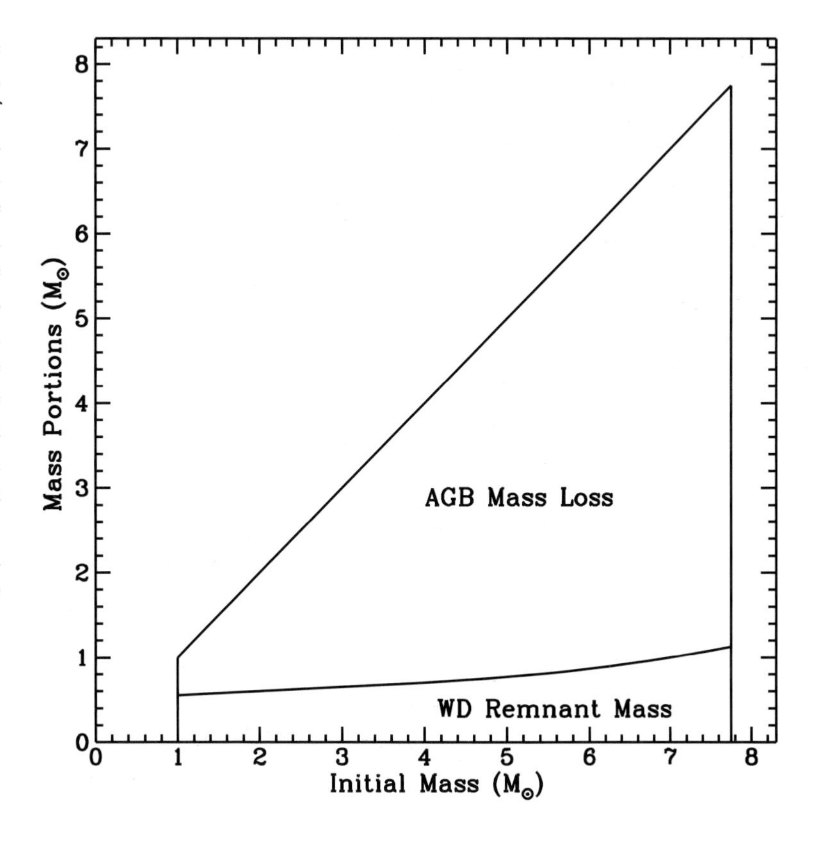

Figure 12.11 The lower curve shows the final mass of white dwarfs as a function of initial mass. The difference between the initial mass and the white dwarf mass is returned to the ISM, primarily while the star was on the asymptotic giant branch. Stars with $M_* \gtrsim 8 M_\odot$ do not lead to white dwarfs. (Adapted from Weidemann, 1987)

half of the mass loss in the solar neighborhood. Jura and Kleinman (1990) extended the survey to stars with distances between 1.0 and 2.5 kpc, and reached the additional conclusion that mass losing carbon stars dominate the return of AGB stars in the outer Galaxy, while closer to the center the carbon stars are less important. Knapp *et al.* (1990) estimate that for the whole Galaxy the mass return rate from red giants, AGB stars and planetary nebulae is about 0.3 to 0.6 $M_\odot$ yr^{-1}. This range is consistent with Weidemann's empirical relation regarding the progenitors of white dwarfs.

It is not yet clear whether the dust that forms in the winds of the AGB stars is still in the form of grains when the material merges with the ISM. Destruction of grains can occur in the wind-wind interaction shocks of planetary neblae.

12.7 Conclusions

We have described the interaction between stellar winds and the interstellar or circumstellar material. We have focussed on two types of models: (1) winds that interact with the interstellar medium, and (2) fast winds that interact with a previously ejected slow wind from the same star. The first model can explain the 'interstellar bubbles' observed around hot stars, ring nebulae around Wolf-Rayet stars, and ultra-compact H II regions. The second model can explain the planetary nebulae.

The time evolution of the interaction zones leads to specific predictions about the sizes, velocities, temperatures and compositions that can be compared with observations. For winds interacting with the ISM, we found that the interacting region quickly reaches the 'snowplow phase', where the swept-up ISM is piled up in front of the expanding interstellar bubble. So almost all observed interstellar bubbles or ring nebulae should be in the snowplow phase. The snowplow phase can exist in two forms: an energy conserving snowplow where the interacting shell is hot, and a momentum conserving snowplow which has a cool and dense interaction shell. The phase of an observed ring nebulae can easily be determined by comparing the momentum and energy of the nebula with that of the stellar wind.

Planetary nebulae are formed by the interaction of a fast wind with a previously ejected slower wind. We have described the structure and evolution of such an interaction model by a simple spherical model. The predictions of this model agree reasonably well with the observed parameters of planetary nebulae.

Actual wind-wind interaction models are more complicated than the simple one dimensional pictures we have studied here. For example, high resolution images of planetary nebulae from the *Hubble Space Telescope* show large numbers of clumps that lead to cometary structures in the inner regions of planetary nebulae. These indicate that two additional effects should be accounted for. First, the evaporation of the clumps leads to 'mass loading', which can slow down an outflow. Second, the oblique shocks from the cometary structures can produce temperatures significantly below those associated with radial-flow shocks. There should be observational conseqences of each of these modifications to the 1-D spherical models. The *HST* images of planetary nebulae also show that many nebulae are not spherical, but have a bipolar structure (Frank and Mellema, 1994). This structure is thought to owe its origin to the rotation of the AGB stars, which leads to higher density mass loss in the equatorial regions, as discussed in Chapter 11.

We have investigated the energy and mass contributions of winds to the ISM. The winds from the very massive stars can produce a larger input of energy integrated over the life of the star than would a supernova. In the case of stars that are too low in mass to become supernovae, the material from the stars primarily re-enters the ISM in the form of slow, dense winds. So for these stars the energy contribution to the ISM dynamics is small, but the injection of chemically enriched material and perhaps grain rich material can have important effects on the life cycles of stars in the galaxy.

12.8 Suggested reading

McCray, R. 1983, 'Stellar Winds and the Interstellar Medium', *Highlights of Astronomy* **6**, 565
(Provides a good discussion of basic wind ISM interaction model used in this chapter.)

Dyson, J.E., & Williams, D.A. 1980, *Physics of the Interstellar Medium* (New York: Wiley)
(This book has useful chapters describing the shock jump conditions and the properties of wind-ISM shocks.)

Kwok, S. 1983, 'Effects of Stellar Mass Loss on the Formation of Planetary Nebulae', in *Planetary Nebulae* IAU Symp 103, ed. D. R. Flower (Dordrecht: Reidel) p. 293
(Describes the wind interaction model for Planetary Nebulae.)

Frank, A. & Mellema, G. 1994, 'The Radiation Gasdynamics of Planetary Nebulae IV. From the Owl to the Eskimo', *Ap. J.* **430**, 800
(Discusses the asymmetries in planetary nebulae that result from non-spherical winds.)

Wood, D.O.S. & Churchwell, E. 1989, 'The Morphologies and Physical Properties of Ultracompact H II regions', *Ap. J. Sup.* **69**, 831
(Describes observations of the wind-ISM interactions around very young and massive stars.)

13 The effects of mass loss on stellar evolution

Mass loss has a profound effect on the evolution of stars. In the case of stars with initial masses greater than about 30 $M_\odot$, mass loss occurs at a considerable rate throughout their whole life. So it affects their evolution from the beginning to the end. In the case of lower mass stars, mass loss is only important in the late stages of their evolution. For those stars only their late evolution is changed dramatically by mass loss. In this chapter we discuss some of the important effects of mass loss on the evolution of the stars. We first discuss the effects in general terms. Later we discuss the evolution of massive stars and of low mass stars under the influence of mass loss. We describe two characteristic examples in some detail: the evolution of a massive star of 60 $M_\odot$ in § 13.2 and of a low mass star of 3 $M_\odot$ in § 13.3. The effect of mass loss on stellar evolution has been described in several reviews: e.g. Iben and Renzini (1983), Chiosi and Maeder (1986) and at several conferences: e.g. Mennessier and Omont (1990) and Leitherer *et al.* (1996).

13.1 The main effects of mass loss

13.1.1 Changes in the surface composition

The outer layers of stars are peeled off by mass loss. Nuclear fusion occurs in the interior of stars. This nuclear fusion changes the chemical composition and the abundance ratios of the elements in the layers where the fusion occurs. If the interior of the star is convective, the convection will mix the products of nuclear burning with the original elements throughout the convective region. When mass loss has peeled off the original unchanged hydrogen-rich layers, the nuclear products will appear at the surface of the star. This results in drastic changes in the abundances of the chemical elements in the stellar photosphere.

This explains for instance
(a) the C-rich atmospheres of low mass stars on the asymptotic giant branch, whereas the stars were initially O-rich,
(b) the ON stars, which are early type stars with a high N and a low C abundance, whereas the stars originally had a higher C than N abundance,
(c) the Wolf-Rayet stars, which have a high He abundance and a low H abundance at their surface. WN stars have a high N surface abundance due to the products of the H-burning via the CN cycle and WC stars have a high C surface abundance due to the products of He-burning.

13.1.2 Changes in the luminosity and main sequence lifetime

The luminosity of a star during the core H-burning depends strongly on the mass of the star. Mass loss reduces the mass of the stars. So for massive stars, which suffer high mass loss already during their main sequence phase, the luminosity of the stars is smaller than for a star with the same *initial mass* but without mass loss. Because of this lower luminosity the main sequence lifetime of stars with mass loss is larger than for stars without mass loss. Figure (13.1) shows this effect schematically for a star with an initial mass of 30 $M_\odot$ with different mass loss rates during the core H-burning phase. The mass loss is parametrized by $\dot{M} = NL_*/c^2$. (The observed mass loss rate

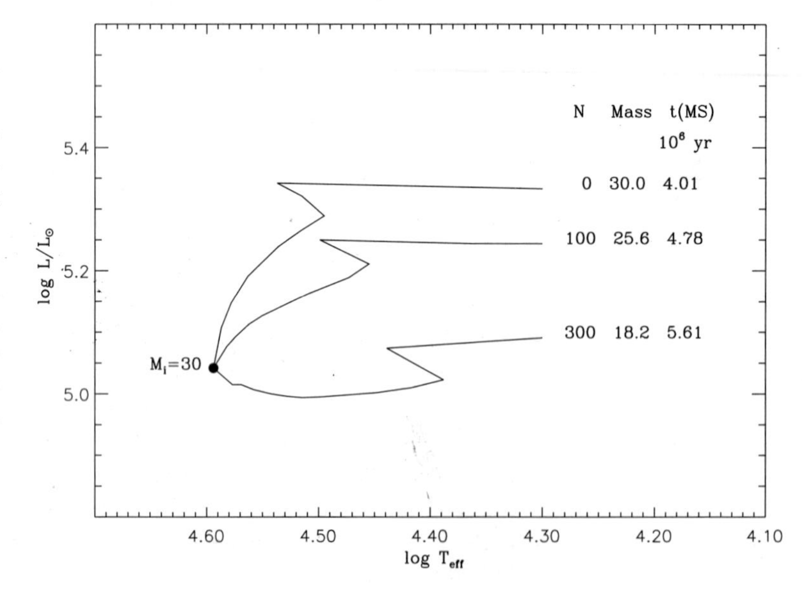

Figure 13.1 Evolutionary tracks of a star of 30 $M_\odot$ with different mass loss rates $\dot{M} = NL_*/c^2$ during the main sequence. The table gives the remaining mass at the end of the main sequence and the main-sequence lifetime. (After de Loore *et al.* 1978)

corresponds to about $N \simeq 100$.) Notice that the luminosity of the star decreases as the mass loss increases. The figure also gives the mass of the star after the main sequence phase. This is obviously lower than the initial mass if $N > 0$. The main sequence lifetime is larger for higher mass loss rates because the luminosity is lower.

13.1.3 The lack of luminous red supergiants

All stars tend to expand their radius and move to the right in the Hertzsprung-Russell diagram (HRD) at the end of their core H-burning phase, because they develop a convective outer envelope. When mass loss has decreased the mass of the H-rich convective envelope below some critical minimum, this envelope can no longer be in convective equilibrium but contracts into a thinner envelope that is in radiative equilibrium. This results in a decrease of the stellar radius and a blueward motion of the star in the HRD. The most massive stars lose so much mass during their main sequence phase and shortly thereafter as luminous blue variables that their envelopes have insufficient mass to become convective. So these stars cannot become red supergiants. This explains the lack of red supergiants brighter than about $L_* > 5 \times 10^5 \ L_\odot$, i.e. at $M_{\rm bol} < -9.^{\rm m}5$. The observed upper limit of the distribution of luminous stars in the HRD is called the 'Humphreys-Davidson limit' (Humphreys and Davidson, 1979). This upper limit is a line that slopes downwards from $T_{\rm eff} \simeq 50\,000$ K to $T_{\rm eff} \simeq 10\,000$ K, and is horizontal for cooler stars. Figure (13.2) shows the upper part of the HRD with the observed Humpreys-Davidson upper limit.

13.1.4 The formation of planetary nebulae

Low mass stars with initial masses of less than about 8 $M_\odot$ suffer severe mass loss during their asymptotic giant branch (AGB) phase. This has two effects: it results in a large amount of circumstellar material expanding with a velocity of about 10 km s^{-1} around the star after the AGB phase, and it makes the star contract when the remaining H-envelope has insufficient mass to sustain a large convective envelope. The star then moves to the left in the HRD with increasing effective temperature. When $T_{\rm eff} \gtrsim 30\,000$ K the radiation ionizes the circumstellar material. At the same time the star starts to blow a fast low density wind that interacts with the slow high density wind from the AGB phase. This results in a planetary nebula (§ 12.5).

13.1.5 The formation of white dwarfs

The fate of stars at the end of their life depends on their mass: stars with a final mass below 1.4 $M_\odot$ become white dwarfs, whereas stars with a final mass above about 1.4 $M_\odot$ end their life in a supernova explosion that leaves either a neutron star or a black hole behind. The amount of mass that a star has at the end of its life is largely determined by the amount of mass that it has lost during its evolution. If low mass stars did not lose a significant fraction of their mass during their late evolution phases, a much larger fraction of the stars would end up as supernovae than observed, and a much smaller number of stars would end up as white dwarfs. The presence of white dwarfs in clusters with a main sequence turn-off mass up to 6 or 8 $M_\odot$ shows that white dwarfs can be formed from stars with initial masses as high as about 8 $M_\odot$. Since the mass of the white dwarfs is less than 1.4 $M_\odot$, these stars must have lost up to about 6.6 $M_\odot$ during their evolution to prevent them from producing supernovae.

13.2 The evolution of a 60 $M_\odot$ star with mass loss

The evolution of a star with initial mass of 60 $M_\odot$, calculated by Maeder and Meynet (1987), is shown in Fig. (13.3).

The upper part (a) of Fig. (13.3) shows the evolutionary track in the HRD. The lower part (b) shows the internal composition of the star during this evolution. The internal composition is shown as a function

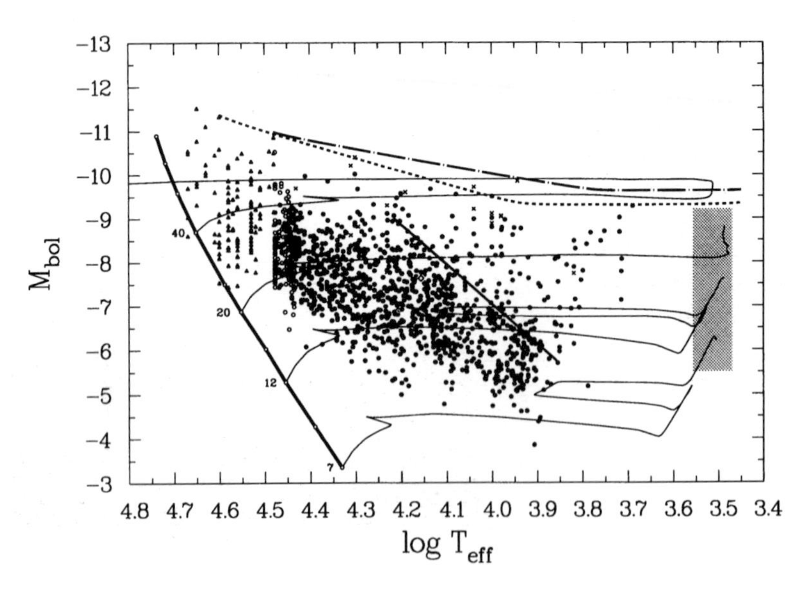

Figure 13.2 The observed 'Humphreys-Davidson' luminosity upper limit for luminous stars in the Large Magellanic Cloud. The high mass loss prevents the formation of luminous red supergiants at $M_{bol} < -9.^m5$. The dotted line is the upper limit according to Fitzpatrick and Garmany (1990) and the dash-dotted line is from Humphreys (1987). Thin lines are evolutionary tracks. The grey area is occupied by red supergiants and the diagonal thick line separates regions of high and low number densities of stars in this plot. (From Fitzpatrick and Garmany, 1990)

of time, with the center of the star at $M/M_\odot = 0$ and the photosphere of the star at the upper boundary. The figure shows the changes in composition for different mass zones within the star: a vertical cut through this figure gives the composition of the star at any given time. Cloudy regions indicate convection zones. Heavy diagonal lines indicate the zones where the nuclear energy production is larger than 10^3 erg g^{-1}s^{-1}. The vertically hatched regions are the zones of variable surface composition.

The upper boundary of Fig. (13.3b) shows the decreasing mass of the star due to mass loss. During the core H-burning phase (from A to C) that lasts 3.7×10^6 years, H is converted into He via the CN-cycle. This results in an increase in He and ^{14}N and a decrease in H and ^{12}C throughout the convective core. Meanwhile the stellar mass decreases

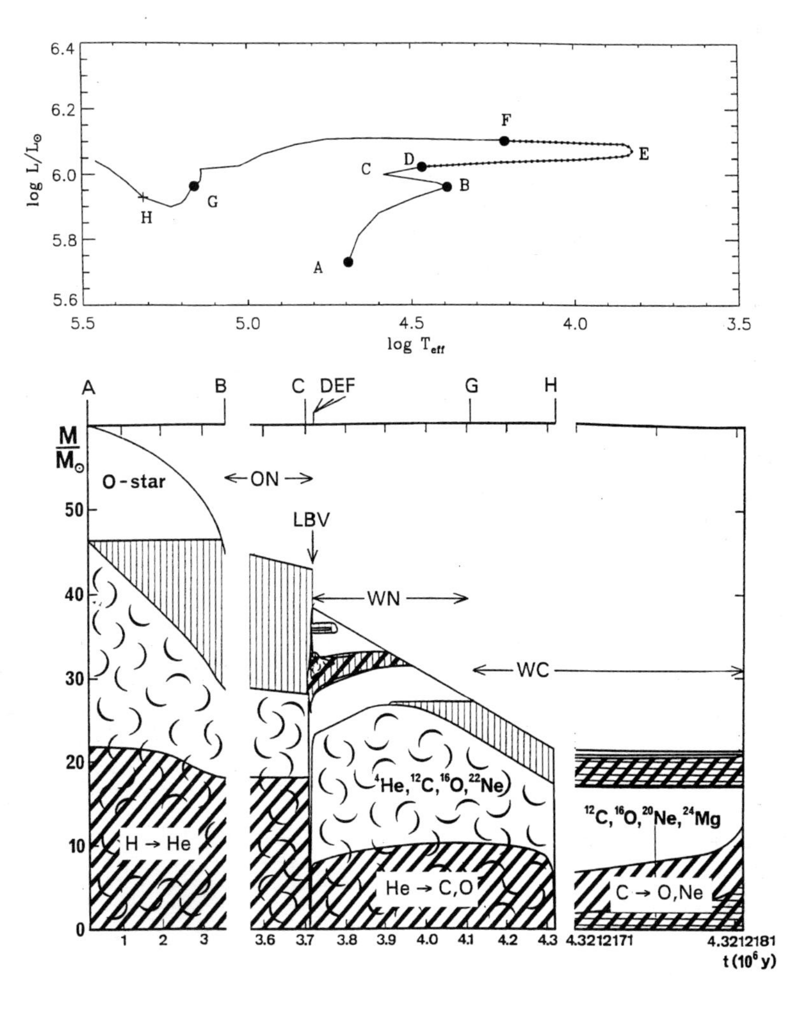

Figure 13.3 Upper figure (a): The evolutionary track of a star with an initial mass of 60 $M_\odot$ in the HRD. Letters indicate specific phases that are discussed in the text. Lower figure (b): The internal structure as a function of time. The phases of different spectral classes are indicated. Cloudy regions are convective. Diagonal lines indicate the regions where most of the nucleur fusion occurs. The time-axis is split into three parts. (From Maeder and Meynet, 1987)

due to mass loss. At first the mass loss rate is $1.4 \times 10^{-6}\ M_{\odot}\ \mathrm{yr}^{-1}$ but this increases to $7.0 \times 10^{-6}\ M_{\odot}\ \mathrm{yr}^{-1}$ at the end of the main sequence phase.

Near the end of the main sequence phase, in phase B, the products of the H fusion by the CN-cycle in the core, that were mixed into the higher layers by convection during the early main sequence, appear at the surface. The surface of the star has reached the vertically dashed zone at phase B. So at that time, the N and He abundances at the stellar surface increase and the H and C abundances decrease: the star becomes a N-rich ON star in phase B.

When the H is exhausted in the core, the star contracts and moves briefly to the left in the HRD, until the H-fusion is ignited in the H-shell surrounding the core in phase C. This is shown by the appearance of a diagonally dashed zone near 30 $M_{\odot}$ in Fig. (13.3b). The star then expands and moves to the right in the HRD. Shortly thereafter, the star becomes unstable and suffers a very high mass loss rate of about $5\ 10^{-4}\ M_{\odot}\ \mathrm{yr}^{-1}$ as a luminous blue variable (phases D,E,F in Fig. (13.3a). The star loses about 5 $M_{\odot}$ in 10^4 years as an LBV. This is shown by the steep drop in the stellar mass between the phases D, E and F in the Fig. (13.3b). Due to this high mass loss the He-enriched layers appear at the surface. This explains why most LBVs have an enhanced He/H ratio of about 0.4 in their photospheres.

The star was starting a post-main sequence expansion and moving to the right in the HRD, but the sudden decrease of the envelope mass in the LBV phase halts the expansion. The envelope then contracts which makes the radius of the star decrease, so the star moves to the left in the HRD (after phase E). The star now becomes a luminous small but hot star with a He-rich and N-rich atmosphere: a Wolf-Rayet star of type WN in phases F to G.

The Wolf-Rayet star gets its energy by converting He into C in the core and from H-fusion in a shell above the He-core. The mass loss rate during the Wolf-Rayet phase is high, of the order of 3×10^{-5} $M_{\odot}\ \mathrm{yr}^{-1}$. At time 4.1×10^6 years (phase G) the star has been peeled-off so far that the C-rich layers appear at the surface. This C-rich material was formed by He-fusion in the core and brought to higher layers by convection: the 'cloudy' region in Fig. (13.3b) between 3.7 and 4.3 $\times 10^6$ years. The star now becomes a Wolf-Rayet star of type WC, i.e. with a C-rich and He-rich atmosphere.

Meanwhile the star keeps contracting at about constant luminosity (phases G to H) until it reaches a radius of about 0.8 $R_{\odot}$ and an effective temperature of 2×10^5 K. This is not the temperature nor the radius that the observer will see. The high mass loss rate makes the wind optically thick in the near-UV, the visual and the IR. So the radiation that escapes from the star comes from the wind where the temperature is of order 3×10^4 K and the radius is about 10 $R_{\odot}$. So the observed WR stars have lower effective temperatures and larger radii than predicted by these evolutionary calculations. (New evolutionary calculations by Schaerer *et al.* (1996) have taken this effect into account.)

The core He-burning phase (phases D to H) lasts about 6×10^5 yrs. The phase of C-burning lasts 2×10^3 years and the later phases of O-burning etc. last less than about a year before the star explodes as a supernova. The core will end up as a black hole. The ejecta will freely expand into the bubble that was blown by the stellar wind during the previous evolution. Eventually the ejecta will catch up with the edge of the bubble and the He-rich and H-rich layers will be mixed.

During its evolution prior to the supernova explosion, the star has ejected 38 $M_{\odot}$ of gas into the interstellar medium: 29 $M_{\odot}$ of H, partly enriched with ^{14}N, and 8 $M_{\odot}$ of He and 1 $M_{\odot}$ of C and O (Chiosi and Maeder, 1986). The interaction of the ejected matter with the surrounding interstellar medium produces a WR ring nebula (see Fig. 12.5). The total wind energy input into the interstellar medium, prior to the supernova explosion, is about $E_w = \int 0.5\dot{M}(t)v_{\infty}{}^2(t)dt \simeq \Delta M v_{\infty}{}^2/2 \simeq 8 \times 10^{50}$ erg for a mean terminal velocity of 2000 km s^{-1}.

13.3 The evolution of a 3 $M_{\odot}$ star with mass loss

We have seen above that mass loss dominates the evolution of a very massive star throughout its evolution. This is not the case for low mass stars, because the mass loss is only important during the later evolutionary stages. In this section we briefly discuss the effects of mass loss on the evolution of low mass stars in the range of $1 \lesssim M_* \lesssim 6\ M_{\odot}$. The evolution of a star with an initial mass of 3 $M_{\odot}$ is taken as an example. The evolutionary track is from Maeder and Meynet (1988) for the early evolution and from Schönberner (1981) for the late evolution.

Figures (13.4) a and b show the evolutionary track and the internal structure of a star of 3 $M_{\odot}$. Letters along the track indicate specific phases that will be discussed. The changes in the internal structure are shown in Fig. (13.4b) only up to the end of the AGB phase (D in Fig. 13.4a).

The mass loss rate of low mass stars during the core H-burning phase (A to B) is very small, of the order of only 10^{-12} to 10^{-10} $M_{\odot}$ yr^{-1}. With a main sequence lifetime of 4×10^8 years, the mass removed during the main sequence phase is therefore negligible.

When the core He-burning and the shell H-burning starts in phase B, the outer layers become convective and start mixing the material. Since the convection extends into the zone where the products of the core H-burning via the CN-cycle are present, the outer convection brings the N-enriched material to the surface. This is called the '*first dredge-up*'. When the star is almost fully convective, shortly after phase B at time 4.00×10^8 years, the star enters the red giant branch (RGB).

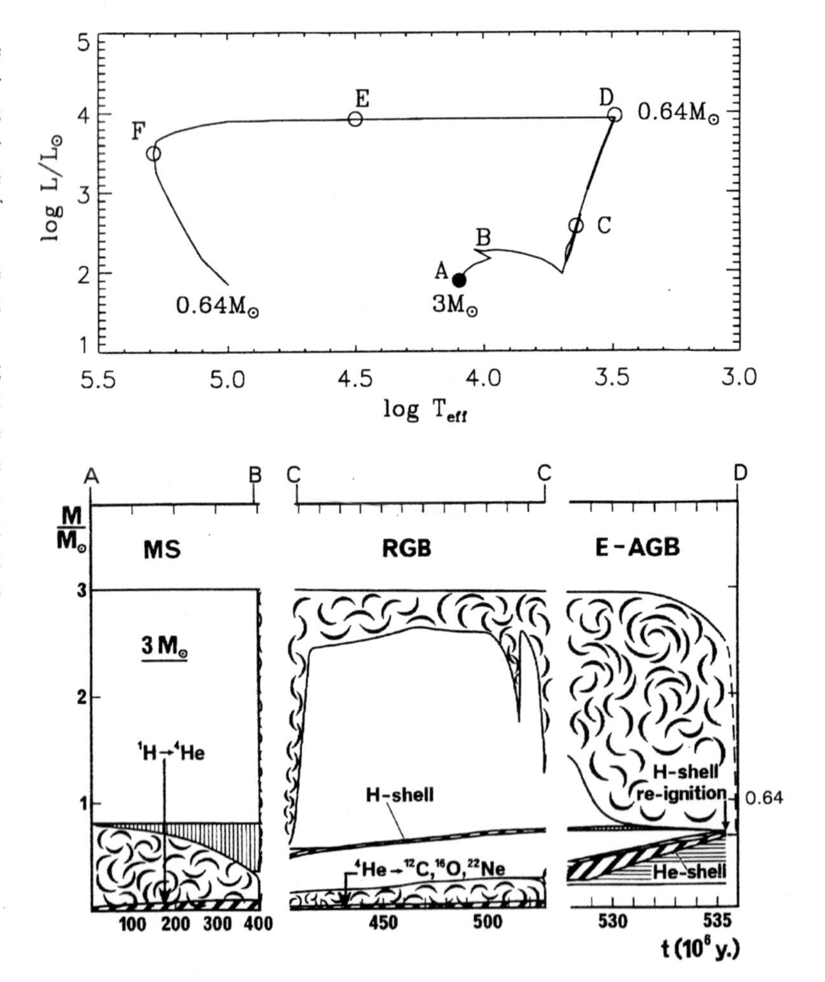

Figure 13.4 The evolution of a star with an initial mass of 3 $M_\odot$. Upper figure: the evolutionary track of the star in the HRD. Lower figure: the internal structure as a function of time. Corresponding phases are indicated in both figures. The symbols are the same as in Fig. (13.3). (From Maeder and Meynet, 1988 and Schönberner, 1981)

In phase C at time 4.03×10^8 years the core He-burning phase starts. During this phase the track makes a downward loop in the HRD. (Evolutionary calculations by others often show a blueward loop.) The mass loss during the red giant phase is about 10^{-10} to 10^{-9} $M_\odot$ yr^{-1}. Since the core He-burning phase lasts 1.30×10^8 years, the mass decreases by only a few percent. This is shown by the very slow decrease of the upper boundary in Fig. (13.4b).

When helium is exhausted in the core, at time 5.27×10^8 years, again at point C in (Fig. 13.4a), the star enters the double shell burning phase for which about 10 percent of the energy is generated by a helium shell around an inert degenerate C core and 90 percent of the energy is generated in a H shell above the inert helium zone. The outer envelope of the star is almost completely convective. The star is now on the early AGB (E-AGB) phase. From now on the evolution is dominated by mass loss.

During the AGB phase, the luminosity of the star is strictly related to the mass of the core by the 'Paczynski-relation'

$$L_*/L_\odot \simeq 5.2 \times 10^4 \left(\frac{M_{\rm core}}{M_\odot} - 0.456 \right) \qquad (13.1)$$

(Boothroyd and Sackmann, 1988) which is valid for core masses of 0.52 to 0.70 $M_\odot$ or luminosities of 3×10^3 to 1.2×10^4 $L_\odot$. As the evolution progresses, the nuclear burning adds material to the core mass, which is the region below the He-burning shell in Fig. (13.4b). As the core mass increases, so does the luminosity according to Eq. (13.1). So the star increases its luminosity and moves up along the AGB. At the same time the mass loss rate increases drastically from about 10^{-9} $M_\odot$ yr^{-1} when the star entered the AGB phase to about 10^{-5} $M_\odot$ yr^{-1} at the tip of the AGB, i.e. phase D (see § 2.8.2).

During the AGB phase the chemical composition at the stellar surface changes gradually. This so-called '*second dredge-up*' occurs during the AGB phase when the convection brings material from the H-burning shell to the surface. This increases the He and N abundance and decreases the C and O abundance.

While the star moves up along the AGB, it will undergo thermal pulses. This is because the thin H-burning layer extinguises after it has burned outwards for some time. The shell He-burning continues, so it moves outwards in mass. Eventually it will reach the H-rich layers. This occurs for the first time in Fig. (13.4b) at $t = 5.35 \times 10^8$ years. When this happens, fresh H-rich material mixes into the very hot 10^8 K He-burning zone, so shell H-burning is suddenly restarted. This produces a thermal pulse in the star. After the thermal pulse the star again has double shell burning for some time. But when the H-burning extinguishes again, the process repeats itself. This leads to a series of thermal pulses, at time intervals of typically 10^4 years, during the AGB phase of the star. The convection brings the products of the thermal pulses to the surface. These products now contain a large fraction of C, formed in the He-burning shell, and also neutron-rich isotopes which are produced during the thermal pulses. This is called the '*third dredge-up*'. It explains the presence of C-rich stars and of AGB stars with a high abundance of neutron-rich isotopes, such as Ba and Sr, in their atmospheres.

The mass of the convective envelope during the AGB phase decreases steadily by two effects: firstly, because the shell burning removes mass from the inside of the envelope by adding it to the core, and secondly, because mass loss from the star removes material from the outside of the envelope. The rate at which nuclear fusion removes mass from the envelope into the core can easily be calculated because the rate of nuclear burning is strictly related to the luminosity of the star. The increase in core mass and the corresponding decrease of the envelope mass is

$$\dot{M}_c = L_*/E_H = 1.0 \times 10^{-11} L_*/L_\odot \qquad (13.2)$$

in $M_\odot\,\mathrm{yr}^{-1}$, where $E_H = 6 \times 10^{18}$ erg g^{-1} is the energy production per gram by H-burning. So an AGB star with $L_* \simeq 3 \times 10^3 L_\odot$ adds about 3×10^{-8} $M_\odot\,\mathrm{yr}^{-1}$ to its core which means that the luminosity, which is connected to the core mass via Eq. (13.1), rises at a rate of about 1.5×10^{-3} $L_\odot$ yr^{-1}. Since the luminosity depends linearly on the core mass and dM_c/dt depends linearly on L_*, we find that $dL_*/dt \sim L_*$ so the luminosity will increase exponentially with time.

The rate at which mass loss removes mass from the envelope is more difficult to estimate because it depends on the mass loss mechanism. Observations of AGB stars suggest that the mass loss rate increases rapidly as the star climbs the AGB (see § 2.8.2). Suppose we adopt a mass loss rate predicted by the Reimers relation, Eq. (2.41), although this is not strictly valid for AGB stars,

$$\dot{M}_{\mathrm{wind}} = 1 \times 10^{-12} \frac{(L_*/L_\odot)(R_*/R_\odot)}{M_*/M_\odot} \qquad (13.3)$$

in $M_\odot\,\mathrm{yr}^{-1}$. During the AGB phase the effective temperature of the star is about constant at $T_{\mathrm{eff}} \simeq 3000$ K, so the radius scales as $R_* \sim L_*^{0.5}$. As the luminosity increases exponentially, the mass loss rate increases almost exponentially. At $L_* = 10^4$ $L_\odot$, $R_* = 400$ $R_\odot$ and

$M_* \simeq M_\odot$ we find a mass loss rate of about 4×10^{-6} $M_\odot\,\mathrm{yr}^{-1}$. Due to the high mass loss rate, the mass of the star decreases very rapidly. This is shown in Fig. (13.4b) by the decrease of the upper boundary. We see that the core mass is about 0.6 $M_\odot$ when the rapid mass loss starts.

The very high mass loss rate near the tip of the AGB is probably due to a pulsating dust driven wind, as described in Chapter 7. This very high mass loss rate produces the OH/IR stars, for which the wind is optically so thick that the emergent radiation is emitted by the dust at a temperature of a few times 10^2 K.

The decrease of the envelope mass on the AGB is

$$\dot{M}_{\mathrm{env}} = |\dot{M}_c| + \dot{M}_{\mathrm{wind}} \tag{13.4}$$

When the mass of the envelope has reached some minimum value, about 10^{-3} $M_\odot$, there is no longer sufficient material to sustain the large convective envelope and so the envelope starts to contract into radiative equilibrium. The star will then move to the left in the HRD. This marks the end of the AGB-phase: phase D in Fig. (13.4a). Calculations show that for a star with an initial mass of 3 $M_\odot$ the core mass at the tip of the AGB, which is almost equal to the total mass of the star at that time, is about 0.64 $M_\odot$. This depends strongly on the adopted mass loss rate during the AGB phase.

The star leaves the AGB when the minimum envelope mass, $M^{\mathrm{env}}_{\mathrm{min}}$, is of order 10^{-3} $M_\odot$ for a core mass of 0.64 $M_\odot$. The star entered the AGB phase with a core mass of $M_{\mathrm{core}} = 0.4$ $M_\odot$ and an envelope mass of $M_{\mathrm{env}} = 2.5$ $M_\odot$. The duration of the AGB phase can thus easily be calculated by combining the Eqs. (13.1), (13.2) and (13.3) and integrating the mass loss rate until an envelope mass of $M^{\mathrm{env}}_{\mathrm{min}}$ is reached. Since the mass loss rate due to the wind is much larger than $\dot{M}_c$, the duration of the AGB phase, the core mass and the corresponding maximum luminosity at the tip of the AGB are essentially determined by the mass loss rate!

After the star has left the AGB, it contracts and moves to the left in the HRD, see Fig. (13.4a). (This phase is not shown in Fig. (13.4b).) The star still satisfies the Paczynski relation between core-mass and luminosity. Since the crossing to the left of the HRD is fast, typically within 10^4 years, the core-mass hardly changes during the contraction and so the luminosity remains constant. So the crossing occurs along a horizontal line in the HRD. The location of the star along this line is determined by the amount of mass in the envelope. This very small envelope mass decreases steadily because the fusion in the shells removes material from the envelope and adds it into the core, and because of a stellar wind, probably of order 10^{-9} $M_\odot\,\mathrm{yr}^{-1}$ in this phase. When the envelope mass is reduced to about 10^{-4} $M_\odot$, the star has an effective temperature of about 30 000 K. This is point E in Fig. (13.4a).

The speed with which the star moves to the left of the HRD, the 'crossing time', is defined as the time it takes for the star to increase its effective temperature from $T_{\mathrm{eff}} \simeq 3000$ K to $T_{\mathrm{eff}} \simeq 30\,000$ K. At this high temperature, the star, which is now small with $R_* \simeq 4$ $R_\odot$, will develop a line driven wind with a mass loss rate of 10^{-9} to 10^{-7} $M_\odot\,\mathrm{yr}^{-1}$ and a velocity of about 1000 $\mathrm{km\,s}^{-1}$. This fast wind will overtake the slow wind of $v_\infty \simeq 10$ $\mathrm{km\,s}^{-1}$ from the AGB phase and the interaction of the two winds will produce a planetary nebula as described in § 12.5. The star is now the central star of a PN (phases E to F in Fig. 13.4a).

The crossing time is determined by the decrease in envelope mass from $M^{\mathrm{env}}_{\mathrm{min}}$ to almost zero. If there is no mass loss from the star during the crossing, this time can be calculated from Eq. (13.2). We find that for a star of $M_c = 0.64$ $M_\odot$ and $L_* = 1 \times 10^4$ $L_\odot$ the crossing time is 3000 years. This agrees with the typical dynamical age of the planetary nebulae, defined as the observed radius divided by the expansion speed of typically 30 to 50 $\mathrm{km\,s}^{-1}$, $\tau_{\mathrm{dyn}} = R/v \simeq 10^3$ to 10^4 yr.

For planetary nebulae with a central star of lower luminosity, $L_{\mathrm{CS}} \lesssim 3\,10^2 L_\odot$, the predicted crossing time is much longer than the measured dynamical ages of the PN. This can only be reached if there is mass loss *after* the AGB phase. Therefore it is assumed that there is a short 'superwind phase' immediately after the AGB phase.

The central stars of the planetary nebulae have a significant mass loss by a radiation driven wind, typically 10^{-9} to 10^{-7} $M_\odot\,\mathrm{yr}^{-1}$ (see § 2.7.3). This reduces the mass of the envelope to an even lower value. When the envelope mass is too low, typically $< 10^{-5}$ $M_\odot$, the shell burning will stop and the star will cool down to become a degenerate white dwarf. This occurs in phase F in Fig. (13.4a). The mass of the white dwarf is almost equal to the core mass when the star left the AGB (apart from a very tiny increase of the core mass during the crossing of the HRD). The mass of the white dwarf is 0.64 $M_\odot$, whereas the initial mass of the star was 3.0 $M_\odot$. So this star has lost 2.36 $M_\odot$.

The mass of white dwarfs is typically between 0.5 and 0.7 $M_\odot$. Stars with an initial mass of up to 8 $M_\odot$ can form white dwarfs. So these stars must have lost a very large fraction of their mass in the form of a stellar wind during their evolution (see Fig. 12.11).

13.4 Conclusions

Mass loss has a significant effect on the evolution of stars of all masses. For massive stars the mass loss is important throughout their evolution. For low mass stars the mass loss is important after the main sequence phase, i.e. in the red giant and AGB phase. The main effects of mass loss are:

(1) It can change the surface chemistry (e.g. high mass ON stars and Wolf-Rayet stars of types WN and WC; low mass C-stars and stars with s-process elements).

(2) It can change the lifetimes of certain evolutionary phases (e.g. longer main sequence life times for massive stars, a shorter AGB-phase for low mass stars).

(3) It explains the occurence of circumstellar nebulae (e.g. planetary nebulae around low mass stars, and ring nebulae around high mass luminous blue variables and Wolf-Rayet stars).

(4) It changes the evolutionary tracks, e.g. the post-AGB evolution of low mass stars and the post main sequence evolution of massive stars.

(5) It explains the lack of luminous red supergiants and the formation of Wolf-Rayet stars from massive stars.

(6) It determines the end products of stellar evolution (e.g white dwarfs from stars of $1.4 < M_* \lesssim 8\ M_\odot$ that would otherwise end as supernovae).

13.5 Suggested reading

Chiosi, C. & Maeder, A. 1986, 'The evolution of massive stars with mass loss', *An. Rev. Astr. Ap.* **24**, 329
(A review on the evolution of massive stars with mass loss)

Humphreys, R.M. & Davidson, K. 1994, 'Luminous blue variables: astrophysical geysers', *Pub. Astr. Soc. Pacific* **106**, 1025
(On the mass loss and evolution of luminous blue variables)

Iben, I. & Renzini, A. 1983, 'Asymptotic giant branch evolution and beyond', *Ann. Rev. Astr. Ap.* **21**, 271
(A review on the evolution of low mass stars)

Mennessier, M.O. & Omont, A. (eds.) 1990, *From Miras to Planetary Nebulae: which path for stellar evolution?*, (Editions Frontieres)
(Proceedings of a conference on the late evolution of low mass stars)

14 Problems

Problem 2.1
Calculate the typical value of r_0 in the expression for the β-law (Eq. 2.3) in units of R_*, for an O star of 40 000 K with a terminal velocity of v_∞ = 2500 km s^{-1} and with a wind velocity of $\beta = 0.8$. The initial velocity is $v_0 \simeq a = \sqrt{kT/\mu m_H}$, where μm_H is the mean mass of the particles (atoms plus electrons) with $\mu = 0.60$.

Problem 2.2
Show that the geometrical dilution factor $W(r)$ (Eq. 2.8) converges to $W = 0.5$ for $r \rightarrow r_{\rm min}$, and to $W(r) \simeq (r/2r_{\rm min})^2$ for $r > 3\ r_{\rm min}$.

Problem 2.3
(a) Derive expressions for the continuum optical depth of an ionized stellar wind with an electron scattering opacity σ_e as a function of radius r and of normalized velocity $w = v/v_\infty$, for a β-law velocity with $\beta = 0.5$, 1, and $\beta > 1$. Notice the very steep dependence on v_0 or r_0 for small values of β. Explain this in physical terms.
(b) For which combination of stellar parameters R_*, $\dot{M}$, v_∞ and v_0 is the wind optically thick $\tau > 0.3$ for electron scattering?

Problem 2.4
(a) Derive expressions for the emission measure of a fully ionized wind for the same set of velocity laws as in the previous problem.
(b) Derive expressions for the total amount of mass in the wind for the same models as above, and for different values of the maximum radius of the wind $r_{\rm max}$ of the wind. Explain why this value diverges if $r_{\rm max} \rightarrow \infty$.

Problem 2.5
The UV spectrum of α Cam (O9.5 Ia) shows strongly saturated P Cygni profiles of the CIV resonance lines near 1500 Å (see Table 2.1). The terminal velocity can be measured from the extent of the 'black' part of the blue component of the doublet, which extends to 1540.5 Å. The star has a luminosity of $L_* = 6 \times 10^5\ L_\odot$, a radius of 28 $R_\odot$ and a mass of 43 $M_\odot$. The effective mass, corrected for radiation pressure by electron scattering is $M_{\rm eff} = M_*(1 - \Gamma_e)$ with $\Gamma_e = 0.34$ (see Eq. 2.36). Determine the effective escape velocity, the terminal velocity v_∞, and the ratio $v_\infty/v_{\rm esc}$.

Problem 2.6
The absorption part of the unsaturated P Cygni profile of the Si IV line at 1393 Å in the spectrum of the star α Problemis has an equivalent width of 3 Å. The Si/H ratio is 3.5×10^{-5} by

number, and the predicted mean ionization fraction in the wind is Si IV/Si = 3.0×10^{-4}. The stellar radius is 30 $R_\odot$. The velocity is a β-law (Eq. 2.3) with $\beta = 1$, $v_\infty = 2000$ km s^{-1}, and $v_0 = 20$ km s^{-1}.

(a) Derive an expression for the column densities of H and Si IV in a column from the observer through the wind ($v > v_0$), as a function of $\dot{M}$.

(b) For unsaturated P Cygni profiles the equivalent width of the absorption (W_λ), depends linearly on the column density N_i of the absorbing ions as

$$\frac{W_\lambda}{\lambda} \simeq 8.8 \times 10^{-21} \; N_i \; f \; \lambda(\text{Å})$$

where f is the oscillator strength of the line (see Table 2.1).

(c) Derive the column density of Si IV and use the ratio Si IV/H to determine the column density of Hydrogen in the wind. Use this and the expression derived in (a) to determine the mass loss rate of the star in $M_\odot$ yr^{-1}.

Problem 2.7

A star with a radius of 10 $R_\odot$ has a wind with a temperature of 40 000 K and a velocity law with $\beta = 1$, $v_\infty = 1500$ km s^{-1} and $v_0 = 30$ km s^{-1}. The wind consists of pure hydrogen ($n_H/\rho = 6.0 \times 10^{23}$ g^{-1}) that is fully ionized.

(a) Express the emission measure $\int n_e n_p dV$ in terms of the mass loss rate.

(b) Calculate the luminosity of the Hα line, in units of $L_\odot$, for mass loss rates of 10^{-6} and 10^{-7} $M_\odot$ yr^{-1}. Adopt for simplicity a mean escape factor of $\langle 1 - W \rangle = 0.75$.

(c) The monochromatic luminosity in the continuum near Hα is $L_\lambda = 2.4 \times 10^{32}$ erg s^{-1} Å^{-1}. Calculate the equivalent width of the Hα emission in Å.

Problem 2.8

The star P Cygni has a radius of 76 $R_\odot$, an effective temperature of 20 000 K, a mass loss rate of 2×10^{-5} $M_\odot$ yr^{-1} and a wind velocity of $v_\infty = 220$ km s^{-1}. The temperature of the wind is about 15 000 K.

(a) Derive an expression for the optical depth of the wind for free-free absorption as a function of wavelength and distance. Assume for simplicity a constant wind velocity of $v(r) \simeq v_\infty$ and a constant gaunt factor of $g = 2.0$. Assume that the wind consists of pure H which is fully ionized, so $n_e = n_p = 6.0 \times 10^{23}$ ρ.

(b) Calculate the radius of the wind where the radial optical depth is $\tau_\lambda = 1/3$ for wavelengths of 10 μm, 1 mm and 10 cm.

(c) Calculate the radio luminosity at 1 mm and at 10 cm using the zero order approximation of Eq. (2.11) and the Rayleigh-Jeans limit for long wavelengths: $\pi B_\nu = 2\pi k T \lambda^{-2}$. Verify that $L_\nu \sim \nu^{2/3} \sim \lambda^{-2/3}$.

(d) Compare this with the flux expected from the photosphere at these wavelengths and express the excess in magnitudes.

Problem 2.9

An AGB star of $L_* = 10^4$ $L_\odot$, $T_{\rm eff}$=2500 K and R_*=532 $R_\odot$ at a distance of 1 kpc has a mass loss rate of 10^{-5} $M_\odot$ yr^{-1} and a wind velocity of 15 km s^{-1}. Radio images of the CO lines show that the wind has an angular diameter of 3 arcmin.

(a) Calculate the duration of the wind phase.

(b) Calculate the total number of CO atoms in the wind, if the abundance ratios by numbers are H : C : O = 1 : 10^{-4} : 10^{-3} and all available C is locked in CO molecules. ($n_H/\rho = 4.4 \times 10^{23}$ g^{-1})

(c) Calculate the luminosity of the CO line at 2.6 mm in terms of the solar luminosity if the wind is isothermal at $T_w = 100$ K, and the excitation of the CO molecules is in LTE ($g_u = 3$, $g_l = 1$, $A_{ul} = 7.2 \times 10^{-8}$ s^{-1} and $n_l \simeq 0.2 n_{CO}$).

Problem 2.10

A red supergiant of $L_* = 2.3 \times 10^5$ $L_\odot$, $T_{\rm eff}$=4000 K and R_*= 1000 $R_\odot$ has a large IR excess due to dust emission. The IR emission is a blackbody spectrum with a temperature of 1200 K and a luminosity of $L_{\rm dust} = 4 \times 10^4$ $L_\odot$.

(a) Assume that the dust is isothermal and consists of grains with a radius of 1 μm and an internal density of 3 g cm^{-3}. Calculate the total emitting surface of the dust grains, the dust mass and the total number of grains.
(b) Assume that the dust, which contributes to the IR emission component of 1200 K, only occurs within a distance of $r_c < r < 2\,r_c$, where the condensation radius is $r_c = 2\,R_*$. The outflow velocity of the dust is 10 km s^{-1}. Calculate the dust mass loss rate and the total mass loss rate if the gas to dust ratio is 200.

Problem 3.1
Derive the expression for the slope of the velocity law at the critical point (Eq. 3.15) from the momentum equation.

Problem 3.2
The solar wind has a mean coronal temperature of about 1.5×10^6 K and a mass loss rate of $2 \times 10^{-14}\ M_\odot\,\mathrm{yr}^{-1}$. The bottom of the corona is at $r_0 \simeq 1.003\ R_\odot$, where the density is $\rho(r_0) = 1 \times 10^{-14}$ g cm^{-3}.
(a) Calculate the potential energy, the kinetic energy and the enthalpy of the gas at r_0.
(b) Calculate the same quantities at the critical point. Which of these energies has absorbed the largest fraction of the energy input?

Problem 3.3
A star with $T_{\rm eff}$=3200 K, R_*=30 $R_\odot$, L_*=85 $L_\odot$ and M_*=6 $M_\odot$ has an isothermal corona of $T = 1. \times 10^6$ K, with a density at the lower boundary of 1×10^{-13} g cm^{-3}.
(a) Calculate the energy per unit mass at the bottom of the corona at $r_0 = R_*$.
(b) Calculate the location of the critical point, r_c, and the mass loss rate.
(c) Calculate the energy per gram gained by the wind between r_0 and r_c. What fraction of the stellar luminosity is used to drive the wind up to the critical point?
(d) Suppose the wind remains isothermal up to a distance of 5 R_*, where $v = 2.2\ a$, and cools adiabatically at $r > 5\ R_*$. What fraction of the stellar luminosity is used to drive the wind?

Problem 3.4
The numerical solution of the momentum equation (Fig. 3.2) shows that the density in the subsonic region of an isothermal corona can be approximated to a good accuracy by Eq. (3.29). Use this expression to derive an approximation for the velocity law in the subsonic region of the wind.

Problem 3.5
Calculate the increase of mass loss rate of a star somewhat like the sun, with an escape velocity of $v_{\rm esc}(r_0) = 700$ km s^{-1} and an isothermal corona of 1.5×10^6 K, if the coronal temperature were increased by only 10 percent. Assume that ρ_0 remains unchanged and that $\mu = 0.60$.

Problem 3.6
The momentum equation of an isothermal wind with an extra force $f = B.v\ dv/dr$ is given by Eq. (3.46). Use this equation and the conditions at the critical point to show that the density distribution is independent of the value of B, i.e. it does not change with the introduction of the extra force.

Problem 4.1
Derive the Bernouilli equation (4.5) for a non-isothermal wind from the first law of thermodynamics and the energy equation.

Problem 4.2
The mean temperature of the stellar wind of the O star ζ Pup is about 30 000 K. The escape velocity at the photosphere is 850 km s^{-1} and the terminal velocity of the wind is 2200 km s^{-1}. The mass loss rate is $3 \times 10^{-6}\ M_\odot\,\mathrm{yr}^{-1}$ and the luminosity is $8 \times 10^5\ L_\odot$.
(a) Calculate the work done by the force due to radiation pressure in the wind per unit mass.
(b) Calculate the total amount of work done by

the force on the wind per second and compare this with the stellar luminosity.

Problem 4.3

The temperature of the solar corona at $r = 1.2\ R_\odot$ is about 1.5×10^6 K and the escape velocity is about 560 km s^{-1}. The velocity of the solar wind at the distance of the earth is 350 km s^{-1}, the temperature is 2.7×10^5 K and the mass loss rate is $2 \times 10^{-14}\ M_\odot\ \mathrm{yr}^{-1}$.
(a) Calculate the energy input into the solar wind per gram between r=1.2 $R_\odot$ and the distance of the earth.
(b) Compare the total energy input into the wind per second with the solar luminosity.

Problem 4.4

Verify the statements a, b and c in § 4.1.5 about the effect of momentum input and heat input on the mass loss from the star, by considering the momentum equation in the form of Eq. (4.24) and requiring that the wind has to go through the critical point where $M = 1$.

Problem 4.5

The general form of the momentum equation of isothermal and non-isothermal winds with a force or energy input that is only a function of distance can be written in the form of Eqs. (4.65) and (4.66) respectively. Show that the left hand sides of these equations reach a minimum value when $M = 1$, which means that the critical point coincides with a minimum of $n(r)$.

Problem 5.1

Derive the momentum equation (5.10) and the energy equation (5.11) in dimensionless units of $\Psi(z)$ and $\tau(z)$.

Problem 5.2

Derive the expression (5.16) for the kinetic energy $\Psi(0)$ at $r \to \infty$ from the conditions at the critical point.

Problem 5.3

Calculate the terminal velocity of a coronal wind of a red giant with $M_* = 4\ M_\odot$, $R_* = 10\ R_\odot$ with a lower boundary of $r_0 = 1.2\ R_*$, $T_0 = 1 \times 10^6$ K and $\mu = 0.60$ as a function of the ratio $L_c(r_0)/\dot{M}$.

Problem 5.4

(a) What is the minimum value of the conductive luminosity $L_c(r_0)$ for a coronal wind for a star of 1 $M_\odot$ if the temperature at the base of the wind at $r_0 = 1.2\ R_\odot$ is $T_0 = 1.5 \times 10^6$ K and the terminal velocity is at least as high as the value of the solar wind speed measured at the earth, i.e. 350 km s^{-1}? Assume a mass loss rate of $\dot{M} = 2 \times 10^{-14}\ M_\odot\ \mathrm{yr}^{-1}$.
(b) What is the ratio between the minimum conductive luminosity and the radiative luminosity of $L_* = L_\odot$?

Problem 6.1

An AGB star with an effective temperature of 3000 K, a radius of 100 $R_\odot$, a luminosity of 730 $L_\odot$ and a mass of 2 $M_\odot$ has a density in the photosphere of 1×10^{-10} g cm^{-3}. The acoustic waves at the lower boundary of the wind with T = 2500 K have an amplitude of δv=0.1 a. Calculate the acoustic luminosity and compare this with the radiative luminosity. Assume a mean particle weight of $\mu = 1$.

Problem 6.2

Calculate the potential energy of the material at the lower boundary of the wind of the AGB star of Problem 6.1. Calculate the maximum mass loss rate that can be driven by the sound wave pressure on the basis of energy considerations only. Compare it with the typical observed mass loss rates at the tip of the AGB stars of about $10^{-5}\ M_\odot\ \mathrm{yr}^{-1}$.

Problem 6.3

Show that the maximum velocity at the lower boundary of a wave driven wind that agrees with the assumption of no dissipation in the subsonic region is given by the Mach number $M_0 < 2(\delta v/a)_0^2$. This implies a second upper limit for the mass loss rate of the AGB star of Problem 6.1. Calculate this upper limit.

Problem 6.4
Show that expression (6.43) for the Mach number at the lower boundary of a wave driven wind converges to expression (3.23) for an isothermal wind without sound waves, if the amplitude of the waves is very small. How small should the amplitude be to find about the same value of M_0 within about 10 percent?

Problem 7.1
Confirm for the case of light coming from a point source, that true absorption or isotropic scattering of incident photons lead to the same transfer of momentum from the radiation field to the gas. Absorption means that all the photon momentum $h\nu/c$ is transfered to the gas. Isotropic scattering means that incident photons are scattered equally into all 4π directions, so the number of photons per solid angle is a constant. To demonstrate this similarity between the absorption and the scattering of radiation
(a) show that the transfer of momentum associated with scattering of an initially radial ($\theta = 0$) photon through an angle θ is $(h\nu/c)(1 - \cos\theta)$, and
(b) then, assuming that the scattering is isotropic, integrate over all 4π steradians. From the result, demonstrate that both isotropic scattering and true absorption lead to a average transfer of momentum, $h\nu/c$, to the particles interacting with a radiation field.

Problem 7.2
Here we derive limits on optically thick continuum driven winds, but rather than assuming that the stellar luminosity is constant we allow for a a transfer of the luminosity from the radiation field to the wind (Ivezic and Elitzur, 1995). This transfer gives the energy equation

$$L_* = L_{\rm rad}(r) + \frac{1}{2}\dot{M}v^2 = \text{constant} \equiv \frac{1}{2}\dot{M}v_L^2.$$

The last equality expresses stellar luminosity in terms of the velocity v_L, which the wind would reach if all the stellar luminosity were converted to wind kinetic energy.

(a) Ignore gravity and gas pressure and show from the momentum equation that

$$v\frac{dv}{dr} = \frac{k_{\rm rp}L_{\rm rad}}{4\pi c r^2} = \frac{k_{\rm rp}\frac{1}{2}\dot{M}(v_L^2 - v^2)}{4\pi c r^2}$$

(b) Gather terms involving v, and integrate over dm, as was done in Eq. (7.5), and thereby show that the terminal velocity of a thick continuum driven wind is

$$v_\infty = v_L\frac{1 - \exp(-v_L\tau_W/c)}{1 + \exp(-v_L\tau_W/c)}$$

where $\tau_W = \int_{r_s}^{\infty} k_{\rm rp}\rho\, dr$.
(c) Now consider the limit where $(v_L/c)\tau_W \ll 1$, and recover the earlier result, Eq. (7.9), for the terminal momentum flux, but now with an upper limit on the optical depth of the wind that is much greater than unity

$$\dot{M}v_\infty = (L/c)\tau_W \quad \text{for} \quad \tau_W \ll c/v_L.$$

Problem 7.3
Derive the ratio of the Planck means given in Eq. (7.35), and derive from the result the expressions for the grain temperature, Eq. (7.36), and condensation radius, Eq. (7.37), for the case $p = 1$.

Problem 7.4
Derive Eq. (7.53) for the mass loss rate of a dust driven wind.

Problem 7.5
In deriving Eq. (7.53), we used some drastic simplifications. Calculate the mass loss rate more accurately, by using Eq. (7.34) for the condensation radius. In finding $Q_{\rm P}^{\rm A}(T)$ assume that the grains are of radius a=0.05 μm, and are composed of silicate material with $T_c = 1500$ K, and with Planck means given in Fig. (7.2). Assume a density in the subsonic region as given by Eq. (7.51) with Δv=4 km s^{-1}, and consider a star with $T_* = 2500$ K, $R_* = 500\ R_\odot$ and M_*=3 $M_\odot$. Find r_c, $\rho(r_c)$, and $\dot{M}$.

Problem 7.6

Carry out the steps in the derivations leading to
(a) the minimal mass loss rate, Eqs. (7.67) and (7.68), and
(b) the maximal mass loss rate (Eq. 7.73).

Problem 8.1

A typical A-type supergiant with a radius of 200 $R_\odot$ and a mass of 20 $M_\odot$ has a wind with a β-type velocity law with $\beta \simeq 1$ and v_∞=200 km s^{-1} and a mass loss rate of about 2×10^{-7} $M_\odot$ yr^{-1}. In this excercise we approximate the velocity with a power law $v = v_0\ (r/R_*)^{1.25}$ with $v_0 = 10$ km s^{-1} for $r < 10.98\ R_*$.
(a) Calculate the time it takes for the gas to be accelerated from v_0 to 0.9 v_∞.
(b) Calculate the acceleration at the distance where $v = v_\infty/2$.
(c) Most of the line driving is done by metal lines with a mean atomic weight of about 50 m_H at a mean wavelength of 2000 Å. Assume a ratio of $N_{\rm metal}/N_H = 4 \times 10^{-5}$. Calculate the number of photon scatterings per second per metal atom needed to accelerate the wind at $v = v_\infty/2$. (Do not forget the gravity!)

Problem 8.2

(a) Calculate the number of effectively optically thick lines required to drive the winds of a typical A-type supergiant with $L_* = 10^5\ L_\odot$, $v_\infty = 200$ km s^{-1} and $\dot{M} = 2 \times 10^{-7}\ M_\odot$ yr^{-1}.
(b) The same for a typical Wolf-Rayet star of $L_* = 3 \times 10^4\ L_\odot$, $v_\infty = 1500$ km s^{-1} and $\dot{M} = 4 \times 10^{-5}\ M_\odot$ yr^{-1}. Compare the result with the maximum value of $N_{\rm eff}$ derived in § 8.2. What can you conclude from this?

Problem 8.3

Calculate the line interaction region for a photon emitted at $\lambda = 796$ Å that is emitted radially from the star and that can be absorbed by an atom with a line at rest wavelength 800 Å in the wind. The velocity is a β-law with $\beta = 1$ and v_∞ = 3000 km s^{-1}. Assume that the absorption profile of the atom is rectangular with a full width of 1 Å.

Problem 8.4

Calculate the shape of the line interaction region at a distance of 2.0 and 10 R_* for a star with a constant velocity wind, $v(r) = v_\infty$, and an absorption line with a rectangular profile with a full width of 0.01 v_∞.

Problem 8.5

Derive the expression for $(dz/d\Delta v)_{r_S}$ of Eq. (8.44).

Problem 8.6

Derive the expression for σ of Eq. (8.47).

Problem 8.7

Calculate the radial Sobolev optical depth, Eq. (8.48), as a function of velocity for the C IV line at 1548.195 Å (Table 2.1) in the spectrum of the star ζ Pup, with parameters given in Table (8.1) and using a velocity law with $\beta = 1.0$.
(a) Assume that all C is in the form of C IV.
(b) Assume an ionization ratio of C IV/C $= 1 \times 10^{-3}$. Will the line be saturated?

Problem 8.8

(a) Calculate the value of the optical depth parameter t (Eq. 8.82) in the wind of the star ϵ Orionis, with the parameters given in Table (8.1) at a distance of $r = 1.5\ R_*$. Assume $\sigma_e = 0.325$ cm^2 g^{-1}, a wind temperature of 20 000 K, and a velocity law with $\beta = 1$.
(b) Calculate the radiative acceleration due to lines at $r = 1.5\ R_*$ (Eq. 8.87) from the data of Table (8.2) for $T_{\rm eff}$=30 000 K. Ignore the weak dependence on δ.
(c) Compare the calculated acceleration with the one derived from the empirical velocity law and the gravity. What could be the reason for the discrepancy?

Problem 8.9

Derive the expression (8.105) for the mass loss

rate of radiation driven winds in the simple way described in § 8.7.1. Start from the momentum equation (8.98) and derive the intermediate equations.

Problem 8.10
The observations of O stars show that $v_\infty/v_{esc} \simeq 2.6$ for O stars (Fig. 2.20). Calculate the value of α that would give this ratio in the point source limit. Assume that the correction for the finite disk is a factor of 1.85, as suggested for the models shown in Fig. (8.17). Calculate the value of α that would result in $v_\infty = 2.6\ v_{esc}$.

Problem 8.11
The wind blanketing can be expressed by means of the parameter f_{refl}, the reflectance of the wind. Calculate the reflectance for the stars listed in Table (8.1).

Problem 9.1
Derive an expression for the Lorentz force, Eq. (9.14), from Eq. (9.13) using a vector identity in spherical coordinates for $(\nabla\times\mathbf{B})\times\mathbf{B}$.

Problem 9.2
Carry out the derivation of the energy constant ε, Eq. (9.26), starting from the radial momentum equation (9.16), and verify Eq. (9.24).

Problem 9.3
We can use properties of the solar wind measured at the earth's orbit, r_E, to estimate the location of the Alfvén radius for the sun. Start with the Alfvén radius condition given in Eq. (9.29), and derive and expression for r_A in terms of the constants, $\mathscr{F}_B$, $\mathscr{F}_m$, and V_A. Assume $\dot{M} = 2\times10^{-14}\ M_\odot\,yr^{-1}$, $v_r(r_E) = 300\ km\,s^{-1}$, $B_r(r_E) = 4\times10^{-5}$ G, Assume also for this problem that $V_A \approx v_r(r_E)$ (ie. the velocity law from r_A to r_E is flat). Evaluate the constants, $\mathscr{F}_B$, $\mathscr{F}_m$, $\mathscr{L}$. Use these constants to find r_A and the spin down time of the present day sun.

Problem 9.4
Derive the maximal surface field Eq. (9.105) for a star with a known mass loss rate and terminal velocity, and discuss the assumptions used in your derivation. Consider a Be star which has the star and wind parameters given in Table (9.1). Assume that line profile analyses show that the star has a projected rotation rate of $v_\phi \sin i = 400\ km\,s^{-1}$, where i is the angle of inclination between the rotation axis and the line of sight. Find ω and the maximal field for this star's wind.

Problem 9.5
Estimate the mass loss rate, $\dot{M}_E$, and terminal velocity of a centrifugal magnetic rotator Be star. Use the stellar parameters from Table (9.1), and assume $B = 10\ B_{min}$, $Z_p = 6$, and $\omega = 0.7$. Equation (9.96) gives the mass loss rate, and assume the terminal velocity is equal to V_M. The sonic radius can be found from Eq. (9.81).

Problem 9.6
Calculate and compare the spin down times of the stars in Table (9.1) for the cases with and without a magnetic field. (Note the table also has $\dot{M}$ and terminal velocities for both non-magnetic and magnetic rotator winds.)

Problem 10.1
Complete the steps in the derivation of Eq. (10.40) and the wave constant, W, (Eq. 10.41), starting from Eq. (10.37).

Problem 10.2
Derive the expression for the mass loss rate of an Alfvén wave driven wind, Eq. (10.72).

Problem 10.3
Verify Eq. (10.77) and find the mass loss rate of a cold wind from a red giant star for which $R=300\ R_\odot$, and with $f_{w,5} = 1$, $B_0 = 1$, $M_* = 1$. (Note that this mass loss rate is comparable with those of dust driven winds.)

Problem 10.4
Carry out the steps leading to the derivation of

the magnetic pressure gradient, Eqs. (10.81) to (10.84), starting with the wave constant, W, of Eq. (10.41).

Problem 10.5
Consider a Wolf-Rayet star, with stellar parameters as given in Table (9.1). Assume that the star has a surface field of $B_0 = 3000$ Gauss and $\alpha = 0.1$, in Eq. (10.78).
(a) Find the Poynting energy flux at the base of the wind. Find the 'stellar Alfvén wave luminosity', Eq. (10.98), and compare it with the Wolf-Rayet radiative luminosity.
(b) Find the mass loss rate, and the terminal velocity. Compare the resultant momentum fluxes with the single scattering limit for the momentum flux of radiatively driven winds, § 7.1 and § 8.2.

Problem 11.1
Starting with Eq. (11.3), derive the cubic equation (11.4). Show that in the limiting case of 'critical rotation', in which the centrifugal acceleration at the star's equator balances gravity, the equatorial radius is 50 % larger than the polar radius, ie. $R_{eq} = 3/2R_{pole}$.

Problem 11.2
The hydrogen optical depth through the wind is given by $\tau_L = \int_R^\infty N_H a_L dr$, where a_L is the value of the photoionization cross section at the Lyman edge of hydrogen. The neutral hydrogen number density is determined by recombination in the wind and thus is given by the proportion $N_H \propto (r/R)^2\rho(r)^2$, where ρ is the density in the wind and is determined by the mass continuity equation, so $\rho \propto \dot{M}/v(r)r^2$. Assume that the velocity in the wind is given by the relation $v(r) = v_\infty(1-R/r)$. Show that $\tau_L \propto (\dot{M}/v_\infty)^2 R^{-1} \propto (\mathscr{F}_m/v_\infty)^2 R^3$. This shows that τ_L is very sensitive to the mass flux rate and flow velocity.

Problem 11.3
Starting with the set of equations in Eqs. (11.18), derive Eq. (11.20) for ϕ' for the simplified case in which $b = 1$. Also for this case, derive from Eq. (11.20) an expression for the maximal displacement angle ϕ'_{max}, Eq. (11.21).

Problem 11.4
Starting with the mass conservaton equation along a flow tube, Eq. (11.26), derive Eq. (11.28). Assume for simplicity that the wind straight out of the pole has the density distribution of a non-rotating star, and that $\mathscr{F}_m{}'$ is independent of $\theta_\circ$. Show that the density contrast from pole to equator is given by Eq. (11.29).

Problem 12.1
Consider a fast, cold wind entering a shock with a speed relative to the shock of $u_1 = v_\infty - v_s$. Since the wind is cold assume that the wind is highly supersonic, so that $u_1^2 \gg \frac{5}{2}\frac{p_1}{\rho_1}$. From the ratio of C_p/C_m show that $p_2/\rho_2 = u_2u_1 - u_2^2$, and from the ratio C_e/C_m show that $(u_1-4u_2)(u_1-u_2) = 0$. This admits two solutions for u_2 versus u_1, only one of which corresponds to a discontinuous jump. Derive this solution.

Problem 12.2
Use the results of Problem (12.1) to show that for the case of 'strong shock', where the Mach number is large, the conservation equations give the jump conditions of Eq. (12.2).

Problem 12.3
Isothermal shock jump conditions. As in Problem (12.1), again consider a fast wind entering a shock with a speed relative to the shock of u_1, which is much faster than the wind isothermal sound speed $a = \sqrt{p_1/\rho_1}$, thus the isothermal Mach number is $m_1 = u_1/a \gg 1$. Consider the jump to the distance behind the shock where the gas has cooled radiatively and the sound speed has returned to its pre-shock value, a. Let the speed relative to the shock at this point be u_3, and let ρ_3 and p_3 be the density and pressure.
(a) Using just the mass and momentum constants C_m and C_p, show that the jump solution for the speeds is $u_3u_1 = a^2$. This means that

$u_3 = u_1/m_1^2 \ll 1$, so the cooled gas moves at the same speed as the shock front.
(b) Now, show from C_m that $\rho_3 = m_1^2 \rho_1$. This tells us that the density in the cooled region is very much larger than the pre-shocked wind density.
(c) Finally, show that the gas pressure at the cooled region is $p_3 = \rho_1 u_1^2(1 - 1/m_1^2) \approx \rho_1 u_1^2$.

Problem 12.4
Find the solution to the differential equation (12.9) that describes the expansion of a wind bubble in the adiabatic phase, and find the solutions for $R(t)$ (Eqs. 12.10, 12.11), and $v(t)$ (Eq. 12.12).

Problem 12.5
Use the parameters for the two winds given in the first row of Table (12.1), and assume that the pressure is determined by the ram pressure of the fast wind $\frac{3}{4}\ \rho_{cs} u_{cs}^2$ (see Eq. 12.2).
(a) Find the mass of the nebula at $t = 3000$ yr using Eq. (12.21).
(b) Find the average density in the nebula assuming that it is isothermal at $T = 10^4$ K and is isobaric at the value determined by the ram pressure.
(c) Find the thickness ΔR of the nebula, and the fractional thickness $\Delta R/R$.

Problem 13.1
The mass-luminosity relation for massive stars on the main sequence is approximately $\log(L_*/L_\odot) \simeq 0.781 + 2.760 \times \log(M_i/M_\odot)$, where M_i is the initial mass. The mass loss rate of massive stars can roughly be approximated by $\log \dot{M} = -12.76 + 1.30 \times \log(L_*/L_\odot)$. The duration of the main sequence phase is $\log \tau_{\rm MS} \simeq 7.719 - 0.655 \times \log(M_i/M_\odot)$, with $\tau_{\rm MS}$ in years.

Calculate the fraction of the mass that is lost by massive stars with $M_i = 25$, 40, 60, 85 and 120 $M_\odot$ during the main sequence phase.

Problem 13.2
A star with an initial mass of 85 $M_\odot$ on the zero age main sequence has a convective core that contains 82 percent of the mass.

Use the mass loss rate and the mass-luminosity relation of Problem (13.1) to calculate the time at which the products of nuclear burning (mainly He and N) will appear at the surface.

Problem 13.3
Evolutionary calculations by Maeder and Meynet (1988) show that the redward evolution of a massive star can be halted due to the high mass loss rate. The track of a star of $M_i = 40\ M_\odot$ will turn leftward if the mass of the star has decreased to 24 $M_\odot$.
(a) Calculate the amount of mass that the star has to lose after the main sequence phase before it reaches this critical value.
(b) The mass loss rate during the redward crossing of the HRD (from main sequence to red supergiant) is about a factor of 3 higher than that given by the mass loss formula of Problem (13.1). The luminosity is about the same as during the main sequence phase. The crossing takes 5×10^3 yr. Calculate the mass of the star when it reaches the red supergiant phase.
(c) Suppose that the mass loss rate during the RSG phase is $10^{-4}\ M_\odot\ {\rm yr}^{-1}$. Calculate the duration of this phase, before the star moves to the left of the HRD and becomes a Wolf-Rayet star.

Problem 13.4
The luminosity of stars on the AGB is related to the core mass via the Paczynski relation $L_*/L_\odot = 5.2 \times 10^4(M_c/M_\odot - 0.456)$ (Eq. 13.1). The nuclear burning in the H- and He-burning shells add matter to the core at a rate of $\dot{M}_c = 1.0 \times 10^{-11}(L_*/L_\odot)$ (Eq. 13.2). Assume that a star enters the AGB with a luminosity of $1.0 \times 10^3\ L_\odot$ and a total mass of 2.0 $M_\odot$.
(a) Derive an expression for the luminosity as a function of time after the star entered the AGB phase.
(b) Assume that $T_{\rm eff}$ remains constant at 3000 K. Derive an expression for the radius as a function of time.

(c) Derive an expression for the core mass as a function of time.

Problem 13.5

This problem shows that the masses of white dwarfs and the luminosity at the tip of the AGB is completely determined by the mass loss rate during the AGB phase. Assume that the mass loss rate of the AGB phase can be approximated by the Reimers relation $\dot{M} = 4 \times 10^{-13} \eta_R (L_*/L_\odot)(R_*/R_\odot)(M_\odot/M_*)$ (Eq. 2.41), with $\eta_R = 3$ for AGB stars. Assume that a star entered the AGB phase with a total mass of 2.0 $M_\odot$ and a luminosity of 1.0×10^3 $L_\odot$.

(a) Derive an expression for the mass of the star as a function of time, using $L(t)$ and $R(t)$ from Problem (13.4). (Hint: $-\dot{M} \times M_* = 0.5(dM_*^2/dt)$)

(b) Use this expression and the one for $M_c(t)$ from Problem (13.4) to derive: the time when the star leaves the AGB ($M_{\rm env} \simeq 0$), the luminosity at the tip of the AGB and the mass of the resulting white dwarf. (This requires a numerical solution of a simple equation.)

(c) Derive the same quatities in the cases when the mass loss rate on the AGB was three times as large, $\eta_R = 9$, and three times as small, $\eta_R = 1$.

14.1 Answers

2.1 $v_0 = 23.5$ km s^{-1}; $r_0 = 0.997$

2.5 $v_{\rm esc} = 622$ km s^{-1}: $v_\infty = 1490$ km s^{-1}; $v_\infty/v_{\rm esc} = 2.40$

2.6a $N_H = \frac{n_H}{\rho} \cdot \frac{\dot{M}}{4\pi R_* v_\infty} - \ln\left(\frac{v_0}{v_\infty}\right)$

2.6b $N_H = 3.86 \times 10^{20} \dot{M}$ with $\dot{M}$ in $M_\odot$ yr^{-1}

2.6c $\dot{M} = 8.9 \times 10^{-7}$ $M_\odot$ yr^{-1}

2.7a Emission measure $EM = 2.7 \times 10^{71} \dot{M}^2$ with $\dot{M}$ in $M_\odot$ yr^{-1}

2.7b $L(H\alpha) = 7.0 \times 10^{-2}$ $L_\odot$ and 7.0 $L_\odot$

2.7c $W_\lambda = 1.1$ Å and $W_\lambda = 110$ Å

2.8a $\tau_\nu = 3.3 \times 10^{26} \nu^{-2} x^{-3}$ where $x = r/R_*$

2.8b $x(\tau_\nu = 1/3) = 1.029$ at 10 μm; 22.2 at 1 mm; 478 at 10 cm

2.8c $L_\nu = 2.3 \times 10^{20}$ erg s^{-1} Hz^{-1} at 1 mm and 1.0×10^{19} at 10 cm

2.8d The excess is $6.^{\rm m}4$ at 1 mm and $13.^{\rm m}1$ at 10 cm

2.9a Duration is 2.8×10^4 years

2.9b 2.5×10^{52} CO molecules

2.9c $L(2.6{\rm mm}) = 2.1 \times 10^{-4}$ $L_\odot$

2.10a Total emitting surface is 1.3×10^{30} cm^2

2.10b $M_d = 9.9 \times 10^{-9}$ $M_\odot$ yr^{-1}; $\dot{M} = 2.0 \times 10^{-6}$ $M_\odot$ yr^{-1}

3.2a $E_{\rm pot} = -1.95 \times 10^{15}$, $E_{\rm kin} = 5.36 \times 10^5$, enthalpy=$5.19 \times 10^{14}$ erg g^{-1}

3.2b $E_{\rm pot} = -4.24 \times 10^{14}$, $E_{\rm kin} = 1.04 \times 10^{14}$, enthalpy=$5.19 \times 10^{14}$ erg g^{-1}

3.3a $e(r_0) = 3.5 \times 10^{13}$ erg g^{-1}

3.3b $\dot{M} = 7.3 \times 10^{-8}$ $M_\odot$ yr^{-1}

3.3c $\Delta e = 1.7 \times 10^{14}$ erg g^{-1}; $\Delta L_w(r_c) = 2.5 \times 10^{-3} L_*$

3.3d $\Delta L_w = 9.1 \times 10^{-3} L_*$

3.5 $\dot{M}$ increases by a factor of 2.7

4.2a Work $= 2.8 \times 10^{16}$ erg g^{-1}

4.2b Total work $= 5.3 \times 10^{36}$ erg s^{-1} $= 1.8 \times 10^{-3}\ L_*$
4.3a $\Delta e = 2.4 \times 10^{15}$ erg g^{-1}
4.3b $\Delta e \times \dot{M} = 3.1 \times 10^{27}$ erg s^{-1} $= 8.0 \times 10^{-7}\ L_*$
5.4 $L_c(\text{min}) = 2.1 \times 10^{27}$ erg s^{-1} $= 5.5 \times 10^{-7} L_*$
6.1 $L_{ac} = 2.9 \times 10^{31}$ erg s^{-1} $= 1.0 \times 10^{-5} L_*$
6.2 $\dot{M}_{max} = 1.2 \times 10^{-8}\ M_\odot$ yr^{-1}
7.5 $r_c = 1.85 R_*$, $\rho(r_c) = 9.7 \times 10^{-15}$ g cm^{-3}, $\dot{M} = 2.5 \times 10^{-6}\ M_\odot$ yr^{-1}
8.1a $t = 2.5 \times 10^7$ s
8.1b $dv/dt = 1.42$ cm s^{-2}
8.1c 110 scatterings per metal atom per second
8.2a $N_{eff} = 30$ and $N_{max} = 1.5 \times 10^3$
8.2b $N_{eff} = 2 \times 10^4$ and $N_{max} = 200$
8.7a $\tau = 2.6 \times 10^3 (v_\infty / v)$
8.7b $\tau = 2.6(v_\infty / v)$
8.8a $t = 4.1 \times 10^4 =$ constant
8.8b $g_L = 0.19$ cm s^{-2}
8.8c $dv/dt = 1.4 \times 10^3$ cm s^{-2}
8.10 $\alpha = 0.871$ in point source limit; $\alpha = 0.664$ for a finite disk
9.3 $r_A = 21\ R_\odot$
9.4 $\omega \geq 0.83$, $B_{r,0} \leq 13.7$ Gauss
9.5 $\dot{M} = 2.6 \times 10^{-9}\ M_\odot$ yr^{-1}, $v_\infty = 2600$ km s^{-1}
9.6 For Be stars: τ_J $(B = 0)$ $= 4.8 \times 10^9$ yr, τ_J (FMR) $= 1.8 \times 10^7$ yr
10.3 $\dot{M} = 8 \times 10^{-7}\ M_\odot$ yr^{-1}
10.5 $L_{Poynting} = 1.0 \times 10^{35}$ erg s^{-1}, $\dot{M} = 1.0 \times 10^{-7}\ M_\odot$ yr^{-1}, $v_\infty = 5600$ km s^{-1} (note: the approximation that $\delta B/B$ breaks down well before this speed is reached)
12.5 $M_{neb} = 0.051\ M_\odot$, $R_{neb} = 0.083$ pc, $\rho = 1.3 \times 10^{-21}$ g cm^{-3}, $\Delta R/R = .38$
13.1 5, 12, 25, 50 and 99 percent
13.2 $t = 1.0 \times 10^6$ yr
13.3 Mass lost during MS $= 4.7\ M_\odot$. Mass lost during crossing is $0.015\ M_\odot$. Mass to be lost as RSG is $11.3\ M_\odot$. Duration of the RSG phase is 1.1×10^5 yr
13.4 $L(t) = L(0) \exp(t/1.92 \times 10^6)$; $R(t) = R(0) \exp(t/3.84 \times 10^6)$; $M_c(t) = 0.456 + 0.0192 \exp(t/1.92 \times 10^6)$
13.5 $M_*{}^2 = M_0^2 + 0.360(\exp(t/1.28 \times 10^6) - 1)$; $t = 2.89 \times 10^6$ yr, $L = 4.5 \times 10^3 L_\odot$, $M_c = 0.542 M_\odot$

APPENDIX I

The chronology of stellar wind studies

Table A1.1 *Chronology of stellar wind studies*

I	THE EARLY HISTORY : THE SIMILARITIES BETWEEN SPECTRA OF NOVAE AND LUMINOUS STARS 1572 – 1910
1572	A nova (actually a supernova) B Cas is observed by Tycho Brahe, and its light variations are followed for 16 months. (Brahe 1573)
1600	Blaeu discovers the 'nova' P Cygni (actually a luminous blue variable) that suddenly increased in brightness to third magnitude. (de Groot, 1985)
1837	η Carinae is observed to have a four magnitude increase in brightness. This star is also a luminous blue variable. (Beals, 1931)
1867	Wolf and Rayet discover with an optical prism, three stars in Cygnus that have 'broad emission bands'. (Wolf and Rayet, 1867)
1876	Copeland notes that Nova Cyg 1876 shows an optical spectrum like that of a planetary nebula. This suggests that the nova is surrounded by a shell which may have been ejected from the star.
1891-1894	Campbell observes Nova Aurigae 1891 and obtains the first high quality photographic spectra which show line profiles consisting of a violet shifted absorption and red shifted emission. (Such profiles are now known as P Cygni lines). (Campbell, 1892)

Table A1.1 *Chronology of stellar wind studies (cont.)*

1892-1894	The Doppler formula is used to interpret the lines of Nova Aurigae in terms of high speed motions. The initial interpretation is that the emission component of the P Cygni line comes from a 'bright lined star' receding away from us at 370 km s^{-1}, while the absorption component comes from a 'dark lined star' moving toward us at 520 km s^{-1}. (Clerke, 1903). The spectrum of Nova Aurigae also shows broad emission lines, which were referred to as 'Wolf-Rayet' bands. The observations by Campbell (1892) establish a link between the formation of P Cygni profiles, Wolf-Rayet bands, and outflows.
1918	Barnard uses photographs to measure the increasing angular diameter of Nova Aquilae 1918.
1920	Adams and Burwell suggest that the emission bands (i.e. the redward side of the P Cygni profiles) arise in the expanding shell and the violet shifted absorptions arise from the portion of the shell directly in front of the star.
1927	Hubble and Duncan combine the observed shell expansion rate (1.0 arcsecond per year), with the radial velocity information from the profile (1700 km s^{-1}), to derive a distance (360 pc) to Nova Aquilae.
II	DIAGNOSTICS OF THE STRUCTURE OF THE OUTER ATMOSPHERES OF THE SUN AND STARS 1913 – 1947
1913	Mitchell measures emission lines obtained during an eclipse of the sun, and concludes that the chromosphere extends to many scale heights above the photosphere.
1919	Saha shows that if quantum theory is used in place of classical electromagnetic wave theory, atoms can stop the flow of radiation and the 'active' atoms could be given sufficient levity to explain the presence of matter at the tops of solar prominences and in the solar corona.
1924-1925	Milne (1924) and Johnson (1925) show that *selective aborption* of line radiation could extend the chromosphere, accelerate ions and heat the solar corona. These papers lay the foundation for the line driven wind theory.
1926	Milne explains that an increasing velocity causes an atom to aborb at steadily shortening wavelength, exposing the atom to a 'much fiercer' radiation field. This will lead to a velocity of material from the sun of about 1600 km s^{-1}. 'Line locking' is suggested as a possible occurence, and Milne derives a velocity law for radiation driven outflows. He finds a β-law with $\beta = 1/2$ (see Eq. 2.2).

Table A1.1 *Chronology of stellar wind studies (cont.)*

1929	Chapman and Ferraro discuss the neutral electrical state of the streams of solar corpuscles (mostly protons and electrons) reaching the orbit of the earth.
1929	Beals presents a sequence of papers classifying P Cygni profiles, and he proposes that a 'continuous ejection of gaseous material from the star' can explain the nova-like line profiles.
1934	Chandrasekhar provides a rigorous radiative transfer explanation for line profiles originating in expanding atmospheres. He also uses 'iso-velocity surfaces' to interpret line profiles and to explain features such as flat-topped emission lines.
1934	Kosirev uses extended photosphere radiation transfer effects to explain the anomalously red colors of P Cygni stars and WR stars. The stars are redder than normal stars that show similar ionizations stages in their spectra. Kosirev derives the 'annual loss of mass from WR stars' of 10^{-5} $M_\odot$ yr^{-1} and a terminal wind velocity of 1000 km s^{-1}. (These are to be compared with modern estimates of 3×10^{-5} $M_\odot$ yr^{-1} and $v_\infty = 2500$ km s^{-1}.) He also suggests that many stars must go through the Wolf-Rayet phase of evolution.
1935	McCrea shows that the densities in the expanding atmospheres of WR stars and novae are sufficiently large to prevent different elements from slipping past one another. Thus he deduces that the gases 'drag each other along' in their motion.
1935	Adams and MacCormack show that the Na D and Mg h and k resonance lines in the spectra of M supergiants are shortward shifted by about 5 km s^{-1}. They suggest that the stars are losing mass in the form of 'gradually expanding envelopes'.
1939	Spitzer argues on the basis of the observed lines in the spectra of α Her, and α Ori, that the cool giants have a 'fountain flow' rather than expansion.
1942	Edlén explains the solar 'coronium' lines at 5303 Å and 6375 Å, observed during eclipses, as lines of the high ion stages Fe XIV and Fe X. This shows that the solar corona has a temperature of about 10^6 K.
1946	Biermann (1946), Alfvén (1947), Schwarzschild (1948) show that the solar corona could be heated by acoustic waves or magneto-hydrodynamic waves, that are generated in the convection zone, as the waves steepen and form shocks. Biermann concludes that the high coronal temperature of the sun does not violate the second law of thermodynamics.
1947	Sobolev develops the escape probability theory for the transfer of line radiation in expanding atmospheres. (Sobolev, 1960)

Table A1.1 *Chronology of stellar wind studies (cont.)*

III	THE DEVELOPMENT OF THE SOLAR WIND THEORY, FURTHER EVIDENCE FOR OUTFLOWS 1947 – 1967
1949	Underhill finds that it is not possible to compute static model atmospheres for O stars because these stars have layers in which the outward force due to radiation pressure exceeds the inward force of gravity.
1949	The idea that Wolf-Rayet stars have fast winds with large mass loss rates is challenged by Thomas (1949) who argues that the emission lines can be formed by a process like that occuring in the solar chromosphere.
1951	Biermann interprets the comet tail deflections observed by Hoffmeister (1943) in terms of a steady flow of 'corpuscular radiation' from the sun with a speed of about 400 $km\,s^{-1}$.
1952	Bondi shows the first plot with an X-type flow topology with a critical point, for the velocity stucture of stellar winds and stellar accretion. He focuses on the infall branch of the solution.
1956	Deutsch develops the first solid evidence that the low speed expansion of the M5 II star α Her is in fact produced in matter escaping from the star. This is because the matter is flowing beyond the orbit of the G star companion at a speed greater than the escape speed at that distance. The rate of mass loss is estimated to be $10^{-7}\ M_{\odot}\ yr^{-1}$.
1960	In a series of papers, Parker and Chamberlain debate whether the hot solar corona leads to a fast solar 'wind' or a slow solar 'breeze'.
1960	Wilson identifies that the major problem of cool star winds lies in explaining why the wind velocity is much less than the surface escape speed.
1960	Parker introduces the name 'stellar wind' and considers applications of solar wind theory to other stars.
1961	Clauser points out the analogy between wind outflow from the solar gravitational field and the supersonic flow through a de Laval nozzle of a rocket.
1962	The Mariner 2 probe to Venus finds that there is a fast wind from the sun occuring at all times. This settles the debate between Parker and Chamberlain in favor of Parker's wind model. (Neugebauer & Snyder, 1962)
1962	Hoyle and Wickramasinghe suggest that grains may form in the atmospheres of carbon stars and can then be driven out by radiation forces. This leads to a renewal of interest in radiation driven wind theory as an explanation for mass loss from cool giant stars.

Table A1.1 *Chronology of stellar wind studies (cont.)*

1964	Kuhi finds evidence for mass loss from pre-main sequence T Tauri stars with speed of about 300 km s^{-1}.
1966-1967	Wilson (1966) and Kraft (1967) show that angular rotation speeds of stars decrease strongly toward spectral types later than F5V. They also show from the study of stars in clusters that chromospheric activity and rotation both diminish sharply as a star ages. This points to a substantial loss of angular momentum in late type main sequence stars.
1967	Weber and Davis develop the basic equations of the magnetic rotator theory and apply it to the problem of the slow rotation rate of the sun.
1968	Coleman studies *Mariner 2* measurements of fluctuations in the solar wind and deduces that transverse Alfvén waves propagate through the wind.
IV	ROCKET AND EARLY SATELLITE OBSERVATIONS OF STELLAR WINDS 1967 – 1982
1967	Morton observes broad P Cygni profiles ($v > 2000$ km s^{-1}) of the C IV, N V and Si IV resonance lines in the ultraviolet spectra of several O and B supergiants, obtained with a rocket launched spectrometer. He estimates the stars have mass loss rates of a few $\times 10^{-6}$ $M_\odot$ yr^{-1}.
1970	Lucy and Solomon develop a radiation driven wind model to explain the high speed outflow from hot stars observed by Morton.
1969-1971	Woolf and Ney (1969) and Gehrz and Woolf (1971) explain the large IR excesses of red supergiants as being caused by dust driven winds.
1972	Gilman develops the idea of 'momentum coupling' in dust driven winds, in which the dust is accelerated by radiation pressure, and subsequent collisions drag the gas along.
1975-1977	It is pointed out in several papers that the mass loss rates from hot stars can be measured accurately with radio measurements of free-free emission (Panagia and Felli, 1975; Wright and Barlow, 1975) or from IR observations of the winds (Barlow and Cohen, 1977).
1975	Castor, Abbott & Klein develop the multi-line radiation driven wind theory, now called the CAK theory. The predicted mass loss rates and terminal velocities agree with the observations within factors of about two.
1975	Castor, McCray & Weaver develop the basic theory for 'wind bubbles' in the interstellar medium.

Table A1.1 *Chronology of stellar wind studies (cont.)*

1976	Snow and Morton publish an atlas of UV P Cygni profiles observed with the *Copernicus* satellite and show that all early type stars with a luminosity in excess of 2×10^4 $L_{\odot}$ have a high mass loss rate.
1976	Lamers, Rogerson & Morton show that the winds of hot stars are superionized, because the UV spectra observed with the *Copernicus* satellite shows P Cygni profiles of high ionization species such as N V and O VI. They make the first detailed empirical models for the winds of hot stars. (Lamers and Morton, 1976; Lamers and Rogerson, 1978)
1979	Linsky and Haisch find evidence for the existence of a 'dividing line' for late type stars in the Hertzsprung-Russell diagram which separates stars with solar type chromospheres from stars with strong winds, based on ultraviolet *IUE* spectra.
1978	Kwok, Purton & FitzGerald propose the 'wind interaction model', in which a planetary nebula is explained as resulting from the collision of the fast wind from the hot central star with the slow dense wind produced while the star was on the asymptotic giant branch (AGB).
1979	Cassinelli and Olson explain the superionization of the winds of early type stars as being due to Auger ionization caused by the presence of X-rays in the winds.
1979	The *Einstein* satellite confirms that essentially all O stars are X-ray sources, and that there is a dividing line in the Hertzsprung-Russell diagram for cool stars with or without an X-ray flux. (Seward *et al.*, 1979)
1980	Lucy and White show that X-rays from hot stars can arise from shocks produced by instabilities in line driven winds.
1982	Abbott finds that wind bubbles can be as important as supernovae in adding energy and momentum into the interstellar medium.
1987-1993	Kudritzki and colleagues develop unified atmosphere radiation driven wind models, and include effects of multiple scattering and rotation. They study the dependence of mass loss on metallicity and show the mass loss of luminous stars in other galaxies, such as the LMC and SMC, will be smaller than for galactic stars. (Kudritzki, Pauldrach and Puls, 1987)
V	INSTABILITIES AND NON-SPHERICAL EFFECTS IN WINDS 1980 – PRESENT
1985	Willson and Bowen show that dust driven winds can occur if pulsation leads to levitation of the underlying atmospheres of AGB stars.

Table A1.1 *Chronology of stellar wind studies (cont.)*

1985	The *IRAS* satellite finds large IR excesses in Be stars due to rotationally distorted atmospheres and evidence for dust in C-rich Wolf-Rayet stars. (Waters, Coté & Lamers, 1987; Williams, van der Hucht & Thé, 1987)
1988	Owocki, Castor & Rybicki develop the theory for the generation of shocks in the winds of hot stars.
1992-1994	*IUE* observations of hot stars indicate that aborption components in the UV spectra appear and disappear with the rotational periods of the stars. This suggests the presence of co-rotating structures in the winds; possibly due to magnetic fields. (Kaper & Henrichs, 1994)
1993-1994	Bjorkman & Cassinelli (1993) and Owocki, Cranmer & Blondin (1994) show that stars with even moderate rotation speeds have winds that converge in the equatorial regions and form wind compressed disks.
1991-95	Conti (1991) identifies 'Wolf-Rayet galaxies', and Leitherer & Heckman (1995) study 'collective-wind' phenomena in galaxies.

APPENDIX 2

Elements of thermodynamics

The first law of thermodynamics relates the internal energy u to the heat energy Q and the work $p\,dV$ done by a gas when it expands. (p = pressure, V = specific volume = volume per unit mass.)

$$dQ = du + p\,dV \tag{A2.1}$$

where Q, u and $p\,dV$ are expressed in erg per unit mass. This equation states that when heat is added to a gas, part of it will be used for expansion and the remainder will increase the internal energy.

The specific heat c of a gas is defined by

$$c = dQ/dT \tag{A2.2}$$

so c (erg g^{-1} K^{-1}) is the heat needed to raise the temperature of a unit mass by one degree. If the gas is kept at a constant volume V the heat needed for $\Delta T = 1$ K is

$$c_v = (dQ/dT)_v \tag{A2.3}$$

If the gas is at constant pressure, the heat needed for $\Delta T = 1$ K is

$$c_p = (dQ/dT)_p \tag{A2.4}$$

The value of c_p will be larger than that of c_v because the gas at constant pressure will expand when heat is added, so it needs more heat input to raise T by 1 K. The first law of thermodynamics implies that

$$dQ = p\,dV + du = \left(p + \frac{\partial u}{\partial V}\right) dV + \frac{\partial u}{\partial T} dT \tag{A2.5}$$

where we have explicitly used the fact that the internal energy, u, of the gas depends on T and V. Equation (A2.5) shows that

$$c_v = (dQ/dT)_v = \partial u/\partial T \tag{A2.6}$$

and

$$c_p = \left(p + \frac{\partial u}{\partial V}\right)\left(\frac{\partial V}{\partial T}\right)_p + \left(\frac{\partial u}{\partial T}\right) = \left(p + \frac{\partial u}{\partial V}\right)\left(\frac{\partial V}{\partial T}\right)_p + c_v \quad \text{(A2.7)}$$

For a perfect gas the internal energy u depends only on temperature, so $\partial u/\partial V = 0$ and the pressure is related to density ρ and temperature as

$$p = \mathcal{R}\rho T/\mu \quad \text{(A2.8)}$$

where $\mathcal{R}$ is the gas constant and μ is the mean weight of the particles in units of proton mass. The specific volume V is related to the density by $V = \rho^{-1}$ and so

$$\left(\frac{\partial V}{\partial T}\right)_p = \left(\frac{\partial \rho^{-1}}{\partial T}\right)_p = \left(\frac{\partial \mathcal{R}T/\mu p}{\partial T}\right)_p = \frac{\mathcal{R}}{\mu}\frac{1}{p} \quad \text{(A2.9)}$$

This reduces c_p in Eq. (A2.7) to

$$c_p = c_v + \mathcal{R}/\mu \quad \text{(A2.10)}$$

The *enthalpy* $h(T, p)$ of a gas is defined as

$$h = u + pV = u + \frac{p}{\rho} \quad \text{(A2.11)}$$

It is the internal energy plus the potential of the gas to do work by adiabatic expansion. For a perfect gas, which we will consider only, the enthalpy is a function of temperature only,

$$h(T) = u(T) + p/\rho = u(T) + \mathcal{R}/\mu \quad \text{(A2.12)}$$

The enthalpy is related to the heat by the first law of thermodynamics (A2.1)

$$dh = du + p\ dV + V\ dp = dQ + V\ dp \quad \text{(A2.13)}$$

Taking the derivative at constant pressure, we find an expression for c_p

$$\frac{dh}{dT} = \left(\frac{\partial Q}{\partial T}\right)_p = c_p \quad \text{(A2.14)}$$

So c_v is the temperature derivative of the internal energy and c_p is the temperature derivative of the enthalpy. Below we will show why the enthalpy is a basic quantity in stellar wind theories.

For adiabatic reversible processes, i.e., without gain or loss of heat, $dQ = 0$, the first law implies $du = -p\ dV$ and $dh = V\ dp$. The derivatives of $u(T)$ and $h(T)$ are

$$(du/dT)dT = c_v dT = -p\ dV = -(\mathscr{R}T/\mu V)dV = -(\mathscr{R}\rho T/\mu)d\rho^{-1} \tag{A2.15}$$

$$(dh/dT)dT = c_p dT = V\ dp = (\mathscr{R}T/\mu p)dp \tag{A2.16}$$

From these equations the logarithmic derivatives of T, p, V can be expressed in terms of c_p and c_v or their ratio

$$\gamma \equiv c_p/c_v \tag{A2.17}$$

$$\frac{d\ \ln T}{d\ \ln V} = -\left(\mathscr{R}/\mu\right) c_v = -\frac{c_p - c_v}{c_v} = 1 - \gamma \tag{A2.18}$$

$$\frac{d\ \ln T}{d\ \ln p} = -\left(\mathscr{R}/\mu\right) c_p = \frac{c_p - c_v}{c_p} = \frac{\gamma - 1}{\gamma} \tag{A2.19}$$

$$\frac{d\ \ln P}{d\ \ln V} = \frac{c_p}{c_v} = \gamma \tag{A2.20}$$

In the stellar wind theory, the specific volume is usually expressed as inverse density, $\rho = V^{-1}$, resulting in

$$\frac{d\ \ln T}{d\ \ln \rho} = \gamma - 1 \tag{A2.21}$$

$$\frac{d\ \ln p}{d\ \ln \rho} = \gamma \tag{A2.22}$$

The relations (A2.18) up to (A2.23) describe the relative variations of p, ρ and T of an adiabatically expanding gas,

$$p \sim \rho^{\gamma},\ T \sim \rho^{\gamma-1},\ T \sim p^{(\gamma-1)/\gamma} \tag{A2.23}$$

The proportionality factors of these equations depend on the entropy of the gas. They are derived below.

For a mono-atomic gas consisting of neutral atoms, ions and electrons which have only three translational degrees of freedom, the internal energy is $u = (3/2)\mathscr{R}T/\mu$ and so $c_v = (3/2)\mathscr{R}/\mu, c_p = (5/2)\mathscr{R}/\mu$ and $\gamma = 5/3$. For multi-atomic gases, such as molecules, which have more degrees of freedom because of their rotation and vibration, the internal energy is more complicated, $u = (3/2)\mathscr{R}T/\mu + u_{\rm vibr} + u_{\rm rot}$. In this case c_v is larger than $(3/2)\mathscr{R}/\mu$ and γ is smaller than 5/3. The same is true for an atomic gas which is partially ionized. For a partially ionized gas the value of u is larger than $(3/2)\mathscr{R}T/\mu$ because a considerable fraction of the energy is in the excitation and ionization. Partially ionized gases have $c_v > (3/2)\mathscr{R}/\mu$ and $4/3 < \gamma < 5/3$.

The importance of the concept of enthalpy for stellar winds can be demonstrated in a simple example. Consider the momentum equation of a stationary wind, moving with a velocity v, in which the outward force is provided by the gas pressure.

$$\frac{dv}{dt} = v\frac{dv}{dr} = -\frac{GM_*}{r^2} - \frac{1}{\rho}\frac{dp}{dr} \tag{A2.24}$$

The gradient of the enthalpy is

$$\frac{dh}{dr} = \frac{du}{dr} + p\frac{d1/\rho}{dr} + \frac{1}{\rho}\frac{dp}{dr} = \frac{dQ}{dr} + \frac{1}{\rho}\frac{dp}{dr} \tag{A2.25}$$

where Q is the heat content per gram and dQ/dr is the heat input per unit mass and per unit distance. The momentum equation can thus be written as

$$v\frac{dv}{dr} + \frac{GM_*}{r^2} + \frac{dh}{dr} - \frac{dQ}{dr} = \frac{d}{dr}\left\{\frac{v^2}{2} - \frac{GM_*}{r} + h - Q\right\} = \frac{de}{dr} = 0 \tag{A2.26}$$

where e is the total energy per unit mass, which is the sum of the kinetic energy and the potential energy plus the enthalpy minus the heat content. This equation shows that the energy e per unit mass is constant in the wind, if the gas pressure and gravity provide the only forces. So in an adiabatic wind where $dQ = 0$, the quantity

$$\frac{v^2}{2} - \frac{GM_*}{r} + h(r) = \text{ constant} \tag{A2.27}$$

with $h = c_p T = (5/2)\mathcal{R}T/\mu$ for a perfect gas. In a wind with energy deposition $dQ(r)/dr > 0$, one finds from Eq. (A2.26) that

$$\frac{v^2}{2} - \frac{GM_*}{r} + h(r) = Q(r) + \text{ constant} \tag{A2.28}$$

This means that the energy, $Q(r)$, deposited in the wind is used for increasing the kinetic energy, the potential energy and the enthalpy.

The *entropy* S of a gas is defined by

$$dQ = TdS \tag{A2.29}$$

where S is the entropy per unit mass in erg g^{-1} K^{-1}. This definition shows that an adiabatic wind, $dQ = 0$, is also isentropic (constant entropy) because $dS = 0$.

The entropy is related to the equation of state which describes the relation between p, ρ, T. The first law expressed in entropy is

$$TdS = du + pd(1/\rho) \tag{A2.30}$$

For a perfect gas this can be written as

$$\begin{aligned} dS &= \frac{du}{T} + \frac{p}{T}d\rho^{-1} = \frac{du}{dT}d\ln T - \frac{\mathcal{R}}{\mu}d\ln\rho \\ &= \frac{du}{dT}(d\ln p - d\ln\rho) - \frac{\mathcal{R}}{\mu}d\ln\rho \end{aligned} \tag{A2.31}$$

Substituting $du/dT = c_v$ and $\mathscr{R}/\mu = c_p - c_v$ we find

$$\frac{dS}{c_v} = d\ \ln\ p - \gamma d\ \ln\ \rho = d\ \ln(p/\rho^\gamma) \qquad \text{(A2.32)}$$

Integration of this equation gives the equation of state for constant values of γ and c_v

$$p = e^{S/c_v}\rho^\gamma = K\rho^\gamma \qquad \text{(A2.33)}$$

where K is a constant for an adiabatic wind.

The *speed of sound* c_s is defined by the relation

$$c_s^2 \equiv (\partial p/\partial \rho)_s \qquad \text{(A2.34)}$$

The adiabatic speed of sound, which is the propagation speed of adiabatic (isentropic) waves, is defined by the derivative at constant entropy

$$c_s^2 \equiv (\partial p/\partial \rho)_s = \exp(S/c_v)\gamma\rho^{\gamma-1} = \gamma p/\rho \qquad \text{(A2.35)}$$

For a perfect monatomic gas the adiabatic sound speed is

$$c_s^2 = \frac{5}{3}\frac{\mathscr{R}T}{\mu} \qquad \text{(A2.36)}$$

The isothermal speed of sound, which is indicated by '*a*' throughout this book, is for a perfect gas

$$a^2 = (\partial p/\partial \rho)_T = \mathscr{R}T/\mu = p/\rho \qquad \text{(A2.37)}$$

so

$$c_s^2 = \gamma a^2 \qquad \text{(A2.38)}$$

The isothermal speed of sound is the propagation velocity of sound waves which remain isothermal during their compression and rarefaction. This condition is usually only met by waves of very long wavelength when radiative cooling during the compression prevents the adiabatic temperature increase, and radiative heating by the surrounding material prevents the adiabatic temperature decrease during the rarefaction. In general, sound waves will be adiabatic and propagate with a velocity $\sqrt{\gamma a^2}$ even if the gas through which they travel is isothermal. For instance, a sound wave propagates horizontally through the earth's atmosphere with the adiabatic sound speed even though it moves through an almost isothermal gas.

APPENDIX 3

De l'Hopital's rule for equations with a singular point

The rule of de l'Hopital provides an expression for the derivative of a first order differential equation with a singular point at that singular point.

Suppose that the function $y(x)$ is described by the differential equation

$$\frac{dy}{dx} = \frac{f(x)}{g(x)} \tag{A3.1}$$

This equation has a singular point at x_c where $f(x_c) = 0$ and $g(x_c) = 0$. The derivative dy/dx at x_c can then be found by developing $f(x)$ and $g(x)$ in a Taylor sequence around x_c.

$$\left(\frac{dy}{dx}\right)_{x_c} = \frac{f(x_c) + (x - x_c)f'(x_c) + 0.5(x - x_c)^2 f''(x_c) + \ldots}{g(x_c) + (x - x_c)g'(x_c) + 0.5(x - x_c)^2 g''(x_c) + \ldots} \tag{A3.2}$$

If the first derivatives of f and g at x_c are both unequal to zero then

$$\left(\frac{dy}{dx}\right)_{x_c} = \frac{f'(x_c)}{g'(x_c)} \tag{A3.3}$$

If only one of the two derivatives $f'(x_c)$ or $g'(x_c)$ is equal to zero, then $(dy/dx)_{x_c}$ is either 0 or $\pm\infty$. If both $f'(x_c)$ and $g'(x_c)$ are zero, then

$$\left(\frac{dy}{dx}\right)_{x_c} = f''(x_c)/g''(x_c) \tag{A3.4}$$

etc.

This rule is very useful for determining the solution of a stellar wind through the critical point of the momentum equation. The momentum equation often has a form

$$\frac{1}{v}\frac{dv}{dr} = \left\{\frac{2\gamma a^2}{r} - \frac{GM_*}{r^2} + f(r)\right\} / \left\{v^2 - \gamma a^2\right\} \tag{A3.5}$$

where $v(r)$ is the velocity, γa^2 is the speed of sound and the function $f(r)$

describes the forces other than gas pressure and gravity in the wind. The singular or critical point of this equation occurs at r_c where the numerator and denominator are zero. So

$$v^2(r_c) = \gamma a^2(r_c) \tag{A3.6}$$

and

$$\frac{GM_*}{r_c^2} - f(r_c) = \frac{2\gamma a^2(r_c)}{r_c} = \frac{2v^2(r_c)}{r_c} \tag{A3.7}$$

The application of de l'Hopital's rule to Eq. (A3.5) for an isothermal wind with $\gamma = 1$ and $a =$ constant gives

$$\frac{1}{v_c}\left(\frac{dv}{dr}\right)_{r_c} = \left\{-\frac{2a^2}{r_c^2} + \frac{2GM_*}{r_c^3} + \left(\frac{df}{dr}\right)_{r_c}\right\} \Big/ \left\{2v_c\left(\frac{dv}{dr}\right)_{r_c}\right\} \tag{A3.8}$$

with $v_c \equiv v(r_c) = a$. This results in a quadratic expression for $(dv/dr)_{r_c}$ with a root

$$\left(\frac{dv}{dr}\right)_{r_c} = \pm\sqrt{-\frac{a^2}{r_c^2} + \frac{GM_*}{r_c^3} + \frac{1}{2}\left(\frac{df}{dr}\right)_{r_c}} \tag{A3.9}$$

The velocity derivative at the critical point of an isothermal wind can have a positive or a negative value. For a wind with an outward increasing velocity the positive gradient should be adopted.

The determination of the velocity gradient through the critical point by de l'Hopital's rule is essential for numerical solutions of the momentum equation because it allows a smooth transition through a region where the numerical calculation will be unstable as dv/dr may vary between large positive or negative values in a small distance interval.

APPENDIX 4

Physical and astronomical constants

Symbol	Value	Units
c	3.00×10^{10}	cm s^{-1}
G	6.67×10^{-8}	dyn cm^2 g^{-1}
h	6.63×10^{-27}	erg s
k	1.38×10^{-16}	erg K^{-1}
m_H	1.66×10^{-24}	g
m_e	9.11×10^{-28}	g
$\mathcal{R}$	8.31×10^{7}	erg K^{-1} mole
σ	5.67×10^{-5}	erg cm^{-2} K^{-4} s^{-1}
$\pi e^2/m_e c$	2.65×10^{-2}	cm^2
$1\ eV$	1.60×10^{-12}	erg
$L_\odot$	3.83×10^{33}	erg s^{-1}
$R_\odot$	6.96×10^{10}	cm
$M_\odot$	1.99×10^{33}	g
pc	3.09×10^{18}	cm
AU	1.50×10^{13}	cm
$1\ M_\odot\ yr^{-1}$	6.30×10^{25}	g s^{-1}
1 yr	3.16×10^{7}	s
n_H/ρ	4.43×10^{23}	H-atoms/g for Pop I stars

Bibliography

Abbott, D.C. 1982a, *Ap. J.* **259**, 282

Abbott, D.C. 1982b, *Ap. J.* **263**, 723

Abbott, D.C., Bieging, J.H. & Churchwell, E. 1981, *Ap. J.* **250**, 645

Abbott, D.C., Bieging, J.H., Churchwell, E. & Torres, A.V. 1986, *Ap. J.* **303**, 239

Abbott, D.C. & Hummer, D.G. 1985, *Ap. J.* **294**, 286

Abbott, D.C. & Lucy, L.B. 1985, *Ap. J.* **288**, 679

Abbott, D.C., Telesco, C.M. & Wolf, S.C. 1984, *Ap. J.* **279**, 225

Achmad, L., Lamers, H.J.G.L.M. & Pasquini, L. 1997, *A & A* **320**, 196

Adams, W.S. & Burwell, C.G. 1920, *Ap. J.* **51**, 121

Adams, W.S. & MacCormack, E. 1935, *Ap. J.* **81**, 119

Alazraki, G. & Couturier, P. 1971, *A & A* **13**, 380

Alfvén, H. 1947, *MNRAS* **107**, 211

Allen, C.W. 1973, *Astrophysical Quantities* (London: The Athlone Press)

Babel, J. 1995, *A & A* **301**, 823

Balona, L.A., Henrichs, H.F. & Le Contel, J.M. (eds) 1994, *Pulsation, Rotation and Mass Loss in Early Type Stars* (Dordrecht: Kluwer)

Barlow, M.J. & Cohen, M. 1977, *Ap. J.* **213**, 737

Barnard, E.E. 1918, *Ap. J.* **49**, 199

Beals, C.S. 1929, *MNRAS* **90**, 202

Beals, C.S. 1931, *Pub. Dominion Astrophysical Obs.* **4**, 271

Bedijn, P.J. 1987, *A & A* **186**, 136

Belcher, J.W. 1971, *Ap. J.* **168**, 509

Belcher, J.W. & MacGregor, K.B. 1976, *Ap. J.* **210**, 498

Bieging, J.H., Abbott, D.C. & Churchwell, E.B. 1989, *Ap. J.* **340**, 518

Biermann, L. 1946, *Naturwiss.* **33**, 118

Biermann, L. 1951, *Zs. f. Astrophys.* **29**, 274

Bjorkman, J.E., & Cassinelli, J.P. 1993, *Ap. J.* **409**, 429

Blöcker, T. 1995, *A & A* **297**, 727

Bondi, H. 1952, *MNRAS* **112**, 195

Boothroyd, A.I. & Sackmann, I.J. 1988, *Ap. J.* **328**, 653

Bowen, G.H. 1988, *Ap. J.* **329**, 299
Bowen, G.H. & Willson, L.A. 1991, *Ap. J.* **375**, L53
Brahe, Tychonis 1573, *De Nova Stella* (Hafniae: Lorentz Benedict)
Brandt, J.C. 1970, *Introduction to the Solar Wind* (San Francisco: Freeman)
Bretherton, F.P. 1970, *Lect. Appl. Math. (Am. Math. Soc.)* **13**, 61
Campbell, W.W. 1892, *A & A* **11**, 799
Cassinelli, J.P. 1979, *Ann. Rev. Astr. Ap* **17**, 275
Cassinelli, J.P. & Haisch, B.M. 1974, *Ap. J.* **188**, 101
Cassinelli, J.P. & Hartmann, L. 1977, *Ap. J.* **212**, 488
Cassinelli, J.P. & Lamers, H.J.G.L.M. 1987 in *Exploring the Universe with the IUE satellite*, eds. Y. Kondo *et al.* (Dordrecht: Reidel) p. 139
Cassinelli, J.P. & Olson, G.D. 1979, *Ap. J.* **229**, 304
Castor, J.I. 1970, *MNRAS* **149**, 111
Castor, J.I. 1987, in *Instabilities in Luminous Early Type Stars*, eds. H.J.G.L.M. Lamers & C.W.H. de Loore (Dordrecht: Reidel) p. 159
Castor, J.I. 1993, in *Massive Stars: Their Lives in the Interstellar Medium* ASP Conf. Series 35, eds. J. Cassinelli & E. Churchwell (ASP: San Francisco) p. 297
Castor, J.I., Abbott, D.C. & Klein, R.I. 1975, *Ap. J.* **195**, 157
Castor, J.I., McCray, R. & Weaver, R. 1975, *Ap. J.* **200**, L107
Chamberlain, J.W. 1961, *Ap. J.* **133**, 675
Chandrasekhar, S. 1934a, *MNRAS* **94**, 444
Chandrasekhar, S. 1934b, *MNRAS* **94**, 522
Chapman, S. & Ferraro, V.C.A. 1929, *MNRAS* **89**, 470
Chiosi, C. & Maeder, A. 1986, *Ann. Rev. Astr. Ap.* **24**, 329
Chu Y. -H. 1991 in *Wolf Rayet Stars and Their Interrelations with Other Massive Stars in Galaxies*, eds. K. A. van der Hucht & B. Hidayat (Dordrecht: Kluwer) p. 349
Churchwell, E. 1990, *A & A Reviews* **2**, 79
Clauser, T. 1961, *Suppl. Nuovo Cimiento* **22**, # 1
Clayton, D. D. 1968, *Principles of Stellar Evolution and Nucleosynthesis* (New York: McGraw Hill)
Clerke, Agnes M. 1903, *Problems in Astrophysics* (London: Adams & Black)
Coleman, P. J. 1968, *Ap. J.* **153**, 371
Collins, G.W. 1989, *Fundamentals of Stellar Astrophysics* (New York: Freeman)
Conti, P.S. 1988, in *Wolf Rayet Stars: Observations, Physics, Evolution*, eds. C. de Loore & A. Willis (Dordrecht: Reidel) p. 3
Conti, P.S. 1991, *Ap. J.* **77**, 115
Cram, L.E. 1985, in *Relations between Chromospheric-coronal Heating and Mass Loss in Stars*, eds. R. Stalio & J.B. Zirker, (Trieste: Osservatorio Astronomico) p. 93
Crowther, P.A. & Bohannan, B. 1996, in *Wolf-Rayet Stars in the Framework of Stellar Evolution*, 33rd Liège Astroph. Coll., eds. Vreux *et al.* (Liège: Univ. Liège)
Crowther, P.A., Hillier, D.J. & Smith, L.J. 1995, *A & A* **293**, 403

Davidson, K. & Harwit, M. 1967 *Ap. J.* **148**, 443

de Groot, M. 1985, *Irish Astr. J.* **17**, 263

de Jager, C. 1980, *The Brightest Stars* (Dordrecht: Reidel)

de Loore, C., de Greve, J.P. & Vanbeveren, D. 1978, *A & A Sup.* **34**, 363

Deutsch, A.J. 1956, *Ap. J.* **123**, 210

Deutsch, A.J. 1960, in *Stellar Atmospheres*, Stars and Stellar Systems Vol VI, ed. J. Greenstein (Chicago: Univ. Chicago Press) p. 543

Deutsch, A.J. 1968, in *Mass Loss from Stars*, ed. M. Hack, (Dordrecht: Reidel), p. 1

Dorschner, J. & Henning, T. 1995, *A & A Reviews* **6**, 271

Dos Santos, L.C., Jatenco-Periera, V. & Opher, R. 1993, *Ap. J.* **410**, 732

Draine, B.T. 1981, *Ap. J.* **245**, 880

Draine, B.T. & Lee, H.M. 1984, *Ap. J.* **285**, 89

Dupree, A.K. 1986, *Ann. Rev. Astr. Ap* **24**, 377

Dupree, A.K. & Reimers, D. 1987, in *Exploring the Universe with the IUE Satellite*, eds. Y. Kondo *et al.* (Dordrecht: Reidel) p. 321

Dyson, J.D. & Williams, D.A. 1980 *Physics of the Interstellar Medium* (New York: Wiley)

Edlén, B. 1942, *Zs. f. Astrophys.* **22**, 30

Fitzpatrick, E.L. & Garmany, C.D. 1990, *Ap. J.* **363**, 119

Frank, A. & Mellema, G. 1994, *Ap. J.* **430**, 800

Friend, D.B. & Abbott, D.C. 1986, *Ap. J.* **311**, 701

Friend, D.B. & Castor, J.I. 1983, *Ap. J.* **272**, 259

Friend, D.B. & MacGregor, K. B. 1984, *Ap. J.* **282**, 591

Gail, H.-P. & Sedlmayr, E. 1987a, *A & A* **171**, 197

Gail, H.-P. & Sedlmayr, E. 1987b, *A & A* **206**, 153

Garmany, C. D., Conti, P. S. & Chiosi, C. 1982, *Ap. J.* **263**, 777

Gayley, K.C., Owocki, S.P. & Cranmer, S.R. 1995, *Ap. J.* **442**, 296

Gehrz, R.D. 1972, *Ap. J.* **178**, 715

Gehrz, R.D. & Hackwell, J.A. 1976, *Ap. J.* **206**, L161

Gehrz, R.D. & Woolf, N.J. 1971, *Ap. J.* **165**, 285

Gilman, R.C. 1969, *Ap. J.* **155**, L185

Gilman, R.C. 1972, *Ap. J.* **178**, 423

Goldberg, L. 1986, in *The M-Type Stars*, eds. H.R. Johnson and F.R. Quercy, NASA SP-492, p. 245

Groenewegen, M.A.T. & Lamers, H.J.G.L.M. 1991, *A & A* **43**, 429

Groenewegen, M.A.T., Lamers, H.J.G.L.M. & Pauldrach, A.W.A. 1989, *A & A* **221**, 78

Habing, H.J. 1996, *A & A Reviews* **7**, 97

Habing, H.J., Tignon, J. & Tielens, A.G.G.M. 1994, *A & A* **286**, 523

Hamann, W.-R. 1995, in *Wolf-Rayet Stars: Binaries, Colliding Winds, Evolution*, eds. K.A. van der Hucht & P.M. Williams (Dordrecht: Kluwer) p. 105

Hamann, W.-R., Koesterke, L. & Wessolowski, U. 1995, *A & A* **299**, 151

Hammer, R. 1982, *Ap. J.* **259**, 767

Hanner, M. S. 1988, *NASA Conf. Pub.* **3004**, p. 22
Hartle, R.A. & Sturrock, P.A. 1968, *Ap. J.* **151**, 1155
Hartmann, L.W. & MacGregor, K.B. 1980, *Ap. J.* **242**, 260
Hartmann, L.W. & MacGregor, K.B. 1982, *Ap. J.* **259**, 180
Hecht, J.H., Holm, A.V., Donn, B. & Wu, C.-C. 1984, *Ap. J.* **280**, 228
Heckman, T.M., Lehnert, M.D. & Armus, L. 1993, in *The Environment and Evolution of Galaxies*, eds. J.M. Shull & H. A. Thronson (Dordrecht: Kluwer) p. 455
Hillier, D.J. 1984, *Ap. J.* **280**, 744
Hoffmeister, C. 1943, *Zs. f. Astrophys.* **23**, 265
Holweg, J.V. 1973, *Ap. J.* **181**, 547
Holzer, T.E. & Axford, W. 1970, *Ann. Rev. Astr. Ap* **8**, 31
Holzer, T.E., Flå, T. & Leer, E. 1983, *Ap. J.* **275**, 808
Howarth, I.D. & Prinja, R.K. 1989, *Ap. J.* **69**, 527
Hoyle, F. & Wickramasinghe, N. C. 1962, *MNRAS* **124**, 417
Hubble, E. & Duncan, J. C. 1927, *Ap. J.* **66**, 59
Humphreys, R.M. 1987, in *Instabilities in Luminous Early Type Stars* eds. H.J.G.L.M. Lamers & C.W.H. de Loore (Dordrecht: Reidel) p. 3
Humphreys, R.M. & Davidson, K. 1979, *Ap. J.* **232**, 409
Humphreys, R.M. & Davidson, K. 1994, *Publ. Astr. Soc. Pacific* **106**, 1025
Hundhausen, A.J. 1972 *Solar Wind and Coronal Expansion* (Heidelberg: Springer Verlag)
Iben, I & Renzini, A. 1983, *Ann. Rev. Astr Ap.* **21**, 271
Ignace, R., Cassinelli, J.P. & Bjorkman, J.E. 1996, *Ap. J.* **459**, 671
Ivezic, Z. & Elitzur, M. 1995, *Ap. J.* **445**, 415
Johnson, M.C. 1925, *MNRAS* **85**, 813
Jordan, S. 1981, *The Sun as a Star*, NASA-SP 450
Jura, M. 1987, *Ap. J.* **313**, 743
Jura, M. & Kleinmann, S.G. 1989, *Ap. J.* **341**, 359
Jura, M. & Kleinmann S.G. 1990, *Ap. J.* **364**, 663
Kaper, L. & Henrichs, H. F. 1994, in *Instability and Variability of Hot Star Winds* eds. A. F. J. Moffat *et al.* (Kluwer: Dordrecht) p. 115
Klein, R.I. & Castor, J.I. 1978, *Ap. J.* **220**, 902
Knapp, G.R. & Morris, M. 1985, *Ap. J.* **292**, 640
Knapp, G.R., Rauch, K.P. & Wilcots, E.M. 1990, in *Evolution of the ISM*, ed. L. Blitz, ASP Conf. Series 12 (San Francisco: ASP) p. 151
Koninx, J.P.M. & Pijpers, F.P. 1992, *A & A* **265**, 183
Kosirev, N.A. 1934, *MNRAS* **94**, 430
Kraft, R.P. 1967, *Ap. J.* **150**, 551
Kudritzki, R.P. 1988, in *Radiation in Moving Gaseous Media*, eds. Y. Chmielewski & T. Lanz (Geneva Observatory) p. 1
Kudritzki, R.P., Pauldrach, A.W.A. & Puls, J. 1987, *A & A* **173**, 293
Kudritzki, R.P., Pauldrach, A.W.A., Puls, J. & Abbott, D.C. 1989, *A & A* **219**, 205

Kudritzki, R.P., Lennon, D.J. & Puls, J. 1995, in *Science with the VLT*, eds. J.R. Walsh & I.J. Danziger (Berlin: Springer) p. 246

Kudritzki, R.P., Mendez, R.H., Puls, J. & McCarthy, J.K. 1997, in *Planetary Nebulae, IAU Symp 180*, eds. H.J. Habing & H.J.G.L.M. Lamers, (Dordrecht: Kluwer) p. 64

Kuhi, L.V. 1964, *Ap. J.* **140**, 1409

Kurucz, R.L. 1979, *Ap. J. Sup.* **40**, 1

Kwok, S. 1975, *Ap. J.* **198**, 583

Kwok, S. 1983 in *Planetary Nebulae* IAU Symp 103, ed. D. R. Flower (Dordrecht: Reidel) p. 293

Kwok, S. & Pottasch, S.R. (eds) 1987, *Late Stages of Stellar Evolution* (Dordrecht: Reidel)

Kwok, S., Purton, C.R. & FitzGerald, P.M. 1978, *Ap. J.* **219**, L125

Lamers, H.J.G.L.M. 1997, in *Stellar Atmospheres: Theory and Observations*, eds. J.P. de Greve, R. Blomme & H. Hensberge (Lecture Notes in Physics) (Berlin:Springer) p. 69

Lamers, H.J.G.L.M. & Cassinelli, J.P. 1996, in *From Stars to Galaxies: the Impact of Stellar Physics on Galaxy Evolution*, ASP Conf. Series, eds. C. Leitherer *et al.* (San Francisco: ASP) p. 162

Lamers, H.J.G.L.M. & Leitherer, C. 1993, *Ap. J.* **412**, 771

Lamers, H.J.G.L.M. & Morton, D.C. 1976, *Ap. J. Sup.* **32**, 715

Lamers, H.J.G.L.M. & Pauldrach, A.W.A. 1991, *A & A* **244**, L5

Lamers, H.J.G.L.M. & Rogerson, J.B. 1978, *A & A* **66**, 417

Lamers, H.J.G.L.M., Vink, J.S., de Koter, A. & Cassinelli, J.P. 1999, in *Nonspherical and Variable Winds from Luminous Hot Stars*, eds. O. Stahl & B. Wolf (in press)

Lamers, H.J.G.L.M. & Waters, L.B.F.M. 1984, *A & A* **136**, 37

Lamers, H.J.G.L.M., de Groot, M. & Cassatella, A. 1983, *A & A* **128**, 299

Lamers, H.J.G.L.M., Maeder, A., Schmutz, W. & Cassinelli, J.P. 1991, *Ap. J.* **368**, 538

Lamers, H.J.G.L.M., Snow, T.P. & Lindholm, D.M. 1995, *Ap. J.* **455**, 269

Lamers, H.J.G.L.M., Haser, S., de Koter, A. & Leitherer, C. 1998, *Ap. J.* (in press)

Landau, L.D & Lifshitz, E.M. 1959, *Mechanics* (London: Pergamon) p. 93

Lang, K.R. 1992, *Astrophysical Data: Planets and Stars* (Berlin: Springer Verlag) p. 133

Leitherer, C. 1988, *Ap. J.* **326**, 356

Leitherer C. & Heckman, T.M. 1995, *Ap. J. Sup.* **96**, 9

Leitherer, C., Chapman, J.M. & Koribalski, B. 1995, *Ap. J.* **450**, 289

Leitherer, C., Fritze-von-Alvensleben, U. & Huchra, J. (eds) 1996, *From Stars to Galaxies: The Impact of Stellar Physics on Galaxy Evolution*, ASP Conf. Series (San Francisco: ASP)

Leitherer, C., Chapman, J.M. & Koribalski, B. 1997, *Ap. J.* **481**, 898

Linsky, J.L. & Haisch, B.M. 1979, *Ap. J.* **229**, L33

Loup, C., Forveille, T., Omont, A. & Paul, J.F. 1993, *A & A Sup.* **99**, 291

Lucy, L.B. & Solomon, P.M. 1970, *Ap. J.* **159**, 879
Lucy, L.B. & White, R.L. 1980, *Ap. J.* **241**, 300
MacGregor, K. B. & Charbonneau, P. 1994, *Ap. J.* **430**, 387
MacGregor, K. B. & Charbonneau, P. 1997, in *Cosmic Winds* (Arizona: Tucson)
MacLow, M.M. & McCray, R. 1988, *Ap. J.* **324**, 776
Maeder, A. & Meynet, G. 1987, *A & A* **182**, 243
Maeder, A. & Meynet, G. 1988, *A & A Sup.* **76**, 411
Maeder, A. & Meynet, G. 1989, *A & A* **210**, 155
Maheswaran, M. & Cassinelli, J.P. 1992, *Ap. J.* **386**, 695
Marston, A.P. 1995, *Astr. J.* **109**, 1839
Martin, P.G & Rogers, C. 1987, *Ap. J.* **322**, 374
Mathis, J.S. & Lamers, H.J.G.L.M. 1992, *A & A* **259**, L39
Mathis, J.S., Rumpl, W. & Nordsieck, K.H. 1977 (MRN), *Ap. J.* **217**, 425
McCray, R. 1983, *Highlights of Astronomy* **6**, 565
McCrea, W.H. 1935, *MNRAS* **95**, 509
Mendez, R.H., Kudritzki, R.P & Herrero, A. 1992, *A & A* **260**, 329
Mennessier, M.O. & Omont, A. (eds.) 1990, *From Miras to Planetary Nebulae: Which Path for Stellar Evolution?* (Editions Frontières)
Michel, F.C. 1969, *Ap. J.* **158**, 727
Mihalas, D. 1978, *Stellar Atmospheres* (San Francisco: Freeman)
Mihalas, D., Kunasz, P. & Hummer, D. 1975, *Ap. J.* **202**, 465
Milne, E.A. 1924, *MNRAS* **84**, 354
Milne, E.A. 1926, *MNRAS* **86**, 459
Mitchell, S.A. 1913, *Ap. J.* **38**, 407
Morris, M. 1980, *Ap. J.* **236**, 823
Morton, D.C. 1967, *Ap. J.* **147**, 1017
Morton, D.C. 1969, in *Mass Loss from Stars*, ed. M. Hack, (Dordrecht: Reidel) p. 36
Nerney, S.F. 1980, *Ap. J.* **242**, 723
Netzer, N. & Elitzur, M. 1993, *Ap. J.* **410**, 701
Neugebauer, M. & Snyder, C.W. 1962, *Science* **138**, 1095
Noble, L.M. & Scarfe, F.L. 1963, *Ap. J.* **138**, 1169
Osterbrock, D.E. 1989, in *Astrophysics of Gaseous Nebulae and Active Galactic Nuclei* (San Francisco: Freeman) p. 80
Owocki, S.P. 1994, *Ap. & Space Sci.* **221**, 3
Owocki, S.P., Castor, J.I. & Rybicki, G.B. 1988, *Ap. J.* **335**, 914
Owocki, S.P., Cranmer, S.R. & Blondin, J.M. 1994, *Ap. J.* **424**, 887
Owocki, S.P., Cranmer, S.R. & Gayley, K.G. 1996, *Ap. J.* **472**, L115
Owocki, S.P., Cranmer, S.R. & Gayley, K.G. 1998 in *$*B[e]$ Stars*, eds. A.M. Hubert & C. Jaschek (Dordrecht: Kluwer) p. 205
Panagia, N. & Felli, M. 1975, *A & A* **39**, 1
Parker, E.N. 1958, *Ap. J.* **128**, 664
Parker, E N. 1960, *Ap. J.* **132**, 821
Parker, E.N. 1965, *Space Sci. Rev.* **4**, 666

Parker, E.N. 1966, *Ap. J.* **143**, 32

Parker, E.N. 1971, in *Solar Wind*, NASA-SP 308, p. 161

Pauldrach, A.W.A. 1987, *A & A* **183**, 295

Pauldrach, A.W.A. & Puls, J. 1990, *A & A* **237**, 409

Pauldrach, A.W.A., Puls, J. & Kudritzki, R.P. 1986, *A & A* **164**, 86

Pauldrach, A.W.A., Kudritzki, R.P., Puls, J., Butler, K., & Hunsinger, J. 1994, *A & A* **283**, 525

Perinotto, M. 1993, in *Planetary Nebulae, IAU Symp. 155*, eds. R. Weinberger & A. Acker (Dordrecht: Kluwer) p. 57

Petrenz, P. & Puls, J. 1996, *A & A* **312**, 195

Pijpers, F.P. & Hearn, A.G. 1989, *A & A* **209**, 198

Poe, C.H. & Friend, D.B. 1986, *Ap. J.* **311**, 317

Poe, C.H., Friend, D.B. & Cassinelli, J.P. 1989, *Ap. J.* **337**, 888

Prinja, R.K., Barlow, M.J. & Howarth, I.D. 1990, *Ap. J.* **361**, 607

Puls, J. 1986, *A & A* **184**, 227

Puls, J., Kudritzki, R.P., Herrero, A., Pauldrach, A.W.A., Haser, S.M., Lennon, D.J., Gabler, R., Voels, S.A., Vilchez, J.M., Wachter, S. & Feldmeier, A. 1995, *A & A* **305**, 171

Quirrenbach, A., Bjorkman, K.S., Bjorkman, J.E., Hummel, C.A., Buscher, D.F., Armstrong, J.T., Mozurkewich, D., Elias, N.M. & Babler, B.L. 1997, *Ap. J.* **479**, 477

Reimers, D. 1975, *Mem. Soc. Roy. Sci. Liège, 6e Ser.* **8**, 369

Renzini, A. 1983, in *Planetary Nebulae*, ed. D. R. Flower (Dordrecht: Reidel) p. 267

Roche, P.F. 1989, in *Infrared Spectroscopy in Astronomy*, ESA SP-290, p. 79

Rowan-Robinson, M. & Harris, S. 1983, *MNRAS* **202**, 767

Rowan-Robinson, M., Lock, J.D., Walker, D.W. & Harris, S. 1986, *MNRAS* **222**, 273

Saha, M.N. 1919, *Ap. J.* **50**, 220

Schaerer, D., de Koter, A., Schmutz, W. & Maeder, A. 1996, *A & A* **310**, 837

Schaller, D., Schaerer, G., Meynet, G. & Maeder, A. 1992, *A & A Sup.* **96**, 269

Schmutz, W., Hamann, W.R. & Wessolowski, U. 1989, *A & A* **210**, 236

Schönberner, D. 1981, *A & A* **103**, 119

Schwarzschild, M. 1948, *Ap. J.* **107**, 1

Scuderi, S., Bonanno, G., Di Benedetto, R., Spadara, D. & Panagia, N. 1992, *Ap. J.* **392**, 201

Seward, F.D, Forman, W.R., Giaconi, R., Griffith, R.B., Harnden, F.R., Jones, C. & Pye, J.P. 1979, *Ap. J.* **234**, L51

Shimada, M.R., Ito, M., Hirata, R. & Horaguchi, T. 1994, in *Pulsation, Rotation and Mass Loss in Early Type Stars*, eds. L. Balona *et al.* (Dordrecht: Kluwer) p. 487

Snow, T.P., Lamers, H.J.G.L.M., Lindholm, D.M. & Odell, A.P. 1994, *Ap. J. Sup.* **95**, 163

Snow, T.P. & Morton, D.C. 1976, *Ap. J. Sup.* **32**, 429

Sobolev, V. 1960, *Moving Envelopes of Stars* (Cambridge MA: Harvard Univ. Press) [Russian Edition, 1947]

Spitzer, L. 1939, *Ap. J.* **90**, 294

Spitzer, L. 1962 *Physics of Fully Ionized Gases* (New York: Interscience Publishers)

Springman, U.W.B. & Pauldrach, A.W.A. 1993, *A & A* **262**, 515

Thomas, R.N. 1949, *Ap. J.* **109**, 500

Underhill, A.B. 1949, *MNRAS* **109**, 562

Underhill, A.B. 1966, *The Early Type Stars* (Dordrecht: Reidel)

van der Hucht, K.A. & Williams, P.M. (eds) 1995, *Wolf Rayet Stars: Binaries, Colliding Winds and Evolution* (Dordrecht: Kluwer)

van de Hulst, H. C. 1957, *Light Scattering by Small Particles* (New York: Dover)

van der Veen, W.E.C.J. 1989, *A & A* **210**, 127

van der Veen, W.E.C.J. & Olofsson, H. 1990 in *From Miras to Planetary Nebulae: Which Path for Stellar Evolution?*, eds. M.O. Mennessier & A. Omont (Editions Frontières) p. 139

Vassiliades, E. & Wood, P.R. 1993, *Ap. J.* **413**, 641

von Zeipel, H. 1924, *MNRAS* **84**, 665

Waters, L.B.F.M., Coté, J. & Lamers H.J.G.L.M. 1987, *A & A* **185**, 206

Waters, L.B.F.M., Trams, N.R. & Waelkens, C. 1992, *A & A* **262**, L37

Waters, L.B.F.M. & Wesselius, P.R. 1986, *A & A* **155**, 104

Weber, E.J. & Davis, L. 1967, *Ap. J.* **148**, 217

Weidemann, V. 1987, *A & A* **188**, 74

Weidemann, V. 1993, in *Mass Loss on the AGB and Beyond.* ESO conference and Workshop Proceedings No. 46, ed. H.E. Schwarz (Garching: ESO) p. 55

Whitelock, P., Menzies, J., Feast, M., Marang, F., Carter, B., Roberts, G., Catchpole, R. & Chapman, J. 1994, *MNRAS* **67**, 711

Whittet, D.C.B. 1992, in *Dust in the Galactic Environment* (Bristol: Inst. of Physics. Publ.)

Whitworth, A.P. 1975, *Ap. & Space Sci.* **34**, 155.

Williams, P.M., van der Hucht, K.A. & Thé, P.S. 1987, *A & A* **182**, 91

Willson. L.A. & Bowen, G. 1986, *Irish Astr. J.* **17**, 249

Wilson, O.C. 1960a, *Ap. J.* **131**, 75

Wilson, O.C. 1960b, *Ap. J.* **132**, 136

Wilson, O.C. 1966, *Ap. J.* **144**, 695

Wolf, C.J.E. & Rayet, G.A.P. 1867, *Compte Rendu* **65**, 291

Wolfire, M.G. & Cassinelli J.P. 1986, *Ap. J.* **310**, 207

Wolfire, M.G. & Cassinelli J.P. 1987, *Ap. J.* **319**, 850

Wood, D.O.S. & Churchwell, E. 1989, *Ap. J. Sup.* **69**, 831

Wood, K., Bjorkman, K.S. & Bjorkman, J.E. 1997, *Ap. J.* **477**, 926

Wood, P.R. 1990, in *From Miras to Planetary Nebulae: Which Path for Stellar Evolution?*, eds. M.O. Mennessier & A. Omont (Editions Frontières) p. 196

Woolf, N.J. & Ney, E.P. 1969, *Ap. J.* **155**, L181

Wright, A.E. & Barlow, M.J. 1975, *MNRAS* **170**, 41

Zickgraf, F.-J. 1992, in *Non-isotropic and Variable Outflows from Stars* ASP Conf. Series, eds. L. Drissen, C. Leitherer & A. Nota (San Francisco: ASP) p. 75

Zickgraf, F.-J., Wolf, B., Leitherer, C., Appenzeller, I. & Stahl, O. 1986, *A & A*, **163**, 119

Object index

Index